FREE SPACE OPTICAL SYSTEMS ENGINEERING

WILEY SERIES IN PURE AND APPLIED OPTICS

The Wiley Series in Pure and Applied Optics publishes authoritative treatments of foundational areas central to optics as well as research monographs in hot-topic emerging technology areas. Existing volumes in the series are such well known books as Gaskill's Linear Systems, Fourier Transforms, and Optics, Goodman's Statistical Optic, and Saleh & Teich's Fundamentals of Optics, among others.

A complete list of titles in this series appears at the end of the volume.

FREE SPACE OPTICAL SYSTEMS ENGINEERING

Design and Analysis

LARRY B. STOTTS

WILEY

Published by John Wiley & Sons, Inc., Hoboken, New Jersey

Published simultaneously in Canada

Library of Congress Cataloging-in-Publication Data

Names: Stotts, Larry B., author.
Title: Free space optical systems engineering : design and analysis / Larry B. Stotts.
Description: Hoboken, New Jersey : John Wiley & Sons, Inc., [2017] | Series: Wiley series in pure and applied optics | Includes bibliographical references and index.
Identifiers: LCCN 2016043633 (print) | LCCN 2016053936 (ebook) | ISBN 9781119279020 (cloth) | ISBN 9781119279037 (pdf) | ISBN 9781119279044 (epub)
Subjects: LCSH: Optical engineering. | Optical communications.
Classification: LCC TA1520 .S76 2017 (print) | LCC TA1520 (ebook) | DDC 621.36–dc23
LC record available at https://lccn.loc.gov/20160436330

Cover Image Credit: Courtesy of the author
Cover Design: Wiley

Set in 10/12pt, TimesLTStd by SPi Global, Chennai, India

Printed in the United States of America

10 9 8 7 6 5 4 3 2 1

CONTENTS

PREFACE

Just before graduating from the college, I took a job at a Navy Laboratory because there were not any research grants to help pay for my future graduate school work. As it turned out, it was a very rewarding experience, allowing me to work on many fascinating projects during my time there. They ranged from fiber optic communications, integrated optics in II–VI compounds, optical signal processing, data storage in electro-optical crystals, laser communications, and atmospheric and space remote sensing. Although I had a very good education in undergraduate applied physics from UCSD, many of these projects involved new optical technologies, as well as engineering concepts, that I had not been exposed to previously. My first two years in graduate school concentrated on graduate physics, which also did not cover these areas. Consequently, I had to spend a large amount of time in the library reading books and papers in order to come up to speed in these areas. I often wished that UCSD offered undergraduate and graduate classes in optical system engineering with an accompanying textbook(s) covering the breadth of the engineering basics necessary to tackle these various engineering areas. As it turned out, the mathematical foundations of each area were common, but many times the definitions and concept descriptions of one area masked its commonality with other topics in optics. In addition, the details were often absent and/or hard to find. In the absence of classes, it would have been nice to have an introductory reference to use to review the basic foundation concepts in optics and to find some of the original key references with the derivations of important equations at the time. This would have made it easier to move among a plethora of ever changing engineering projects.

Since then, several comprehensive books on optics have been written, for example, *Fundamentals of Photonics* by Saleh and Teich, the SPIE *Encyclopedia of Optics*, *Electro-Optics Handbook* by Waynant and Ediger. Although excellent in their content, these are written for a conversant researcher who has done graduate work, and/or been working, in optics for several years in order to fill in the blanks or to understand the nuances contained within the text. Unfortunately, this leaves junior, senior, and first/second year graduate students behind the power curve, requiring additional time, work, and consultation with their advisor or seasoned colleague, to understand what is written. Even with the Internet, this can be a formidable task. Thus, it appears that they are in the same situation as I was at the beginning of my career. In looking across the literature, there also are introductory textbooks focused on certain aspects of optics such as lens design, lasers, detectors, optical communications, and remote sensing, but none of which seem to encompass the breadth of free space optical systems engineering at a more basic level.

This textbook is an attempt to fulfill this need. It is intended to be the reference book for the engineer changing fields, and at the same time, to be an introduction to the field of electro-optics for upper division undergraduates and/or graduate students. Many of the original papers for the field are referenced, and an (comprehensive) introduction and overview of the topic has been attempted. Presentation and integration of physical (quantum mechanical), mathematical, and technological concepts, where possible, hopefully assists in the students' understanding.

It has been suggested that this material is too advanced for upper division undergraduate students. I think not for two reasons. First, today's students have been exposed to advanced subjects since middle/junior high school, for example, calculus and differential equations. They are used to being challenged. Second, and more importantly, the book provides the details of complex calculations in the many examples and discussions, so the students can become comfortable with complex mathematical manipulations. I never thought the concepts and calculations described by professors as "obvious to the most casual observer" were, and my fellow students and I struggled because of our lack of confidence, experience, and familiarity in figuring complex things out. Professors Booker and Lohmann independently taught me that if I understood the mathematical details, it would be easier to understand experimental results and to invent and explain complex concepts. This has helped me greatly over my career. However, getting this understanding sooner over a broader range of subjects in optics would have benefited me a lot and I hope to achieve that for readers of this book. I also believe students will be better prepared for graduate school and jobs by seeing complex subjects with this foundation. This book breaks down as follows:

Chapter 1 provides the background mathematics for the rest of the book. Specific topics include linear algebra, Fourier series, Fourier transforms, Dirac Delta function, and probability theory.

In Chapter 2, we discuss Fourier Optics, which includes sections on (1) Maxwell Equations, (2) Rayleigh–Sommerfeld–Debye Theory of Diffraction, (3) The Huygens–Fresnel–Kirchhoff Theory of Diffraction, (4) Fresnel Diffraction, and (5) Fraunhofer Diffraction. The Huygens–Fresnel–Kirchhoff formalism is the workhorse of laser propagation analysis, as the reader will soon find out. Examples and comments are also provided, so the reader gets insights on the application of Fourier Optics in typical engineering problems.

Geometrical Optics uses the concept of rays, which have direction and position but no phase information, to model the way light travels through space or an optical system. This is the subject of Chapter 3. In this chapter, we focus on imaging systems, which cover a broad class of engineering applications. We begin by summarizing the first-order lens design approaches. Key concept and definitions are explained, so the student can understand the key aspects of lens design. We also discuss the basic elements in an optical system such as windows, stops, baffles, and pupils that are sometimes confusing to the new optical engineer.

In Chapter 4, we outline the field of Radiometry, which is the characterization of the distribution of the optical power or energy in space. It is distinct from the quantum processes such as photon counting because this theory uses "ray tracing" as its means for depicting optical radiation transfer from one point to another. It ignores the dual nature of light.

Chapter 5 deals with the convolutional theory of image formation. Specifically, the reader will find that convolution process can characterize the effects of an imperfect optical system or those of an optical channel such as the optical scatter channel or turbulent channel on an input distribution. Most of the engineering analyses one finds in the literature exploit this mathematical theory.

Chapter 6 focuses on partial coherence theory. It covers the situation where the resulting light interference is barely visible, exhibiting only low contrast effects. Partial coherence theory is considered the most difficult subject in optics. Much is written on this subject; sometimes successfully, sometimes not. This chapter looks at this theory from the most basic level, clarifying the definitions and concepts with examples, so the reader will better understand the theory, compared to others, after completing this chapter.

In Chapter 7, we address the characterization of optical channel effects. We begin a discussion of radiative transfer through particulate media, then move to the development of the mutual coherence function (MCF) for aerosols and molecules, and then turbulence. Finally, we provide a set of engineering equations useful in understanding and characterizing light propagation in those same channels.

In Chapter 8, we provide a first-order overview of the various optical detector mechanisms and devices. We next look at the possible noise sources in an optical receiver that influence the quality of signal reception. When these detector and noise mechanisms are combined with the received signal, we obtain the arguably key parameter in detection theory, the electrical signal-to-noise ratio (SNR). The construction of this particular metric connects it with RF engineering, so the synergism between the two areas can be easily exploited. Finally, we discuss the various forms of SNR and include some detection sensor/receiver examples to illustrate their variation.

Chapter 9 reviews the classical statistical detection theory and then shows its applicability to optical communications and remote sensing. This is not found in many introductory optics books and discusses two of the key detection concepts: the probabilities of detection and false alarm. Both the signal-plus-additive-noise and replacement model hypothesis testing approaches are discussed. Examples of the theory's application to communications and remote sensing system are given.

In the early chapters, we emphasized blackbody sources, which are based on the concept of the spontaneous emission of light from materials such as gases and solids. An alternative source concept was proposed in 1917 by Albert Einstein.

It is called the simulated emission of light. Although it was almost 60 years before it became a reality, the laser, which is an acronym for Light Amplification by Stimulated Emission of Radiation, has revolutionized optical system design. The final chapter of this textbook, Chapter 10, provides an overview of the fundamentals of laser theory, the key source for all optical communications and remote sensing applications.

During my career, I have had the great fortune to have worked under the guidance of some key leaders in Electrical Engineering and Optics. Alphabetically, they were H.G. Booker (RF propagation), S.Q. Duntley (Visibility/Underwater Optics), R.M. Gagliardi, (Optical Communications), C.W. Helstrom (RF and Optical Detection and Estimation theory), S. Karp (Optical Communications), A.W. Lohmann (Optical Signal Processing), and I.S. Reed (RF/Optical Remote Sensing and Communications, and Forward Error Correction). These minds have served as both inspiration and mentors in my studies and projects, and I am forever grateful to have passed through their lives.

In addition, I would like to thank my many colleagues who helped me over the years succeed in my optical systems projects and gain the insights I discuss in this book. They also have influenced me greatly. In writing this book, I want to acknowledge Larry Andrews, Guy Beaghler, David Buck, Ralph Burnham, Barry Hunt, Skip Hoff, Juan Juarez, Ron Phillips, H. Alan Pike, Gary Roloson, and Ed Winter for their help and assistance.

I also want to thank Mr. Jim Meni, Science and Technology Associates (STA), for his friendship and for his support of my little project. He kindly let me use an office and his staff at STA to facilitate this textbook's development. It would have been much harder without his strong support.

Finally, I want to recognize the biggest contributors to this book, my partner Debra, who has shown both understanding and patience during the preparation of this manuscript. Anyone who has imbedded themselves in an all-consuming goal knows the importance of receiving those facets from loved ones.

Although I stand upon the shoulders of all of these individuals, any deficiencies in this textbook are attributable to me alone. I request your feedback regarding all improvements that might help this text, both in content, style, and perhaps most importantly, in its ability to serve as the desired introduction to upper division undergraduate students. Please pass along any and all suggestions to attain this goal.

Larry B. Stotts
Arlington, Virginia
September 2016

ABOUT THE COMPANION WEBSITE

This book is accompanied by a companion website:

www.wiley.com/go/stotts/free_space_optical_systems_engineering

The website includes:

- Solutions Manual

1

MATHEMATICAL PRELIMINARIES

1.1 INTRODUCTION

Free space optical systems engineering addresses how light energy is created, manipulated, transferred, changed, processed, or any combination of these entities, for use in atmospheric and space remote sensing and communications applications. To understand the material in this book, some basic mathematical concepts and relationships are needed. Because our audience is envisioned to be junior and senior undergraduates, it is not possible to require the readers to have had exposure to these topics at this stage of their education. Normally, they will have a working knowledge of algebra, geometry, and differential and integral calculus by now, but little else.

In this chapter, we provide a concise, but informative, summary of the additional mathematical concepts and relationships needed to perform optical systems engineering. The vision is to form a strong foundation for understanding what follows in subsequent chapters.

The intent is to establish the lexicon and mathematical basis of the various topics, and what they really mean in very simple, straightforward means, establishing that envisioned foundation.

1.2 LINEAR ALGEBRA

Imaging sensors create pictures stored as two-dimensional, discrete element arrays, that is, matrices. It is often convenient to convert these image arrays into a vector form

Free Space Optical Systems Engineering: Design and Analysis, First Edition. Larry B. Stotts.

Companion website: www.wiley.com\go\stotts\free_space_optical_systems_engineering

by column (or row) scanning the matrix, and then stringing the elements together in a long vector; that is, a lexicographic form [1]. This is called "vectorization." This means that the optical engineers need to be adept in linear algebra to take on problems in optical signal processing and target detection. This section reviews the notational conventions and basics of linear algebra following Bar-Shalom and Fortmann, Appendix A [2]. A more extensive review can be found in books on linear algebra.

1.2.1 Matrices and Vectors

A *matrix*, $\boldsymbol{A}$, is a two-dimensional array that can be mathematically written as

$$\boldsymbol{A} = [a_{ij}] = \begin{bmatrix} a_{11} & \cdots & a_{1J} \\ \vdots & \ddots & \vdots \\ a_{I1} & \cdots & a_{IJ} \end{bmatrix}. \tag{1.1}$$

The first index in the matrix element indicates the row number and the second index, the column number. The dimensions of the matrix are $I \times J$. The *transpose* of the above matrix is written as

$$\boldsymbol{A}^T = \begin{bmatrix} a_{11} & \cdots & a_{1I} \\ \vdots & \ddots & \vdots \\ a_{J1} & \cdots & a_{JI} \end{bmatrix}. \tag{1.2}$$

A *square matrix* is said to be symmetric, which means that

$$\boldsymbol{A} = \boldsymbol{A}^T, \tag{1.3}$$

which means that $a_{ij} = a_{ji} \;\; \forall i, j$.

A *vector* is a one-dimensional matrix array, which is written as

$$\boldsymbol{a} = \text{column}(a_i) = \begin{bmatrix} a_1 \\ \vdots \\ a_I \end{bmatrix}. \tag{1.4}$$

The column vector has dimension I in this case. By convention, we assume all vectors are column vectors. The transpose of a column vector is a row vector and the transpose of Eq. (1.4) can be written as

$$\boldsymbol{a}^T = \text{row}(a_i) = \begin{bmatrix} a_1 & \cdots & a_I \end{bmatrix}. \tag{1.5}$$

Comparing Eqs. (1.4) and (1.5), it is clear that

$$\boldsymbol{a} = \begin{bmatrix} a_1 & \cdots & a_I \end{bmatrix}^T. \tag{1.6}$$

1.2.2 Linear Operations

The *addition* of matrices and *multiplication* of a matrix by a scalar are given by the following equation:

$$\boldsymbol{C} = r\boldsymbol{A} + s\boldsymbol{B}, \tag{1.7}$$

where

$$c_{ij} = ra_{ij} + sb_{ij} \tag{1.8}$$

for $\{i = 1, \ldots, I; j = 1, \ldots, J\}$. Obviously, all three matrices have the same dimensions.

The *product* of two matrices is written in general as

$$\boldsymbol{C} = \boldsymbol{AB}, \tag{1.9}$$

where

$$c_{ip} = \sum_{j=1}^{J} a_{ij} b_{jp} \tag{1.10}$$

for $\{i = 1, \ldots, I; p = 1, \ldots, P\}$. Here, $\boldsymbol{A}$ is a $I \times J$ matrix, $\boldsymbol{B}$ is a $J \times P$ matrix, and $\boldsymbol{C}$ is a $I \times P$ matrix. In general, matrix products are not commutative, that is, $\boldsymbol{AB} \neq \boldsymbol{BA}$.

The transpose of a product is

$$\boldsymbol{C}^T = (\boldsymbol{AB})^T = \boldsymbol{B}^T \boldsymbol{A}^T. \tag{1.11}$$

Equation (1.10) implies that if the matrix–vector product is written as

$$\boldsymbol{Ab} = \boldsymbol{c}, \tag{1.12}$$

where $\boldsymbol{A}$ is a $I \times J$ matrix, $\boldsymbol{b}$ is a $J \times 1$ vector, and $\boldsymbol{c}$ is a $I \times 1$ vector, then its transpose is equal to

$$\boldsymbol{c}^T = \boldsymbol{b}^T \boldsymbol{A} \tag{1.13}$$

with $\boldsymbol{c}^T$ being a $1 \times I$ (row) vector, $\boldsymbol{b}^T$ being a $1 \times J$ vector, and $\boldsymbol{A}$ still being a $I \times J$ matrix.

1.2.3 Traces, Determinants, and Inverses

The *trace* of a $I \times I$ matrix $\boldsymbol{A}$ is defined as

$$\mathrm{tr}(\boldsymbol{A}) = \sum_{i=1}^{I} a_{ii} = \mathrm{tr}(\boldsymbol{A}^T), \tag{1.14}$$

which implies that

$$\mathrm{tr}(\boldsymbol{AB}) = \mathrm{tr}(\boldsymbol{BA}). \tag{1.15}$$

The *determinant* of a $I \times I$ matrix $\boldsymbol{A}$ is defined as

$$|\boldsymbol{A}| = a_{11}c_{11} + a_{12}c_{12} + \cdots + a_{1I}c_{1I} = |\boldsymbol{A}^T|, \tag{1.16}$$

where

$$c_{ij} = (-1)^{i+j} |\boldsymbol{A}^{ij}|; \quad i,j = 1, \ldots, I. \tag{1.17}$$

The parameter set $\{c_{ij}\}$ are called the *cofactors* of $\boldsymbol{A}$ and $\boldsymbol{A}^{ij}$ is the $(I-1)\times(I-1)$ matrix formed by deleting the ith row and jth column from $\boldsymbol{A}$. The determinant of a scalar is defined as the scalar itself since essentially is a 1×1 matrix. This implies that the determinant of a matrix multiplied by a scalar is given by

$$|r\boldsymbol{A}| = r^I|\boldsymbol{A}|, \tag{1.18}$$

and the determinant of a product of two matrices is written as

$$|\boldsymbol{AB}| = |\boldsymbol{BA}|. \tag{1.19}$$

Example 1.1

(a) The determinant of a 2×2 matrix is given by

$$\begin{vmatrix} a_{11} & a_{12} \\ a_{21} & a_{22} \end{vmatrix} = a_{11}a_{22} - a_{21}a_{12}. \tag{1.20}$$

(b) The determinant of a 3×3 matrix is given by

$$\begin{vmatrix} a_{11} & a_{12} & a_{13} \\ a_{21} & a_{22} & a_{23} \\ a_{31} & a_{32} & a_{33} \end{vmatrix} = a_{11}\begin{vmatrix} a_{22} & a_{23} \\ a_{32} & a_{33} \end{vmatrix} - a_{12}\begin{vmatrix} a_{21} & a_{23} \\ a_{31} & a_{33} \end{vmatrix} + a_{13}\begin{vmatrix} a_{21} & a_{22} \\ a_{31} & a_{32} \end{vmatrix} \tag{1.21}$$

$$= a_{11}(a_{22}a_{33} - a_{23}a_{32}) - a_{12}(a_{21}a_{33} - a_{23}a_{31}) + a_{13}(a_{21}a_{32} - a_{22}a_{31}). \tag{1.22}$$

Example 1.2 Let us now look at the solution for Eq. (1.12), where matrix $\boldsymbol{A}$ is a 3×3 matrix. Multiplying Eq. (1.12) out, we have three simultaneous equations:

$$\begin{aligned} a_{11}x + a_{12}y + a_{13}z &= u, \\ a_{21}x + a_{22}y + a_{23}z &= v, \\ a_{31}x + a_{32}y + a_{33}z &= w. \end{aligned} \tag{1.23}$$

The solutions to these equations are:

$$x = \frac{\begin{vmatrix} u & a_{12} & a_{13} \\ v & a_{22} & a_{23} \\ w & a_{32} & a_{33} \end{vmatrix}}{\begin{vmatrix} a_{11} & a_{12} & a_{13} \\ a_{21} & a_{22} & a_{23} \\ a_{31} & a_{32} & a_{33} \end{vmatrix}}, \tag{1.24}$$

$$y = \frac{\begin{vmatrix} a_{11} & u & a_{13} \\ a_{21} & v & a_{23} \\ a_{31} & w & a_{33} \end{vmatrix}}{\begin{vmatrix} a_{11} & a_{12} & a_{13} \\ a_{21} & a_{22} & a_{23} \\ a_{31} & a_{32} & a_{33} \end{vmatrix}}, \tag{1.25}$$

and

$$z = \frac{\begin{vmatrix} a_{11} & a_{12} & u \\ a_{21} & a_{22} & v \\ a_{31} & a_{32} & w \end{vmatrix}}{\begin{vmatrix} a_{11} & a_{12} & a_{13} \\ a_{21} & a_{22} & a_{23} \\ a_{31} & a_{32} & a_{33} \end{vmatrix}}, \tag{1.26}$$

assuming the determinant of matrix $\boldsymbol{A}$ is not zero.

The *inverse* $\boldsymbol{A}^{-1}$ of a $I \times I$ matrix $\boldsymbol{A}$ (if it exists) can be expressed as

$$\boldsymbol{A}^{-1}\boldsymbol{A} = \boldsymbol{A}\boldsymbol{A}^{-1} = \begin{bmatrix} 1 & \cdots & 0 \\ \vdots & \ddots & \vdots \\ 0 & \cdots & 1 \end{bmatrix} = \boldsymbol{I}. \tag{1.27}$$

In Eq. (1.20), the $I \times I$ matrix $\boldsymbol{I}$ is called the *identity matrix*, which has 1's down the diagonal and 0's everywhere else. The inverse is given by the equation

$$\boldsymbol{A}^{-1} = \frac{1}{|\boldsymbol{A}|}\boldsymbol{C}^T, \tag{1.28}$$

where $\boldsymbol{C}$ are the cofactors of $\boldsymbol{A}$. The matrix $\boldsymbol{C}^T$ is called the *adjugate* of $\boldsymbol{A}$. A matrix is considered *invertible* or *nonsingular* if and only if its determinant is nonzero; otherwise, it is said to be singular. Let us discuss these points a little more.

The inverse of a matrix exists if and only if the columns of the matrix (or its rows) are linearly independent. This means that

$$\sum_{i=1}^{m} r_i \boldsymbol{a}_i = \boldsymbol{0} \rightarrow r_i = 0 \quad \text{for} \quad i = 1, \ldots, m, \tag{1.29}$$

where $\boldsymbol{0}$ is the zero vector.

A general $m \times m$ matrix can be inverted using methods such as the Cayley–Hamilton (CH) method, Gauss–Jordan elimination, Gaussian elimination, or LU decomposition.

Example 1.3 The cofactor equation given in Eq. (1.17) gives the following expression for the inverse of a 2×2 matrix:

$$\begin{bmatrix} a_{11} & a_{12} \\ a_{21} & a_{22} \end{bmatrix}^{-1} = \frac{1}{\begin{vmatrix} a_{11} & a_{12} \\ a_{21} & a_{22} \end{vmatrix}} \begin{bmatrix} a_{22} & -a_{21} \\ -a_{12} & a_{11} \end{bmatrix}. \tag{1.30}$$

The CH method gives the solution

$$A^{-1} = \frac{1}{|A|}[(\text{tr}(A)\, I - A)]. \tag{1.31}$$

Example 1.4 The inverse of a 3×3 matrix is given by

$$A^{-1} = \begin{bmatrix} a_{11} & a_{12} & a_{13} \\ a_{21} & a_{22} & a_{23} \\ a_{31} & a_{32} & a_{33} \end{bmatrix}^{-1} = \frac{1}{|A|}\begin{bmatrix} d_{11} & d_{12} & d_{13} \\ d_{21} & d_{22} & d_{23} \\ d_{31} & d_{32} & d_{33} \end{bmatrix}^{T} = \frac{1}{|A|}\begin{bmatrix} d_{11} & d_{21} & d_{31} \\ d_{12} & d_{22} & d_{32} \\ d_{13} & d_{23} & d_{33} \end{bmatrix}, \tag{1.32}$$

where

$$\begin{array}{lll} d_{11} = a_{22}a_{33} - a_{23}a_{32} & d_{12} = -(a_{21}a_{33} - a_{23}a_{31}) & d_{13} = (a_{21}a_{32} - a_{22}a_{31}) \\ d_{21} = -(a_{12}a_{33} - a_{13}a_{32}) & d_{22} = -(a_{11}a_{33} - a_{13}a_{31}) & d_{23} = -(a_{11}a_{32} - a_{12}a_{31}). \\ d_{31} = a_{12}a_{23} - a_{13}a_{22} & d_{32} = -(a_{11}a_{23} - a_{13}a_{21}) & d_{33} = (a_{21}a_{22} - a_{12}a_{21}) \end{array}$$

The CH method gives the solution

$$A^{-1} = \frac{1}{|A|}\left[\frac{1}{2}[(\text{tr}(A))^2 - \text{tr}(A^2)]\, I - A\, \text{tr}(A) + A^2\right]. \tag{1.33}$$

Example 1.5 With increasing dimensions, expressions for A^{-1} becomes complicated. However, for $m = 4$, the CH method yields

$$A^{-1} = \frac{1}{|A|}\left[\begin{array}{c} \frac{1}{6}[(\text{tr}(A))^3 - 3\text{tr}(A)\text{tr}(A^2) + \text{tr}(A^3)] \\ I - \frac{1}{2}A[(\text{tr}(A))^2 - \text{tr}(A^2)] - A^2\, \text{tr}(A) - A^3 \end{array}\right]. \tag{1.34}$$

Example 1.6 The inverse of a (nonsingular) partitioned $I \times I$ matrix also can be shown to be given by

$$\begin{bmatrix} A & B \\ C & D \end{bmatrix}^{-1} = \begin{bmatrix} E & F \\ G & M \end{bmatrix}, \tag{1.35}$$

where A is a $I_1 \times I_1$ matrix, B is a $I_1 \times I_2$ matrix, C is a $I_2 \times I_1$ matrix, D is a $I_2 \times I_2$ matrix, and $I_1 + I_2 = I$. In the above,

$$E = A^{-1} + A^{-1}BMCA^{-1} = (A - BD^{-1}C)^{-1} \tag{1.36}$$

$$F = A^{-1}BM = -EBD^{-1} \tag{1.37}$$

$$G = -MCA^{-1} = -D^{-1}CE \tag{1.38}$$

and

$$M = (D - CA^{-1}B)^{-1} = D^{-1} + D^{-1}CEBD^{-1}. \tag{1.39}$$

If $\boldsymbol{R} = -\boldsymbol{A}$, $\boldsymbol{P} = \boldsymbol{D}^{-1}$, and $\boldsymbol{H} = \boldsymbol{B} = \boldsymbol{C}^T$, then the following matrix equation

$$(\boldsymbol{P}^{-1} + \boldsymbol{H}^T\boldsymbol{R}^{-1}\boldsymbol{H})^T = \boldsymbol{P} - \boldsymbol{P}\boldsymbol{H}^T(\boldsymbol{H}\boldsymbol{P}\boldsymbol{H}^T + \boldsymbol{R})^{-1}\boldsymbol{H}\boldsymbol{P} \tag{1.40}$$

can be rewritten as

$$(\boldsymbol{R} + \boldsymbol{H}^T\boldsymbol{P}\boldsymbol{H})^T = \boldsymbol{R}^{-1} - \boldsymbol{R}^{-1}\boldsymbol{H}(\boldsymbol{P}^{-1} + \boldsymbol{H}^T\boldsymbol{R}^{-1}\boldsymbol{H} + \boldsymbol{R})^{-1}\boldsymbol{H}^T\boldsymbol{R}^{-1}. \tag{1.41}$$

The above is known as *matrix inversion formula.*

It is easy to show that

$$(\boldsymbol{A}\boldsymbol{B})^{-1} = \boldsymbol{B}^{-1}\boldsymbol{A}^{-1}. \tag{1.42}$$

1.2.4 Inner Products, Norms, and Orthogonality

The inner product of two arbitrary vectors of the same dimension is given by

$$\boldsymbol{a}^T\boldsymbol{b} = \sum_{i=1}^{I} a_i b_i. \tag{1.43}$$

If $\boldsymbol{a} = \boldsymbol{b}$, then we write

$$\boldsymbol{a}^T\boldsymbol{a} = \boldsymbol{a}^2 = \sum_{i=1}^{I} a_i^2. \tag{1.44}$$

Equation (1.44) is called the squared *norm* of the vector $\boldsymbol{a}$. The *Schwartz inequality* states that

$$|\boldsymbol{a}^T\boldsymbol{b}| \leq \boldsymbol{a}\boldsymbol{b}. \tag{1.45}$$

Two vectors are defined to be orthogonal ($\boldsymbol{a} \perp \boldsymbol{b}$) if

$$\boldsymbol{a}^T\boldsymbol{b} = \boldsymbol{0}. \tag{1.46}$$

The orthogonal projection of the vector $\boldsymbol{a}$ onto $\boldsymbol{b}$ is

$$\Pi_b(\boldsymbol{a}) = \frac{\boldsymbol{a}^T\boldsymbol{b}}{\boldsymbol{b}^2}\,\boldsymbol{b}, \tag{1.47}$$

and the difference between it and the original vector $\boldsymbol{a}$ is orthogonal to $\boldsymbol{b}$. That is, we have

$$[\boldsymbol{a} - \Pi_b(\boldsymbol{a})] \perp \boldsymbol{b}. \tag{1.48}$$

Finally, the outer product of the vectors $\boldsymbol{a}$ and $\boldsymbol{b}$ is the matrix $\boldsymbol{C}$, that is,

$$\boldsymbol{a}\boldsymbol{b}^T = \boldsymbol{C}. \tag{1.49}$$

1.2.5 Eigenvalues, Eigenvectors, and Rank

The *eigenvalues* of a square matrix are the scalars λ_i such that

$$\boldsymbol{A}\boldsymbol{x}_i = \lambda_i \boldsymbol{x}_i, \tag{1.50}$$

where the vectors $\boldsymbol{x}_i$ are the corresponding *eigenvectors*. Equation (1.50) has the following properties:

- A matrix $\boldsymbol{A}$ is nonsingular if and only if all its eigenvalues are nonzero.
- The ***rank*** of the matrix $\boldsymbol{A}$ is equal to the number of its nonzero eigenvalues. A nonsingular matrix is said to be of full rank.
- The eigenvalues of a real matrix can be either real or complex, but a symmetric matrix only has eigenvalues that are real.
- The trace of a matrix $\boldsymbol{A}$ is equal to the sum of its eigenvalues.
- The determinant of a matrix $\boldsymbol{A}$ is equal to the product of its eigenvalues.

1.2.6 Quadratic Forms and Positive Definite Matrices

The scalar equation

$$q(\boldsymbol{x}) = \boldsymbol{x}\boldsymbol{A}\boldsymbol{x} \tag{1.51}$$

is called a *quadratic form*. This relationship is true when the matrix $\boldsymbol{A}$ is symmetric. The equation of this form is called *positive definite* if

$$q(\boldsymbol{x}) = \boldsymbol{x}\boldsymbol{A}\boldsymbol{x} > \boldsymbol{0} \quad \forall\, \boldsymbol{x} \neq \boldsymbol{0}. \tag{1.52}$$

The matrix $\boldsymbol{A}$ also is called positive definite, which we denote by $\boldsymbol{A} > \boldsymbol{0}$. It goes without saying that a matrix is positive definite if and only if its eigenvalues are positive. If the inequality in Eq. (1.52) is nonnegative rather than positive, then the matrix is referred to as positive semidefinite or nonnegative definite.

The quadratic equation given in Eq. (1.51) is the squared weighted norm of the vector $\boldsymbol{x}$ where the weighting is in accordance with the matrix $\boldsymbol{A}$.

1.2.7 Gradients, Jacobians, and Hessians

The *gradient operator* is written as

$$\nabla_x = \begin{bmatrix} \dfrac{\partial}{\partial x_1} \\ \vdots \\ \dfrac{\partial}{\partial x_I} \end{bmatrix}, \tag{1.53}$$

with the properties

$$\nabla_x \boldsymbol{x}^T = \boldsymbol{I} \tag{1.54}$$

and

$$\nabla_x(\boldsymbol{x}^T\boldsymbol{A}\boldsymbol{x}) = 2\boldsymbol{A}\boldsymbol{x}. \tag{1.55}$$

The gradient of a vector-valued function, say $\boldsymbol{f}(\boldsymbol{x})$, is equal to

$$\nabla_x \boldsymbol{f}^T(\boldsymbol{x}) = \begin{bmatrix} \frac{\partial}{\partial x_1} \\ \vdots \\ \frac{\partial}{\partial x_I} \end{bmatrix} \{ f_1(\boldsymbol{x}) \ \cdots \ f_I(\boldsymbol{x}) \} = \begin{bmatrix} \frac{\partial f_1(\boldsymbol{x})}{\partial x_1} & \cdots & \frac{\partial f_I(\boldsymbol{x})}{\partial x_1} \\ \vdots & \ddots & \vdots \\ \frac{\partial f_1(\boldsymbol{x})}{\partial x_I} & \cdots & \frac{\partial f_I(\boldsymbol{x})}{\partial x_I} \end{bmatrix}. \tag{1.56}$$

The transpose of Eq. (1.56) is called the *Jacobian* and is defined as

$$\boldsymbol{f}_x(\boldsymbol{x}) = \frac{\partial \boldsymbol{f}(\boldsymbol{x})}{\partial \boldsymbol{x}} = [\nabla_x \boldsymbol{f}^T(\boldsymbol{x})]^T. \tag{1.57}$$

The Hessian of the scalar function $f(\boldsymbol{x})$ is defined to be given by

$$\boldsymbol{f}_{xx}(\boldsymbol{x}) = \frac{\partial^2 \boldsymbol{f}(\boldsymbol{x})}{\partial \boldsymbol{x}^2} = \nabla_x \nabla_x{}^T f(\boldsymbol{x}) = \begin{bmatrix} \frac{\partial^2 f_1(\boldsymbol{x})}{\partial x_1 \partial x_1} & \cdots & \frac{\partial^2 f_I(\boldsymbol{x})}{\partial x_1 \partial x_I} \\ \vdots & \ddots & \vdots \\ \frac{\partial^2 f_1(\boldsymbol{x})}{\partial x_I \partial x_1} & \cdots & \frac{\partial^2 f_I(\boldsymbol{x})}{\partial x_I \partial x_I} \end{bmatrix}, \tag{1.58}$$

which is a symmetric matrix.

1.3 FOURIER SERIES

One of the main tools used in physics and engineering are Fourier series and Fourier integrals, which were published by Jean Baptiste Fourier in 1822, almost 200 years ago. This section reviews the basics of these important entities [3–5] as they are the foundation for a lot of the mathematical concepts used in optics.

1.3.1 Real Fourier Series

A *Fourier series* is used to represent a periodic function $f(x)$. The definition of a periodic function is

$$f(x) = f(x + md); \quad \text{where } |m| = 0, 1, 2, 3, \ldots \tag{1.59}$$

where d is the length of the period (Figure 1.1) and $\nu = \frac{1}{d}$ is the fundamental spatial frequency in one dimension. The following expression is the usual way for writing a Fourier series representing the periodic function $f(x)$:

$$f(x) = \sum_{n=0}^{\infty} [a_n \cos(2\pi n \nu x) + b_n \sin(2\pi n \nu x)]. \tag{1.60}$$

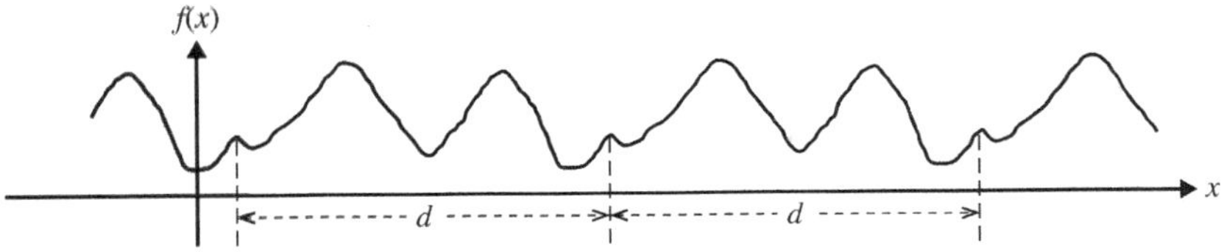

FIGURE 1.1 Example of a periodic function $f(x)$.

In short, a Fourier series is an expansion of a periodic function $f(x)$ in terms of sines and cosines. The coefficients for this series are given by:

$$a_0 = \frac{1}{\pi}\int_{-\pi}^{\pi} f(x)\,dx, \tag{1.61}$$

$$a_n = \frac{1}{\pi}\int_{-\pi}^{\pi} f(x)\,\cos(nx)dx, \tag{1.62}$$

and

$$b_n = \frac{1}{\pi}\int_{-\pi}^{\pi} f(x)\,\sin(nx)dx \tag{1.63}$$

for $n = 1, 2, \ldots, \infty$. It is noted that this basis set is orthonormal, as seen by the following the integral identities:

$$\int_{-\pi}^{\pi} \sin(mx)\,\sin(nx)dx = \pi\delta_{mn}, \tag{1.64}$$

$$\int_{-\pi}^{\pi} \cos(mx)\,\cos(nx)dx = \pi\delta_{mn}, \tag{1.65}$$

$$\int_{-\pi}^{\pi} \sin(mx)\,\cos(nx)dx = 0, \tag{1.66}$$

$$\int_{-\pi}^{\pi} \sin(mx)\,dx = 0, \tag{1.67}$$

and

$$\int_{-\pi}^{\pi} \cos(mx)\,dx = 0 \tag{1.68}$$

for $m, n \neq 0$. In the first two equations, δ_{mn} denotes the Kronecker delta function.

1.3.2 Complex Fourier Series

The natural extension of the above is to express a Fourier series in terms of complex coefficients. Consider a real-valued function $f(x)$. In this case, we have

$$f(x) = \sum_{n=-\infty}^{\infty} C_n\,e^{+inx}, \tag{1.69}$$

where

$$C_n = \frac{1}{2\pi}\int_{-\pi}^{\pi} f(x)\, e^{-inx} dx. \tag{1.70}$$

The above coefficients can be expressed in terms of those in the Fourier series given in Eq. (1.60), namely,

$$C_n = \begin{cases} \frac{1}{2\pi}\int_{-\pi}^{\pi} f(x)[\cos(nx) + i\,\sin(nx)]dx & \text{for } n < 0 \\ \frac{1}{2\pi}\int_{-\pi}^{\pi} f(x)\, dx & \text{for } n < 0 \\ \frac{1}{2\pi}\int_{-\pi}^{\pi} f(x)\,[\cos(nx) - i\,\sin(nx)]dx & \text{for } n > 0 \end{cases} \tag{1.71}$$

$$= \begin{cases} \frac{1}{2\pi}(a_n + i b_n) & \text{for } n < 0 \\ \frac{1}{2\pi}\, a_0 & \text{for } n = 0\ . \\ \frac{1}{2\pi}(a_n - i\, b_n) & \text{for } n < 0 \end{cases} \tag{1.72}$$

For a function that is periodic in $\left\{-\frac{d}{2}, \frac{d}{2}\right\}$, this complex series become

$$f(x) = \sum_{n=-\infty}^{\infty} C_n e^{+2\pi i n \nu x}, \tag{1.73}$$

where

$$C_n = \frac{1}{d}\int_{-\frac{d}{2}}^{\frac{d}{2}} f(x) e^{-2\pi i n \nu x} dx \tag{1.74}$$

and $\nu = \frac{1}{d}$. These equations are the basis for the extremely important Fourier transform, which is obtained by transforming C_n from a discrete variable to a continuous one as the length $d \to \infty$. We will discuss Fourier Transforms shortly.

1.3.3 Effects of Finite Fourier Series Use

Sometimes, it is necessary not to have an infinite series, but a finite series to represent the periodic function $f(x)$. In this case, we write

$$f(x) = \sum_{n=0}^{N} [a_n \cos(2\pi n \nu x) + b_n \sin(2\pi n \nu x)] = S_N(x). \tag{1.75}$$

What is the best choice of coefficients, $\{a_n\}$ and $\{b_n\}$? Let us define an error function

$$E_N(x) = f(x) - S_N(x). \tag{1.76}$$

Rewriting Eq. (1.75), we find that

$$S_N(x) = \sum_{n=0}^{\infty}[a_n \cos(2\pi n \nu x) + b_n \sin(2\pi n \nu x)]$$

$$= a_0 + \sum_{n=1}^{\infty}\left[\frac{(a_n - ib_n)}{2}\right] e^{+2\pi n \nu x} + \sum_{n=1}^{\infty}\left[\frac{(a_n + ib_n)}{2}\right] e^{-2\pi n \nu x}. \quad (1.77)$$

Manipulating the terms in Eq. (1.77), we find that

$$S_N(x) = \sum_{n=-N}^{N} C_n e^{+2\pi i n \nu x}; \quad C_n = f(x) \cdot e^{-2\pi i n \nu x}. \quad (1.78)$$

Now let us return to the questions: "How good is our approximation?" or "Does $\sigma_N^2 \to 0$ as $N \to \infty$?" Using our complex notation, the error σ_N^2 can be rewritten as

$$\sigma_N^2 = \left(f - \sum_{n=-N}^{N} C_n e^{+2\pi i n \nu x}\right) \cdot \left(f^* - \sum_{n=-N}^{N} C_n^* e^{-2\pi i n \nu x}\right) \quad (1.79)$$

$$= f \cdot f^* - \sum_{m=-N}^{N} |C_m|^2. \quad (1.80)$$

This equation implies the following desirable feature, $\sigma_{N+1}^2 - \sigma_N^2 = -2|C_{N+1}|^2 \leq 0$. This means that the error σ_N^2 can only decrease if more terms are added to the Fourier series as in $S_N \to S_{N+1}$. If $\sigma_N^2 \to 0$, then

$$\sum_{m=-N}^{N} |C_m|^2 \overset{\text{yields}}{\to} f \cdot f^*.$$

The ultimate result, which holds for all healthy functions (and also for some strange ones too), is the *completeness relationship*,

$$f \cdot f^* = \sum_{m=-\infty}^{\infty} |C_m|^2. \quad (1.81)$$

However, this relationship is not completely satisfactory because it always cannot easily be checked and $f(x)$ can be quite different from $\sum_{n=-\infty}^{\infty} C_n e^{+2\pi n \nu x}$ for a finite number of points.

Example 1.7 Let

$$f(x) = \begin{cases} 1 & \text{for } 0 \leq x \leq \frac{d}{2} \\ -1 & \text{for } -\frac{d}{2} < x < 0 \end{cases}, \quad (1.82)$$

which implies that

$$C_n = \begin{cases} 0 & \text{if } n \text{ is even} \\ \frac{2}{\pi i n} & \text{if } n \text{ is odd} \end{cases}. \tag{1.83}$$

Equation (1.82) is depicted in Figure 1.2a.

From Eq. (1.83), we see that the Fourier series for the function in Eq. (1.70) can be written as

$$\sum_{n=-\infty}^{\infty} C_n e^{+2\pi i n \nu x} = \sum_{n=1,3,5,\ldots} \frac{4}{\pi n} \sin(2\pi n \nu x) = S_{\infty(x)}. \tag{1.84}$$

It is clear that the series in Eq. (1.84) certainly is zero at $x = 0$, while $f(x)$ is +1 for $x = 0$ (Figure 1.2b).

For functions with a finite number of discontinuities, Dirichlet has shown that at discontinuous point of $f(x)$ the infinite Fourier series assumes the arithmetic means of the right- and left-hand limits (Figure 1.2):

$$\sum_{m=-\infty}^{\infty} C_n e^{+2\pi i n \nu x} = \frac{1}{2} \lim_{\varepsilon \to 0} [f(x+\varepsilon) + f(x-\varepsilon)]. \tag{1.85}$$

Figure 1.2b shows how a square wave is perfectly represented by the infinite series, except at points of discontinuity, where the series assumes the mean value.

Another effect occurs when one tries to approximate a discontinuous function $f(x)$ by the *finite* series Fourier series $S_N(x)$. Figure 1.3 shows a rectangular function $f(x)$ and examples of the series, $S_N(x)$, for $N = 9$, 15, and 25, for the interval $0 \le x \le \frac{d}{2}$. These graphs clearly show there is overshoot and ringing occurring in the S_N curves for $N > 1$. This is the so-called Gibbs effect. Looking closely at these three plots, we see that as you add more sinusoids to the series, the width of the overshoot decreases, but the amplitude of the overshoot peak remains about the same, $\frac{4}{\pi}$. These characteristics continue as $N \to \infty$. What is most interesting is that the overshoot is still present with an infinite number of sinusoids, but it has zero width; hence, no energy. Exactly

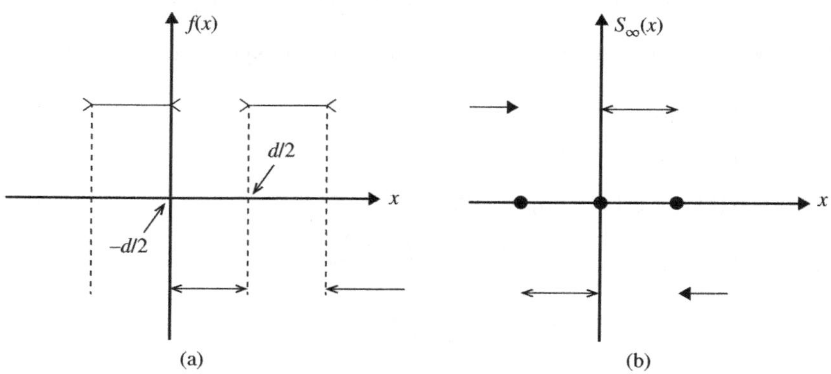

FIGURE 1.2 Dirichlet examples of (a) $f(x)$ and (b) $S_\infty(x)$.

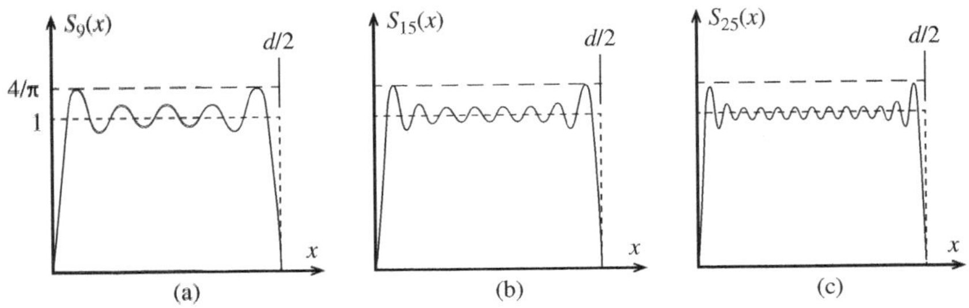

FIGURE 1.3 Plots of (a) $S_9(x)$, (b) $S_{15}(x)$, and (c) $S_{25}(x)$ as a function of x.

at the discontinuity the value of the reconstructed signal converges to the midpoint of the step. The result is the rectangle of amplitude 1. In other words, the summation converges to the signal in the sense that the error between the two has zero energy (Gibbs).

1.3.4 Some Useful Properties of Fourier Series

This section provides some of the interesting properties for Fourier series. If the function $f(x)$ has the Fourier coefficients $\{A_n\}$, $g(x)$ have the coefficients $\{B_n\}$ and both $f(x)$ and $g(x)$ are periodic in $|x| \leq \frac{d}{2}$, then the following properties in both Fourier domains are equivalent:

$$[f(x) + g(x)] \leftrightarrow [A_n + B_n] \tag{1.86}$$

$$f(x) = a\,g(x) \leftrightarrow A_n = aB_n \tag{1.87}$$

$$f(x) = g(Mx) \leftrightarrow A_{Mn} = B_n \text{ (M is a fixed integer > 0).} \tag{1.88}$$

The following two properties, which are quite important, are called the "shift theorem":

$$f(x) = g(x + c) \leftrightarrow A_n = B_n e^{+2\pi\, in\nu c} \tag{1.89}$$

$$f(x) = g(x)e^{+2\pi\, iM\nu c}(M, \text{ an integer}) \leftrightarrow A_n = B_{n-M}. \tag{1.90}$$

The next two properties are sometimes called "reality symmetry" or Hermitian:

$$f(x) = g^*(x) \leftrightarrow A_n = B_n^* \tag{1.91}$$

$$f(x) = f^*(x) \leftrightarrow A_n = A_n^* \tag{1.92}$$

$$f(x) = -g^*(x) \leftrightarrow A_n = -B_n^* \tag{1.93}$$

$$f(x) = g_1(x)g_2(x) \leftrightarrow A_n = \sum_m B_m^{(1)} B_{n-m}^{(2)} \tag{1.94}$$

$$f(x) = g(x)g^*(x) \leftrightarrow A_n = \sum_m B_m B_{m-n}^* \tag{1.95}$$

$$f(x) = \frac{dg(x)}{dx} \leftrightarrow A_n = (2\pi i n\nu)B_n \tag{1.96}$$

$$f(x) = \int_{-x}^{x} g(x')dx' \leftrightarrow A_n = \frac{(B_n + B_{-n})}{(2\pi i n v)} \quad [B_0 = 0 \text{ assumed}] \tag{1.97}$$

$$\underbrace{f(x) = \left(\frac{1}{d}\right) \int_{-d/2}^{d/2} g_1(x')g_2(x' - x)dx' \leftrightarrow A_n = B_n^{(1)} B_{-n}^{(2)}}_{\text{[Cross-Correlation]}} \tag{1.98}$$

$$\underbrace{f(x) = \left(\frac{1}{d}\right) \int_{-d/2}^{d/2} g_1(x')g_2(x - x')dx' \leftrightarrow A_n = B_n^{(1)} B_n^{(2)}}_{\text{[Convolution]}} \tag{1.99}$$

$$\underbrace{f(x) = \left(\frac{1}{d}\right) \int_{-d/2}^{d/2} g(x')g^*(x - x')dx' \leftrightarrow A_n = |B_n|^2}_{\text{[Autocorrelation]}} \tag{1.100}$$

$$\underbrace{f(x) = \left(\frac{1}{d}\right) \int_{-d/2}^{d/2} g(x')g^*(x - x')dx' \leftrightarrow A_n = {B_n}^2.}_{\text{[Autoconvolution]}} \tag{1.101}$$

1.4 FOURIER TRANSFORMS

As noted earlier, the Fourier transform is a generalization of the complex Fourier series in the limit as $d \rightarrow \infty$. We first replace the discrete C_n with the continuous $\hat{f}(v)dv$ while letting $\frac{n}{d} \rightarrow v$, and then change the sum to an integral. The result is that Fourier series representation for $f(x)$ becomes

$$f(x) = \int_{-\infty}^{\infty} \hat{f}(v)e^{2\pi i v x} dx. \tag{1.102}$$

Its counterpart, the inverse Fourier transform, is given by

$$\hat{f}(k) = \int_{-\infty}^{\infty} f(x)e^{-2\pi i v x} dx. \tag{1.103}$$

1.4.1 Some General Properties

Since we generated the Fourier-integral transform as an extension of the Fourier-series transform for periodic functions, we can expect some of the same properties. In particular the Fourier-integral representation is optimized in the Gaussian sense:

$$\int \left| f(x) - \int_{-\infty}^{\infty} \hat{f}(v)e^{2\pi i v x} dx \right|^2 dx \rightarrow \text{Minimum.} \tag{1.104}$$

Hence, it is plausible that the Fourier integral also exhibits the Dirichlet effect. Here are some other properties:

$$f(x) + g(x) \leftrightarrow \widehat{f}(\nu) + \widehat{g}(\nu) \tag{1.105}$$

$$f(x) = ag(x) \leftrightarrow \widehat{f}(\nu) = a\widehat{g}(\nu) \tag{1.106}$$

$$f(x) = g(mx) \leftrightarrow \widehat{f}(\nu) = \frac{\widehat{g}\left(\frac{\nu}{m}\right)}{|m|} \tag{1.107}$$

$$f(x) = g(x + c) \leftrightarrow \widehat{f}(\nu) = \widehat{g}(\nu)e^{2\pi i\nu c} \quad \text{[Shift Theorem]} \tag{1.108}$$

$$f(x) = g(x)e^{2\pi i\nu_0 c} \leftrightarrow \widehat{f}(\nu) = \widehat{g}(\nu - \nu_0) \tag{1.109}$$

$$f(x) = g(-x) \leftrightarrow \widehat{f}(\nu) = \widehat{g}(-\nu) \tag{1.110}$$

$$f(x) = -g(-x) \leftrightarrow \widehat{f}(\nu) = \widehat{g}(\nu) \tag{1.111}$$

$$f(x) = f^*(x) \leftrightarrow \widehat{f}(\nu) = \widehat{f}^*(-\nu) \quad \text{[Reality Symmetry]} \tag{1.112}$$

$$f(x) = -f^*(x) \leftrightarrow \widehat{f}(\nu) = -\widehat{f}^*(-\nu) \tag{1.113}$$

$$f(x) = g_1(x)g_2(x) \leftrightarrow \widehat{f}(\nu) = \int \widehat{g}_1(\mu)\widehat{g}_2(\nu - \mu)d\mu \quad \text{[Convolution]} \tag{1.114}$$

$$\underbrace{f(x) = g(x)g^*(x)}_{\text{Real nonnegative}} \leftrightarrow \widehat{f}(\nu) = \int \widehat{g}(\mu)\widehat{g}^*(\nu - \mu)d\mu \quad \text{[Auto-correlation]} \tag{1.115}$$

$$f(x) = \frac{dg(x)}{dx} \leftrightarrow \widehat{f}(\nu) = 2\pi i\nu\widehat{g}(\nu) \tag{1.116}$$

$$f(x) = \int_{-x}^{x} g(x')dx' \leftrightarrow \widehat{f}(\nu) = \frac{\widehat{g}(\nu) + \widehat{g}(-\nu)}{2\pi i\nu} \tag{1.117}$$

$$f(x) = \int g_1(x')g_2(x' - x)dx' \leftrightarrow \widehat{f}(\nu) = \widehat{g}_1(\nu)\widehat{g}_2(-\nu) \tag{1.118}$$

$$f(x) = \int g_1(x')g_2(x - x')dx' \leftrightarrow \widehat{f}(\nu) = \widehat{g}_1(\nu)\widehat{g}_2(\nu) \tag{1.119}$$

$$f(x) = \int g(x')g^*(x' - x)dx' \leftrightarrow \widehat{f}(\nu) = |\widehat{g}(\nu)|^2 \tag{1.120}$$

$$f(x) = \int g(x')g^*(x - x')dx' \leftrightarrow \widehat{f}(\nu) = \widehat{g}^2(\nu) \tag{1.121}$$

$$f(x, y) = g(x + x_0, y + y_0) \leftrightarrow \widehat{f}(\nu, \mu) = \widehat{g}(\nu, \mu)e^{2\pi i(\nu x_0 + \mu y_0)} \tag{1.122}$$

$$f(x, y) = g(-x, -y) \leftrightarrow \widehat{f}(\nu, \mu) = (-\nu, -\mu) \tag{1.123}$$

$$\underbrace{f(x, y) = g(-x, -y)}_{\text{Inversion around the } y\text{-axis}} \leftrightarrow \widehat{f}(\nu, \mu) = \widehat{g}(-\nu, +\mu) \tag{1.124}$$

$$f(x,y)=f^*(x,y)\leftrightarrow \hat{f},\mu=\hat{f}^*(-\nu,-\mu) \tag{1.125}$$

$$f(x,y)=\iint g_1(x',y')g_2(x'-x,y'-y)dx'dy' \leftrightarrow \hat{f}(\nu,\mu)=\hat{g}_1(\nu,\mu)\hat{g}_2^*(\nu,\mu). \tag{1.126}$$

One of the most important properties of Fourier Transform is the following:

$$\iint f(x',y')g^*(x',y')dx'dy' = \iint \hat{f}(\nu,\mu)\hat{g}^*(\nu,\mu)d\nu d\mu \tag{1.127}$$

or

$$f\cdot g^* = \hat{f}\cdot\hat{g}^*, \tag{1.128}$$

that many other formulae can be derived from it. For example, if $g(x',y')=f(x',y')$, then we have

$$\iint |f(x',y')|^2dx'dy' = \iint |\hat{f}(\nu,\mu)|^2 d\nu d\mu. \tag{1.129}$$

Equation (1.129) comes from *Parseval's Theorem.* In addition, the *Wiener–Khinchin Formula*

$$\iint |f(x',y')|^2 e^{-2\pi i(\nu'x'+\mu'y')}dx'dy' = \iint \hat{f}(\nu,\mu)\hat{f}^*(\nu-\nu',\mu-\mu')d\nu d\mu \tag{1.130}$$

also can be created via Eq. (1.127) by setting $g(x',y')=f(x',y')e^{2\pi i(\nu'x'+\mu'y')}$ and $\hat{g}^*(\nu,\mu)=\hat{f}^*(\nu-\nu',\mu-\mu')$, using previously stated properties.

Before leaving this topic, let us review the Fourier series in polar coordinates. Mathematically, we can write the series as

$$f(r,\varphi)=\sum f_m(r)\,e^{im\varphi}, \tag{1.131}$$

where the coefficients can be written as

$$f_m(r)=\frac{1}{2\pi}\int_0^{2\pi} f(r,\varphi)\,e^{im\varphi}\,d\varphi \tag{1.132}$$

or

$$f_m(r)=\int_{-\infty}^{\infty}\hat{f}_m(\vartheta)\,e^{2\pi i r\vartheta}d\vartheta. \tag{1.133}$$

This last equation implies that Eq. (1.131) can be rewritten as

$$f(r,\varphi)=\sum e^{im\varphi}\int_{-\infty}^{\infty}\hat{f}_m(\vartheta)e^{2\pi i r\vartheta}d\vartheta. \tag{1.134}$$

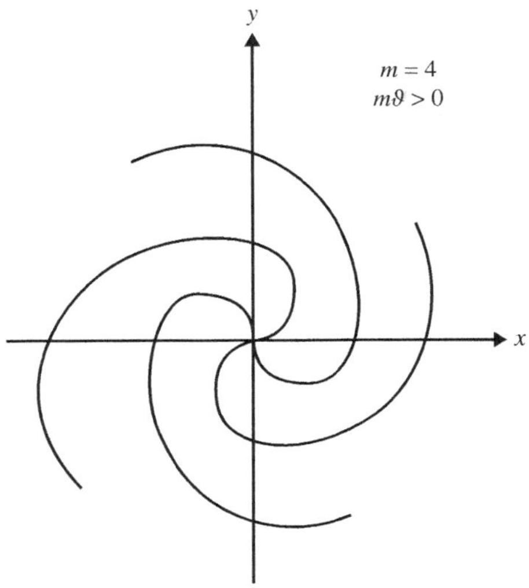

FIGURE 1.4 Plots of polar spirals.

The real part of $e^{im\varphi+2\pi ir\vartheta}$ is $\cos\{m\varphi + 2\pi r\vartheta\}$ and has maximum curves that form as set of m spirals that obey the equation

$$m\varphi + 2\pi r\vartheta = W, \quad \text{for } W = 0, \pm 1, \pm 2, \ldots \tag{1.135}$$

This suggests that Eq. (1.134) represents a superposition of spirals following this expression. The parameter $\frac{2\pi\vartheta}{m}$ specifies the extent of the spirals as illustrated in Figure 1.4.

An alternative to the above is the two-dimensional version of the Fourier transform in polar coordinates. Expressing the Cartesian coordinates in terms of polar coordinates, we have

$$\begin{aligned} x &= r\cos\{\varphi\} & y &= r\sin\{\varphi\}, \\ \nu &= \rho\cos\{\theta\} & \mu &= \rho\sin\{\theta\}, \\ x\nu &= r\rho\cos\{\varphi\}\cos\{\theta\} & y\mu &= r\rho\sin\{\varphi\}\sin\{\theta\}, \end{aligned}$$

and

$$\cos\{\varphi\}\cos\{\theta\} + \sin\{\varphi\}\sin\{\theta\} = \cos\{\varphi - \theta\}. \tag{1.136}$$

This implies that

$$f(x, y) = \int\int_{-\infty}^{\infty} \hat{f}(\mu, \nu)e^{-2\pi i(\nu x+\mu y)}\,d\nu d\mu \tag{1.137a}$$

$$= \int_0^{2\pi}\int_0^{\infty} \hat{f}_{\text{pol}}(\varrho, \theta)e^{2\pi ir\varrho\cos\{\varphi-\theta\}}\varrho d\varrho d\theta = f_{\text{pol}}(r, \varphi). \tag{1.137b}$$

Comparing Eqs. (1.137a) and (1.137b), we see that the transformation to polar coordinates changes the functions $f(x,y)$ and $\hat{f}(\mu,\nu)$ to $f_{\text{pol}}(r,\varphi)$ and $\hat{f}_{\text{pol}}(\varrho,\theta)$, respectively, even though they represent the same spatial pattern. This is important for the reader to recognize and not assume it is just a change of variables in the cited functions.

Rewriting Eq. (1.137b) into the following form:

$$\begin{aligned} f_{\text{pol}}(r,\varphi) &= \int_0^{2\pi}\int_0^{\infty} \hat{f}_{\text{pol}}(\varrho,\theta) e^{2\pi i r\varrho\cos\{\varphi-\theta\}}\varrho d\varrho d\theta \\ &= \int_0^{2\pi}\int_0^{\infty} \hat{f}_{\text{pol}}(\varrho,\theta) e^{2\pi i r\varrho\sin\left\{\frac{\pi}{2}+\varphi-\theta\right\}}\varrho d\varrho d\theta, \end{aligned} \tag{1.138}$$

one can use the following Bessel Function Series

$$e^{A\sin(\alpha)} = \sum_{n=-\infty}^{\infty} J_n(A) e^{in\alpha} \tag{1.139}$$

to yield

$$f_{\text{pol}}(r,\varphi) = \sum_{n=-\infty}^{\infty} \int_0^{\infty} J_n(2\pi r\rho) \int_0^{2\pi} \hat{f}_{\text{pol}}(),\theta) e^{in\left\{\frac{\pi}{2}+\varphi-\theta\right\}} \rho d\rho d\theta. \tag{1.140}$$

Expressing $\hat{f}_{\text{pol}}(\rho,\theta)$ as a Fourier series, we obtain

$$\begin{aligned} f_{\text{pol}}(r,\varphi) &= \sum_{n=-\infty}^{\infty} \int_0^{\infty} J_n(2\pi r\rho) \sum_{m=-\infty}^{\infty} \hat{f}_m^{\text{pol}}(\rho) \int_0^{2\pi} e^{i(m-n)\theta+in\left\{\frac{\pi}{2}+\varphi\right\}} \rho d\rho d\theta \\ &= \sum_{n=-\infty}^{\infty} \int_0^{\infty} J_n(2\pi r\rho) \sum_{m=-\infty}^{\infty} \hat{f}_m^{\text{pol}}(\rho) e^{in\left\{\frac{\pi}{2}+\varphi\right\}} \int_0^{2\pi} e^{i(m-n)\theta} \rho d\rho d\theta. \end{aligned} \tag{1.141}$$

$$\begin{aligned} &= \sum_{n=-\infty}^{\infty} \int_0^{\infty} J_n(2\pi r\rho) \sum_{n=-\infty}^{\infty} \hat{f}_m^{\text{pol}}(\rho) e^{in\left\{\frac{\pi}{2}+\varphi\right\}} (2\pi\delta_{mn}) \rho d\rho. \\ &= 2\pi \sum_{n=-\infty}^{\infty} \int_0^{\infty} J_n(2\pi r\rho) \hat{f}_n^{\text{pol}}(\rho) e^{in\left\{\frac{\pi}{2}+\varphi\right\}} \rho d\rho. \end{aligned} \tag{1.142}$$

Example 1.8 If $f_{\text{pol}}(r,\varphi)$ does not depend on φ, then all the coefficients $\{\hat{f}_n^{\text{pol}}(\rho)\}$ are zero, except for $\hat{f}_0^{\text{pol}}(\rho)$. If we set $2\pi\hat{f}_n^{\text{pol}}(\rho) = \hat{f}_{\text{pol}}(\rho)$, we obtain the Bessel Transformations given in the following equations:

$$f_{\text{pol}}(r) = \int_0^{\infty} J_0(2\pi r\rho) \hat{f}_{\text{pol}}(\rho) \rho d\rho \tag{1.143}$$

and

$$\hat{f}_{pol}(\rho) = \int_0^{\infty} J_0(2\pi r\rho) f_{pol}(r) r dr. \tag{1.144}$$

1.5 DIRAC DELTA FUNCTION

The companion to the Kronecker Delta Function introduced in the previous section is the Dirac Delta function. It is used with continuous functions and has application throughout science, engineering, and other disciplines. Consequently, it warrants a discussion to ensure the reader understands its function and properties.

The *Dirac Delta Function* $\delta(x)$ often is described as infinitely high spike with an infinitesimal width such that the "area" essentially is 1. This is not too satisfying as a mathematical function definition and a better way is to put in terms of limiting function, say like the following:

$$\delta(x) = \lim_{A\to\infty} \begin{cases} A & \text{if } |x| \leq \frac{1}{2A} \\ 0 & \text{otherwise} \end{cases} \tag{1.145a}$$

$$= \lim_{A\to\infty} A \text{ rect}\left[\frac{x}{\frac{1}{A}}\right], \tag{1.145b}$$

where

$$\text{rect}[x] = \begin{cases} 1 & \text{if } |x| \leq \frac{1}{2A} \\ 0 & \text{if } |x| > \frac{1}{2A} \end{cases} \tag{1.146}$$

is the *rectangle function*. This form of the delta function satisfies the implicit definition by keeping the area constant while its height gets larger and its width gets narrower as A moves to infinity. This implies that

$$\int_{-\infty}^{\infty} g(x)\delta(x-c)dx = \lim_{A\to\infty} \int_{-\infty}^{\infty} g(x)\, A \text{ rect}\left[\frac{x-c}{\frac{1}{A}}\right] dx = g(c). \tag{1.147}$$

There are other forms for the delta function than the above that satisfy the implicit definition and normalize to one. However, not all of them mean that $\delta(x-c) = 0$ whenever $x \neq c$. For example, the function

$$A\,\text{sinc}[A(x-c)] = \frac{\text{sinc}[\pi A(x-c)]}{\pi A(x-c)}. \tag{1.148}$$

Equation (1.148) defines the *sinc function*. This means that we hypothesize that

$$\int_{-\infty}^{\infty} g(x)\delta(x-c)dx = \lim_{A\to\infty} \int_{-\infty}^{\infty} g(x)\, A\,\text{sinc}\left[\frac{x-c}{A}\right] dx = g(c). \tag{1.149}$$

Because of the periodic nature of the sine function, there will be points where this sinc function will equal 1 when $x \neq c$. Fortunately, when A is large, it will oscillate rapidly, resulting in the $g(x)$ contributions outside of $x = c$, averaging to zeros because of that fact over the long integration interval. This implies that

$$\delta(x) = \lim_{A\to\infty} A\,\mathrm{sinc}\left[\frac{x}{A}\right] \tag{1.150}$$

is a legitimate representation of the Dirac Delta Function.

One of the properties of this function is that it is the derivative of the *Unit Step Function*. In particular, we have

$$\frac{d}{dx}U(x-c) = \delta(x-c), \tag{1.151}$$

where

$$U(x) = \begin{cases} 1 & \text{if } x > 0 \\ \frac{1}{2} & \text{if } x = 0 \\ 0 & \text{if } x < 0 \end{cases} \tag{1.152}$$

is the Heaviside (or Unit) Step Function.

1.6 PROBABILITY THEORY

Optical detection and estimation theory is based on the concepts derived from probability theory. This theory is fundamental to understanding the performance of any optical remote sensing or communications system. This section provides a brief review of probability theory following Bar-Shalom and Fortmann, Appendix B [2] and Helstrom [6]. It recognizes that this theory is designed to understand the possible outcomes of *chance experiments*. These outcomes are unpredictable, but known with certain knowledge of their relative frequency of occurrence, countable or not. The theory provides a means for creating a strategy for "predicting" the various outcomes, which hopefully maximizes possible gains and minimizes losses. More information can be found in books on probability, for example, by Venkatesh [7] and Helstrom [6].

1.6.1 Axioms of Probability

Let $\boldsymbol{A}$ specify an *event* in a chance experiment whose results come from a specific set of random outcome. This could be "heads" if one was flipping a coin or the number "3" in throwing a die. Let $\boldsymbol{S}$ represent an assured event in the experiment; this would mean "a number between 1 and 6 coming up after throwing a die." The *probability* of an event, denoted by $P\{\boldsymbol{A}\}$, occurring obeys these following axioms:

Axiom 1

It is nonnegative, that is,

$$P\{\boldsymbol{A}\} \geq 0. \tag{1.153}$$

Axiom 2

It is unity for assured event, that is,

$$P\{\boldsymbol{S}\} = 1. \tag{1.154}$$

Axiom 3

It is additive over the union of mutually exclusive events, that is, if the events $\boldsymbol{A}$ and $\boldsymbol{B}$ have no common elements, then

$$\boldsymbol{A} \cap \boldsymbol{B} \equiv \{\boldsymbol{A} \text{ and } \boldsymbol{B}\} = \{\boldsymbol{A}, \boldsymbol{B}\} = \emptyset, \tag{1.155}$$

and

$$P\{\boldsymbol{A} \cup \boldsymbol{B}\} = P\{\boldsymbol{A} \text{ or } \boldsymbol{B}\} = P\{\boldsymbol{A} + \boldsymbol{B}\} = P\{\boldsymbol{A}\} + P\{\boldsymbol{B}\}. \tag{1.156}$$

Axiom 4

If $\boldsymbol{A}_i \cap \boldsymbol{A}_j = \emptyset$ for all i and j, $i \neq j$, then

$$P\{\cup_{i=1}^{\infty} \boldsymbol{A}_i\} = \sum_{i=1}^{\infty} P(\boldsymbol{A}_i). \tag{1.157}$$

An assignment of probabilities to the events consistent with these axioms is known as a *probability measure.*

Corollaries to the above are

Corollary 1

$$P(\emptyset) = 0. \tag{1.158}$$

Corollary 2

$$0 \leq P\{\boldsymbol{A}\} \leq 1. \tag{1.159}$$

Corollary 3

With $\overline{\boldsymbol{A}}$ being the complement of the event $\boldsymbol{A}$, then

$$P\{\overline{\boldsymbol{A}}\} = 1 - P\{\boldsymbol{A}\}, \tag{1.160}$$

using Eqs. (1.154) and (1.157).

Corollary 4

If the n events $\boldsymbol{A}_i$, $\{1 \leq i \leq n\}$, are mutually exclusive, for example, $\boldsymbol{A}_i \cap \boldsymbol{A}_j = \emptyset$ for all i and j, $i \neq j$, then

$$P\{\cup_{i=1}^{n} \boldsymbol{A}_i\} = \sum_{i=1}^{n} P(\boldsymbol{A}_i). \tag{1.161}$$

Corollary 5

$$P\{\boldsymbol{A} \cup \boldsymbol{B}\} = P\{\boldsymbol{A}\} + P\{\boldsymbol{B}\} - P\{\boldsymbol{A} \cap \boldsymbol{B}\}. \tag{1.162}$$

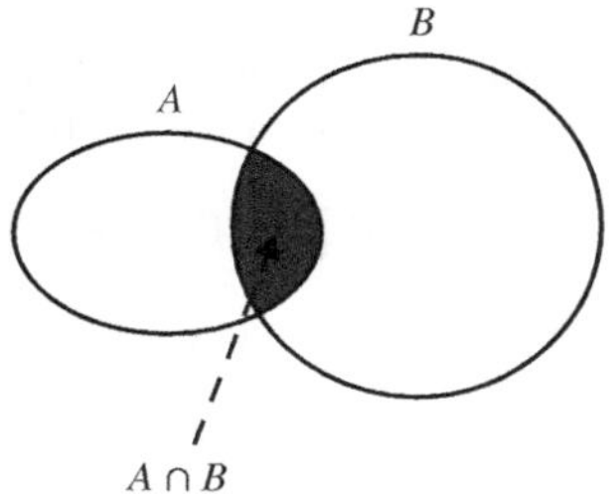

FIGURE 1.5 Venn diagram of events ***A*** and ***B***.

It should be noted that probability theory is essentially computing the new probabilities from an initial probability measure set over the events of the chance experiment. It is a Bookkeeping drill that requires that all the unit quantities of probability be accounted for, no probability is negative, and no excess probability emerges.

1.6.2 Conditional Probabilities

Sometimes the outcomes from two events are common. These type of outcomes form the intersection $\boldsymbol{A} \cap \boldsymbol{B}$, as depicted in Figure 1.5. The conditional probability of the event $\boldsymbol{A}$, given event $\boldsymbol{B}$, is defined as

$$P\{\boldsymbol{A} \mid \boldsymbol{B}\} = \frac{P\{\boldsymbol{A} \cap \boldsymbol{B}\}}{P\{\boldsymbol{B}\}} \tag{1.163}$$

in terms of the probability measure assigned to the event. The probability $P\{\boldsymbol{A} \mid \boldsymbol{B}\}$ can be interpreted as the likelihood of event $\boldsymbol{A}$ occurring when it is known that event $\boldsymbol{B}$ has occurred.

Example 1.9 A coin is tossed three times in succession. The possible outcomes from this chance experiment are

ϑ_1	ϑ_2	ϑ_3	ϑ_4	ϑ_5	ϑ_6	ϑ_7	ϑ_8
$\{HHH\}$	$\{HHT\}$	$\{HTH\}$	$\{HTT\}$	$\{THH\}$	$\{THT\}$	$\{TTH\}$	$\{TTT\}$

It is clear that there are eight elements to the *universal set* of outcomes of this experiment. For illustrative purposes, let us assume that the coin is unbalanced and the individual probabilities of occurrence for the above universal set after flipping the coin many times come out nonuniform and have the probabilities given in Table 1.1.

TABLE 1.1 Probabilities for Possible Outcomes from Flipping a Coin

ϑ_1	ϑ_2	ϑ_3	ϑ_4	ϑ_5	ϑ_6	ϑ_7	ϑ_8
$\{HHH\}$	$\{HHT\}$	$\{HTH\}$	$\{HTT\}$	$\{THH\}$	$\{THT\}$	$\{TTH\}$	$\{TTT\}$
0.07	0.31	0.17	0.05	0.29	0.01	0.06	0.04

(a) What is the probability that three heads will appear, given the coin's first toss is heads? From our table, the possible outcomes beginning with heads are $\boldsymbol{B} = \{\vartheta_1, \vartheta_2, \vartheta_3, \vartheta_4\}$. This conditioning event has probability

$$P\{\boldsymbol{B}\} = 0.07 + 0.31 + 0.17 + 0.05 = 0.60.$$

The probability that three heads come up is

$$P\{HHH\} = 0.07.$$

Using Eq. (1.163), we have

$$P\{HHH\,|\boldsymbol{B}\} = \frac{P\{HHH\}}{P\{\boldsymbol{B}\}} = \frac{0.07}{0.6} \approx 0.11.$$

(b) What is the probability that three heads will appear, given the coin was heads on the first two tosses? In this case, the possible outcomes are $\boldsymbol{B} = \{\vartheta_1, \vartheta_2\}$. This implies that

$$P\{\boldsymbol{B}\} = 0.07 + 0.31 = 0.38,$$

and we have

$$P\{HHH\,|\boldsymbol{B}\} = \frac{0.07}{0.38} \approx 0.18.$$

(c) What is the probability that tails appeared on the second and third tosses, given the number of tails is odd? In this case, the events are $\boldsymbol{A} = \{\vartheta_4, \vartheta_8\}$ and $\boldsymbol{B} = \{\vartheta_2, \vartheta_3, \vartheta_5, \vartheta_8\}$. The intersection of $\boldsymbol{A}$ and $\boldsymbol{B}$ is $\boldsymbol{A} \cap \boldsymbol{B} = \{\vartheta_8\}$ and we find that

$$P\{\boldsymbol{A}\,|\boldsymbol{B}\} = \frac{P\{\boldsymbol{A} \cap \boldsymbol{B}\}}{P\{\boldsymbol{B}\}} = \frac{0.04}{0.31 + 0.17 + 0.29 + 0.04} = \frac{0.04}{0.81} \approx 0.5.$$

Example 1.10 The reliability of a rectifier is such that the probability of it lasting at least t hours is given by $e^{-\alpha t^2}$. What is the probability that the rectifier fails between times t_1 and t_2, given it is still operating after time τ, where $\tau < t_1 < t_2$.

The universal set of time possible goes from zero to infinity. The conditioning event $\boldsymbol{B}$ is that the failure happens in the interval $\tau < t < \infty$. The associated probability of that happening is $e^{-\alpha\tau^2}$, by the hypothesis. Event $\boldsymbol{A}$ is that the rectifier fails within the interval $t_1 < t < t_2$. This implies that

$$P\{\text{failure in } t_1 < t < \infty\} = P\{\text{failure in } t_1 < t < t_2\} + P\{\text{failure in } t_2 < t < \infty\}$$

allows us to write

$$\begin{aligned} P\{\text{failure in } t_1 < t < t_2\} &= P\{\text{failure in } t_2 < t < \infty\} - P\{\text{failure in } t_1 < t < \infty\} \\ &= e^{-\alpha t_2{}^2} - e^{-\alpha t_1{}^2} \end{aligned}$$

using Eq. (1.144). From the above description, we have

$$P\{\boldsymbol{A}\,|\boldsymbol{B}\} = \frac{P\{\boldsymbol{A} \cap \boldsymbol{B}\}}{P\{\boldsymbol{B}\}} = \frac{e^{-\alpha t_2{}^2} - e^{-\alpha t_1{}^2}}{e^{-\alpha\tau^2}}.$$

If one cross multiplies the above conditional probability and use the fact that

$$A \cap B = B \cap A,$$

then the following theorem can be derived:

$$P\{A \cap B\} = P\{B\}\, P\{A\, |B\}.$$

This results in the following corollary:

Corollary 6

For any set of events $\{A_1, A_2, \dots, A_m\}$, we have

$$P\{A_1 \cap A_2, \cap \cdots \cap A_m\}$$
$$= P\{A_1\}P\{A_2\, |A_2\}P\{A_3\, |A_2 \cap A_2\} \cdots P\{A_m\, |A_1 \cap A_2, \cap \cdots \cap A_{m-1}\}$$

The above theorem is known as the *multiplication theorem.*

Example 1.11 A class contains 12 boys and 4 girls. If three students are chosen at random from the class, what is the probability that they are all boys?

The probability that the first student chosen is a boy is $\frac{12}{16}$ since there are 12 boys in a class of 16 students. If the first selection is a boy, then the probability that the next student chosen is a boy is $\frac{11}{15}$. Finally, If the first two selections are boys, then the probability that the final selected student is a boy is $\frac{10}{14}$. Then by the multiplication theorem, the probability that all three students chosen are boys is

$$\left(\frac{12}{16}\right)\left(\frac{11}{15}\right)\left(\frac{10}{14}\right) = \left(\frac{1}{4}\right)\left(\frac{11}{1}\right)\left(\frac{1}{7}\right) = \frac{11}{28}.$$

1.6.3 Probability and Cumulative Density Functions

Nearly all applications of probability to science and engineering are derived from the outcomes from chance experiments, which are associated with specific numbers, for example, voltages, currents, and power. Thus, events are sets of outcomes from chance experiments. A scalar *random variable* is a real-valued parameter that labels an outcome of a chance experiment for a given event with a probability measure assigned to it. Its value is known as its *realization.* It follows that the outcome of experiment can be a single number, a pair or more of numbers, and even a function of the parameter. We now turn to defining the two key functions for characterizing all of the above. They are the *probability density function (PDF)* and *cumulative distribution function (CDF).*

The PDF $p_x(x)$ of a continuous-valued random variable x at $x = \vartheta$ is defined as

$$p_x(\vartheta)d\vartheta = P\{\vartheta \le x \le \vartheta + d\vartheta\} \ge 0. \tag{1.164}$$

For simplicity, the PDF can be shorted to $p(\vartheta)$, where the argument of the function defines it. We will do so for the remainder of this section.

Example 1.12 Let us assume we have a device for counting the number of electrons radiated by some sort of emissive surface during the period of τ seconds. We also assume we have a constant time average rate of occurrence for said electron emissions equal to λ. A stochastic process

$$Z(t) = C\sum_{m=1}^{k} \delta(t - t_m), \tag{1.165}$$

composed of a sequence of electron emissions (impulses) occurring at times t_m in a time period $-\frac{\tau}{2}$ to $\frac{\tau}{2}$ and multiplied by a constant C, is a Poisson process if the probability that the number, X, of impulses occurring in the time period τ is an integer with probability

$$P\{X(t) = k\} = \frac{(\lambda t)^k}{k!} e^{-\lambda t}. \tag{1.166}$$

[8]. In Eq. (1.165), an impulse is represented by $\delta(t)$, the Dirac delta function defined previously.

The PDF for $Z(t)$ is not Poisson, but rather equals

$$P\{Z(t)\} = P\{t_1, t_2, \ldots, t_k \, |k\} P\{X(t) = k\}. \tag{1.167}$$

If the times of the impulse occurrences are independent of one another and independent of the total number of pulses, then Eq. (1.167) can be written as

$$P\{Z(t)\} = \frac{1}{\tau^k} P\{X(t) = k\}. \tag{1.168}$$

Using Eq. (1.164) and Corollary 4, we can write

$$P\{a \le \mathbf{x} \le b\} = \int_a^b p(x)dx. \tag{1.169}$$

The function

$$P\{-\infty \le \mathbf{x} \le b\} \equiv P_x(b) = P\{\mathbf{x} \le b\} = \int_{-\infty}^{b} p(x)dx \tag{1.170}$$

is called the CDF of $\mathbf{x}$ at b. Since the event $S \equiv (\mathbf{x} \le \infty)$ is a sure thing, then

$$P\{\mathbf{x} \le \infty\} = \int_{-\infty}^{\infty} p(x)dx = 1. \tag{1.171}$$

Equation (1.171) is known as the normalization integral for the PDF $p(x)$. This equation must be true for $p(x)$ to be a proper PDF.

The relationship between the PDF and CDF can be derived from Eq. (1.170), namely, we see that

$$p(x) = \frac{d}{d\vartheta} P_{\mathbf{x}}(\vartheta)|_{\vartheta = x}, \tag{1.172}$$

if the derivative exists.

1.6.4 Probability Mass Function

Let us assume that we have discrete random variable $\boldsymbol{x}$ comprising one set of possible countable values $\{\vartheta_1, \vartheta, \ldots, \vartheta_m\}$. (Note: the set could extend to infinity and still be countable.) As a result, their probabilities are written as

$$P\{\boldsymbol{x} = \vartheta_i\} = \mu_{\boldsymbol{x}}(\vartheta_i) = \mu_i, \qquad \text{for } i = 1, 2, \ldots, m. \tag{1.173}$$

The parameter μ_i is the point mass associated with outcome ϑ_i. Similar to Eq. (1.171), we have

$$\sum_{i=1}^{m} \mu_i = 1. \tag{1.174}$$

Using the Dirac Delta Function, we can write the PDF for the above random variable as

$$p(x) = \sum_{i=1}^{m} \mu_i \, \delta(x - \vartheta_i). \tag{1.175}$$

This PDF normalizes properly, so it is a true PDF. The distribution expressed in Eq. (1.175) has jumps in it at values of ϑ_i. Its CDF is a "stair step" or "staircase" function whose derivative is zero everywhere but at the jumps, where it is an impulse.

As a final note, a random variable $\boldsymbol{x}$ can take on values that are continuous, discrete, or a combination of both. The PDF for a mixed random variable has the form

$$p(x) = p_c(x) + \sum_{i=1}^{m} \mu_i \, \delta(x - \vartheta_i). \tag{1.176}$$

where $p_c(x)$ represents the continuous part of the PDF. As one would expect,

$$\begin{aligned} \int p(x)dx &= \int_{x \in X} p_c(x)dx + \sum_{i=1}^{m} \mu_i \int \delta(x - \vartheta_i)dx \\ &= \int_{x \in X} p_c(x)dx + \sum_{i=1}^{m} \mu_i = 1. \end{aligned} \tag{1.177}$$

Example 1.13 Let us again assume that we have discrete random variable $\boldsymbol{x}$ that only takes on the following values $\{\vartheta_1, \vartheta, \ldots, \vartheta_m\}$. Their associated probabilities are $P\{\boldsymbol{x} = \vartheta_i\}; i = 1, \ldots, m$, which normalize properly. Its CDF will have steps of height $P\{\boldsymbol{x} = \vartheta_i\}$ at all points ϑ_i and can be written as

$$P_{\boldsymbol{x}}(x) = \sum_{i=1}^{m} P\{\boldsymbol{x} = \vartheta_i\} \, U(x - \vartheta_i). \tag{1.178}$$

Using Eq. (1.172) for the PDF, we find that

$$p_{\boldsymbol{x}}(x) = \sum_{i=1}^{m} P\{\boldsymbol{x} = \vartheta_i\} \, \delta(x - \vartheta_i) \tag{1.179a}$$

$$= \sum_{i=1}^{m} \mu_i \, \delta(x - \vartheta_i). \tag{1.179b}$$

Equation (1.179b) is the same one we used in Eq. (1.173) for characterizing the PDF in terms of the point masses.

1.6.5 Expectation and Moments of a Scalar Random Variable

The expected value, or *mean* or first moment, of a random variable is defined as

$$\bar{\boldsymbol{x}} \equiv \mathcal{E}\{\boldsymbol{x}\} = \int_{-\infty}^{\infty} x \, p(x) dx. \tag{1.180}$$

The expected value of any function of the random variable therefore is

$$\mathcal{E}\{g(\boldsymbol{x})\} = \int_{-\infty}^{\infty} g(x) \, p(x) dx. \tag{1.181}$$

The *variance*, or second moment, of a random variable is defined as

$$\text{var}(\boldsymbol{x}) = \sigma_{\boldsymbol{x}}^2 \equiv \mathcal{E}\{(\boldsymbol{x} - \bar{\boldsymbol{x}})^2\} = \mathcal{E}\{\boldsymbol{x}^2\} - (\bar{\boldsymbol{x}})^2 = \int_{-\infty}^{\infty} (x - \bar{\boldsymbol{x}})^2 \, p(x) dx. \tag{1.182}$$

The square root of the variance is the called the *standard deviation*. Looking at the forms of Eqs. (1.180) and (1.182) implies that the nth (noncentral) moment of the random variable is given by

$$\mathcal{E}\{\boldsymbol{x}^n\} = \int_{-\infty}^{\infty} x^n \, p(x) dx. \tag{1.183}$$

Example 1.14 Let us assume a random variable $\boldsymbol{x}$ that is uniformly distributed. Its PDF is given by

$$p(x) = \begin{cases} \frac{1}{x_2 - x_1} & \text{for } x_1 \le x \le x_2 \\ 0 & \text{otherwise} \end{cases}. \tag{1.184}$$

The mean for this random variable equals

$$\bar{\boldsymbol{x}} \equiv \mathcal{E}\{\boldsymbol{x}\} = \int_{x_1}^{x_2} \left(\frac{x}{x_2 - x_1} \right) dx = \frac{1}{2} \left(\frac{x^2}{x_2 - x_1} \right) \Bigg|_{x_1}^{x_2} = \frac{1}{2} \left(\frac{{x_2}^2 - {x_1}^2}{x_2 - x_1} \right) = \frac{(x_2 + x_1)}{2}.$$

The covariance equals

$$\begin{aligned} \text{var}(\boldsymbol{x}) &= \int_{x_1}^{x_2} \left(\frac{x^2}{x_2 - x_1} \right) dx - (\bar{\boldsymbol{x}})^2 = \frac{1}{3} \left(\frac{x^3}{x_2 - x_1} \right) \Bigg|_{x_1}^{x_2} - \left(\frac{(x_2 + x_1)}{2} \right)^2 \\ &= \frac{(x_2 - x_1)^2}{12}. \end{aligned}$$

1.6.6 Joint PDF and CDF of Two Random Variables

Given two random variables $\boldsymbol{x}$ and $\boldsymbol{y}$, the joint PDF is defined as the probability of the joint event, which is denoted by the set intersection symbol, and is written as

$$p_{xy}\{\vartheta,\zeta\}d\vartheta\, d\zeta \equiv P_{xy}\{[\vartheta < x < \vartheta + d\vartheta] \cap [\zeta < y < \zeta + d\zeta]\}. \tag{1.185}$$

The function $p_{xy}\{x, y\}$ is called the *join* or *bivariate PDF*. It follows that

$$p_{xy}(x,y) = \frac{d}{d\vartheta\, d\zeta} P_x(\vartheta,\zeta)|_{\vartheta=x,\zeta=y} \tag{1.186}$$

defines the relationship between the bivariate PDF and bivariate CDF, provided that $P_x(x, y)$ is continuous and differentiable. Integrating Eq. (1.185) over one of the variables yields the PDF of the other variable, that is,

$$\int_{-\infty}^{\infty} p_{xy}(\vartheta,\zeta)d\zeta = p_x(\vartheta) \tag{1.187}$$

or

$$\int_{-\infty}^{\infty} p(\vartheta,\zeta)d\zeta = p(\vartheta) \tag{1.188}$$

in reduced notation. The resulting PDF is called the *marginal PDF* or *marginal density*.

The *covariance* of two scalar random variables ϑ_1 and ϑ_2 with means $\overline{\vartheta}_1$ and $\overline{\vartheta}_2$, respectively, is equal to

$$\mathcal{E}\{(\vartheta_1 - \overline{\vartheta}_1)(\vartheta_2 - \overline{\vartheta}_2)\} = \int_{-\infty}^{\infty} (\vartheta_1 - \overline{\vartheta}_1)(\vartheta_2 - \overline{\vartheta}_2)p(\vartheta_1,\vartheta_2)d\vartheta_1 d\vartheta_2 \tag{1.189}$$

$$= \sigma^2_{\vartheta_1\vartheta_2}. \tag{1.190}$$

The correlation coefficient of these two random variables can be written as

$$\rho_{12} = \frac{\sigma^2_{\vartheta_1\vartheta_2}}{\sigma_{\vartheta_1}\sigma_{\vartheta_2}}, \tag{1.191}$$

where σ_{ϑ_i} is the standard deviation of the variable ϑ_i, $i = 1, 2$. The correlation coefficient of any two random variables must obey the following inequality:

$$|\rho_{12}| \le 1. \tag{1.192}$$

1.6.7 Independent Random Variables

If the conditional PDF $p_x(x|y)$ does not depend on y, then we have

$$p_x(x|y) = p_x(x), \tag{1.193}$$

which says that any observation of the random variables $\boldsymbol{x}$ provides no information about the variable $\boldsymbol{y}$. This situation establishes that the two random variables $\boldsymbol{x}$ and $\boldsymbol{y}$ are *statistically independent*. Another way to say this is that two events are independent if the probability of their joint event equals the product of their marginal probabilities, or

$$P\{\boldsymbol{A} \cap \boldsymbol{B}\} = P\{\boldsymbol{A}, \boldsymbol{B}\} = P\{\boldsymbol{A}\}P\{\boldsymbol{B}\}. \tag{1.194}$$

The conclusion from this is that

$$p_{xy}(x, y) = p_x(x)p_y(y) \tag{1.195}$$

and

$$P_{xy}(x, y) = P_x(x)P_y(y), \tag{1.196}$$

both in terms of marginal probabilities.

1.6.8 Vector-Valued Random Variables

The PDF of the vector-valued random variable

$$\boldsymbol{x} = [x_1 \cdots x_m]^T \tag{1.197}$$

is defined as the joint density of its components

$$p_{x_1 \cdots x_m}(\vartheta_1 \cdots \vartheta_m)d\vartheta_1 \cdots d\vartheta_m \equiv p_x(\boldsymbol{\vartheta})d\boldsymbol{\vartheta} = P_x(\cap_{i=1}^m \{\vartheta_i < x_i \le \vartheta_i + d\vartheta_i\}), \tag{1.198}$$

where the set intersection symbol is used to denote the joint event

$$\boldsymbol{A} \cap \boldsymbol{B} = \{\boldsymbol{A} \text{ and } \boldsymbol{B}\} = \{\boldsymbol{A}, \boldsymbol{B}\}. \tag{1.199}$$

The *mean* of $\boldsymbol{x}$ is the result of the m-fold integration

$$\bar{\boldsymbol{x}} = \mathcal{E}\{\boldsymbol{x}\} = \int \boldsymbol{x} p_x(\boldsymbol{x})d\boldsymbol{x}. \tag{1.200}$$

The *covariance matrix* of $\boldsymbol{x}$ is given by

$$\boldsymbol{\Gamma}_{xx} = \mathcal{E}\left\{(\boldsymbol{x} - \bar{\boldsymbol{x}})(\boldsymbol{x} - \bar{\boldsymbol{x}})^T\right\} = \int (\boldsymbol{x} - \bar{\boldsymbol{x}})(\boldsymbol{x} - \bar{\boldsymbol{x}})^T \, p_x(\boldsymbol{x})d\boldsymbol{x}. \tag{1.201}$$

The covariance matrix is a positive definite (or semidefinite) matrix, whose diagonal elements are the variances of the random variable $\boldsymbol{x}$. The off-diagonal elements are the covariances between various components of $\boldsymbol{x}$.

The *characteristic* or *moment generating function* of a vector random variable is defined as

$$M_x(\boldsymbol{S}) = \mathcal{E}\left\{e^{\boldsymbol{S}^T\boldsymbol{x}}\right\} = \int e^{\boldsymbol{S}^T\boldsymbol{x}} p_x(\boldsymbol{x})d\boldsymbol{x}, \tag{1.202}$$

which is the Fourier transform of the PDF. The first moment of $\boldsymbol{x}$ is related to the characteristic function via the relation

$$\mathcal{E}\{\boldsymbol{x}\} = \nabla_S M_x(\boldsymbol{S})|_{\boldsymbol{S}=\mathbf{0}}, \tag{1.203}$$

where ∇_S is the (column) gradient operator defined in Section 1.2.7. Similarly, we have

$$\mathcal{E}\left\{\boldsymbol{x}\boldsymbol{x}^T\right\} = \nabla_S \nabla_S^T M_x(\boldsymbol{S})|_{\boldsymbol{S}=\mathbf{0}}. \tag{1.204}$$

1.6.9 Gaussian Random Variables

The PDF of a *Gaussian or normally distributed* random variable is

$$p(x) = N(x, \overline{x}, \sigma^2) = \frac{1}{\sqrt{2\pi\sigma^2}} \exp\left\{-\frac{(x-\overline{x})^2}{2\sigma^2}\right\}, \tag{1.205}$$

where $N(x, \overline{x}, \sigma^2)$ denotes the normal PDF, $\overline{x}$ is the mean, and σ^2 its variance. These first two moments totally characterize the Gaussian random variance and are known as its statistics.

A vector-valued Gaussian random variable has a PDF of the form

$$p(\boldsymbol{x}) = N(\boldsymbol{x}, \overline{\boldsymbol{x}}, \boldsymbol{\Gamma}) = \frac{1}{\sqrt{2\pi\boldsymbol{\Gamma}}} \exp\{-(\boldsymbol{x}-\overline{\boldsymbol{x}})^T \boldsymbol{\Gamma}^{-1}(\boldsymbol{x}-\overline{\boldsymbol{x}})\}, \tag{1.206}$$

where $\overline{\boldsymbol{x}}$ is the mean of the vector $\boldsymbol{x}$ and $\boldsymbol{\Gamma}$ is its covariance matrix. If $\boldsymbol{\Gamma}$ is a diagonal matrix, then the elements of vector $\boldsymbol{x}$ are uncorrelated and independent. Consequently, their joint PDF equals the product of their marginal PDFs.

Two vector random vectors $\boldsymbol{x}$ and $\boldsymbol{y}$ are jointly Gaussian if the stacked vector

$$z = \begin{bmatrix} \boldsymbol{x} \\ \boldsymbol{y} \end{bmatrix}. \tag{1.207}$$

Its PDF is given by

$$p(\boldsymbol{x}, \boldsymbol{y}) = p(z) = N(z, \overline{z}, \boldsymbol{\Gamma}_{zz}). \tag{1.208}$$

The mean and covariance matrix of z in terms of those vectors $\boldsymbol{x}$ and $\boldsymbol{y}$ are

$$\overline{z} = \begin{bmatrix} \overline{\boldsymbol{x}} \\ \overline{\boldsymbol{y}} \end{bmatrix} \tag{1.209}$$

and

$$\boldsymbol{\Gamma}_{zz} = \begin{bmatrix} \boldsymbol{\Gamma}_{xx} & \boldsymbol{\Gamma}_{xy} \\ \boldsymbol{\Gamma}_{yx} & \boldsymbol{\Gamma}_{yy} \end{bmatrix}, \tag{1.210}$$

where

$$\boldsymbol{\Gamma}_{xx} = \mathcal{E}\left\{(\boldsymbol{x}-\overline{\boldsymbol{x}})(\boldsymbol{x}-\overline{\boldsymbol{x}})^T\right\}, \tag{1.211}$$

$$\mathbf{\Gamma}_{xy} = \mathcal{E}\left\{(\boldsymbol{x} - \overline{\boldsymbol{x}})(\boldsymbol{y} - \overline{\boldsymbol{y}})^T\right\}, \tag{1.212}$$

$$\mathbf{\Gamma}_{yx} = \mathcal{E}\left\{(\boldsymbol{y} - \overline{\boldsymbol{y}})(\boldsymbol{x} - \overline{\boldsymbol{x}})^T\right\}, \tag{1.213}$$

and

$$\mathbf{\Gamma}_{yy} = \mathcal{E}\left\{(\boldsymbol{y} - \overline{\boldsymbol{y}})(\boldsymbol{y} - \overline{\boldsymbol{y}})^T\right\}. \tag{1.214}$$

The conditional probability for $\boldsymbol{x}$, given $\boldsymbol{y}$, is written as

$$p(\boldsymbol{x}|\boldsymbol{y}) = \frac{p(\boldsymbol{x}, \boldsymbol{y})}{p(\boldsymbol{y})}. \tag{1.215}$$

Let

$$\boldsymbol{\xi} = (\boldsymbol{x} - \overline{\boldsymbol{x}}) \quad \text{and} \quad \boldsymbol{\eta} = (\boldsymbol{y} - \overline{\boldsymbol{y}}).$$

Substituting Eqs. (1.206) and (1.208) into Eq. (1.215), we obtain for the argument in the resulting exponent the following equation:

$$q = \begin{bmatrix} \boldsymbol{\xi} \\ \boldsymbol{\eta} \end{bmatrix}^T \begin{bmatrix} \mathbf{\Gamma}_{xx} & \mathbf{\Gamma}_{xy} \\ \mathbf{\Gamma}_{yx} & \mathbf{\Gamma}_{yy} \end{bmatrix}^{-1} \begin{bmatrix} \boldsymbol{\xi} \\ \boldsymbol{\eta} \end{bmatrix} - \boldsymbol{\eta}^T \mathbf{\Gamma}_{yy}{}^{-1} \boldsymbol{\eta} \tag{1.216}$$

$$= \begin{bmatrix} \boldsymbol{\xi} \\ \boldsymbol{\eta} \end{bmatrix}^T \begin{bmatrix} \mathbf{\Upsilon}_{xx} & \mathbf{\Upsilon}_{xy} \\ \mathbf{\Upsilon}_{yx} & \mathbf{\Upsilon}_{yy} \end{bmatrix} \begin{bmatrix} \boldsymbol{\xi} \\ \boldsymbol{\eta} \end{bmatrix} - \boldsymbol{\eta}^T \mathbf{\Gamma}_{yy}{}^{-1} \boldsymbol{\eta}, \tag{1.217}$$

using the results in Section 1.3. In particular, we find for the partitions that

$$\mathbf{\Upsilon}_{xx}{}^{-1} = \mathbf{\Gamma}_{xx} - \mathbf{\Gamma}_{xy} \mathbf{\Gamma}_{yy}{}^{-1} \mathbf{\Gamma}_{yx}, \tag{1.218}$$

$$\mathbf{\Gamma}_{yy}{}^{-1} = \mathbf{\Upsilon}_{yy} - \mathbf{\Upsilon}_{yx} \mathbf{\Upsilon}_{xx}{}^{-1} \mathbf{\Upsilon}_{xy}, \tag{1.219}$$

and

$$\mathbf{\Upsilon}_{xx}{}^{-1} \mathbf{\Upsilon}_{xy} = -\mathbf{\Gamma}_{xy} \mathbf{\Gamma}_{yy}{}^{-1}, \tag{1.220}$$

which allows us to write Eq. (1.217) [after a multiplication of −2] as

$$\begin{aligned} q &= \boldsymbol{\xi}^T \mathbf{\Upsilon}_{xx} \boldsymbol{\xi} + \boldsymbol{\xi}^T \mathbf{\Upsilon}_{xy} \boldsymbol{\eta} + \boldsymbol{\eta}^T \mathbf{\Upsilon}_{yx} \boldsymbol{\xi} + \boldsymbol{\eta}^T \mathbf{\Upsilon}_{yy} \boldsymbol{\eta} - \boldsymbol{\eta}^T \mathbf{\Gamma}_{yy}{}^{-1} \boldsymbol{\eta} \\ &= (\boldsymbol{\xi} + \mathbf{\Upsilon}_{xx}{}^{-1} \mathbf{\Upsilon}_{xy} \boldsymbol{\eta})^T \mathbf{\Upsilon}_{xx} (\boldsymbol{\xi} + \mathbf{\Upsilon}_{xx}{}^{-1} \mathbf{\Upsilon}_{xy} \boldsymbol{\eta}) + \boldsymbol{\eta}^T (\mathbf{\Upsilon}_{yy} - \mathbf{\Upsilon}_{yx} \mathbf{\Upsilon}_{xx}{}^{-1} \mathbf{\Upsilon}_{xy}) \boldsymbol{\eta} - \boldsymbol{\eta}^T \mathbf{\Gamma}_{yy}{}^{-1} \boldsymbol{\eta} \\ &= (\boldsymbol{\xi} + \mathbf{\Upsilon}_{xx}{}^{-1} \mathbf{\Upsilon}_{xy} \boldsymbol{\eta})^T \mathbf{\Upsilon}_{xx} (\boldsymbol{\xi} + \mathbf{\Upsilon}_{xx}{}^{-1} \mathbf{\Upsilon}_{xy} \boldsymbol{\eta}), \end{aligned} \tag{1.221}$$

using the partitioning results. It is clear that Eq. (1.221) has a quadratic form, which means that the conditional PDF of $\boldsymbol{x}$, given $\boldsymbol{y}$, also is Gaussian.

Using Eq. (1.220), we find that

$$\boldsymbol{\xi} + \mathbf{\Upsilon}_{xx}{}^{-1} \mathbf{\Upsilon}_{xy} \boldsymbol{\eta} = (\boldsymbol{x} - \overline{\boldsymbol{x}}) - \mathbf{\Gamma}_{xy} \mathbf{\Gamma}_{yy}{}^{-1} (\boldsymbol{y} - \overline{\boldsymbol{y}}). \tag{1.222}$$

The conditional mean of $\boldsymbol{x}$, given $\boldsymbol{y}$, is

$$\hat{\boldsymbol{x}} = \mathcal{E}\{\boldsymbol{x}|\boldsymbol{y}\} = \overline{\boldsymbol{x}} - \mathbf{\Gamma}_{xy} \mathbf{\Gamma}_{yy}{}^{-1} (\boldsymbol{y} - \overline{\boldsymbol{y}}), \tag{1.223}$$

and its covariance matrix

$$\boldsymbol{\Gamma}_{xx|y} = \mathcal{E}\{(\boldsymbol{x}-\hat{\boldsymbol{x}})(\boldsymbol{x}-\hat{\boldsymbol{x}})^T|\boldsymbol{y}\} = \boldsymbol{\Upsilon}_{xx}{}^{-1} = \boldsymbol{\Gamma}_{xx} - \boldsymbol{\Gamma}_{xy}\boldsymbol{\Gamma}_{yy}{}^{-1}\boldsymbol{\Gamma}_{yx}, \tag{1.224}$$

using Eq. (1.218).

As a final comment, Gaussian random variables remain Gaussian even under linear or affine transformations.

1.6.10 Quadratic and Quartic Forms

The expected value of quadratic and quartic forms of Gaussian random variables can be derived as follows. Let us assume that

$$p(\boldsymbol{x}) = N(\boldsymbol{x}, \overline{\boldsymbol{x}}, \boldsymbol{\Gamma}).$$

Then the characteristic function for the above is

$$M_x(\boldsymbol{S}) = \mathcal{E}\left\{e^{\boldsymbol{S}^T\boldsymbol{x}}\right\} = e^{\frac{1}{2}\boldsymbol{S}^T\boldsymbol{\Gamma}\,\boldsymbol{S}+\boldsymbol{S}^T\overline{\boldsymbol{x}}}. \tag{1.225}$$

For convenience and without any loss of generality, let us assume $\overline{\boldsymbol{x}} = 0$.

Given the vector random variable $\boldsymbol{x}$, we can write the following general equation:

$$\mathcal{E}\left\{\boldsymbol{x}^T\boldsymbol{A}\,\boldsymbol{x}\right\} = \mathcal{E}\left\{\mathrm{tr}[\boldsymbol{A}\,\boldsymbol{x}\boldsymbol{x}^T]\right\} = \mathrm{tr}[\boldsymbol{A}\,\boldsymbol{\Gamma}] \tag{1.226}$$

for any arbitrary matrix $\boldsymbol{A}$. The same result can be obtained using the characteristic function; specifically, we have

$$\begin{aligned}\mathcal{E}\left\{\boldsymbol{x}^T\boldsymbol{A}\,\boldsymbol{x}\right\} &= \mathcal{E}\left\{\nabla_S{}^T e^{\boldsymbol{S}^T\boldsymbol{x}}\boldsymbol{A}\,\boldsymbol{x}\right\}\Big|_{\boldsymbol{S}=\boldsymbol{0}} = \mathcal{E}\left\{\nabla_S{}^T\boldsymbol{A}\,\boldsymbol{x}e^{\boldsymbol{S}^T\boldsymbol{x}}\right\}\Big|_{\boldsymbol{S}=\boldsymbol{0}}\\ &= \nabla_S{}^T\boldsymbol{A}\,\mathcal{E}\left\{\boldsymbol{x}e^{\boldsymbol{S}^T\boldsymbol{x}}\right\}\Big|_{\boldsymbol{S}=\boldsymbol{0}} = \nabla_S{}^T\boldsymbol{A}\,\mathcal{E}\left\{\nabla_S e^{\boldsymbol{S}^T\boldsymbol{x}}\right\}\Big|_{\boldsymbol{S}=\boldsymbol{0}}\\ &= \nabla_S{}^T\boldsymbol{A}\,\nabla_S M_x(\boldsymbol{S})|_{\boldsymbol{S}=\boldsymbol{0}}.\end{aligned} \tag{1.227}$$

Substituting Eq. (1.225) into Eq. (1.227), with $\overline{\boldsymbol{x}} = 0$, will give us Eq. (1.226). This same procedure can be applied to the quartic form.

Specifically, we find that

$$\begin{aligned}\mathcal{E}\{\boldsymbol{x}^T\boldsymbol{A}\,\boldsymbol{x}\boldsymbol{x}^T\boldsymbol{B}\,\boldsymbol{x}\} &= \nabla_S{}^T\boldsymbol{A}\,\nabla_S\nabla_S{}^T\boldsymbol{B}\,\nabla_S M_x(\boldsymbol{S})|_{\boldsymbol{S}=\boldsymbol{0}} &(1.228)\\ &= \mathrm{tr}[\boldsymbol{A}\,\boldsymbol{\Gamma}]\mathrm{tr}[\boldsymbol{B}\,\boldsymbol{\Gamma}] + 2\mathrm{tr}[\boldsymbol{A}\,\boldsymbol{\Gamma}\,\boldsymbol{B}\,\boldsymbol{\Gamma}]. &(1.229)\end{aligned}$$

Example 1.15 If $\boldsymbol{A}$ and $\boldsymbol{B}$ equal the scalar 1 and $\boldsymbol{\Gamma}$ equals the scalar σ^2, then the scalar counterpart for Eq. (1.224) is the well-known expression for the fourth moment of a Gaussian random variable given by

$$\mathcal{E}\left\{x^4\right\} = 3\sigma^4. \tag{1.230}$$

Using Eqs. (1.226) and (1.229), the covariance of two quartic forms is equal to

$$\begin{aligned}\mathcal{E}\left\{\left(x^TAx-\mathcal{E}\left\{x^TAx\right\}\right)\left(x^TBx-\mathcal{E}\left\{x^TBx\right\}\right)\right\} &= \mathcal{E}\left\{x^TAxx^TBx\right\}\\ &\quad-\mathcal{E}\left\{x^TAx\right\}\mathcal{E}\left\{x^TBx\right\}\\ &\quad-\mathcal{E}\left\{x^TBx\right\}\mathcal{E}\left\{x^TAx\right\}\\ &\quad+\mathcal{E}\left\{x^TBx\right\}\mathcal{E}\left\{x^TAx\right\}\\ &=\mathcal{E}\left\{x^TAxx^TBx\right\}\\ &\quad-\mathcal{E}\left\{x^TAx\right\}\mathcal{E}\left\{x^TBx\right\}\\ &=\mathrm{tr}\,[A\Gamma]\,\mathrm{tr}\,[B\Gamma]+2\mathrm{tr}\,[A\Gamma B\Gamma]\\ &\quad-\mathrm{tr}\,[A\Gamma]\,\mathrm{tr}\,[B\Gamma]\\ &=2\mathrm{tr}\,[A\Gamma B\Gamma]\,. \end{aligned} \tag{1.231}$$

1.6.11 Chi-Squared Distributed Random Variable

If an m-dimensional random variable x is Gaussian with mean $\overline{x}$ and covariance matrix Γ, then the scalar random variable of the form

$$q=(x-\overline{x})^T\Gamma^{-1}(x-\overline{x}) \tag{1.232}$$

is the sum of the squares of m independent, zero-mean, unity-variance Gaussian random variables and consequently, has a *chi-squared distribution* with m degrees of freedom. Let's see why this is true.

Define

$$u=\Gamma^{-1/2}\,(x-\overline{x}). \tag{1.233}$$

Obviously, the form of the above implies that u is Gaussian, with

$$\mathcal{E}\{u\}=0 \tag{1.234}$$

and

$$\mathcal{E}\{uu^T\}=\Gamma^{-1/2}\,\mathcal{E}\left\{(x-\overline{x})(x-\overline{x})^T\right\}\,\Gamma^{-1/2}=\Gamma^{-1/2}\Gamma\,\Gamma^{-1/2}=I. \tag{1.235}$$

Because the above covariance matrix is a diagonal matrix, its components are independent. This means that

$$q=u^Tu=\sum_{i=1}^{m}u_i^2, \tag{1.236}$$

where $u_i \sim N(0,1)$.

Given the above, the convention is to write

$$q\sim\chi_m^2. \tag{1.237}$$

The mean for q is

$$\mathcal{E}\{q\} = \mathcal{E}\left\{ \sum_{i=1}^{m} u_i^2 \right\} = m, \tag{1.238}$$

and its variances equals

$$\mathcal{E}\left\{ \left[\sum_{i=1}^{m} (u_i^2 - 1) \right]^2 \right\} = \sum_{i=1}^{m} \mathcal{E}\{(u_i^2 - 1)^2\}$$

$$= \sum_{i=1}^{m} [\mathcal{E}\{u_i^4\} - 2\mathcal{E}\{u_i^2\} + 1] = 2m, \tag{1.239}$$

using Eq. (1.230) for the fourth moment of a Gaussian random variable as $\sigma^4 = \sigma^2 = 1$.

1.6.12 Binomial Distribution

A chance experiment $\mathbb{E}$ is repeated m times. The m trials effectively create an m -fold chance experiment, that is, $\mathbb{E}^m$ defines our new chance experiment. The outcomes of $\mathbb{E}^m$ are m -tuples of outcomes generated by the various outcome combinations from the m trials of $\mathbb{E}$. Let us look how to characterize these experiments statistically by an example.

The convention is to recognize that any event $\boldsymbol{B}$ in the experiment $\mathbb{E}$ has a probability p and its complement $\boldsymbol{B}'$ has probability $(1 - p) = q$. An occurrence of $\boldsymbol{B}$ is a "success" and the occurrence of its complement is a "failure." For simplicity discussion here, let us assume that our experiment involves a biased coin, and "heads" denote "success" and "tails" denote "failure. Clearly, these trials are statistically independent. The universal set has 2^m possible outcomes. To characterize the compound experiments, we seek the probability density for the number of "heads" and "tails" in m trials of chance experiment $\mathbb{E}$, which we interpret as a single experiment, $\mathbb{E}' = \mathbb{E}^m$. $\mathbb{E}'$ is often described as a succession of *Bernoulli trials*.

Let A_k denote the event where k "heads" and $(m - k)$ "tails" both occur after m coin tosses happen. The number of outcomes of $\mathbb{E}'$ in event A_k is equal to

$$\binom{m}{k} = \frac{m!}{r!(m-k)!}, \tag{1.240}$$

which is called the *binominal coefficient* [6, pp. 40–41]. Taking the sum of Eq. (1.240) gives

$$\sum_{k=0}^{m} \binom{m}{k} = \sum_{k=0}^{m} \frac{m!}{r!(m-k)!} = (1+1)^m = 2^m \tag{1.241}$$

using the binomial theorem,

$$(x+y)^n = \sum_{k=0}^{n} \frac{n!}{r!(n-r)!} x^{n-k} y^r. \tag{1.242}$$

Equation (1.241) confirms our earlier claim that we have 2^m possible outcomes in the universal set.

Let us now interpret our compound experiment as an atomic event (single outcome) in $\mathbb{E}'$ consisting of k "successes" and $(m-k)$ "failures" in a particular order. The probability of this event is $p^k q^{m-k}$ because each trial is independent of the others. The number of such atomic events in A_k is $\binom{m}{k}$ and the probability of A_k is

$$P\{A_k\} = \binom{m}{k} p^k q^{m-k} \tag{1.243}$$

from Corollary 4, Eq. (1.161). Equation (1.243) is known as *binomial distribution*. Taking the sum of Eq. (1.243) yields

$$\sum_{k=0}^{m} P\{A_k\} = \sum_{k=0}^{m} \binom{m}{k} p^k q^{m-k} = (p+q)^m = (1)^m = 1. \tag{1.244}$$

Example 1.16 Let us assume we have a pair of fair dice that we toss 10 times. What is the probability that the dice totals seven points exactly four times? The possible outcomes that total seven points are $(1,6), (2,5), (3,4), (4,3), (5,2)$, and $(6,1)$. The probability of success equals to $p = 6 \times 6^{-2} = \frac{1}{6}$. The probability of failure is $q = 1 - p = \frac{5}{6}$. The probability of the cited event is

$$\binom{10}{4}\left(\frac{1}{6}\right)^4\left(\frac{5}{6}\right)^6 = (210)(0.000771605)(0.334897977) = 0.0543.$$

The probability that at most k successes occur in m trials is defined as

$$B(k, m; p) = \sum_{r=0}^{k} \binom{m}{r} p^r q^{m-r}, \tag{1.245}$$

and its results comprise what is known as the *cumulative binomial distribution*.

The mean of the binomial distribution can be calculated as follows:

$$\begin{aligned}
\bar{k} = \mathcal{E}\{k\} &= \sum_{k=0}^{m} k\left(\frac{m!}{k!(m-k)!}\right) p^k q^{m-k} = \sum_{k=1}^{m}\left(\frac{m!}{(k-1)!(m-k)!}\right) p^k q^{m-k} \\
&= \sum_{l=0}^{m-1}\left(\frac{m!}{l!(m-1-l)!}\right) p^{l+1} q^{m-1-l} = mp\sum_{l=0}^{m-1}\left(\frac{(m-1)!}{l!(m-1-l)!}\right) p^l q^{m-1-l} \\
&= mp\sum_{l=0}^{n}\left(\frac{n!}{l!(n-l)!}\right) p^l q^{n-l} = mp.
\end{aligned} \tag{1.246}$$

The variance of the binomial distribution can be derived as follows:

$$\sigma^2 = \mathcal{E}\{k^2\} - [\mathcal{E}\{k\}]^2 = \mathcal{E}\{k^2\} - \mathcal{E}\{k\} + \mathcal{E}\{k\} - [\mathcal{E}\{k\}]^2$$
$$= \mathcal{E}\{k(k-1)\} + \mathcal{E}\{k\} - [\mathcal{E}\{k\}]^2. \quad (1.247)$$

Now,

$$\mathcal{E}\{k(k-1)\} = \sum_{k=0}^{m} k(k-1)\left(\frac{m!}{k!(m-k)!}\right)p^k q^{m-k} = \sum_{k=2}^{m}\left(\frac{m!}{(k-2)!(m-k)!}\right)p^k q^{m-k}$$
$$= \sum_{l=0}^{m-2}\left(\frac{m!}{l!(m-2-l)!}\right)p^{l+2}q^{m-2-l}$$
$$= m(m-1)p^2\sum_{l=0}^{m-1}\left(\frac{(m-2)!}{l!(m-2-l)!}\right)p^l q^{m-2-l}$$
$$= m(m-1)p^2\sum_{l=0}^{n}\left(\frac{n!}{l!(n-l)!}\right)p^l q^{n-l} = m(m-1)p^2. \quad (1.248)$$

Substituting Eqs. (1.246) and (1.248) into Eq. (1.247) yields

$$\sigma^2 = \mathcal{E}\{k(k-1)\} + \mathcal{E}\{k\} - [\mathcal{E}\{k\}]^2 = m(m-1)p^2 + mp - (mp)^2$$
$$= m^2p^2 - mp^2 + mp - m^2p^2 = mp - mp^2 = mp(1-p). \quad (1.249)$$

Example 1.17 The probability that a certain diode fails before operating 1000 h is 0.15. The probability that among 20 such diodes, at least 5 diodes will fail before reaching 1000 h is

$$\sum_{r=5}^{20}\binom{20}{r}(0.15)^m(0.85)^{m-r} = 1 - B(4,20;0.15) \approx 0.17.$$

1.6.13 Poisson Distribution

When the number of Bernoulli trial is large and the probability p gets small, the binomial distribution approximates another probability distribution, namely, the Poisson distribution. We show that as follows:

Recall from the last section that the mean of the binomial distribution is mp, which we now will call λ. This implies that

$$p = \frac{\lambda}{m}. \quad (1.250)$$

This implies that

$$\left(\frac{m!}{k!(m-k)!}\right)p^k q^{m-k} = \left(\frac{m!}{k!(m-k)!}\right)\left(\frac{\lambda}{m}\right)^k\left(1-\frac{\lambda}{m}\right)^{m-k}$$

$$= \left(\frac{m(m-1)\cdots(m-k+1)}{k!}\right)\left(\frac{\lambda}{m}\right)^k\left(1-\frac{\lambda}{m}\right)^{m-k}$$

$$= \left(\frac{m}{m}\right)\left(\frac{m-1}{m}\right)\cdots\left(\frac{(m-k+1)}{m}\right)\frac{\lambda^k}{k!}\left(1-\frac{\lambda}{m}\right)^{m-k}$$

$$= \left(\frac{m}{m}\right)\left(\frac{m-1}{m}\right)\cdots\left(\frac{(m-k+1)}{m}\right)$$

$$\frac{\lambda^k}{k!}\left(1-\frac{\lambda}{m}\right)^m\left(1-\frac{\lambda}{m}\right)^{-k}. \tag{1.251}$$

Now if we let m get very large, then all but the last three factors go to one. Looking at the second to last factor, if m is getting large while keeping k and λ fixed, then

$$\left(1-\frac{\lambda}{m}\right)^m \approx \left(\sum_{i=0}^{\infty}\frac{1}{i!}\left(-\frac{\lambda}{m}\right)^i\right)^m = \left(e^{-\frac{\lambda}{m}}\right)^m = e^{-\lambda}. \tag{1.252}$$

The last factor goes to 1 as m is getting large. The result is that

$$\lim_{m\to\infty}P\{A_k\} = \frac{\lambda^k}{k!}e^{-\lambda}. \tag{1.253}$$

The function on the right side of Eq. (1.253) is called *Poisson distribution*. The Poisson distribution has a mean equal to $\lambda = mp$ and its variance also is its mean.

Example 1.18 Suppose that the expected number of electrons counted in $(0, \tau)$ is $m = 4.5$. Assume the electron count follows a Poisson distribution. What is the probability that exactly two electrons are counted? The probability in this case is given by

$$p(2, 4.5) = \frac{(4.5)^2}{2!}e^{-4.5} = 0.1125.$$

What is the probability that at least six electrons are counted? The probability in this case equals

$$P\{k \geq 6\} = 1 - \sum_{k=0}^{5}p(k, 4.5) = \sum_{k=0}^{5}\frac{(4.5)^k}{k!}e^{-4.5} = 0.2971.$$

1.6.14 Random Processes

In the previous sections, we dealt with the random variable $\boldsymbol{x}$ that is a real number defined by the outcome ϑ of a chance experiment, that is,

$$\boldsymbol{x} = \boldsymbol{x}(\vartheta). \tag{1.254}$$

A *random or stochastic process* is function of time determined by the outcome of that chance experiment. That is, we rewrite Eq. (1.254) as

$$\boldsymbol{x}(t) = \boldsymbol{x}(t, \vartheta), \tag{1.255}$$

which now represents a family of functions in time, ever changing for each outcome ϑ. Let us now define its properties.

The mean of the random process is written as

$$\overline{\boldsymbol{x}}(t) = \mathcal{E}\{\boldsymbol{x}(t)\}, \tag{1.256}$$

while its *autocorrelation* is given as

$$\boldsymbol{R}(t_1, t_2) = \mathcal{E}\{\boldsymbol{x}(t_1)\boldsymbol{x}(t_2)\}. \tag{1.257}$$

The autocovariance of the random process is defined as

$$\boldsymbol{\Gamma}(t_1, t_2) = \mathcal{E}\{[\boldsymbol{x}(t_1) - \overline{\boldsymbol{x}}(t_1)][\boldsymbol{x}(t_2) - \overline{\boldsymbol{x}}(t_2)]\} = \boldsymbol{R}(t_1, t_2) - \overline{\boldsymbol{x}}(t_1)\overline{\boldsymbol{x}}(t_2). \tag{1.258}$$

Looking at the above two equations, it is clear that $\boldsymbol{R}(t_1, t_2)$ is a joint noncentral moment as its mean of the process has not been removed, while $\boldsymbol{\Gamma}(t_1, t_2)$ is joint central moment because it had its mean removed. A zero-mean random process makes them equivalent.

When a random process has a mean that is time independent, its autocorrelation depends on the length of the time interval between t_1 and t_2, not their specific values. In other words, we write

$$\boldsymbol{R}(t_1, t_2) = \boldsymbol{R}(\tau), \tag{1.259}$$

where

$$\tau = t_2 - t_1. \tag{1.260}$$

This type of random process is called *stationary*. (Its counterpart where the process depends on the specific times used is called *nonstationary*.)

Let us now define the power spectrum of a stationary random process. The Fourier transform of the autocovariance function is equal to

$$\int_{-\infty}^{\infty} \boldsymbol{\Gamma}(\xi) e^{-i\omega\xi} d\xi = \int_{-\infty}^{\infty} \boldsymbol{R}(\xi) e^{-i\omega\xi} d\xi + \int_{-\infty}^{\infty} \overline{\boldsymbol{x}}^2 e^{-i\omega\xi} d\xi \tag{1.261a}$$

$$= \int_{-\infty}^{\infty} \boldsymbol{R}(\xi) e^{-i\omega\xi} d\xi + \overline{\boldsymbol{x}}^2 \delta(\omega), \tag{1.261b}$$

where ω denotes radial (temporal) frequency. The second term on the right of Eq. (1.261b) is the "dc" component of the power spectrum ($\omega = 0$). It has no bandwidth, but has a magnitude equal to the square of the mean. On the other hand, the first term on the right of Eq. (1.261b) refers to the AC terms and has bandwidth. We define this term as the *power spectrum* of the random process, that is,

$$\boldsymbol{S}(\omega) = \int_{-\infty}^{\infty} \boldsymbol{R}(\xi) e^{-i\omega\xi} d\xi. \tag{1.262}$$

In other words, it is the Fourier transform of the autocorrelation function. This result is the Wiener–Khinchin Theorem. The integral of $S(\omega)$ is related to the variance, which can be seen as follows:

$$\left.\int_{-\infty}^{\infty} S(\omega)e^{i\omega\xi}d\omega\right|_{\tau=0} = \int_{-\infty}^{\infty} S(\omega)d\omega = \boldsymbol{R}(0) \tag{1.263a}$$

$$= \mathcal{E}\{(x(t_1))^2\} = \sigma_x^2 + \overline{x}^2. \tag{1.263b}$$

For a zero-mean random process, it is the variance.

An important random process found in both radio frequency (RF) and optical systems is *white noise*. This is where the random process has both a zero mean and zero autocorrelation. Mathematically, this is characterized by the equation

$$\boldsymbol{R}(t_1, t_2) = \boldsymbol{R}(\tau) = \delta(\tau). \tag{1.264}$$

This means that $S(\omega) = 1$, which implies that we have a constant power spectrum. The term "white noise" comes from the analogy with white light possessing all wavelengths (frequencies) of light.

1.7 DECIBELS

The decibel, or dB, for short, is one of the most widely used mathematical tools for those involved with system engineering. It is related to the common (base 10) logarithm and has the following mathematical form:

$$X, \text{ measure in dB} = 10 \log_{10} X\text{dB}. \tag{1.265}$$

The decibel, named after Alexander Graham Bell, was created to measure the ratio of two power levels; the most typical use is the ratio of power out of systems over its input power. The parameter X typically is the ratio of two power levels P_1 and P_2, or

$$\frac{\text{Gain}}{\text{Loss (dB)}} = 10 \log_{10}\left(\frac{P_2}{P_1}\right) \text{ dB}. \tag{1.266}$$

One of the key reasons that the decibel is used in engineering is that it can transform a multiplicative equation into an additive equation. However, one must be careful in its use since there are various units involved. Let us look at an example to show how to do it properly.

Consider the range equation we will define in Chapter 4:

$$P_{rx} \approx \gamma_{tx} P_{tx} \frac{A_{tx} A_{rx}}{(\lambda R)^2}, \tag{1.267}$$

where P_{rx} is the received power, γ_{tx} the transmitter optics transmittance, P_{tx} the laser transmitter power, λ the wavelength of light, R the distance between the transmitter and receiver, and A_{tx} and A_{rx} the areas of the transmitter and receiver apertures, respectively. Applying Eq. (1.265) to Eq. (1.267) yields

$$10\log_{10}(P_{rx}) = 10\log_{10}(\gamma_{tx}) + 10\log_{10}(P_{tx}) + 10\log_{10}\left(\frac{A_{tx}A_{rx}}{(\lambda R)^2}\right). \tag{1.268}$$

Equation (1.268) has a unit problem. The elements for the two power levels have unit of watts, while everything else is unitless. To remedy this problem, we can subtract $10\log_{10}(1\text{ W})$ to each side of Eq. (1.268) without affecting the correctness of the equations and obtain the following new equation:

$$10\log_{10}\left(\frac{P_{rx}}{1\text{ W}}\right) = 10\log_{10}(\gamma_{tx}) + 10\log_{10}\left(\frac{P_{tx}}{1\text{ W}}\right) + 10\log_{10}\left(\frac{A_{tx}A_{rx}}{(\lambda R)^2}\right). \tag{1.269}$$

This corrects the problem. If we define

$$P'_{tx}(\text{dB W}) = 10\log_{10}\left(\frac{P_{rx}}{1\text{ W}}\right) \tag{1.270}$$

and

$$P'_{rx}(\text{dB W}) = 10\log_{10}\left(\frac{P_{rx}}{1\text{ W}}\right), \tag{1.271}$$

then Eq. (1.268) becomes

$$P'_{rx}(\text{dB W}) = 10\log_{10}(\gamma_{tx})dB + P'_{tx}(\text{dB W}) + 10\log_{10}\left(\frac{A_{tx}A_{rx}}{(\lambda R)^2}\right)\text{ dB}. \tag{1.272}$$

If we had specified 1 mW rather than 1 W in the above derivation, Eq. (1.272) would be written as

$$P'_{rx}(\text{dB mW}) = 10\log_{10}(\gamma_{tx})\text{ dB} + P'_{tx}(\text{dB mW}) + 10\log_{10}\left(\frac{A_{tx}A_{rx}}{(\lambda R)^2}\right)\text{ dB}. \tag{1.273}$$

Clearly, we could write the last term in Eq. (1.272) as

$$\begin{aligned} 10\log_{10}\left(\frac{A_{tx}A_{rx}}{(\lambda R)^2}\right) &= \text{dB}\,10\log_{10}\left(\frac{A_{tx}}{1\text{ m}}\right)\text{dB} \\ &\quad + 10\log_{10}\left(\frac{A_{rx}}{1\text{ m}}\right)\text{dB} - 20\log_{10}(\lambda R)\,\text{dB} \\ &= A'_{tx}(\text{dB m}) + A'_{rx}(\text{dB m}) - 20\log_{10}(\lambda R)\text{ dB} \end{aligned} \tag{1.274}$$

$$= A'_{tx}(\text{dB m}) + A'_{rx}(\text{dB m}) - 2\lambda'(\text{dB m}) - 2R'(\text{dB m}) \tag{1.275}$$

TABLE 1.2 Equivalent Power Ratios

Decibel Value	Positive dB Power Ratios	Negative dB Power Ratios
0	1	1
1	1.3	0.79
2	1.6	0.63
3	2.0	0.50
4	2.5	0.40
5	3.2	0.32
6	4.0	0.25
7	5.0	0.20
8	6.3	0.16
9	7.9	0.13

following the same procedure. This implies that

$$P'_{rx}(\text{dB W}) = 10\log_{10}(\gamma_{tx})\ \text{dB} + P'_{tx}(\text{dB W}) + = A'_{tx}(\text{dB m}) + A'_{rx}(\text{dB m}) - 2\lambda'(\text{dB m}) \\ - 2R'(\text{dB m}). \tag{1.276}$$

Now, the reader seeing Eq. (1.276) for the first time may think we again have a unit problem, mixing watts and meters since they have learned that quantities that have different units should not add. However, logarithmic units are different and entities with different logarithmic or decibel units do add. The reason is that all dB m element units cancel, leaving only the unitless $10\log_{10}(\gamma_{tx})$ dB and (dB W) terms left, agreeing with the right side of the equation. This may be hard to get used to at first, but the benefit will pay off when dealing with complex equations that must be having several parameters varied in a trade study. In addition, adding and subtracting numbers is easier to do in one's head than multiplying or dividing numbers; at least for most people. This is an easy bookkeeping approach for engineers.

Some common increments in decibels should be memorized by the reader so they are proficient in using decibels in link budgets. A factor of 1 equals 0 dB, a factor of 10 equals 10 dB, a factor of 100 is 20 dB, and so on. However, the workhorse engineering number to remember is 3 dB. Its +3 dB value is 1.995, which basically is 2, and its −3 dB number is 0.5012, which essentially is 0.5. Therefore, if one increases the signal power by 2, we say we have a 3 dB increase in power; if we cut the signal power by half, we decrease the power by −3 dB. Table 1.2 provides a list of the real number equivalents for decibels between 1 and 9.

1.8 PROBLEMS

Problem 1.1. Let

$$\boldsymbol{A} = \begin{bmatrix} 1 & 2 & 3 \\ 0 & 1 & 4 \end{bmatrix} \text{ and } \boldsymbol{B} = \begin{bmatrix} 2 & 3 & 0 \\ -1 & 2 & 5 \end{bmatrix}.$$

(a) what is $A + B$?

(b) What is $A - B$?

Problem 1.2. Compute the determinant

$$|A| = \begin{vmatrix} 1 & 2 & 10 \\ 2 & 3 & 9 \\ 4 & 5 & 11 \end{vmatrix}.$$

Problem 1.3. Compute the determinant

$$|A| = \begin{vmatrix} +2 & +3 & -2 & +4 \\ +3 & -2 & +1 & +2 \\ +3 & +2 & +3 & +4 \\ -2 & +4 & 0 & +5 \end{vmatrix}.$$

Problem 1.4. Compute the determinant

$$|A| = \begin{vmatrix} +2 & +3 & -2 & +4 \\ +7 & +4 & -3 & +10 \\ +3 & +2 & +3 & +4 \\ -2 & +4 & 0 & +5 \end{vmatrix}.$$

Problem 1.5. Show that the cofactor of each element of

$$A = \begin{bmatrix} -\frac{1}{3} & -\frac{2}{3} & -\frac{2}{3} \\ +\frac{2}{3} & +\frac{1}{3} & -\frac{2}{3} \\ +\frac{2}{3} & -\frac{2}{3} & +\frac{1}{3} \end{bmatrix}$$

is that element.

Problem 1.6. Show that the cofactor of an element of any row of

$$A = \begin{bmatrix} -4 & -3 & -3 \\ +1 & 0 & +1 \\ +4 & +3 & +3 \end{bmatrix}$$

is the corresponding element of the same numbered column.

Problem 1.7. Find the inverse of

$$A = \begin{vmatrix} 2 & 3 \\ 1 & 4 \end{vmatrix}.$$

Problem 1.8. Find the inverse of

$$A = \begin{vmatrix} 2 & 3 & 1 \\ 1 & 2 & 3 \\ 3 & 1 & 2 \end{vmatrix}.$$

Problem 1.9. Find the inverse of

$$A = \begin{vmatrix} +2 & +4 & +3 & +2 \\ +3 & +6 & +5 & +2 \\ +2 & +5 & +2 & -3 \\ +4 & +5 & +14 & +14 \end{vmatrix}.$$

Problem 1.10. Calculate the Fourier series coefficients C_n for the following periodic function:

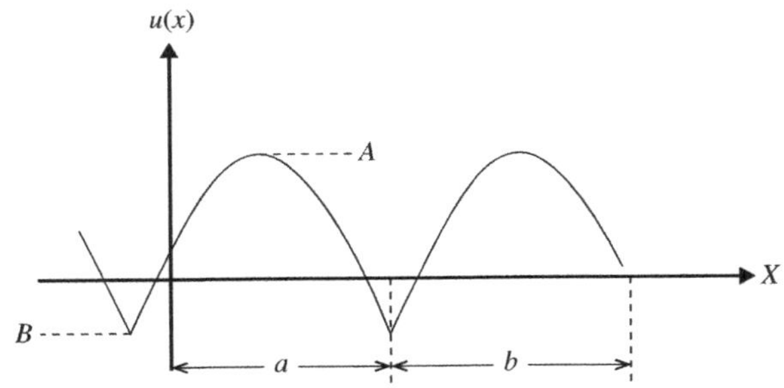

NOTE: The curve essentially follows the positive part of the sine function with period $2b$.

Problem 1.11. Calculate the Fourier series coefficients C_n for the function $u(x)$ plotted as follows, but do so exploiting some of the properties contained in Eqs. (1.4)–(1.24) to provide a solution derived from the coefficient calculation of a square wave. The slopes up and down are equally steep, and that B depicts a negative number in the graph.

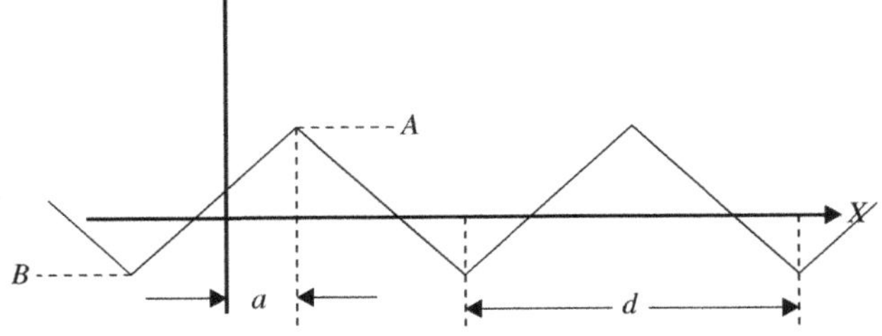

Problem 1.12. Calculate the Fourier series coefficients C_n for the function $u(x)$ plotted as follows, but do so exploiting some of the properties contained in Eqs. (1.4)–(1.24) to provide a solution derived from the coefficient calculation of a square wave. The slopes up and down are equally steep, and that B depicts a negative number in the graph.

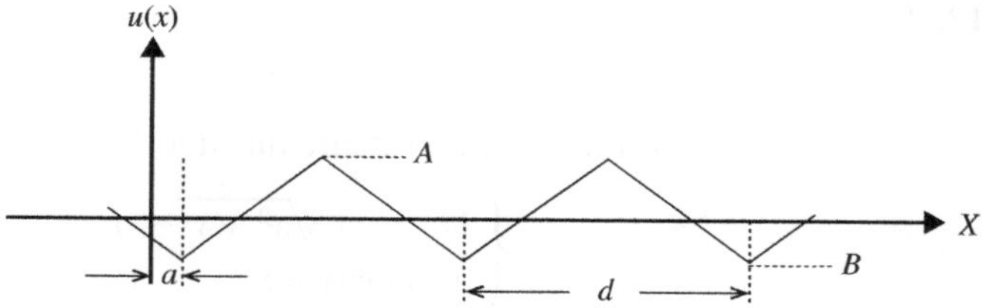

Problem 1.13. Calculate the Fourier series coefficients C_n for $u(x) = A(x)e^{i\varphi(x)}$. $\varphi(x) = \varphi_0 + 10\pi\frac{x}{b}$. The curve shown in the following figure is like the lower portion of the cosine function.

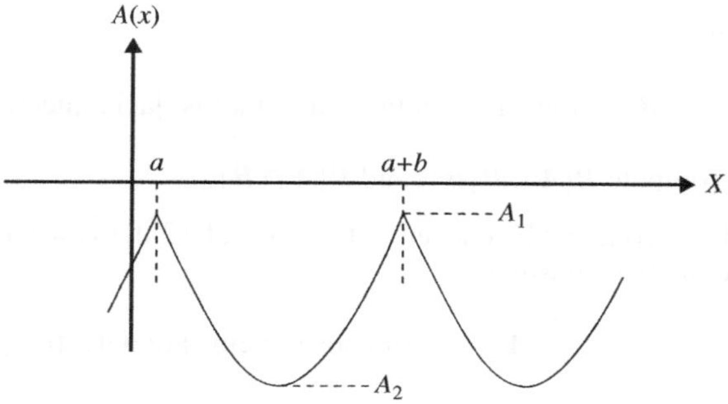

Problem 1.14. Let

$$f(x) = e^{-\left(\frac{x}{a}\right)^2}.$$

Prove that its Fourier transform is

$$\hat{f}(\nu) = \sqrt{\pi}a\, e^{-(\pi\nu a)^2}.$$

Problem 1.15. Let

$$f(x) = \text{rect}\left(\frac{x}{a}\right) = \begin{cases} 1 & \text{if } |x| \le \frac{a}{2} \\ 0 & \text{otherwise} \end{cases}.$$

Prove that its Fourier transform is

$$\hat{f}(\nu) = \frac{\sin(a\nu)}{a\nu} \equiv \text{sinc}(a\nu).$$

Problem 1.16. Let

$$f(x, y) = \text{rect}\left(\frac{x}{a}\right)\text{rect}\left(\frac{y}{b}\right).$$

What is its two-dimensional Fourier transform?

Problem 1.17. Let

$$f(x, y) = \mathrm{Circ}(r) \equiv \text{circ function}$$
$$= \begin{cases} 1 & |r| = \sqrt{x^2 + y^2} \le 1 \\ 0 & \text{otherwise} \end{cases}.$$

What is its two-dimensional Fourier transform?

Problem 1.18. Assume a card selected out of an ordinary deck of 52 cards. Let

$$\boldsymbol{A} = \{\text{the card is a spade}\}$$

and

$$\boldsymbol{B} = \{\text{the card is a face card, that is, jack, queen, or king}\}.$$

Compute $P\{\boldsymbol{A}\}$, $P\{\boldsymbol{B}\}$, and $P\{\boldsymbol{A} \cap \boldsymbol{B}\}$.

Problem 1.19. Let two items be chosen out of a lot of 12 items where 4 of them are defective. Assume

$$\boldsymbol{A} = \{\text{both chosen items are defective}\}$$

and

$$\boldsymbol{B} = \{\text{both chosen items are not defective}\}.$$

Compute $P\{\boldsymbol{A}\}$ and $P\{\boldsymbol{B}\}$.

Problem 1.20. Given the problem laid out in Problem 1.19. Assume now that

$$\boldsymbol{C} = \{\text{At least one chosen item is defective}\}.$$

What is the probability that event $\boldsymbol{C}$ occurs?

Problem 1.21. Let a pair of fair dice be tossed. If the sum is 6, what is the probability that one of the dice is a 2? In other words, we have

$$\boldsymbol{A} = \{\text{sum is } 6\}$$

and

$$\boldsymbol{B} = \{\text{a 2 appears on at least one die}\}.$$

Find $P\{\boldsymbol{B}|\boldsymbol{A}\}$.

Problem 1.22. In a certain college, 25% of the students fail in mathematics, 15% of the students fail in chemistry, and 10% of the students fail both in mathematics and chemistry. A student is selected at random.

(a) If the student failed in chemistry, what is the probability that the student also failed in mathematics?

(b) If he failed in mathematics, what is the probability that he failed in chemistry too?

(c) What is the probability that he failed both in mathematics and chemistry?

Problem 1.23. Let $\boldsymbol{A}$ and $\boldsymbol{B}$ be events with $P\{\boldsymbol{A}\} = 1/2, P\{\boldsymbol{B}\} = 1/3$, and $P\{\boldsymbol{A} \cap \boldsymbol{B}\} = 1/4$. Find (a) $P\{\boldsymbol{B}|\boldsymbol{A}\}$, (b) $P\{\boldsymbol{A}|\boldsymbol{B}\}$, and (c) $P\{\boldsymbol{A} \cup \boldsymbol{B}\}, P\{\boldsymbol{A}_c|\boldsymbol{B}_c\}$, and $P\{\boldsymbol{B}_c|\boldsymbol{A}_c\}$. Here, $\boldsymbol{A}_c$ and $\boldsymbol{B}_c$ are the complements of $\boldsymbol{A}$ and $\boldsymbol{B}$, respectively.

Problem 1.24. A lot contains 12 items of which 4 are defective. Three items are drawn at random from that lot one after another. Find the probability that all three are nondefective.

Problem 1.25. A card player is dealt 5 cards one right after another from an ordinary deck of 52 cards. What is the probability that they are all spades?

Problem 1.26. Let $\varphi(t)$ be the standard normal distribution (i.e., mean equals zero and variance equals to unity)? Find $\varphi(t)$ for (a) $t = 1.63$, (b) $t = -0.75$ and $t = -2.08$.

HINT: You may need a standard normal distribution table to solve this problem.

Problem 1.27. Let $\boldsymbol{x}$ be a random variable with a standard normal distribution $\varphi(t)$. Find

(a)

$$P\{0 \leq \boldsymbol{x} \leq 1.42\}.$$

(b)

$$P\{-0.73 \leq \boldsymbol{x} \leq 0\}.$$

(c)

$$P\{-1.37 \leq \boldsymbol{x} \leq 2.01\}.$$

(d)

$$P\{0.65 \leq \boldsymbol{x} \leq 1.26\}.$$

(e)

$$P\{-1.79 \leq \boldsymbol{x} \leq -0.54\}.$$

(f)

$$P\{\boldsymbol{x} \geq 1.13\}.$$

(g)

$$P\{|\boldsymbol{x}| \leq 0.5\}.$$

HINT: You may need a standard normal distribution table to solve this problem.

Problem 1.28. A fair die is tossed seven times. Let us assume that success occurs if a 5 or 6 appear. Let $n = 7$, $p = P\{5, 6\} = \frac{1}{3}$, and $q = 1 - p = \frac{2}{3}$.

(a) What is the probability that a 5 or a 6 occurs exactly three times (i.e., $k = 3$)?

(b) What is the probability that a 5 or a 6 occurs at least once?

Problem 1.29. A fair coin is tossed six times. Let us assume that success is a heads. Let $n = 6$ and $p = q = \frac{1}{2}$.

(a) What is the probability that exactly two heads occur (i.e., $k = 2$)?

(b) What is the probability of getting at least four heads (i.e., $k = 4,\ 5,\ \text{and } 6$)?

(c) What is the probability that at least one head occurs?

Problem 1.30. For a Poisson distribution

$$p(k, \lambda) = \frac{\lambda^k}{k!} e^{-\lambda},$$

find (a) $p(2, 1)$, (b) $p\left(3, \frac{1}{2}\right)$, and (c) $p(2, 7)$.

Problem 1.31. Suppose 300 misprints are randomly distributed throughout a book of 500 pages. Find the probability that a given page contains (a) exactly 2 misprints, (b) 2 or more misprints.

HINT: You may want to consider the number of misprints on one page as the number of successes in a sequence of Bernoulli trials. Note that we are dealing with large numbers.

Problem 1.32. Suppose 2% of the items made by a factory are defective. Find the probability that there are 3 defective items in a sample of 100 items.

Problem 1.33. Given

$$X(\text{dB}) = 10 \log_{10} X,$$

derive an equation for X in terms of $X(\text{dB})$.

Problem 1.34. Given

$$X(\text{dB}) = 10 \log_{10} X,$$

find $X(\text{dB})$ for (a) $X = 63^2$, (b) $X = 4000$, and (c) $X = \frac{1}{2500}$.

REFERENCES

1. Pratt, W.K. (2007) *Digital Image Processing*, 4th edn, John Wiley and Sons, New York, pp. 121–123.
2. Bar-Shalom, Y. and Fortmann, T.E. (1988) *Tracking and Data Association*, Academic Press, New York, 353 pages.
3. Lohmann, A.W. (2006) in *Optical Information Processing* (ed. S. Sinzinger), Universitätsverlag, Ilmenau, Germany, ISBN 4-939474-00-6

4. Goodman, J.W. (2004) *Introduction to Fourier Optics*, 3rd edn, Roberts and Company, Englewood, CO.
5. Papoulis, A. (1968) *Systems and Transforms With Applications in Optics*, McGraw-Hill Series in Systems Science, Edition reprint, Robert Krieger Publishing Company, Malabar, FL, 474 pages, ISBN 0898743583, 9780898743586
6. Helstrom, C.W. (1991) *Probability and Stochastic Processes for Engineers*, 2nd edn, Macmillan Publishing Company, New York, 610 pages.
7. Venkatesh, S.S. (2013) *The Theory of Probability: Exploration and Applications*, Cambridge University Press, Cambridge, UK, 805 pages.
8. Fry, W. (1928) *Probability and Its Engineering Uses*, Von Nostrand, New York.

2

FOURIER OPTICS BASICS

2.1 INTRODUCTION

The underlining concept in the characterization of light propagation through a vacuum, channel, or system is *linear superposition*. That is, the combined response from a collection of electromagnetic waves into specific medium or systems is just the linear sum of individual responses created by each individual electromagnetic wave structure going through said medium or system. Booker and Clemmow were one of the first to apply it to modern RF system analysis in their seminal paper in 1950 on *Angular Spectrum* [1], highlighting its benefit over other conventional techniques at the time. Authors like Lohmann [2], Goodman [3], Papoulis [4], and Born and Wolf [5] elaborated on it in their respective textbooks and showed the basic property of *linearity* greatly simplifies the mathematics for describing optical signal processing concepts, and the effects of the turbulent [6–10] and particulate channels [9, 10] on incoming signals as well.

In this and the following chapter, we look into the mathematical theories used to describe the linear superposition approach to optical propagations following Lohmann [2] and others, hopefully providing some insights how they are used in optical systems characterization. The areas covered in this chapter are often referred to as *Fourier Optics*. We begin with the four major theoretical views of this approach:

- Rayleigh–Sommerfeld–Debye Theory of Diffraction
- Huygens–Fresnel–Kirchhoff Theory of Diffraction
- Fresnel Diffraction
- Fraunhofer Diffraction.

Free Space Optical Systems Engineering: Design and Analysis, First Edition. Larry B. Stotts.

Companion website: www.wiley.com\go\stotts\free_space_optical_systems_engineering

Examples and comments also are provided, so that the students get insights on the application of Fourier Optics in optical engineering analyses.

2.2 THE MAXWELL EQUATIONS

In early 1860s, the laws of electromagnetism were unified into one complete theory by James Clerk Maxwell. The result was the first complete description of how electric and magnetic fields are generated and altered by each other, and by charges and currents. His marriage of the Laws of Gauss, Ampere, Faraday, and Lorentz force law into an interrelated set of differential equations formed the foundation of classical electrodynamics, classical optics, and electric circuit analytics for more than a century and a half. However today, it has been understood that the Maxwell equations are not exact, but are a classical field theory approximation to the more accurate and fundamental theory of quantum electrodynamics. These equations do not cover processes such as photon–photon scattering, quantum optics, and many other phenomena related to photons or virtual photons. Fortunately, for most practical instances in optical systems engineering, deviation from the Maxwell equations is immeasurably small and will not affect the development to follow in this book. (Exceptions will be photon counting applications, but we will treat them as separate topics.)

Light is a transverse electromagnetic wave. The electric $\boldsymbol{E}$ and magnetic $\boldsymbol{H}$ fields are perpendicular to each other and to the propagation vector $\boldsymbol{k}$, as shown in Figure 2.1. Both fields obey the Maxwell equations, which are:

$$\nabla \times \boldsymbol{H} = \frac{\partial \boldsymbol{D}}{\partial t} + \boldsymbol{I}, \tag{2.1}$$

$$\nabla \times \boldsymbol{E} = -\frac{\partial \boldsymbol{B}}{\partial t}, \tag{2.2}$$

$$\nabla \cdot \boldsymbol{B} = 0, \tag{2.3}$$

and

$$\nabla \cdot \boldsymbol{D} = \rho \tag{2.4}$$

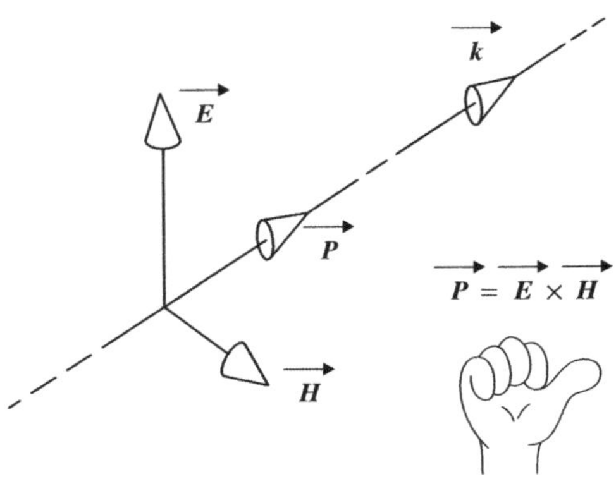

FIGURE 2.1 Right-hand rule for electromagnetic waves. Source: Reproduced with permission of Newport Corporation.

with

$$\boldsymbol{D} = \varepsilon \boldsymbol{E}, \tag{2.5}$$

$$\boldsymbol{I} = \sigma \boldsymbol{E}, \tag{2.6}$$

and

$$\boldsymbol{B} = \mu \boldsymbol{H}. \tag{2.7}$$

In the above, ε is the medium's permittivity, μ the medium's permeability, and σ the medium's conductivity. The refractive index of the medium is defined as

$$n = \sqrt{\varepsilon \mu}. \tag{2.8}$$

In free space, $\varepsilon = \varepsilon_0$ and $\mu = \mu_0$ are the free space permittivity and permeability, respectively, and define the speed of light in free space via the equation

$$v_{light} = \frac{c}{\sqrt{\varepsilon_0 \mu_0}} = c = 2.99792258 \times 10^8 \text{ m/s} \tag{2.9}$$

is the speed of light in free space. The phase velocity in the medium related to the free space speed of light, which is given by

$$v_p = \frac{c}{n}. \tag{2.10}$$

The *Poynting vector* represents the magnitude and direction of the flow of energy in electromagnetic waves. Mathematically, it is written as

$$\boldsymbol{S} = \boldsymbol{E} \times \boldsymbol{H}. \tag{2.11}$$

Its SI unit is watt per square meter (w/m^2). You can easily remember the directions if you "curl" $\boldsymbol{E}$ into $\boldsymbol{H}$ with the fingers of the right hand: your thumb points in the direction of propagation.

If the "material's constants" $\{\varepsilon, \mu, \rho, \sigma\}$ do not depend on $\boldsymbol{E}$ and $\boldsymbol{H}$, then the Maxwell equations are linear. This means that if $\boldsymbol{E}_1$ and $\boldsymbol{B}_1$ are one solution set and $\boldsymbol{E}_2$ and $\boldsymbol{B}_2$ are different solution set for the Maxwell equations, then $\boldsymbol{E}_1 + \boldsymbol{E}_2$ and $\boldsymbol{B}_1 + \boldsymbol{B}_2$ also is a valid solution set. This last relation is known as *linear superposition*.

If there are no charges or currents in the medium of interest, and we specify that ε is a constant and $\mu = \mu_0$, then the Maxwell equations reduce to

$$\nabla \times \boldsymbol{H} = \varepsilon \frac{\partial \boldsymbol{E}}{\partial t}, \tag{2.12}$$

$$\nabla \times \boldsymbol{E} = -\mu_0 \frac{\partial \boldsymbol{H}}{\partial t}, \tag{2.13}$$

$$\nabla \cdot \boldsymbol{B} = 0, \tag{2.14}$$

and

$$\nabla \cdot \boldsymbol{E} = 0. \tag{2.15}$$

If we apply $\mu_0 \frac{\partial}{\partial t}$ to Eq. (2.12) and $\nabla \times$ to Eq. (2.13), then we have

$$\mu_0 \nabla \times \frac{\partial \boldsymbol{H}}{\partial t} = \varepsilon \mu_0 \frac{\partial^2 \boldsymbol{E}}{\partial t^2}$$

and

$$\nabla \times \nabla \times \boldsymbol{E} = -\mu_0 \nabla \times \frac{\partial \boldsymbol{H}}{\partial t}.$$

This implies that

$$\nabla \times \nabla \times \boldsymbol{E} = \varepsilon \mu_0 \frac{\partial^2 \boldsymbol{E}}{\partial t^2}. \tag{2.16}$$

Now the left side of Eq. (2.16) can be written as

$$\nabla \times \nabla \times \boldsymbol{E} = \nabla (\nabla \cdot \boldsymbol{E}) - \nabla \nabla \boldsymbol{E} = -\nabla \nabla \boldsymbol{E} \tag{2.17}$$

using Eq. (2.15), and we find that

$$\nabla^2 \boldsymbol{E} - \varepsilon \mu_0 \frac{\partial^2 \boldsymbol{E}}{\partial t^2} = 0, \tag{2.18}$$

which is a time-dependent wave equation. A similar equation can be derived for $\boldsymbol{H}$.

If we now decompose the electric field into time frequencies, we have

$$\boldsymbol{E}(\boldsymbol{r}, t) = \int_{-\infty}^{\infty} \boldsymbol{u}(\boldsymbol{r}, \gamma)\, e^{-2\pi i \gamma t} d\gamma. \tag{2.19}$$

Substituting Eq. (2.19) into Eq. (2.18) yields

$$\int_{-\infty}^{\infty} \left[\nabla^2 \boldsymbol{u} - (2\pi\delta)^2 \varepsilon \mu_0 \boldsymbol{u}\right] e^{-2\pi i \gamma t} d\delta = 0.$$

The above equation implies that the bracketed term must be zero for all values of time. The result is the following equation must be true:

$$\nabla^2 \boldsymbol{u}(\boldsymbol{r}) - (2\pi\delta)^2 \varepsilon \mu_0 \boldsymbol{u}(\boldsymbol{r}) = \nabla^2 \boldsymbol{u}(\boldsymbol{r}) - k^2 \boldsymbol{u}(\boldsymbol{r}) = 0, \tag{2.20}$$

where $k = \sqrt{(2\pi\delta)^2 \varepsilon \mu_0}$. Equation (2.20) must be true for each vector component individually, so we have a general equation of the form

$$\nabla^2 \boldsymbol{u}(\boldsymbol{r}) - k^2 \boldsymbol{u}(\boldsymbol{r}) = 0. \tag{2.21}$$

Equation (2.21) is the time-independent or stationary wave equation known as the *Helmholtz equation*. This equation will be the basis for the developments in the sections to come.

2.3 THE RAYLEIGH–SOMMERFELD–DEBYE THEORY OF DIFFRACTION

Given an initial source distribution $u(x, y, z_0)$ in the $z = z_0$ plane, what is the distribution $u(x, y, z)$ in some other plane, where $z \geq z_0$. The function $u(x, y, z)$ must satisfy the following four conditions:

i. The Helmholtz equation
ii. The boundary condition $\lim_{z \to z_0} u(x, y, z) \to u(x, y, z_0)$
iii. The irradiation condition
iv. The damping condition.

The first two conditions are straightforward, but the last two need to be explained. That will happen shortly.

Let us begin by defining the two-dimensional Fourier transform pair u and $\hat{u}$ given by

$$u(x, y, z) = \iint_{-\infty}^{\infty} \hat{u}(k_x, k_y; z)\, e^{+2\pi i(k_x x + k_y y)}\, dk_x\, dk_y \tag{2.22}$$

and

$$\hat{u}(k_x, k_y; z) = \iint_{-\infty}^{\infty} u(x, y, z)\, e^{-2\pi i(k_x x + k_y y)}\, dx\, dy. \tag{2.23}$$

Inserting Eq. (2.22) into the wave equations, we obtain the following equation:

$$\left(\frac{\partial^2}{\partial z^2} + k^2 \left[1 - \lambda^2 \left({k_x}^2 + {k_y}^2\right)\right] \right) \hat{u}(k_x, k_y; z) = 0. \tag{2.24}$$

The solution to Eq. (2.24) is

$$\hat{u}(k_x, k_y; z) = A(k_x, k_y)\, e^{ikz\sqrt{[1-\lambda^2({k_x}^2+{k_y}^2)]}} + B(k_x, k_y)\, e^{-ikz\sqrt{[1-\lambda^2({k_x}^2+{k_y}^2)]}}. \tag{2.25}$$

This equation satisfies condition (i). The coefficients $A(k_x, k_y)$ and $B(k_x, k_y)$ also are what satisfy condition (ii), the boundary conditions for $u(x, y, z)$. Specifically, by taking the limit stated in (ii), we have

$$\hat{u}(k_x, k_y; z_0) = A(k_x, k_y)\, e^{ikz_0\sqrt{[1-\lambda^2({k_x}^2+{k_y}^2)]}} + B(k_x, k_y)\, e^{-ikz_0\sqrt{[1-\lambda^2({k_x}^2+{k_y}^2)]}}. \tag{2.26}$$

Unfortunately, we have one equation and two unknowns, A and B. This is reconciled by condition (iii), the *irradiation condition*.

It is clear that besides the coefficient, the big difference between the two components in Eq. (2.25) is the opposite signs of the z-component in the exponent, not the field component comprised solely of the other two wavenumbers. This difference is illustrated in Figure 2.2. From a physical point of view, only the positive propagation field makes sense, since we assume that the object was illuminated from the left or the negative z-axis side. This implies that $B(k_x, k_y) = 0$ for all values of k_x and k_y. By doing this, we satisfy condition (iii), the irradiation condition.

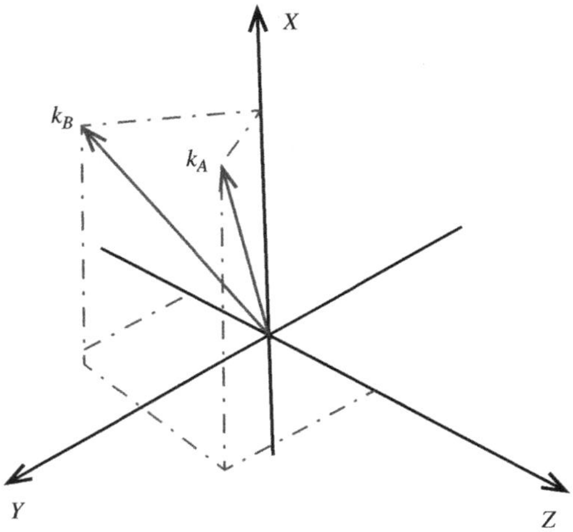

FIGURE 2.2 Wavenumbers for fields A and B in three space.

In Eq. (2.25), it is assumed that $\left[1 - \lambda^2 \left({k_x}^2 + {k_y}^2\right)\right]$ to be real, which means that $\left({k_x}^2 + {k_y}^2\right) \le \lambda^{-2}$. Otherwise, we would have an evanescent or decaying electromagnetic field. What do we mean by that? The integral in Eq. (2.22) ranges over the whole $\left(k_x, k_y\right)$-domain, $(-\infty, \infty)$. Inside the circle, $\left({k_x}^2 + {k_y}^2\right) \le \lambda^{-2}$, the bracketed term above is real. Outside that circle, the bracketed term becomes imaginary, or we have

$$\sqrt{\left[1 - \lambda^2 \left({k_x}^2 + {k_y}^2\right)\right]} = i\sqrt{\left[\lambda^2 \left({k_x}^2 + {k_y}^2\right) - 1\right]},$$

for $\left({k_x}^2 + {k_y}^2\right) > \lambda^{-2}$. Under these conditions, the A and B solutions become

$$A\left(k_x, k_y\right) e^{i\,k\,z\sqrt{\left[1-\lambda^2\left({k_x}^2+{k_y}^2\right)\right]}} = A\left(k_x, k_y\right) e^{-k\,z\,\sqrt{\left[\lambda^2\left({k_x}^2+{k_y}^2\right)-1\right]}}$$

and

$$B\left(k_x, k_y\right) e^{-i\,k\,z\sqrt{\left[1-\lambda^2\left({k_x}^2+{k_y}^2\right)\right]}} = B\left(k_x, k_y\right) e^{+k\,z\sqrt{\left[\lambda^2\left({k_x}^2+{k_y}^2\right)-1\right]}},$$

respectively. The A solution decreases exponentially as the distance z increases. This type of solution is known as an evanescent or decaying electromagnetic field. It is of the form required by condition (iv), that is, the damping condition requirement. On the other hand, the amplitude of the B solution has very large amplitude for large values of z. This is not a realistic expectation as it violates the damping condition, so again we set B to zero.

Thus, we find that the solution to the wave equation that meets all of our original four conditions is given by

$$\hat{u}\left(k_x, k_y; z\right) = A\left(k_x, k_y\right) e^{ikz\sqrt{\left[1-\lambda^2\left({k_x}^2+{k_y}^2\right)\right]}}. \tag{2.27}$$

Recall that because of condition (ii), we must have

$$\hat{u}\left(k_x, k_y; z_0\right) = A\left(k_x, k_y\right) e^{ikz_0\sqrt{[1-\lambda^2(k_x{}^2+k_y{}^2)]}}. \tag{2.28}$$

Solving for $A\left(k_x, k_y\right)$, we have

$$A\left(k_x, k_y\right) = \hat{u}\left(k_x, k_y; z_0\right) e^{-i\,k\,z_0\sqrt{[1-\lambda^2(k_x{}^2+k_y{}^2)]}}. \tag{2.29}$$

This implies that Eq. (2.22) can be rewritten as

$$u(x, y, z) = \iint_{-\infty}^{\infty} A\left(k_x, k_y\right) e^{i\,k\,z\sqrt{[1-\lambda^2(k_x{}^2+k_y{}^2)]}}\, e^{2\pi i(k_x\,x+k_y\,y)}\, dk_x\, dk_y$$

or

$$u(x, y, z) = \iint_{-\infty}^{\infty} \hat{u}\left(k_x, k_y; z_0\right) e^{i\,k\,(z-z_0)\sqrt{[1-\lambda^2(k_x{}^2+k_y{}^2)]}}\, e^{2\pi i(k_x\,x+k_y\,y)}\, dk_x\, dk_y \tag{2.30a}$$

$$= \iint_{-\infty}^{\infty}\iint_{-\infty}^{\infty} u\left(x', y', z_0\right) e^{2\pi i[k_x(x-x')+k_y(y-y')]}$$
$$\times\, e^{ik(z-z_0)\sqrt{1-\lambda^2(k_x{}^2+k_y{}^2)}}\, dk_x dk_y dx' dy' \tag{2.30b}$$

or

$$\hat{u}\left(k_x, k_y; z\right) = \hat{u}\left(k_x, k_y;\ z_0\right) e^{ik(z-z_0)\sqrt{1-\lambda^2(k_x{}^2+k_y{}^2)}} \tag{2.31a}$$

$$= \iint_{-\infty}^{\infty} u\left(x', y', z_0\right) e^{2\pi i[k_x\,x'+k_y\,y']}\, e^{ik(z-z_0)\sqrt{1-\lambda^2(k_x{}^2+k_y{}^2)}}\, dx' dy'. \tag{2.31b}$$

Equations (2.30) and (2.31) are known as the Rayleigh–Sommerfeld–Debye formulae and are useful in different circumstances.

Let us look at the above to derive some important facts. The range of $\left(k_x, k_y\right)$ covers both the range of frequencies in the circle $\left(k_x{}^2 + k_y{}^2\right) \le \lambda^{-2}$ and the region outside this circle where the evanescent field occur. These regions are illustrated in Figure 2.3. The key aspect of this fact is that the spatial frequency spectrum $\hat{u}\left(k_x, k_y; z\right)$ describing $u(x, y, z)$ does not know anything about the super-high frequencies of the object spectrum $\hat{u}\left(k_x, k_y; z_0\right)$ outside the circle $\left(k_x{}^2 + k_y{}^2\right) \le \lambda^{-2}$. We see that this will affect the quality of the profile $u(x, y, z)$ in the observation plane. For example, optical waves cannot reveal object details smaller than a fraction of a wavelength, if the observer is many wavelengths away no matter what the lens, mirror or other long distance sensor you use [3, p. 151]. Super resolution, that is, breaking the fundamental resolution barrier limit, is one of the key technology challenges that have plagued the community for decades. To solve this problem, one must learn to handle the evanescent fields

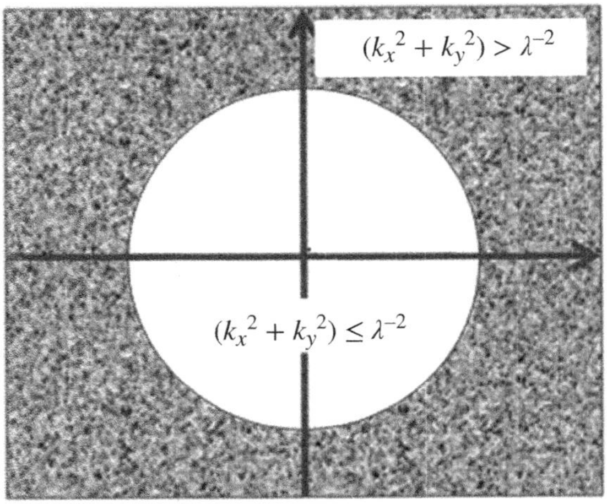

FIGURE 2.3 Wavenumber regimes for real and imaginary arguments.

where the key information lies. One might wonder why this is so hard. The following example illustrates the problem [3].

Example 2.1 What percentage of resolution increase would occur if we were able to measure the field amplitudes at the distance of one wavelength? In this case, the intensity of the observable would be down to 0.14 $\left(= 1/e^2\right)$. If p equals the fractional increase above one wavelength, we will have to sample the additional frequencies, that is,

$$\sqrt{\left(k_x^{\ 2} + k_y^{\ 2}\right)} = \frac{(1+p)}{\lambda}.$$

Substituting this relationship into the bracketed z-term above for the evanescent field condition, we obtain

$$\left(\frac{2\pi z}{\lambda}\right)\sqrt{\lambda^2\left(k_x^{\ 2} + k_y^{\ 2}\right) - 1} = \left(\frac{2\pi z}{\lambda}\right)\sqrt{(1+p)^2 - 1} \approx \left(\frac{2\pi z}{\lambda}\right)\sqrt{2p}$$

for $p \ll 1$. For $z = \lambda$, the damping factor is $\exp\left\{-2\pi\sqrt{2p}\right\}$. The specific fraction p of increased resolution for a $\frac{1}{e}$ damping factor is then given by

$$2\pi\sqrt{2p} = 1 \ \text{ or } \ 4\pi^2(2p) = 1,$$

which results in

$$p = \frac{1}{8\pi^2} = 0.1267.$$

Therefore, in order to improve the resolution, or spatial bandwidth, of object by 1.27%, we need to go as close as one wavelength to the object and measure its intensities that are down 14% of the original value. This does not bode well for improving resolution for most practical applications!

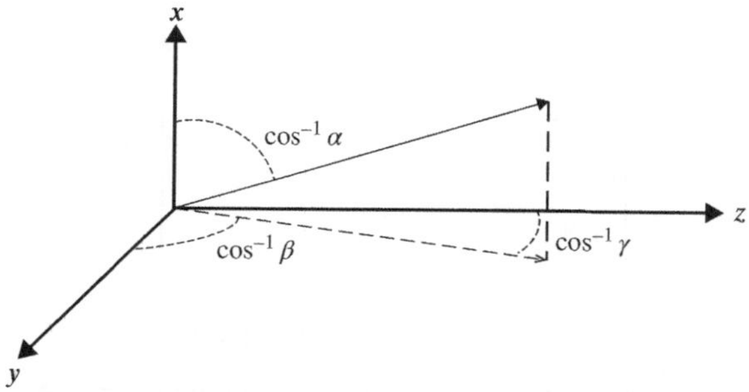

FIGURE 2.4 The wave vector in angular coordinates.

Example 2.2 Referring to Figure 2.4, we can write the components of the wave vector as

$$\alpha = \lambda k_x, \tag{2.32}$$

$$\beta = \lambda k_y, \tag{2.33}$$

and

$$\gamma = \sqrt{1 - (\lambda k_x)^2 - (\lambda k_y)^2} = \sqrt{1 - \alpha^2 - \beta^2}. \tag{2.34}$$

Rewriting Eq. (2.31b) using the above definitions, we have

$$\hat{u}\left(\frac{\alpha}{\lambda}, \frac{\beta}{\lambda}; z\right) = \iint_{-\infty}^{\infty} u\left(x, y, z_0\right) e^{2\pi i\left[\frac{\alpha}{\lambda} x + \frac{\beta}{\lambda} y\right]} e^{ik(z-z_0)\sqrt{1-\lambda^2(\alpha^2+\beta^2)}} dxdy. \tag{2.35}$$

Equation (2.14) is known as the *Angular Spectrum* characterization of the initial field $u\left(x, y, z_0\right)$. Modern authors found that a spatial frequencies characterization of Eq. (2.35) was more useful in describing resulting optical phenomena because it related better to the Fourier transform theory. This became the trend and the area of *Fourier Optics* was created as a result.

2.4 THE HUYGENS–FRESNEL–KIRCHHOFF THEORY OF DIFFRACTION

In the Huygens–Fresnel–Kirchhoff theory of diffraction, each point in the source in the $\left(x, y, z_0\right)$-plane emits a spherical wave and reaches various points in the observation plane (x, y, z). Figure 2.5 illustrates a two-dimensional cut of this situation and Figure 2.6 is its simple interpretation. Recalling Eq. (2.30b), we have

$$u(x, y, z) = \iint_{-\infty}^{\infty} \iint_{-\infty}^{\infty} u\left(x', y', z_0\right) e^{2\pi i\left[k_x(x-x')+k_y(y-y')\right]}$$
$$\times e^{ik(z-z_0)\sqrt{1-\lambda^2\left(k_x^2+k_y^2\right)}} dk_x dk_y dx' dy'$$

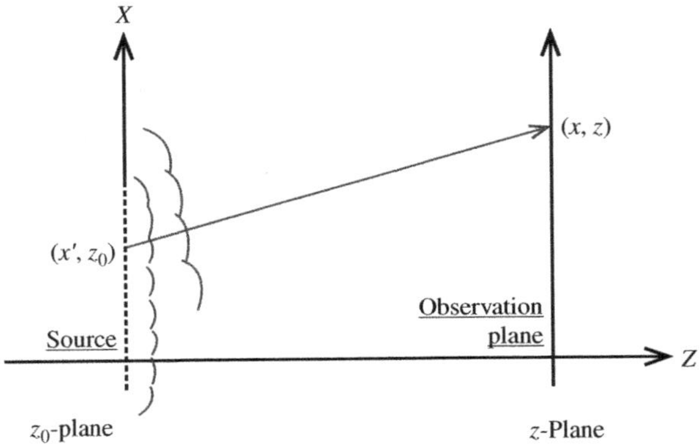

FIGURE 2.5 Spherical wave geometry for Huygens–Fresnel–Kirchhoff theory.

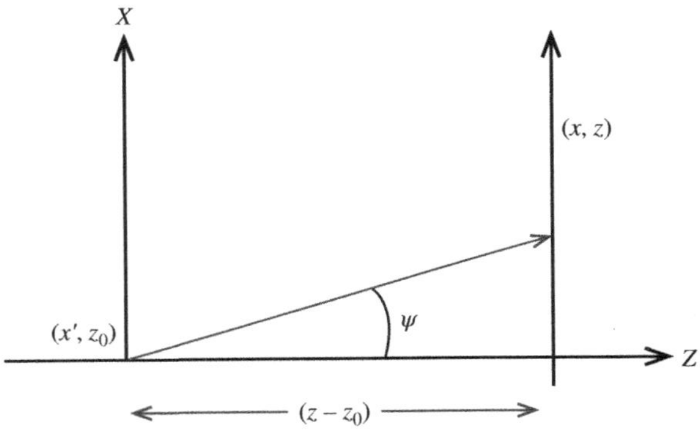

FIGURE 2.6 Angular interpretation of the spherical wave geometry.

$$= \iint_{-\infty}^{\infty} u\left(x', y', z_0\right) \left[\iint_{-\infty}^{\infty} e^{2\pi i\left[k_x(x-x')+k_y(y-y')\right]} \right.$$
$$\left. \times e^{ik(z-z_0)\sqrt{1-\lambda^2\left(k_x{}^2+k_y{}^2\right)}} dk_x dk_y \right] dx' dy'. \tag{2.36}$$

If the Huygens–Fresnel–Kirchhoff theory of diffraction is to be derived from Eq. (2.36), then the integral in brackets over the frequency spectrum must represent a spherical wave emitted from every point (x, y) in the source plane $z = z_0$. To solve Eq. (2.36), we employ the *saddle-point method* (see Appendix A). This method's general solution is given by

$$\int_{A_x}^{B_x} \int_{A_y}^{B_y} g(x, y)\, e^{ikf(x,y)}\, dxdy$$
$$\approx \frac{2\pi\, g\left(x_0, y_0\right) e^{ikf(x_0,\ y_0)+\dfrac{i\pi}{2}}}{k\sqrt{\left(\dfrac{\partial^2 f\left(x_0, y_0\right)}{\partial x^2}\right)\left(\dfrac{\partial^2 f\left(x_0, y_0\right)}{\partial y^2}\right) - \left(\dfrac{\partial^2 f\left(x_0, y_0\right)}{\partial x \partial y}\right)^2}}. \tag{2.37}$$

Looking at Eq. (2.36), we see that

$$kf\left(k_x, k_y\right) = \frac{2\pi}{\lambda}\left[\lambda k_x\left(x - x'\right) + \lambda k_y\left(y - y'\right) + \left(z - z_0\right)\sqrt{1 - \lambda^2\left({k_x}^2 + {k_y}^2\right)}\right] \tag{2.38}$$

Taking the derivatives with respect to k_x and k_y yield

$$\left.\frac{\partial f}{\partial k_x}\right|_{k_x = k_{x0}} = 0 = \left[\lambda\left(x - x'\right) - \frac{\lambda^2 k_x\left(z - z_0\right)}{\sqrt{1 - \lambda^2\left({k_x}^2 + {k_y}^2\right)}}\right]_{k_x = k_{x0}}$$

and

$$\left.\frac{\partial f}{\partial k_y}\right|_{k_y = k_{y0}} = 0 = \left[\lambda\left(y - y'\right) - \frac{\lambda^2 k_y\left(z - z_0\right)}{\sqrt{1 - \lambda^2\left({k_x}^2 + {k_y}^2\right)}}\right]_{k_x = k_{x0}},$$

respectively. These equations imply that

$$\frac{\lambda k_{x0}}{\sqrt{1 - \lambda^2\left({k_{x0}}^2 + {k_{y0}}^2\right)}} = \frac{\left(x - x'\right)}{\left(z - z_0\right)} \tag{2.39}$$

and

$$\frac{\lambda k_{y0}}{\sqrt{1 - \lambda^2\left({k_{x0}}^2 + {k_{y0}}^2\right)}} = \frac{\left(y - y'\right)}{\left(z - z_0\right)}. \tag{2.40}$$

Using Eqs. (2.39) and (2.40), we can write the function $f\left(k_{x0}, k_{y0}\right)$ as

$$\begin{aligned} f\left(k_{x0}, k_{y0}\right) &= \sqrt{1 - \lambda^2\left({k_{x0}}^2 + {k_{y0}}^2\right)}\left(\frac{\left(x - x'\right)^2}{\left(z - z_0\right)} + \frac{\left(y - y'\right)^2}{\left(z - z_0\right)} + \left(z - z_0\right)\right) \\ &= \sqrt{1 - \lambda^2\left({k_{x0}}^2 + {k_{y0}}^2\right)}\left(\frac{r^2}{\left(z - z_0\right)}\right), \end{aligned} \tag{2.41}$$

where

$$r^2 = \left(x - x'\right)^2 + \left(y - y'\right)^2 + \left(z - z_0\right)^2. \tag{2.42}$$

To reduce Eq. (2.41) into an equation solely expressed in just spatial coordinates, we note that

$$\frac{\left(\lambda k_{x0}\right)^2}{1-\lambda^2\left(k_{x0}{}^2+k_{y0}{}^2\right)}+\frac{\left(\lambda k_{y0}\right)^2}{1-\lambda^2\left(k_{x0}{}^2+k_{y0}{}^2\right)}+1=\frac{1}{1-\lambda^2\left(k_{x0}{}^2+k_{y0}{}^2\right)} \tag{2.43}$$

and

$$\frac{\left(\lambda k_{x0}\right)^2}{1-\lambda^2\left(k_{x0}{}^2+k_{y0}{}^2\right)}+\frac{\left(\lambda k_{y0}\right)^2}{1-\lambda^2\left(k_{x0}{}^2+k_{y0}{}^2\right)}+1$$

$$=\left[\frac{(x-x')}{(z-z_0)}\right]^2+\left[\frac{(y-y')}{(z-z_0)}\right]^2+\left[\frac{(z-z_0)}{(z-z_0)}\right]^2 \tag{2.44a}$$

$$=\left(\frac{r^2}{\left(z-z_0\right)^2}\right), \tag{2.44b}$$

which shows that

$$\frac{1}{1-\lambda^2\left(k_{x0}{}^2+k_{y0}{}^2\right)}=\frac{r^2}{\left(z-z_0\right)^2}. \tag{2.45}$$

This means that

$$f\left(k_{x0},k_{y0}\right)=r=\sqrt{(x-x')^2+(y-y')^2+\left(z-z_0\right)^2}. \tag{2.46}$$

Now, let us compute the denominator of the saddle-point solution given in Eq. (2.37).

Taking the first-order differentiated equations above and differentiating again, we obtain

$$\left.\frac{\partial^2 f}{\partial k_x{}^2}\right|_{k_x=k_{x0};k_y=k_{y0}}=\frac{-\lambda^2\left(z-z_0\right)\left(1-\lambda^2k_{y0}^2\right)}{\left[\sqrt{1-\lambda^2\left(k_{x0}{}^2+k_{y0}{}^2\right)}\right]^3}, \tag{2.47}$$

$$\left.\frac{\partial^2 f}{\partial k_y{}^2}\right|_{k_y=k_{y0};k_y=k_{y0}}=\frac{-\lambda^2\left(z-z_0\right)\left(1-\lambda^2k_{y0}^2\right)}{\left[\sqrt{1-\lambda^2\left(k_{x0}{}^2+k_{y0}{}^2\right)}\right]^3}, \tag{2.48}$$

and

$$\left.\frac{\partial^2 f}{\partial k_x\partial k_y}\right|_{k_x=k_{x0};\ k_y=k_{y0}}=\frac{-\lambda^4\left(z-z_0\right)k_xk_y}{\left[\sqrt{1-\lambda^2\left(k_{x0}{}^2+k_{y0}{}^2\right)}\right]^3}, \tag{2.49}$$

which leads to

$$\left(\frac{\partial^2 f\left(x_0, y_0\right)}{\partial x^2}\right)\left(\frac{\partial^2 f\left(x_0, y_0\right)}{\partial y^2}\right)-\left(\frac{\partial^2 f\left(x_0, y_0\right)}{\partial x \partial y}\right)=\frac{\lambda^4\left(z-z_0\right)^2}{\left[1-\lambda^2\left({k_{x0}}^2+{k_{y0}}^2\right)\right]^2}. \tag{2.50}$$

Relooking at the denominator term in the saddle-point solution, we find that Eq. (2.50) allows us to write

$$\left(\frac{\partial^2 f\left(x_0, y_0\right)}{\partial x^2}\right)\left(\frac{\partial^2 f\left(x_0, y_0\right)}{\partial y^2}\right)-\left(\frac{\partial^2 f\left(x_0, y_0\right)}{\partial x \partial y}\right)=\frac{\lambda^4 r^4}{\left(z-z_0\right)^2}, \tag{2.51}$$

and we have

$$\frac{1}{\sqrt{\left(\frac{\partial^2 f(x_0, y_0)}{\partial x^2}\right)\left(\frac{\partial^2 f(x_0, y_0)}{\partial y^2}\right)-\left(\frac{\partial^2 f(x_0, y_0)}{\partial x \partial y}\right)}}=\frac{\left(z-z_0\right)}{\lambda^2 r^2}. \tag{2.52}$$

Referring to Figure 2.6, we can write

$$\cos \Psi=\frac{\left(z-z_0\right)}{r}=\frac{1}{\sqrt{1-\lambda^2\left({k_{x0}}^2+{k_{y0}}^2\right)}}. \tag{2.53}$$

Let us now turn to the integral inside the brackets in Eq. (2.36). Substituting all of the above equations into saddle-point solution for that integral, we find that

$$\iint_{-\infty}^{\infty} e^{2\pi i\left[k_x(x-x')+k_y(y-y')+\frac{1}{\lambda}(z-z_0)\sqrt{1-\lambda^2\left({k_x}^2+{k_y}^2\right)}\right]} dk_x dk_y=\frac{\cos \Psi}{\lambda r} e^{i\left(kr+\frac{\pi}{2}\right)}. \tag{2.54}$$

Substituting into Eq. (2.36) yields

$$u(x, y, z)=\iint_{-\infty}^{\infty} \iint_{-\infty}^{\infty} u\left(x', y', z_0\right) \frac{\cos \Psi}{\lambda r} e^{i\left(kr+\frac{\pi}{2}\right)} dx' dy'. \tag{2.55}$$

Equation (2.55) is known as the *Huygens–Fresnel–Kirchhoff integral* or *Kirchhoff integral* for short.

Something very interesting happens to the previous development if one does not go through the above rigor, but assumes that the spatial frequencies of dominance are the ones close to the optical axis, z, that is, the low spatial frequency domain where $\left({k_{x0}}^2+{k_{y0}}^2\right) \ll \lambda^2$. In this case, we have

$$\cos \Psi=\sqrt{1-\sin^2 \Psi} \approx 1-\frac{1}{2}\sin^2 \Psi \approx \sqrt{1-\lambda^2\left({k_x}^2+{k_y}^2\right)}. \tag{2.56}$$

Let us now use these relationships in developing a different saddle-point solution. Specifically, we now can write

$$\begin{aligned} f\left(k_x, k_y\right) &= \lambda k_x \left(x - x'\right) + \lambda k_y \left(y - y'\right) + \left(z - z_0\right) \sqrt{1 - \lambda^2 \left({k_x}^2 + {k_y}^2\right)} \\ &\approx \lambda k_x \left(x - x'\right) + \lambda k_y \left(y - y'\right) + \left(z - z_0\right) + \left(z - z_0\right) \\ &\quad - \frac{\lambda^2}{2} \left({k_x}^2 + {k_y}^2\right) \left(z - z_0\right), \end{aligned} \tag{2.57}$$

which yields

$$\left.\frac{\partial f}{\partial k_x}\right|_{k_x=k_{x0};\ k_y=k_{y0}} = 0 = \left[\lambda \left(x - x'\right) - \lambda^2 k_{x0} \left(z - z_0\right)\right] \tag{2.58}$$

and

$$\left.\frac{\partial f}{\partial k_y}\right|_{k_x=k_{x0};\ k_y=k_{y0}} = 0 = \left[\lambda \left(y - y'\right) - \lambda^2 k_{y0} \left(z - z_0\right)\right]. \tag{2.59}$$

The evaluation points k_{x0} and k_{y0} are then

$$k_{x0} = \frac{\left(x - x'\right)}{\lambda \left(z - z_0\right)} \tag{2.60}$$

and

$$k_{y0} = \frac{\left(y - y'\right)}{\lambda \left(z - z_0\right)}. \tag{2.61}$$

Continuing with the above, we find that

$$\left.\frac{\partial^2 f}{\partial {k_x}^2}\right|_{k_x=k_{x0};k_y=k_{y0}} = \lambda^2 \left(z - z_0\right) = \left.\frac{\partial^2 f}{\partial {k_y}^2}\right|_{k_y=k_{y0};k_y=k_{y0}} \tag{2.62}$$

and

$$\left.\frac{\partial^2 f}{\partial k_x \partial k_y}\right|_{k_x=k_{x0};\ k_y=k_{y0}} = 0. \tag{2.63}$$

Substituting into the denominator of the saddle-point solution, we obtain

$$\sqrt{\left(\frac{\partial^2 f\left(x_0, y_0\right)}{\partial x^2}\right)\left(\frac{\partial^2 f\left(x_0, y_0\right)}{\partial y^2}\right) - \left(\frac{\partial^2 f\left(x_0, y_0\right)}{\partial x \partial y}\right)} = \lambda^2 \left(z - z_0\right). \tag{2.64}$$

The function $f(k_x, k_y)$ evaluated at (k_{x0}, k_{y0}) then is

$$f(k_{x0}, k_{y0}) = \frac{(x-x')^2}{(z-z_0)} + \frac{(y-y')^2}{(z-z_0)} + (z-z_0) - \frac{(x-x')^2 + (y-y')^2}{2(z-z_0)}$$
$$= (z-z_0) + \frac{(x-x')^2 + (y-y')^2}{2(z-z_0)}. \tag{2.65}$$

Using the above equation, we find that

$$\iint_{-\infty}^{\infty} e^{2\pi i\left[k_x(x-x')+k_y(y-y')+\frac{1}{\lambda}(z-z_0)\sqrt{1-\lambda^2(k_x{}^2+k_y{}^2)}\right]}\, dk_x dk_y$$
$$\approx \frac{\exp\left\{i\pi\frac{(x-x')^2+(y-y')^2}{\lambda(z-z_0)} + i\left[k(z-z_0)+\frac{\pi}{2}\right]\right\}}{\lambda(z-z_0)}. \tag{2.66}$$

Substituting the above into Eq. (2.36), we obtain

$$u(x,y,z) = \frac{\exp\left\{i\left[k(z-z_0)+\frac{\pi}{2}\right]\right\}}{\lambda(z-z_0)} \iint_{-\infty}^{\infty} u(x',y',z_0)$$
$$\times \exp\left\{i\pi\frac{(x-x')^2+(y-y')^2}{\lambda(z-z_0)}\right\} dx'dy'. \tag{2.67}$$

Equation (2.67) is known as the quadratic, or parabolic, approximation of the Huygens–Fresnel–Kirchhoff integral. This approximation basically describes *Fresnel diffraction* by a Fresnel transformation.

Example 2.3 To demonstrate the utility of the Eq. (2.67), we derive the *Lens Formula*. For simplicity, we derive it using only a single-dimension version. Figure 2.7 depicts the system setup. Assume that we have spatial source, say, $u_0(x')$, in the $(z=0)$-plane.

This source will propagate to the $(z=z_1^-)$-plane, resulting in the distribution $u(x'', z_1^-) = u_1(x'')$. Mathematically, we can write

$$u_1(x'') = \int_{-\infty}^{\infty} u_0(x')\, e^{\frac{i\pi(x'-x'')^2}{\lambda z_1}}\, dx', \tag{2.68}$$

using what we learned above.

In the $(z=z_1^-)$-plane, we place a lens and the distribution passes through the lens to create the distribution $v_1(x'')$. This distribution is written as

$$v_1(x') = u_1(x'')\, e^{\frac{-i\pi x''^2}{\lambda f}}, \tag{2.69}$$

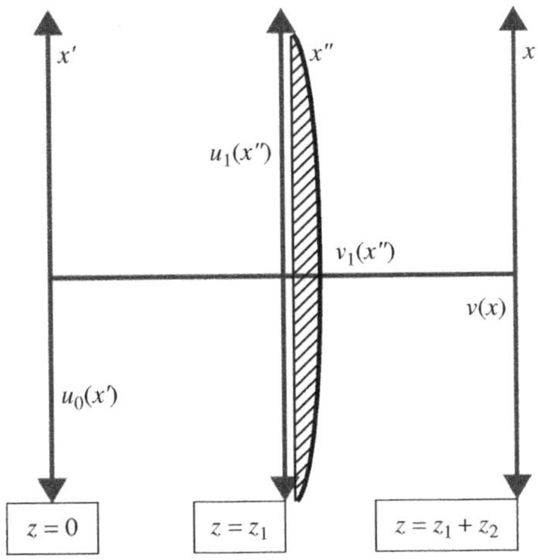

FIGURE 2.7 Geometry for image formation.

where $z = z_1^+ = f$ ≡ the focal length of the lens. This new source then will propagate to the $(z = z_1 + z_2)$-plane, resulting in the distribution $v(x)$ given by

$$v(x) = \int_{-\infty}^{\infty} v_1(x'') \, e^{\frac{i\pi(x-x'')^2}{\lambda z_2}} \, dx''. \tag{2.70}$$

Substituting the previous equations into Eq. (2.70) yields

$$v(x) = \iint_{-\infty}^{\infty} u_0(x') \, e^{\frac{i\pi(x-x'')^2}{\lambda z_2}} \left(e^{\frac{-i\pi(x'')^2}{\lambda f}} \right) e^{\frac{i\pi(x'-x'')^2}{\lambda z_1}} \, dx' dx''. \tag{2.71}$$

Ideally, we wish $v(x)$ to be an image of $u_0(x')$. This implies that the exponential terms in the above integrand should be acting like a delta function to achieve that mapping, that is, changes the x' of $u_0(x')$ into x. Combining the arguments of the exponents into one term, we have

$$\begin{aligned}
&\frac{i\pi\,(x-x'')^2}{\lambda\, z_2} - \frac{i\pi x''^2}{\lambda f} + \frac{i\pi\,(x'-x'')^2}{\lambda\, z_1} \\
&\quad = \frac{i\pi}{\lambda}\left[x''^2 \left[\frac{1}{z_2} - \frac{1}{f} + \frac{1}{z_1} \right] - 2x'' \left(\frac{x}{z_2} + \frac{x'}{z_1} \right) + \frac{x^2}{z_2} + \frac{x'^2}{z_1} \right].
\end{aligned}$$

Since the initial distribution is a function of x', the integration over (x'') must be the one to yield a delta function. That is, the delta function must come from the following integration:

$$\int_{-\infty}^{\infty} e^{\frac{i\pi}{\lambda}\left[x''^2 \left[\frac{1}{z_2} - \frac{1}{f} + \frac{1}{z_1} \right] - 2x'' \left(\frac{x}{z_2} + \frac{x'}{z_1} \right) \right]} \, dx''. \tag{2.72}$$

For that to happen, we need to get rid of the x''^2 term. This implies that we set the output plane at distance such that

$$\frac{1}{z_2} = \frac{1}{f} - \frac{1}{z_1} \tag{2.73a}$$

or

$$\frac{1}{f} = \frac{1}{z_1} + \frac{1}{z_2}. \tag{2.73b}$$

Equation (2.73b) is the *Lens Law*. Using the Lens Law in Eq. (2.72) gives

$$\int_{-\infty}^{\infty} e^{\frac{-2\pi i}{\lambda}x''\left(\frac{x}{z_2}+\frac{x'}{z_1}\right)} dx'' = \delta\left(\frac{x}{\lambda\, z_2} + \frac{x'}{\lambda\, z_1}\right) = \lambda z_1\, \delta\left(x' + \frac{x z_1}{z_2}\right). \tag{2.74}$$

In the last equation, we make use of the *Dirac-Delta or Delta function*, $\delta(x)$, defined by the following integral:

$$\int_{-\infty}^{\infty} e^{2\pi i\, x(v'-v)}\, dx \equiv \delta\left(v' - v\right). \tag{2.75}$$

Properties of this function include:

$$\delta(v) = \delta(-v) = \delta^*(v), \tag{2.76a}$$

$$\delta(v) = \left(\frac{1}{A}\right)\delta\left(\frac{v}{A}\right), \tag{2.76b}$$

$$\delta(A\, v) = \frac{1}{A}\delta(v). \tag{2.76c}$$

This implies that

$$\begin{aligned} v(x) &= \iint_{-\infty}^{\infty} u_0\left(x'\right) e^{\frac{i\pi\left(x-x''\right)^2}{\lambda z_2}} \left(e^{\frac{-i\pi x''^2}{\lambda f}}\right) e^{\frac{i\pi\,\left(x'-x''\right)^2}{\lambda z_1}}\, dx'dx'' \\ &\propto \int_{-\infty}^{\infty} u_0\left(x'\right) e^{\frac{i\pi}{\lambda}\left(\frac{x^2}{z_2}+\frac{x'^2}{z_1}\right)} \delta\left(x' + \frac{x\, z_1}{z_2}\right) dx' = u_0\left(-\frac{x\, z_1}{z_2}\right) e^{\frac{i\pi x^2}{\lambda}\left(\frac{1}{z_2}+\frac{z_1}{z_2}\right)}, \end{aligned} \tag{2.77}$$

ignoring the $\lambda\, z_1$ factor in Eq. (2.67). Squaring Eq. (2.77), we find that the resulting intensity is

$$|v(x)|^2 = \left|u_0\left(-\frac{x\, z_1}{z_2}\right)\right|^2 \tag{2.78a}$$

$$= \left|u_0\left(-\frac{x}{M}\right)\right|^2. \tag{2.78b}$$

In two dimensions, Eq. (2.78) is written as

$$|v(x,y)|^2 = \left|u_0\left(-\frac{x\,z_1}{z_2}, -\frac{y\,z_1}{z_2}\right)\right|^2 \tag{2.79a}$$

$$= \left|u_0\left(-\frac{x}{M}, -\frac{y}{M}\right)\right|^2. \tag{2.79b}$$

From Eqs. (2.78) and (2.79), we see that a "perfect lens" in the right setup creates a perfect production of the initial source field, but with an inverse of said field and a different scaling. The term $M = \frac{z_2}{z_1}$ represents the magnification of the imaged object; the minus sign in the initial distribution's argument implies an image inversion.

2.5 FRAUNHOFER DIFFRACTION

Far-field, or Fraunhofer, diffraction has been the workhorse of all electromagnetic field propagation, for example, optical signal processing, holography, and propagation in a turbulence atmosphere. In this section, we extend the two-dimension discussion in the previous sections to three dimensions to describe this important phenomenon.

Figure 2.8 shows from a source plane at $z = 0$ to an observation plane at $z = R$. Define

$$p = (x_p, y_p, z_p) = (R\sin\alpha, R\cos\beta, R\cos\varepsilon), \tag{2.80}$$

$$u(x,y,0) = u_0(x,y), \tag{2.81}$$

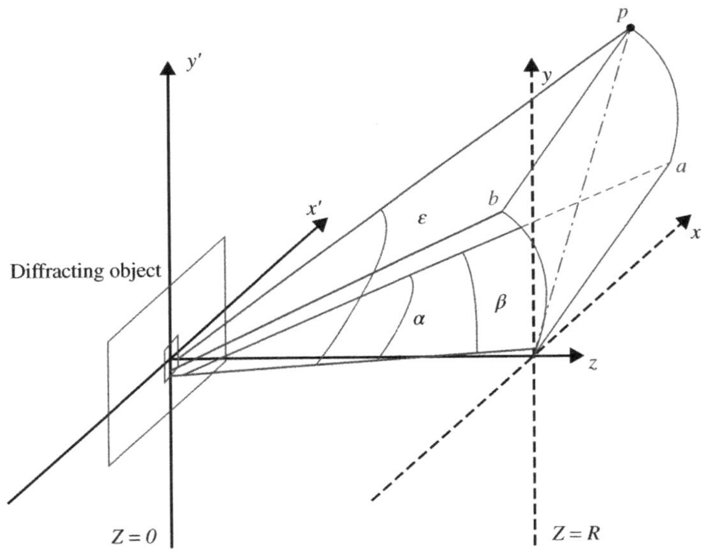

FIGURE 2.8 Fraunhofer diffraction geometry.

and

$$\hat{u}_0\left(k_x, k_x\right) = \iint u_0\left(x', y'\right) e^{2\pi i k_x x' + 2\pi i k_x y'}\, dx' dy', \tag{2.82}$$

where $\cos\mathcal{E} = \sqrt{1 - \sin^2\alpha - \sin^2\beta}$. Using Eq. (2.30a), we write

$$u\left(x_p, y_p, z_p\right) = \iint_{-\infty}^{\infty} \hat{u}_0\left(k_x, k_y\right) e^{ikz_p\sqrt{\left[1-\lambda^2\left(k_x{}^2+k_y{}^2\right)\right]}} e^{2\pi i\left(k_x\, x_p + k_y\, y_p\right)} dk_x dk_y \tag{2.83a}$$

$$= \iint_{-\infty}^{\infty} \hat{u}_0\left(k_x, k_y\right) e^{2\pi i R\cos\mathcal{E}\frac{\sqrt{\left[1-\lambda^2\left(k_x{}^2+k_y{}^2\right)\right]}}{\lambda}}$$

$$\times\, e^{2\pi i\, R\left(k_x\ \sin\alpha + k_y\ R\cos\beta\right)}\, dk_x\, dk_y. \tag{2.83b}$$

Equation (2.83b) represents a collection of plane waves, some propagating in z and others evanescent in z, which are negligible in the Fraunhofer regime because $R \gg \lambda$. Referring to Figure 2.8, there is only one whose wavenumber from the origin located at $z = 0$ that directly projects to point $\boldsymbol{p}$. This single-wave vector implies that associated plane wave dominates the contribution to that point. Rewriting Eq. (2.83b), we obtain

$$u\left(x_p, y_p, z_p\right) = \iint_{-\infty}^{\infty} \hat{u}_0\left(\nu, \mu\right) e^{i2\pi R\, f(\nu,\mu)}\, d\nu\, d\mu, \tag{2.84}$$

where

$$f\left(k_x, k_y\right) = k_x\ \sin\alpha + k_y\ R\cos\beta + \cos\mathcal{E}\frac{\sqrt{\left[1-\lambda^2\left(k_x{}^2+k_y{}^2\right)\right]}}{\lambda}. \tag{2.85}$$

In Eq. (2.84), $2\pi\, R$ replaces k in the exponential's argument and is very large. Let us again apply the saddle-point method to this equation. Using Eq. (2.85), we have

$$\left.\frac{\partial f}{\partial k_x}\right|_{k_x=k_{x0};\ k_y=k_{y0}} = 0 = \sin\alpha - \frac{\lambda^2 k_{x0}\cos\mathcal{E}}{\lambda\sqrt{1-\lambda^2\left(k_{x0}{}^2+k_{y0}{}^2\right)}} \tag{2.86}$$

and

$$\left.\frac{\partial f}{\partial k_y}\right|_{k_x=k_{x0};\ k_y=k_{y0}} = 0 = \sin\beta - \frac{\lambda^2 k_{y0}\cos\mathcal{E}}{\lambda\sqrt{1-\lambda^2\left(k_{x0}{}^2+k_{y0}{}^2\right)}}. \tag{2.87}$$

The result is that we find that

$$\frac{\sin\alpha}{\cos\mathcal{E}} = \frac{\lambda k_{x0}}{\sqrt{1-\lambda^2\left(k_{x0}{}^2+k_{y0}{}^2\right)}} \tag{2.88}$$

and

$$\frac{\sin\beta}{\cos\mathcal{E}} = \frac{\lambda k_{y0}}{\sqrt{1-\lambda^2\left(k_{x0}{}^2+k_{y0}{}^2\right)}}. \tag{2.89}$$

Since $\cos\varepsilon = \sqrt{1 - \sin^2\alpha - \sin^2\beta}$, we see that

$$1 + \left(\frac{\sin\alpha}{\cos\varepsilon}\right)^2 + \left(\frac{\sin\beta}{\cos\varepsilon}\right)^2 = \frac{1}{\cos^2\varepsilon}. \tag{2.90}$$

Equation (2.90) implies that

$$\frac{1 - \lambda^2\left({k_{x0}}^2 + {k_{y0}}^2\right)}{1 - \lambda^2\left({k_{x0}}^2 + {k_{y0}}^2\right)} + \frac{\left(\lambda^2{k_{x0}}^2 + \lambda^2{k_{y0}}^2\right)}{1 - \lambda^2\left({k_{x0}}^2 + {k_{y0}}^2\right)} = \frac{1}{1 - \lambda^2\left({k_{x0}}^2 + {k_{y0}}^2\right)} = \frac{1}{\cos^2\varepsilon}$$

or

$$\cos\varepsilon = \sqrt{1 - \lambda^2\left({k_{x0}}^2 + {k_{y0}}^2\right)}\,. \tag{2.91}$$

Substituting Eq. (2.91) into Eqs. (2.88) and (2.89) implies that

$$\sin\alpha = \lambda k_{x0} \tag{2.92}$$

and

$$\sin\beta = \lambda k_{y0}. \tag{2.93}$$

Continuing on with the saddle-point method derivation, we see that

$$\left.\frac{\partial^2 f}{\partial {k_x}^2}\right|_{k_x \equiv k_{x0}; k_y = k_{y0}} = \frac{-\lambda\cos\varepsilon\left(1 - \lambda^2{k_{y0}}^2\right)}{\left[\sqrt{1 - \lambda^2\left({k_{x0}}^2 + {k_{y0}}^2\right)}\right]^3} = \frac{-\lambda\left(1 - \lambda^2{k_{y0}}^2\right)}{\cos^2\varepsilon}, \tag{2.94}$$

$$\left.\frac{\partial^2 f}{\partial {k_y}^2}\right|_{k_y = k_{y0}; k_y = k_{y0}} = \frac{-\lambda\cos\varepsilon\left(1 - \lambda^2{k_{x0}}^2\right)}{\left[\sqrt{1 - \lambda^2\left({k_{x0}}^2 + {k_{y0}}^2\right)}\right]^3} = \frac{-\lambda\left(1 - \lambda^2{k_{x0}}^2\right)}{\cos^2\varepsilon}, \tag{2.95}$$

and

$$\left.\frac{\partial^2 f}{\partial k_x \partial k_y}\right|_{k_x = k_{x0}; k_y = k_{y0}} = \frac{-\left(\lambda^3 k_{x0} k_{y0}\right)}{\cos^2\varepsilon}. \tag{2.96}$$

Using these equations, we find that

$$\sqrt{\left(\frac{\partial^2 f\left(x_0, y_0\right)}{\partial x^2}\right)\left(\frac{\partial^2 f\left(x_0, y_0\right)}{\partial y^2}\right) - \left(\frac{\partial^2 f\left(x_0, y_0\right)}{\partial x \partial y}\right)^2}$$

$$= \sqrt{\frac{\lambda^2\left(1 - \lambda^2{k_{y0}}^2 - \lambda^2{k_{y0}}^2\right)}{\cos^4\varepsilon}} = \frac{\lambda}{\cos\varepsilon}. \tag{2.97}$$

Putting this all together into the saddle-point method solution gives the following result:

$$u(R\sin\alpha, R\cos\beta, R\cos\mathcal{E}) = \left(\frac{\cos\mathcal{E}}{\lambda R}\right) e^{\left\{2\pi i\frac{R}{\lambda}+\frac{\pi i}{2}\right\}} \hat{u}_0\left(\frac{\sin\alpha}{\lambda}, \frac{\sin\beta}{\lambda}\right) \tag{2.98a}$$

or

$$u\left(x_p, y_p, R\cos\mathcal{E}\right) = \left(\frac{\cos\mathcal{E}}{\lambda R}\right) e^{\left\{2\pi i\frac{R}{\lambda}+\frac{\pi i}{2}\right\}} \hat{u}_0\left(\frac{x_p}{\lambda R}, \frac{y_p}{\lambda R}\right). \tag{2.98b}$$

Aside from the multiplicative factor upfront in the above equations, Eq. (2.98) clearly shows that the received field in the far field, or Fraunhofer regime, is directly proportional to the Fourier transform of the initial planar source distribution, evaluated at the frequencies $k_{x0} = \frac{\sin\alpha}{\lambda} = \frac{x_p}{\lambda R}$ and $k_{y0} = \frac{\sin\beta}{\lambda} = \frac{y_p}{\lambda R}$, respectively. Unlike RF systems, optical engineers generally deal with optical intensity because optical detectors measure current, photoelectrons, not field strength.[1] Hence, all the complex terms disappear in Eq. (2.98) and we are left with the signal of interest being proportional to the square of the Fourier transform of the initial field distribution.

It should be noted that some books use the following equation for describing Fraunhofer diffraction:

$$u\left(x_p, y_p, R\right) = \left(\frac{i}{\lambda R}\right) e^{\frac{2\pi i}{\lambda}\left[R+\frac{(x^2+y^2)}{R}\right]} \hat{u}_0\left(\frac{x_p}{\lambda R}, \frac{y_p}{\lambda R}\right), \tag{2.99}$$

when one is looking at the field very close to the optical axis where $\cos\mathcal{E} \cong 1$. This is the next order correction to Eq. (2.98), but changes nothing in our overall conclusion cited above when we measure optical intensity for the field in Eq. (2.99).

For most applications, the optical receiver is situated on the normal axis in plane $z = R$. The conclusion is that the lowest frequencies are the most important to the measurement of the received signal. This provides some of the motivation in including Example 2.3, where we looked at the diffraction pattern analysis when $\left(k_{x0}{}^2 + k_{y0}{}^2\right) \ll \lambda^2$.

Example 2.4 Assume that a plane wave is illuminating an object, which is a rectangular aperture defined by the function

$$\begin{aligned} \text{rect}\,(w) &\equiv \text{Rectangular Function} \\ &= \begin{cases} 1 & \text{for } |w| \le 0.5 \\ 0 & \text{otherwise} \end{cases} \end{aligned} \tag{2.100}$$

in two dimensions. Here, the observation plane is located at a distance R away from the object.

[1]The received intensity of an optical field is proportional to time-averaged square of the field amplitude u.

The resulting field amplitude in the observation plane will be dependent on the Fourier transform of the rectangular aperture, namely,

$$\begin{aligned}
\hat{u}_0\left(k_x, k_y\right) &= \iint_{-\infty}^{\infty} u_0(x, y)\, e^{-2\pi i\left(k_x\, x + k_y\, y\right)}\, dx\, dy \\
&= \iint_{-\infty}^{\infty} \operatorname{rect}\left(\frac{x}{2a}\right) \operatorname{rect}\left(\frac{y}{2b}\right) e^{-2\pi i\left(k_x\, x + k_y\, y\right)}\, dx\, dy \\
&= (ab)\operatorname{sinc}\left(a\, k_x\right) \operatorname{sinc}\left(b\, k_y\right) \qquad (2.101) \\
&\approx (ab)\operatorname{sinc}\left(\frac{a x_p}{\lambda R}\right) \operatorname{sinc}\left(\frac{b y_p}{\lambda R}\right) = \hat{u}_0\left(x_p, y_p\right) \qquad (2.102)
\end{aligned}$$

since $k_{x0} = \frac{\sin\alpha}{\lambda} = \frac{x_p}{\lambda R}$ and $k_{y0} = \frac{\sin\beta}{\lambda} = \frac{y_p}{\lambda R}$. In the above equation, the Sinc function is defined to be

$$\begin{aligned}
\operatorname{sinc}(w) &\equiv \text{Sinc Function} \\
&= \frac{\sin(\pi\, w)}{\pi\, w}. \qquad (2.103)
\end{aligned}$$

Here, $2a$ is the length of the aperture in the x-direction and $2b$ is the length of the aperture in the y-direction. In Eq. (2.103), $\operatorname{sinc}(x)$ has the property of being zero whenever x is an integer, except for $x = 0$, where $\operatorname{sinc}(0) = 1$.

The received intensity in the observation plane is then

$$|u(x, y\, z)|^2 \propto \frac{(ab)^2}{(\lambda\, R)^2} \operatorname{sinc}^2(x)\left(\frac{ax}{\lambda\, R}\right)^2 \operatorname{sinc}^2(x)\left(\frac{by}{\lambda\, R}\right). \qquad (2.104)$$

Figure 2.9 shows a plot of $\operatorname{sinc}^2(x)$. The width of the main lobe of $\operatorname{sinc}^2(x)$ is 2, which implies that the widths of the Sinc functions in equation Eq. (2.104) are equal to

$$\Delta x_0 = 2\left(\frac{\lambda\, R}{a}\right) \quad \text{and} \quad \Delta y_0 = 2\left(\frac{\lambda\, R}{b}\right),$$

respectively.

Example 2.5 Assume that a plane wave is illuminating an object, u, which is a circular aperture of radius a. The observation plane is located at a distance R away from the object. In this case, the resulting field amplitude in the observation plane is proportional to:

$$\begin{aligned}
\hat{u}_0\left(k_x, k_x\right) &= \iint u_0\left(x', y'\right) e^{2\pi i k_x x' + 2\pi i k_x y'}\, dx' dy' \\
&= \iint \operatorname{circ}\left(\frac{r}{a}\right) e^{2\pi i k_x x' + 2\pi i k_x y'}\, dx' dy' \qquad (2.105) \\
&= \int_0^a \int_0^{2\pi} e^{\frac{2\pi i \rho w \cos(\theta - \vartheta)}{\lambda R}}\, \rho d\rho\theta \qquad (2.106)
\end{aligned}$$

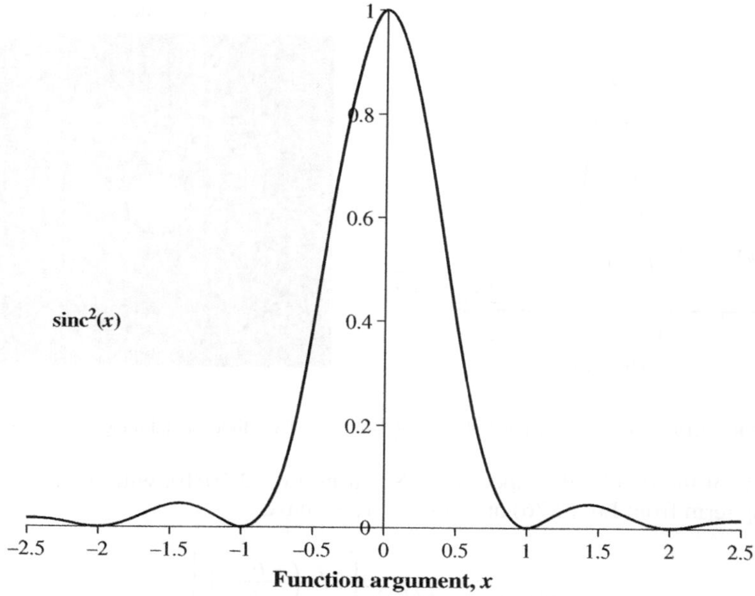

FIGURE 2.9 Plot of $\text{sinc}^2(x) = \frac{\sin^2(\pi x)}{(\pi x)^2}$.

$$= 2\pi \int_0^a J_0\left(\frac{2\pi\,\rho\,w}{\lambda R}\right)\rho\,d\rho \tag{2.107}$$

$$= 2\left(\pi a^2\right)\left[\frac{J_1\left(\frac{2\pi\,a\,w}{\lambda R}\right)}{\left(\frac{2\pi\,a\,w}{\lambda R}\right)}\right] \tag{2.108a}$$

$$= 2\pi\left(\frac{D^2}{4}\right)\left[\frac{J_1\left(\frac{\pi\,D\,w}{\lambda R}\right)}{\left(\frac{\pi\,D\,w}{\lambda R}\right)}\right] = \hat{u}_0\left(x_p, y_p\right), \tag{2.108b}$$

where

$$\text{circ}\,(r) \equiv \text{circ function}$$

$$= \begin{cases} 1 & |r| = \sqrt{x^2+y^2} \le 1 \\ 0 & \text{otherwise} \end{cases} \tag{2.108c}$$

since $k_{x0} = \frac{\sin\alpha}{\lambda} = \frac{x_p}{\lambda R}$ and $k_{y0} = \frac{\sin\beta}{\lambda} = \frac{y_p}{\lambda R}$ and $D = 2a$ is the diameter of the aperture. In Eqs. (2.107) and (2.108), $J_0(\rho)$ and $J_1(\rho)$ are the zeroth and first-order Bessel

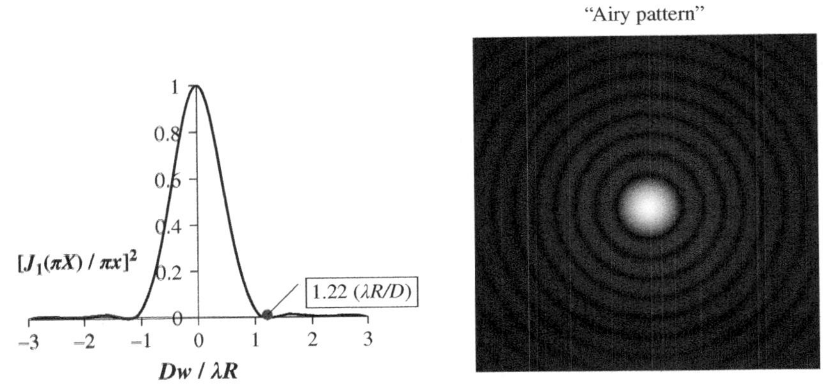

FIGURE 2.10 (a) One-dimensional plot of $\left[\frac{J_1(\pi x)}{(\pi x)}\right]^2$. (b) Two-dimensional image of $\left[\frac{J_1(\pi x)}{(\pi x)}\right]^2$.

functions of the first kind, respectively. Squaring Eq. (2.108b), with the proper normalizing term from Eq. (2.76) and $\cos \varepsilon \approx 1$, we have

$$I\left(\frac{D\,w}{\lambda R}\right) \approx \left(\frac{\pi D^2}{4\lambda R}\right)^2 \left[\frac{2J_1\left(\frac{\pi\,D\,w}{\lambda R}\right)}{\left(\frac{\pi\,D\,w}{\lambda R}\right)}\right]^2, \tag{2.109}$$

which is the celebrated formula first derived in a different form by Airy [2–5]. Figure 2.10a,b shows the one-dimensional plot and two-dimensional image, respectively, of $\left[\frac{J_1(\pi x)}{(\pi x)}\right]^2$.

Figure 2.10b is known as the Airy Pattern and the central bright region (to first-zero ring) is called the "Airy disk."

Integrating $\left[\frac{J_1\left(\frac{2\pi\,a\,w}{\lambda R}\right)}{\left(\frac{2\pi\,a\,w}{\lambda R}\right)}\right]^2$ over 2π and radius w, we obtain

$$L\left(\frac{\pi D w}{\lambda R}\right) = 1 - \left[J_0\left(\frac{\pi D w}{\lambda R}\right)\right]^2 - \left[J_1\left(\frac{\pi D w}{\lambda R}\right)\right]^2, \tag{2.110}$$

which was originally derived by Lord Rayleigh. Figure 2.11 shows this function. It represents the total energy contained in the radius w_0. The fractional energies contained in the first, second, and third rings are about 83%, 91%, and 94%, respectively. Clearly, more than 90% of the total energy in the Airy Pattern is within the second dark ring.

The first minima of the diffraction pattern is located at $x = 1.220$, so the effective radius at that point is

$$\Delta w_0 = 1.22\,\frac{\lambda R}{D}. \tag{2.111}$$

Equation (2.111) is normally known as the angular beam spreading for diffraction-limited optics. However, this is not the radius that should be used in

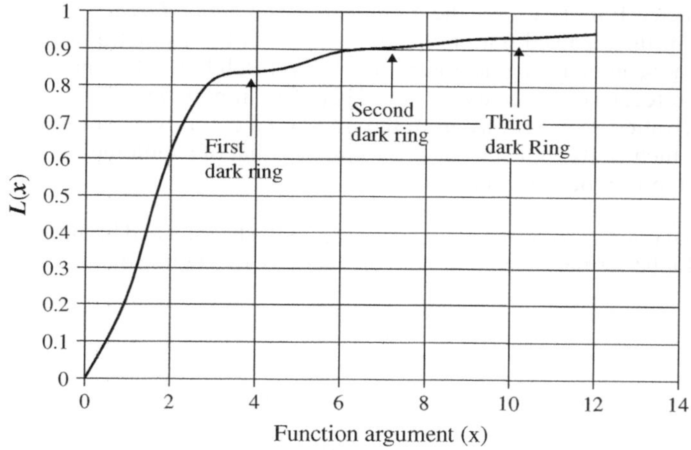

FIGURE 2.11 Fractional energy of $L(x)$ contained in a functional argument $x = \frac{\pi D w}{\lambda R}$.

modern link budget equations. In free space, Fraunhofer diffraction gives the far-field irradiance pattern at the receiver aperture plane. From Figure 2.11, we see that a large portion of the fractional energy, 80%, is obtained by an argument of 2.5. The value of the fractional energy at the first zero, which is at an argument of 3.83, is 83%. This means that one only obtains an additional 3% of energy by increasing the argument by 28% $\left[\left(\frac{3.83}{3}\right) = 1.28\right]$. Not the best estimate for deriving the radius capturing most of the power. Therefore, if we chose the former case as the means for calculating the desired radius, then we find that

$$\Delta w_a \approx \left(\frac{3}{\pi}\right)\frac{\lambda R}{D} = 0.95\frac{\lambda R}{D} \approx \frac{\lambda R}{D}. \tag{2.112}$$

Squaring Eq. (2.112) to create the beam area, the irradiance at the receiver plane would be written as

$$I_{\text{aperture}} \cong \gamma_{tx} P_{tx} \frac{A_{tx}}{(\lambda R)^2}, \tag{2.113}$$

where γ_{tx} is the transmitter optics transmittance, P_{tx} the laser transmitter power, and $A_{tx} = \frac{\pi D_{tx}^2}{4}$ the area of the transmitter. The power captured by the receiver is the above irradiance multiplied by the area of the receiver, or

$$P_{rx} \approx \gamma_{tx}\gamma_{rx} P_{tx} \frac{A_{tx}A_{rx}}{(\lambda R)^2} = \gamma_{tx} P_{tx}\ \text{FSL}, \tag{2.114}$$

where γ_{rx} is the receiver optics transmittance, $A_{rx} = \frac{\pi D_{rx}^2}{4}$ the area of the receiver aperture, and

$$\text{FSL} \approx \frac{A_{tx}A_{rx}}{(\lambda R)^2}. \tag{2.115}$$

The parameter FSL is called the *Fraunhofer Spreading Loss*. Although a crude derivation of the received power equation, Eq. (2.114) was first derived in 1945 by Danish-American radio engineer Harald T. Friis at Bell Labs [11] and is the accepted form of the received power used today [12]. The importance of this equation was that its derivation moved away from the traditional circuit viewpoint of analysis of the time, which deals with current and voltage, to that of a radiometric one that dealt with power, power density, and so on.

As a final point, what is the validity ranges for the Fresnel and Fraunhofer transformations? In the former case (near-field diffraction), the regime of validity is given by

$$R \leq \frac{D^2}{\lambda}. \tag{2.116}$$

The latter case (near-field diffraction) dictates that

$$R \gg \frac{D^2}{\lambda}. \tag{2.117}$$

Let us assume that the wavelength of interest is 1.55 μm and we have our aperture diameter is 4 in. (=10 cm). Then,

$$\frac{D^2}{\lambda} = 6.450 \text{ m}.$$

This is quite a distance between the source and observation plane. Fortunately, this criterion is met for most problems of interest to the working engineer. In the following section, we look at one way of getting the Fraunhofer diffraction result into the near field.

2.6 BRINGING FRAUNHOFER DIFFRACTION INTO THE NEAR FIELD

One of the working definitions of "infinity" is "the place where all parallel lines cross." In optics, "infinity" can be created using a simple lens. In this section, we see how a Fraunhofer diffraction pattern (usually observed at infinity) can be brought into the near field via a lens system.

Figure 2.12 illustrates the basic setup for the derivation. Assume that a pinhole creates monochromatic plane wave, $\exp(ikz)$, that is normally incident of a scattering or diffractive object. The scattering object is located a focal length,[2] f, away from an imaging lens. The observation plane is located at $z = 2f$. At $z = 0$, we have

$$u_0(x, y) = \iint_{-\infty}^{\infty} \hat{u}_0\left(k_x, k_y\right) e^{2\pi i k_x x + 2\pi i k_x y}\, dk_x dk_y. \tag{2.118}$$

For simplicity, we now employ an approximate form of the parabolic approximation of the Huygens–Fresnel–Kirchhoff integral. In this case, we obtain the following

[2]The focal length of an optical system is a measure of how strongly the system converges or diverges light.

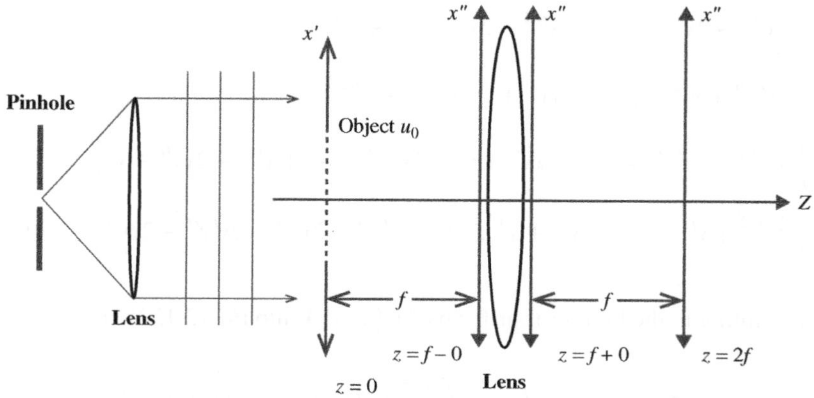

FIGURE 2.12 Plane wave illuminating an object, which is imaged by a simple lens.

field amplitude from the propagation form $z = 0$ to $z = f$–0:

$$u\left(x'', y'', f-0\right) \approx \iint_{-\infty}^{\infty} u_0\left(x', y'\right) e^{i\pi \frac{\left[(x''-x')^2+(y''-y')^2\right]}{\lambda f}} dx'dy'. \tag{2.119}$$

Propagating through the simple lens, we have

$$u\left(x'', y'', f+0\right) = u\left(x'', y'', f-0\right) e^{-i\pi \frac{\left[x''^2+y''^2\right]}{\lambda f}}, \tag{2.120}$$

where the exponential in the above equation is the phase factor induced by the simple lens. Finally, the received field amplitude in the observation plane $z = 2f$ is given by:

$$u(x, y, 2f) \approx \iint_{-\infty}^{\infty} u\left(x'', y'', f+0\right) e^{i\pi \frac{\left[(x-x'')^2+(y-y'')^2\right]}{\lambda f}} dx''dy'' \tag{2.121}$$

$$= \iint_{-\infty}^{\infty} \iint_{-\infty}^{\infty} u\left(x', y', f+0\right) e^{i\frac{\pi}{\lambda} g(x,y;\, x',y';\, x'',y'')} \times dx'dy'dx''dy'' \tag{2.122}$$

with

$$g\left(x, y;\, x', y';\, x'', y''\right) = \frac{\left[(x''-x')^2 + (y''-y')^2\right]}{f} + \frac{\left[(x-x'')^2 + (y-y'')^2\right]}{f} - \frac{\left[x''^2 + y''^2\right]}{f}$$

$$= \frac{1}{f}\left[(x''-x')^2 + (y''-y')^2 + (x-x'')^2 + (y-y'')^2 - x''^2 - y''^2\right]$$

$$= \frac{1}{f}\left[x''^2 - 2x'x'' + x'^2 + x''^2 - 2xx'' + x^2 + y''^2\right.$$
$$\left.-2y'y'' + y'^2 + y''^2 - 2yy'' + y^2 - x''^2 - y''^2\right]$$
$$= \frac{1}{f}\left[x''^2 - 2x'x'' + x'^2 - 2xx'' + x^2 - 2y'y'' + y'^2 + y''^2 - 2yy'' + y^2\right]$$
$$= \frac{1}{f}\left[x''^2 + y''^2 + x'^2 + y'^2 + x^2 + y^2 - 2x'x'' - 2xx'' - 2y'y'' - 2yy''\right]. \quad (2.123)$$

Substituting in the Fourier transform of $u\left(x', y'\right)$ into Eq. (2.122) yields

$$u\,(x, y, 2f) = \iint_{-\infty}^{\infty}\iint_{-\infty}^{\infty}\iint_{-\infty}^{\infty} \hat{u}_0\left(k_x, k_y\right) e^{i\,\frac{\pi}{\lambda}\,g'(x,y;\,x',y';\,x'',y'')}$$
$$\times dx''dy''dx'dy'dk_xdk_y, \quad (2.124)$$

where

$$g'\left(x, y;\; x', y';\; x'', y''\right) = g\left(x, y;\; x', y';\; x'', y''\right) + 2\,\lambda\left(k_x\, x' + k_y\, y'\right) \quad (2.125)$$
$$= \frac{1}{f}\left[x''^2 + y''^2 + x'^2 + y'^2 + x^2 + y^2 - 2x'x'' - 2xx'' - 2y'y'' - 2yy''\right]$$
$$+ 2\,\lambda\left(k_x x' + k_y\, y'\right) \quad (2.126)$$
$$= \frac{1}{f}\left[x''^2 + y''^2 + x'^2 + y'^2 + x^2 + y^2 - 2x''\left(x' + x\right) - 2y''\left(y' + y\right)\right]$$
$$+ 2\,\lambda\left(k_x x' + k_y y'\right)$$
$$= \frac{1}{f}\left[\left(x'' - \left(x' + x\right)\right)^2 - \left(x' + x\right)^2 + \left(y'' - \left(y' + y\right)\right)^2\right.$$
$$\left.-\left(y' + y\right)^2 + x'^2 + y'^2 + x^2 + y^2\right] + 2\,\lambda\left(k_x\, x' + k_y\, y'\right) \quad (2.127)$$

Using x''-part of Eq. (2.127), we find that

$$\int_{-\infty}^{\infty} e^{\left\{\frac{i\pi}{\lambda f}\left(x'' - \left(x' + x\right)\right)^2\right\}} dx'' = \int_{-\infty}^{\infty} e^{iaz^2}\, dz, \quad (2.128)$$

where $a = \frac{\pi}{\lambda f}$.

Example 2.6 Define

$$F = \int_{-\infty}^{\infty} e^{iax^2} dx = \int_{-\infty}^{\infty} \cos\left(ax^2\right) dx + i \int_{-\infty}^{\infty} \sin\left(ax^2\right) dx. \quad (2.129)$$

Referring to Figure 2.13a, the main contribution to the real part of F is equal to

$$F_{\text{real}} = \int_{-\infty}^{\infty} \cos\left(ax^2\right) dx \approx \int_{-\sqrt{\frac{\pi}{2a}}}^{+\sqrt{\frac{\pi}{2a}}} \cos\left(ax^2\right) dx = \sqrt{\frac{\pi}{2a}} \tag{2.130}$$

because the integral is stable over the interval $-\sqrt{\frac{\pi}{2a}} < x < \sqrt{\frac{\pi}{2a}}$ and the integrals outside that interval are essentially zero because of the oscillations of the cosine function. The question now is how good this approximation is to the correct value. Referring to Gradshteyn and Ryzhik [13, Eq. (3.961.1), p. 395], we see that

$$\int_{-\infty}^{\infty} \cos\left(ax^2\right) dx \approx 2\int_{0}^{\infty} \cos\left(ax^2\right) dx = 2\left(\frac{1}{2}\sqrt{\frac{\pi}{2a}}\right) = \sqrt{\frac{\pi}{2a}}$$

because $\cos\left(ax^2\right)$ is a symmetric function (Figure 2.13a). Looks like a very good approximation.

Referring to Figure 2.13b, it is plausible that the imaginary part of F is given by

$$F_{\text{img}} = \int_{-\infty}^{\infty} \sin\left(ax^2\right) dx \approx \int_{-\sqrt{\frac{\pi}{2a}}}^{+\sqrt{\frac{\pi}{2a}}} \sin\left(ax^2\right) dx = \sqrt{\frac{\pi}{2a}} \tag{2.131}$$

over the same interval, $-\sqrt{\frac{\pi}{2a}} < x < \sqrt{\frac{\pi}{2a}}$, with the remaining integration canceling out because of the oscillations of the sine function similar to the cosine case. Again referring to Gradshteyn and Ryzhik [13, Eq. (3.961.1), p. 395], we see that

$$\int_{-\infty}^{\infty} \sin\left(ax^2\right) dx \approx 2\int_{0}^{\infty} \sin\left(ax^2\right) dx = 2\left(\frac{1}{2}\sqrt{\frac{\pi}{2a}}\right) = \sqrt{\frac{\pi}{2a}}$$

because $\sin\left(ax^2\right)$ also is a symmetric function (Figure 2.13b). Very good correlation between the two answers.

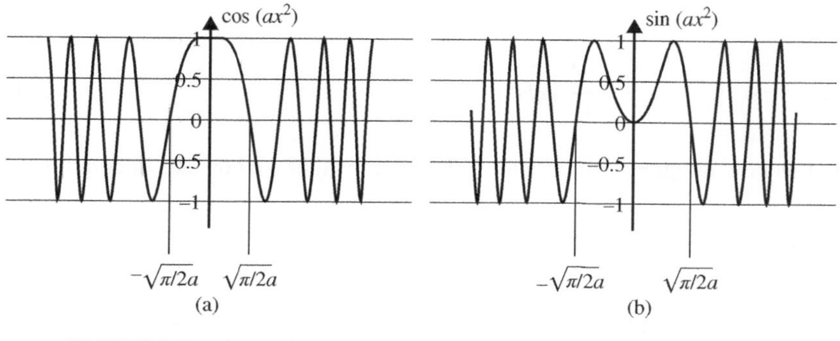

FIGURE 2.13 Oscillatory nature of the (a) cosine function and (b) sine function.

The resulting solution is then

$$F = \int_{-\infty}^{\infty} e^{iax^2} dx \approx \sqrt{\frac{\pi}{2a}}\,(1+i) = \sqrt{\frac{\pi}{a}}\, e^{i\frac{\pi}{4}} = \sqrt{\lambda f}\, e^{i\frac{\pi}{4}}. \tag{2.132}$$

The solution also could be derived from the method of stationary phase used before.

From Example 2.5, we know that

$$\int_{-\infty}^{\infty} e^{\left\{\frac{i\pi}{\lambda f}\left(x''-(x'+x)\right)^2\right\}} dx'' = \int_{-\infty}^{\infty} e^{iaz^2}\, dz = \sqrt{\lambda f}\, e^{i\frac{\pi}{4}}. \tag{2.133}$$

Similarly, we find that the y''-integration yields the same result as Eq. (2.133). Both results are constants. As a result, Eq. (2.124) reduces to

$$\begin{aligned} u(x,y,2f) &= \left[\sqrt{\lambda f}\, e^{i\frac{\pi}{4}}\right]^2 \iint_{-\infty}^{\infty} \iint_{-\infty}^{\infty} \hat{u}_0\left(k_x,k_y\right) e^{i\,\frac{\pi}{\lambda} g'(x,y;\,x',y';\,x'',y'')} \\ &\quad \times dx'dy'\; dk_x dk_y \\ &= [\lambda f]^2 e^{i\frac{\pi}{2}} \iint_{-\infty}^{\infty} \iint_{-\infty}^{\infty} \hat{u}_0\left(k_x,k_y\right) e^{i\,\frac{\pi}{\lambda}\, g'(x,y;\,x',y';\,x'',y'')} \\ &\quad \times dx'dy'\; dk_x dk_y \end{aligned} \tag{2.134}$$

with

$$\begin{aligned} g'\left(x,y;\,x',y'\right) &= \frac{1}{f}\left[-(x'+x)^2 - (y'+y)^2 + x'^2 + y'^2 + x^2 + y^2\right] \\ &\quad + 2\,\lambda\left(k_x\, x' + k_y\, y'\right) \\ &= \frac{1}{f}\left[-\left(x'^2 + 2xx' + x^2\right) - \left(y'^2 + 2yy' + y^2\right)\right. \\ &\quad \left. + x'^2 + y'^2 + x^2 + y^2\right] + 2\,\lambda\left(k_x\, x' + k_y\, y'\right) \\ &= \frac{1}{f}\left[-2xx' - 2yy'\right] + 2\,\lambda\left(k_x\, x' + k_y\, y'\right) = \frac{2x'}{f}\left(\lambda f k_x - x\right) \\ &\quad + \frac{2y'}{f}\left(\lambda f k_y - y\right). \end{aligned} \tag{2.135}$$

Let us now integrate over x' and y'. This means that we now are integrating

$$\begin{aligned} \iint_{-\infty}^{\infty} e^{i\,\frac{\pi}{\lambda}\, g'(x,y;\,x',y';\,x'',y'')} dx'dy' &= \iint_{-\infty}^{\infty} e^{i\frac{\pi}{\lambda}\left(\frac{2x'}{f}\left(\lambda f k_x - x\right) + \frac{2y'}{f}\left(\lambda f k_y - y\right)\right)} dx'dy' \\ &= \delta\left(\lambda f k_x - x\right)\delta\left(\lambda f k_y - y\right) = \frac{1}{(\lambda f)^2}\delta\left(k_x - \frac{x}{\lambda f}\right)\delta\left(k_y - \frac{y}{\lambda f}\right), \end{aligned} \tag{2.136}$$

using Eq. (2.76c). Substituting Eq. (2.136) into Eq. (2.134) yields

$$u(x, y, 2f) = \frac{[\lambda f]^2 e^{i\frac{\pi}{2}}}{(\lambda f)^2} \int\int_{-\infty}^{\infty} \int\int_{-\infty}^{\infty} \hat{u}_0(k_x, k_y)$$
$$\times \delta\left(k_x - \frac{x}{\lambda f}\right) \delta\left(k_y - \frac{y}{\lambda f}\right) dk_x dk_y$$
$$= \hat{u}_0\left(\frac{x}{\lambda f}, \frac{y}{\lambda f}\right) e^{i\frac{\pi}{2}}. \tag{2.137}$$

It is clear that Eq. (2.137) says the electric field in the observation plane is proportional to the Fourier transform of the illuminated object. Focusing our attention on the intensity distribution in the observation plane $z = 2f$, we find that

$$|u(x, y, 2f)|^2 = \left|\hat{u}_0\left(\frac{x}{\lambda f}, \frac{y}{\lambda f}\right)\right|^2. \tag{2.138}$$

Thus, the diffraction pattern using the above lens setup creates the same intensity pattern as the Fraunhofer diffraction does in the far field. There is a magnification, here $M = \lambda f$, but no image inversion like we found in Example 2.2.

Example 2.7 There is another way to accomplish Fraunhofer diffraction with lens [2, p. 237].

Let two lenses be separated by one focal length, f, as illustrated in Figure 2.14. Assume that the source distribution $u(x, y, -0)$ is known. From above, we know that the source distribution after the first lens is given by

$$u(x, y, +0) = u(x, y, -0)\, e^{-\frac{i\pi}{\lambda f}(x^2+y^2)}. \tag{2.139}$$

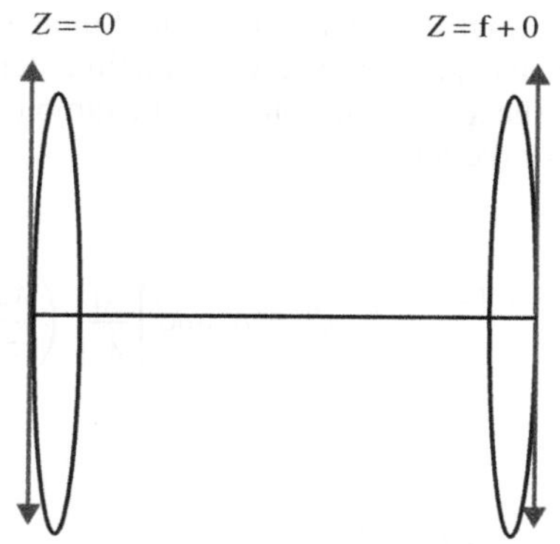

FIGURE 2.14 Two lens geometry setup.

Propagating this distribution over a distance $f - 0$ yields

$$u(x, y, f - 0) = \iint_{-\infty}^{\infty} u\left(x', y', +0\right) e^{\frac{i\pi}{\lambda f}\left[(x-x')^2+(y-y')^2\right]}\, dx'dy'. \tag{2.140}$$

The distribution on the other side of the second lens equals

$$\begin{aligned} u(x, y, f + 0) &= u(x, y, f - 0)\, e^{-\frac{i\pi}{\lambda f}(x^2+y^2)} \\ &= e^{-\frac{i\pi}{\lambda f}(x^2+y^2)} \iint_{-\infty}^{\infty} u\left(x', y', +0\right) \\ &\quad \times e^{-\frac{i\pi}{\lambda f}\frac{\left(x'^2+y'^2\right)}{\lambda f}} e^{\frac{i\pi}{\lambda f}\left[(x-x')^2+(y-y')^2\right]}\, dx'dy' \\ &= \iint_{-\infty}^{\infty} u\left(x', y', +0\right) e^{-\frac{2\pi i}{\lambda f}(xx'+yy')}\, dx'dy' \\ &= \hat{u}\left(\frac{x}{\lambda f}, \frac{y}{\lambda f}, -0\right). \end{aligned} \tag{2.141}$$

Equation (2.141) is the same result we found in Eq. (2.137), except for the inclusion of the constant term in the latter. Fortunately, squaring Eq. (2.141) gives the same intensity as Eq. (2.138), which is the important point.

2.7 IMPERFECT IMAGING

Until this point, we have assumed that a lens creates perfect imaging. This is not true since the lens has a finite diameter and not all light rays from the source can be accepted by the lens. Let us look at one of our previous examples to define effect a finite diameter lens has on image quality, that is, resolution.

Recall Example 2.3 where we established the Lens Formula and that a perfect lens create a perfect image, but inverted and rescaled image of the source field. Let us see how this theory can also predict the effect of a finite lens diameter. This can be accomplished by redoing the x''-integration in that example not from $(-\infty, \infty)$, but just over the finite extent of the lens.

Specifically, we have

$$\int_{-\infty}^{\infty} \mathrm{rect}\left(\frac{x''}{2a}\right) e^{\frac{-2\pi i}{\lambda}x''\left(\frac{x}{z_2}+\frac{x'}{z_1}\right)}\, dx'' = a\, \mathrm{sinc}\left[\frac{a}{\lambda z_1}\left(\frac{xz_1}{z_2}+x'\right)\right]. \tag{2.142}$$

This implies that

$$v(x) \propto \int_{-\infty}^{\infty} u_0\left(x'\right) e^{\frac{i\pi}{\lambda}\left(\frac{x^2}{z_2}+\frac{x'^2}{z_1}\right)} \mathrm{sinc}\left[\frac{a}{\lambda z_1}\left(\frac{xz_1}{z_2}+x'\right)\right] dx'. \tag{2.143}$$

Equation (2.143) shows that the image amplitude $v(x)$ is just a convolution of the propagated object amplitude and a Sinc-function. For a point source, $u_0(x') = \delta(x')$, the received intensity is given by

$$|v(x)|^2 \propto \left| \text{sinc}\left(\frac{ax}{\lambda z_1}\right) \right|^2. \tag{2.144}$$

It is clear from Eq. (2.144) that a finite aperture essentially acts a low pass spatial filter function that degrades spatial resolution, creating a "point source" now having harmonic distortion characterized by fundamental point spread equal to

$$\frac{2\lambda z_1}{a}. \tag{2.145}$$

Unfortunately, this is just the "tip of the iceberg." In addition, a lens also may have defects called "aberration" that modifies the source field going through it in a bad way as well as amplitude defects such as dust and fingerprints that will change the segments of the source field. Figure 2.15 illustrates some typical aberration types. The types of distortion generate additional loss in the specific details of the image besides the blurring from diffraction. In engineering terms, these aberrations create a reduction in both the image resolution and contrast. We address a more general mathematical characterization of imperfect imaging in Chapter 4.

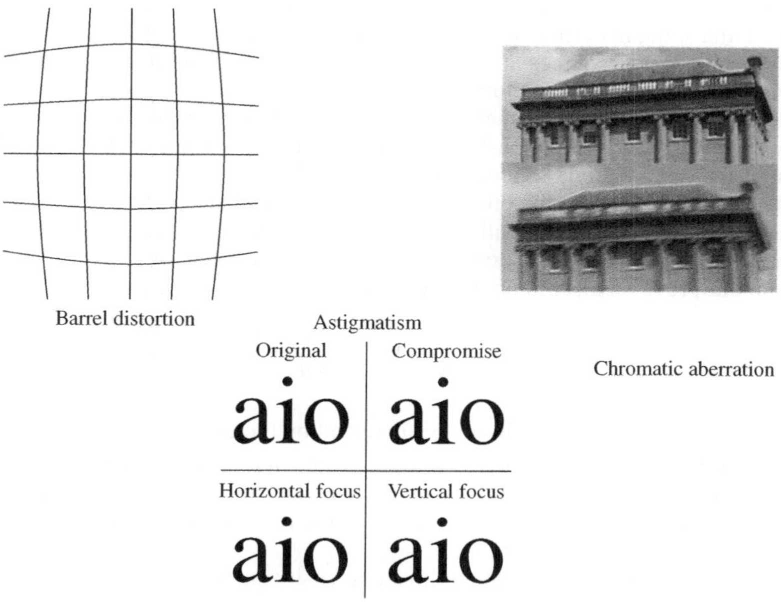

FIGURE 2.15 Examples of optical aberrations; barrel distortion, chromatic aberrations (https://en.wikipedia.org/wiki/Chromatic_aberration), and astigmatism (https://en.wikipedia.org/wiki/Astigmatism).

As a final comment, the sizes of the lens, the observation plane (sometimes called the *focal plane*), and the object distribution are connected, affecting the theory in the previous section. In particular, it can be shown that the following relationship must be true:

$$a + c \leq b \tag{2.146}$$

with a being the diameter of the input source, b the diameter of the lens, and c the diameter of the focal plane [2, p. 236].This is because the bandwidth of spatial frequencies for the above setup is

$$\Delta\nu \leq \frac{(b-a)}{\lambda f}. \tag{2.147}$$

This equation represents the limit of high spatial frequencies content (high angle light rays) that can make through the optical systems unblocked. The bottom line is that the lens must be much larger in diameter than the diameter of the source distribution for the above equations to work.

2.8 THE RAYLEIGH RESOLUTION CRITERION

In the last section, we established that a finite aperture lens is incapable of creating a perfect focus even under the best circumstances. This affects image resolution, which affects our ability to see details in said image. The standard metric for understanding the amount of detail one can see is the distance between two point sources that allow each to be simultaneously seen in the image under distortion. Therefore, what is the accepted image quality criterion one uses to characterize this separation between two closely spaced image sources at the origin? The most widely used criterion is called the *Rayleigh Criterion* for resolution.

The Rayleigh Criterion is where two incoherent closely spaced point sources are barely resolved by a diffraction-limited, circular aperture lenses. This is when the center of the Airy Pattern (Figure 2.10a) created by one of the point sources falls exactly on the first zero of the other point source's Airy Pattern. Figure 2.16 depicts the Rayleigh Criterion graphically. This is what is meant as barely resolved. It is apparent that the central dip is down 27% from the peak intensity.

From our previous discussion, we know that the first zero occurs at

$$r_{\text{first zero}} = \frac{0.61\lambda R}{a} = \frac{0.61\lambda}{n\sin\theta} = \frac{0.61\lambda}{\text{NA}}, \tag{2.148}$$

where $\theta = \sin^{-1}\left(\frac{a}{R}\right)$ is the half-angle subtended by the exit pupil and $\text{NA} = n\sin\theta$ the *Numerical Aperture* of the optical system. Again, n represents the refractive index of the medium where the optical system is located. One now might ask, "Does the use of a coherent optical system increase/improve the resolution of two closely spaced point sources?"

The following cross section of the image intensity answers that question:

$$I(x) = \left|2\,\frac{J_1(\pi[r-0.61])}{r-0.61} - 2e^{i\varphi}\,\frac{J_1(\pi[r+0.61])}{r+0.61}\right|^2, \tag{2.149}$$

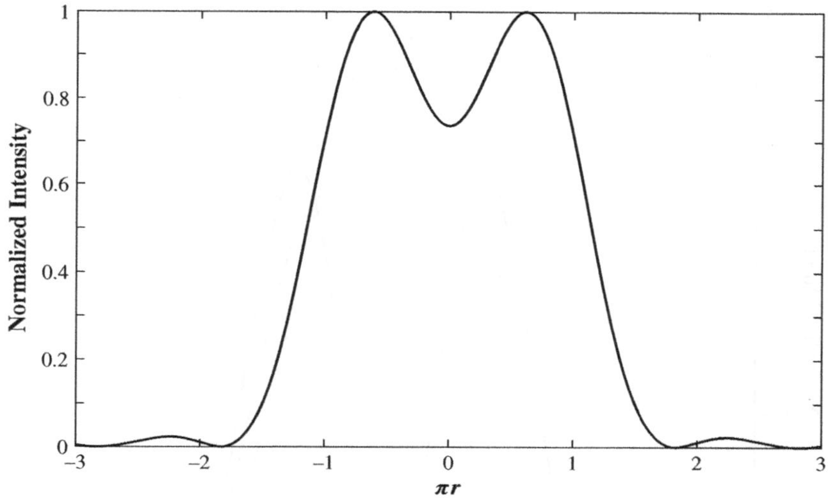

FIGURE 2.16 Two barely resolved point sources (Rayleigh Criterion).

where φ is the relative phase between the two point sources. Figure 2.17 plots Eq. (2.140) for three phase angles, $\varphi = 0, \frac{\pi}{2}$, and π, which shows the following:

- When $\varphi = 0$, the point sources are "in phase" and the dip in Figure 2.17 disappears and the point sources cannot be resolved.
- When $\varphi = \frac{\pi}{2}$, the point sources are "in quadrature" and we obtain the same intensity distribution as shown in Figure 2.16.
- When $\varphi = \pi$, the point sources are "out of phase" and the dip between the two sources goes to zero at $\pi r = 0$, and the left and right intensity peaks have shifted away from $\pi r = \pm 0.61$, respectively.

The above tells us that in-phase coherent sources totally obscured the presence of the two sources and in-quadrature coherent sources provide the same resolving capability as a totally incoherent source. However, totally out-of-phase coherent sources resolved the sources perfectly. The conclusion is that improved two-point resolution depends on the particular phase distributions between the two point sources. Unfortunately, it cannot be generalized to arbitrary phase distributions.

2.9 THE SAMPLING THEOREM

The Sampling Theorem is one of the foundation concepts in engineering systems analysis. The normal approach is to use the periodic nature of the signal of interest to develop that theorem. In real life, most optical radiance distributions $f(x, y)$ are not periodic pattern, but rather a radiance pattern limited in extent, that is, totally completely contained with the area interval $\left\{\pm\frac{d_x}{2}, \pm\frac{d_y}{2}\right\}$. (It goes without saying that there is no periodicity within said function either.) Thus, $f(x, y)$ is aperiodic

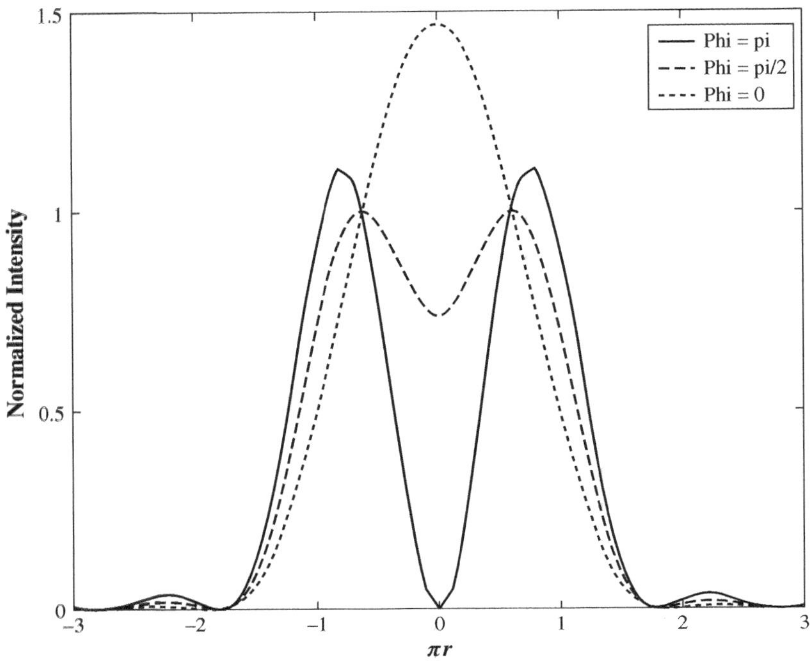

FIGURE 2.17 Two barely resolved point sources (Rayleigh Criterion) with three phase differences between said sources.

function, which cannot be practically characterized by a Fourier Series because of this situation. Therefore, we need to take a different approach to developing this theorem. This section provides a means to do so. For simplicity to begin with, let us limit our derivation to one dimension only.

Given the function description in the introductory paragraph, let us specify our original function to be represented in the following mathematical form:

$$F(x) = f(x)\,\mathrm{rect}\left(\frac{x}{d}\right) = \begin{cases} f(x) & \text{for } -\frac{d}{2} \le x \le \frac{d}{2} \\ 0 & \text{otherwise} \end{cases}. \tag{2.150}$$

If we define a new function,

$$f(x) = \mathrm{rect}\left(\frac{x}{d}\right) F(x) = \mathrm{rect}\left(\frac{x}{d}\right) \sum_{m=-\infty}^{\infty} f(x + md) \tag{2.151}$$

we now obtain a periodic series of the original radiance distribution. Fourier transforming Eq. (2.151), we have

$$\hat{f}\left(k_x\right) = \sum_{n=-\infty}^{\infty} \hat{f}\left(\frac{n}{d}\right) \mathrm{sinc}\left(k_x\, d - n\right), \tag{2.152}$$

where

$$f(x) = \int_{-\infty}^{\infty} \hat{f}(k_x)\, e^{2\pi i k_x x}\, dk_x \text{ and } \hat{f}(k_x) = \int_{-\infty}^{\infty} f(x)\, e^{-2\pi i k_x x}\, dx.$$

These two equations are one version of the well-known *Sampling Theorem*. Its spectrum is infinite. Two of this version's general properties are:

$$\int_{-\infty}^{\infty} \left| f(x) - \int_{-\infty}^{\infty} \hat{f}(k_x)\, e^{2\pi i k_x x}\, dk_x \right|^2 dx \text{ yields}\rightarrow \text{Minimum}$$

and

$$\int_{-\infty}^{\infty} \hat{f}(k_x)\, e^{2\pi i k_x x}\, dk_x = \frac{[f(x-0) + f(x+0)]}{2}.$$

Alternately, if the spectrum of a function is contained within a finite spectrum interval of $\pm\mu_0$ line pairs per spatial dimensions, that is, band-limited, then that function $f(x)$ can be completely determined by the following formula:

$$f(x) = \sum_{n=-\infty}^{\infty} f\left(\frac{n}{2\mu_0}\right) \operatorname{sinc}\left(2\mu_0\left(x - \frac{n}{2\mu_0}\right)\right). \tag{2.153}$$

Equation (2.153) is known as the *Whittaker–Shannon Sampling Theorem*. This equation states that the creation of $f(x)$ comes from the sum of $f(x)$ sampled at locations

$$\left\{\frac{n}{2\mu_0};\ \text{all integer points between } \pm\infty\right\}$$

using an interpolation function $\operatorname{sinc}\left(2\mu_0\left(x - \frac{n}{2\mu_0}\right)\right)$. This is shown graphically in Figure 2.18.

It is clear from this figure that the sample point of the Sinc function at one location occurs at the zeros of the other Sinc functions, which means they do not contaminate the samples. This is because of the periodicity of the Sinc function. In general, the sampling must occur at sampling intervals satisfying the following conditions:

$$\Delta x_n = \frac{n}{2\mu_0};\quad n = 1, 2, 3, \ldots \tag{2.154}$$

This condition ensures we do not have sampling errors. If the sampling interval Δx is greater than $\frac{1}{2\mu_0}$, the series of band-limited spectra are more widely spaced and $f(x)$ still is recoverable even though more sampling locations have been used.

If the sampling interval Δx is less than $\frac{1}{2\mu_0}$, the spectra will overlap and $f(x)$ is not completely recoverable. This degradation due to undersampling is called *Aliasing*. Figure 2.19 illustrates an example of imaging undersampling. In the field of optics, aliasing errors also are known as *Moiré Patterns*. Figure 2.20a shows a perfect image and Figure 2.20b, the resulting Moiré patterns from aliasing [14, p. 103]. The Moiré patterns can be clearly seen in the third ring of (b).

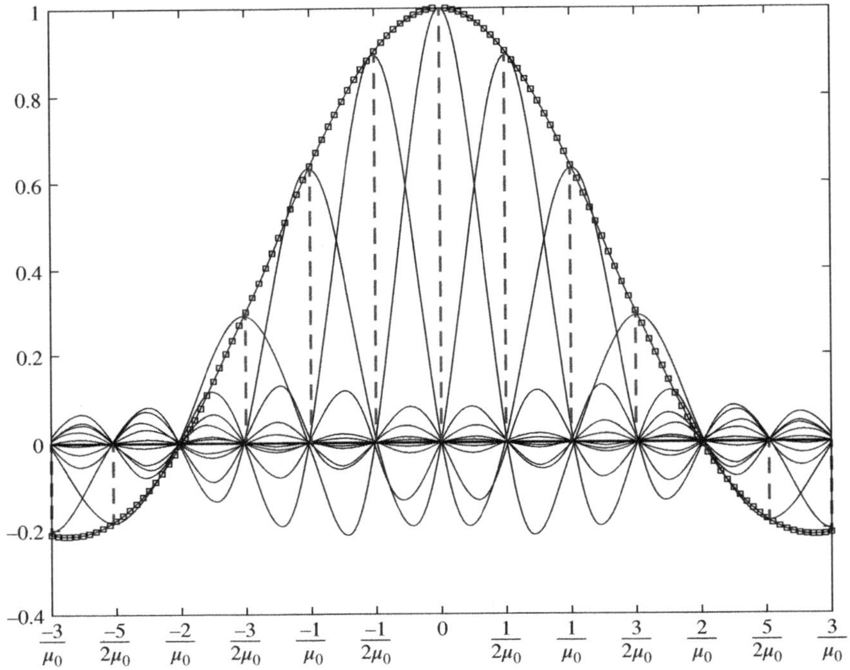

FIGURE 2.18 Plot of the Sampling Theorem.

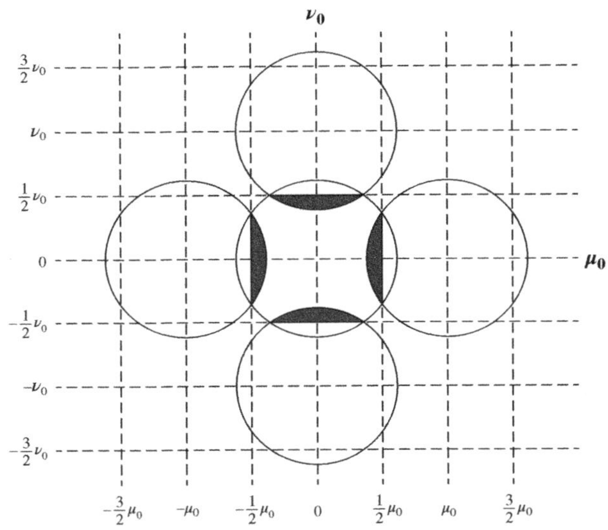

FIGURE 2.19 Depiction of image undersampling. Source: Pratt, 2007 [14]. Reproduced with permission of Wiley.

(a) Original image

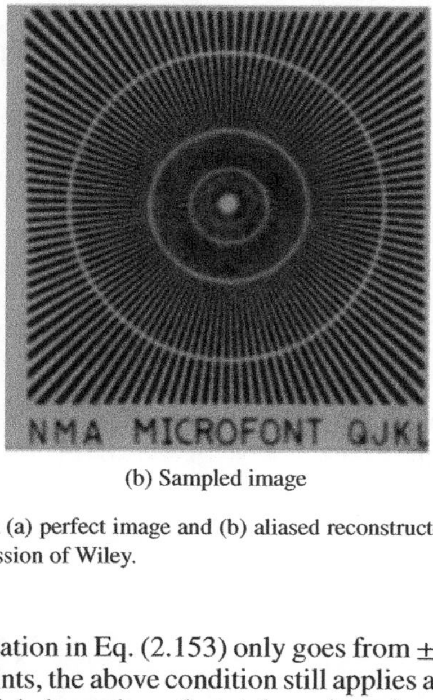

(b) Sampled image

FIGURE 2.20 Example of a (a) perfect image and (b) aliased reconstructed image. Source: Pratt, 2007 [14]. Reproduced with permission of Wiley.

In general, the summation in Eq. (2.153) only goes from $\pm N$ rather than $\pm\infty$. Even with only $2N$ sample points, the above condition still applies at each sample point. The product $(2N)\left(2\mu_0\right)$, which depends on the total number of samples and the bandwidth requirement given above, is called the *Space-Bandwidth Product* (one dimensional), similar in nature to the Time-Bandwidth Product of time series analysis.

2.10 PROBLEMS

Problem 2.1. Solve the following integral:

$$\iint_{-\infty}^{\infty} \int_{i\sqrt{x/2}}^{\sqrt{x/2}} e^{2\pi i\left[v\left(\frac{x-3}{2}\right)+y^2\right]} y\,dy\,dv\,dx.$$

Integrals running from $-\infty$ to $+\infty$ integrate over the parameters v and x, while the integral with finite limits integrates over the parameter y.

HINT: Watch out for the "hidden delta function." Also be careful when changing the sequence of integration.

Problem 2.2. Let us assume that the refractive index of the medium above the (x, y) plane is 1 and that the refractive index of the medium below the (x, y) plane is $n_1 \neq 1$. Also, let us assume a point source at $(0, 0, h)$ on the positive z-axis define the point Q to be an arbitrary point $(x, y, 0)$ on the boundary between the two media and the point B to be the receiver location $(a, 0, h)$. The distance from point P to point Q is R and the distance from point Q to point B, is R'. See the following left figure for this situation, which is not drawn to scale. The change in entry angle from the upper medium to the bottom medium is shown in the lower right-hand image, that is, Snell's Law.

According to Huygens, the spherical wave from P is $\frac{e^{ikR}}{R}$ at $z = 0$ for any (x, y) in that plane. From Q, a secondary spherical wave emerges with a beginning amplitude and a field amplitude at B.

$$\left(\frac{e^{ikR}}{R}\right)\left(\frac{e^{ikn_1R'}}{n_1R'}\right).$$

The complete field amplitude at B from all the incoming wavelets equals

$$u_B = \iint \frac{e^{ikR+ikn_1R'}}{n_1R'R}\,dx\,dy.$$

Compute u_B using the saddle point method.

HINT: Watch out for Snell!

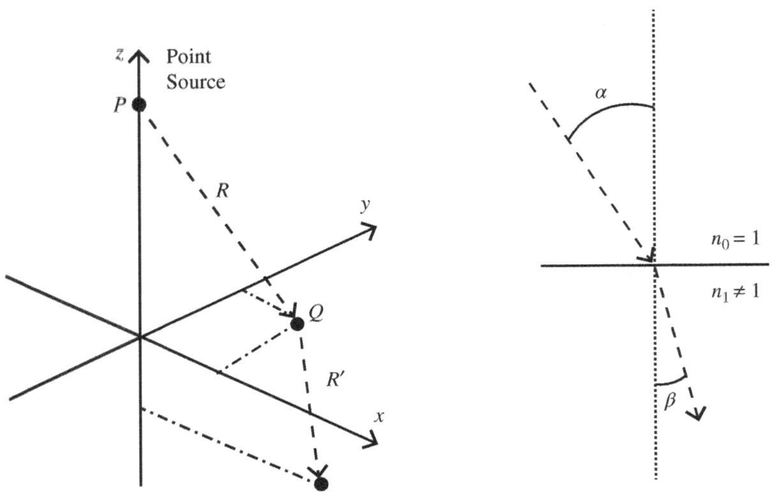

$\sin\alpha = n_1 \sin\beta$

Problem 2.3. Compute

$$I = \int_A^B g(x)\, e^{ikf(x)}\, dx$$

with $A = \frac{5}{2}$, $B = \frac{15}{2}$, $k = \frac{2\pi}{5}$, $f(x) = x(x-5)^2$, and $g(x) = \sin\left(\frac{\pi x}{30}\right)$.

Problem 2.4. We assume $u(x) = 0$ in $|x| \geq \frac{P}{2}$. Hence, $\tilde{u}(v)$ can be sampled at $v_n = \frac{n}{P}$, which is enough to know, $\tilde{u}(v)$ and $u(x)$ complex. Refer to the following figure. Unfortunately, we do not get $\{\tilde{u}(v)\}$, but $\{\tilde{w}(v)\}$, where

$$w\left(\frac{n}{P}\right) = Q^{-1} \int_{\frac{n}{P}-\frac{Q}{2}}^{\frac{n}{P}-\frac{Q}{2}} \tilde{u}(v)\, dv.$$

Is $w(x) = u(x)$, or at least, $w(x) \approx u(x)$?

HINT:

$$\sum_{n=-\infty}^{\infty} e^{-2\pi i n v r} = \sum_{m=-\infty}^{\infty} \delta\left(v - \frac{m}{r}\right)$$

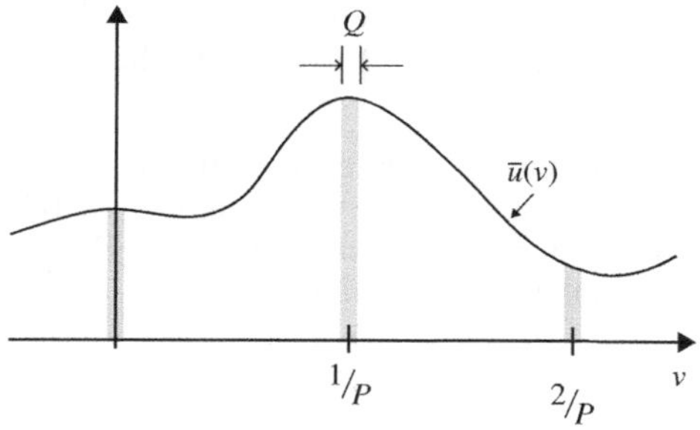

Problem 2.5. The Rayleigh Criterion for resolution states that two point sources are just resolved when the central maximum from one source falls on the first minimum of the diffraction pattern from the other source. Referring to the following figure, derive an equation for the separation between the peaks under this criteria assuming diffraction-limited optics.

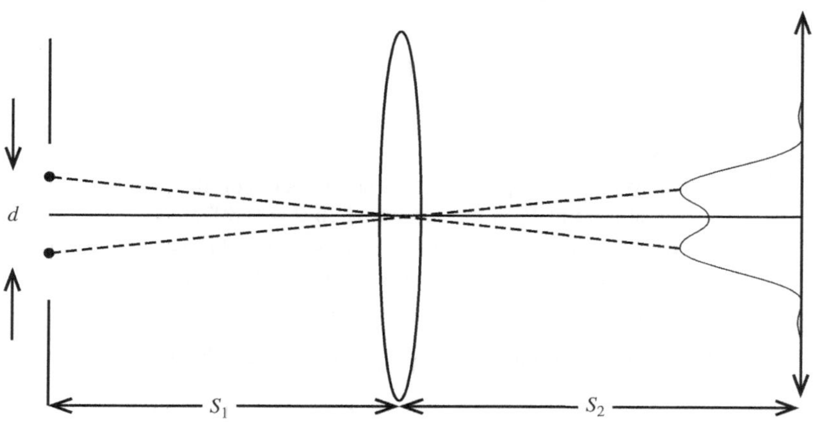

Problem 2.6. Two stars are a distance 1.5×10^8 km apart. At what distance can they be resolved by the unaided eye? Assume that the refractive index and the lens aperture of the eye are 1.34 and 5 mm, respectively. How much range improvement occurs if the eye now is aided by a telescope? Assume that a 200-inches astronomical telescope displaying the image on a TV screen.

Problem 2.7. If a car is approaching you at night with headlights 1 m apart, how far away must you be in order to just resolve them? (Treat the headlights as single slits of width 1 mm and assume that the lamps are monochromatic sodium sources of wavelength 5892.9 Å.)

Problem 2.8. In Young's double-slit experiment, light intensity is a maximum when the two waves interfere constructively. This occurs when

$$d \sin \theta = m\lambda; \quad m = 0, \pm 1, \pm 2, \pm 3, \ldots$$

where d is the separations of the slits, λ the wavelength of light, m the order of the maximum, and θ the scatter angle that the maxima occurs. Assume $d = 0.320$ mm. If a beam of $\lambda = 500$ nm light strikes the two slits, how many maxima will there be in the angular range $45° \leq \theta \leq 45°$?

Problem 2.9. Refer to the introduction to Problem 2.8. Assume the incident light is monochromatic with a wavelength $\lambda = 500$ nm and slit separation. Suppose $d = 0.100$ mm. The Young's experimental geometry is shown in the following figure:

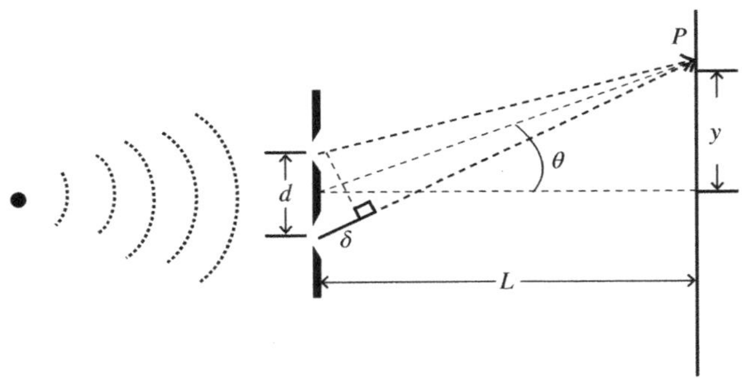

Assume $L = 1$ m.

(a) What is the phase difference φ between the two waves arriving at a point P on the screen when $\theta = 0.80°$?

(b) What is the phase difference between the two waves arriving at a point P on the screen when $y = 4.00$ mm?

(c) If $\varphi = \frac{1}{3}$ rad, what is the value of θ?

(d) If the path difference is $\frac{\lambda}{4}$, what is the value of θ?

Problem 2.10. Refer to the introduction to Problem 2.8 and the figure in Problem 2.9. Using the small angle approximation, develop an expression for wavelength used in this experiment. A pair of screens are placed 13.7 m apart. A third order fringe is seen on the screen 2.50 cm from the central fringe. Does the scatter angle a value consistent with the small angle approximation? If the slits were cut 0.0960 cm apart, determine the wavelength of this light.

REFERENCES

1. Booker, H.G. and Clemmow, P.C. (1950) *The Concept of an Angular Spectrum of Plane Waves, and its Relation to that of Polar Diagram and Aperture Distribution*. Proc. Inst. Electr. Eng. 97, Part III, pp. 11–17.
2. Lohmann, A.W. (2006, ISBN 2-939472-00-6) in *Optical Information Processing* (ed. S. Sinzinger), Universitätsverlag, Ilmenau, Germany.
3. Goodman, J.W. (2004) *Introduction to Fourier Optics*, 3rd edn, Roberts and Company, Englewood, CO.
4. Papoulis, A. (1968, 474 pages, ISBN 0898743583; 9780898743586) *Systems and Transforms with Applications in Optics*, McGraw-Hill Series in Systems Science, Edition reprint, Robert Krieger Publishing Company, Malabar, FL.
5. Born, M. and Wolf, E. (1999) *Principles of Optics: Electromagnetic Theory of Propagation, Interference and Diffraction of Light*, Cambridge University Press, London.
6. Andrews, L.C., Phillips, R.L., Bagley, Z.C. *et al.* (2012) in *Advanced Free Space Optics (FSO) a Systems Approach (Chapter 9)*, Springer Series in Optical Sciences, vol. **186** (ed. W.T. Rhodes), Springer, New York.
7. Andrews, L.C. and Phillips, R.L. (2005) *Laser Beam Propagation through Random Media*, 2nd edn, SPIE Press, Bellingham, WA.
8. Andrews, L.C. (2004) *Field Guide to Atmospheric Optics*, SPIE Press, Bellingham, WA.
9. Karp, S. and Stotts, L.B. (2012) *Fundamentals of Electro-Optic Systems Design: Communications, Lidar, and Imaging*, Cambridge Press, New York.
10. Karp, S., Gagliardi, R.M., Moran, S.E., and Stotts, L.B. (1988) *Optical Channels: Fiber, Atmosphere, Water and Clouds*, Plenum Publishing Corporation, New York.
11. Friis, H.T. (1946) A note on a simple transmission formula. *Proceedings of the IRE*, **34**, 252–256.
12. Shaw, J.A. (2013) Radiometry and the Friis transmission equation. *American Journal of Physics*, **81** (**1**), 32–37.
13. Gradshteyn, I.S. and Ryzhik, I.M. (1965) in *Table of Integrals, Series and Products*, 4th edn (eds Y.V. Geronimus and M.Y. Tseytlin), Translated by Alan Jeffrey, Academic Press, New York.
14. Pratt, W.K. (2007) *Digital Image Processing*, 4th edn, Wiley-Interscience, John Wiley & Sons, Inc., Hoboken, NJ.

3

GEOMETRICAL OPTICS

3.1 INTRODUCTION

There are two basic ways to describe the transfer of energy or power from one point to another, and how it interacts with various media such as lens and apertures. One is Geometrical Optics, which involves tracing light rays emanating from various sources to a point in space or within different media, and is the focus of this chapter. The other is Physical Optics, which we discussed in Chapter 2.

Geometrical Optics uses the concept of rays, which have direction and position but no phase information, to model the way light travels through space or an optical system. It approximates Physical Optics and is generally employed when the smallest dimension of the optical system is much larger than the wavelength of light. If applicable, its strength is that light propagation is simply governed by the principles of refraction and reflection, and a few simple rules about the geometry involved in tracing light rays through various media such as simple and complex lens systems. In practice, geometrical optics is good when the diameter of the optical systems is greater than a 100 wavelengths. However, if the dimensions of the optical components are smaller, tending toward the size of a wavelength, then Physical Optics must be used.

In this chapter, we focus on imaging systems, which cover a broad class of engineering applications. References [1–5] are excellent reference materials for the interested reader, which we will be leveraging in this chapter, especially Smith [3]. We will assume that an object is viewed as a collection of many pinpoint sources of light and that some of them will be captured by the imaging system. Those captured ray

Free Space Optical Systems Engineering: Design and Analysis, First Edition. Larry B. Stotts.

Companion website: www.wiley.com\go\stotts\free_space_optical_systems_engineering

sources will be traced through an optical system to determine what image will be formed. This technique is useful for designing lenses, developing systems of lenses for use in microscopes, telescopes, cameras, and other devices.

3.2 THE FOUNDATIONS OF GEOMETRICAL OPTICS – EIKONAL EQUATION AND FERMAT PRINCIPLE

Born and Wolf stated that in regions that are many wavelengths from a source that the general electric and magnetic fields have the form

$$\boldsymbol{E}_0(\boldsymbol{r}) = \boldsymbol{e}(\boldsymbol{r})\, e^{ik_0\mathcal{L}(\boldsymbol{r})} \tag{3.1}$$

and

$$\boldsymbol{H}_0(\boldsymbol{r}) = \boldsymbol{h}(\boldsymbol{r})\, e^{ik_0\mathcal{L}(\boldsymbol{r})}, \tag{3.2}$$

where $\mathcal{L}(\boldsymbol{r})$, *the optical path*, is a scalar function of position, $k_0 \left(= \frac{2\pi}{\lambda_0}\right)$ the wavenumber, λ_0 the vacuum wavelength of the propagating light, and $\boldsymbol{e}(\boldsymbol{r})$ and $\boldsymbol{h}(\boldsymbol{r})$ the vector functions of position, which may be complex [4, p. 111]. For example, the function $k_0\mathcal{L}(\boldsymbol{r})$ is written as $k_x x + k_y y + k_z z$ for a plane wave and $k_0\sqrt{x^2 + y^2 + z^2}$ for a spherical wave.

Maxwell's Equations lead to a set of relations among $\mathcal{L}(\boldsymbol{r})$, $\boldsymbol{e}(\boldsymbol{r})$, and $\boldsymbol{b}(\boldsymbol{r})$. The one of most interest to us in optics is that if the wavenumber k_0 is large (small λ_0), these relations require that $\mathcal{L}(\boldsymbol{r})$ satisfy a certain differential equation, which is independent of the amplitude vectors $\boldsymbol{e}(\boldsymbol{r})$ and $\boldsymbol{b}(\boldsymbol{r})$ [4, pp. 111–112]. Specifically, the authors showed this equation to be

$$[\nabla\mathcal{L}(\boldsymbol{r})]^2 = [n(\boldsymbol{r})]^2 \tag{3.3a}$$

or in specific terms

$$\left(\frac{\partial\mathcal{L}(x,y,z)}{\partial x}\right)^2 + \left(\frac{\partial\mathcal{L}(x,y,z)}{\partial x}\right)^2 + \left(\frac{\partial\mathcal{L}(x,y,z)}{\partial x}\right)^2 = n^2(x,y,z), \tag{3.3b}$$

where $n(\boldsymbol{r})$ denotes the refractive index of the medium. The function $\mathcal{L}(\boldsymbol{r})$ also is called the *eikonal* and Eq. (3.3) is known as the *eikonal* equation. This is the basic equation of geometrical optics. The surfaces where $\mathcal{L}(\boldsymbol{r}) \equiv$ constant are called the *geometrical wave surfaces* or *geometrical wave fronts*.

Recall from basic electromagnetic theory that the Poynting vector represents the amount of energy that crosses per second a unit area normal to the direction of $\boldsymbol{E}$ and $\boldsymbol{H}$. It can be shown that the average Poynting vector in the direction of these geometrical wave fronts and its magnitude is equal to the average energy density and the velocity $v = \frac{c}{n}$. This allows us to define the geometrical light rays as the orthogonal trajectories to the geometrical wave fronts where $\mathcal{L}(\boldsymbol{r}) \equiv$ constant. This situation is illustrated in Figure 3.1. If $\boldsymbol{r}(s)$ denotes a position vector of a point v on ray, the authors considered it a function of the length of arc s of the ray, and this implied that

$$\frac{d\boldsymbol{r}}{ds} = s. \tag{3.4}$$

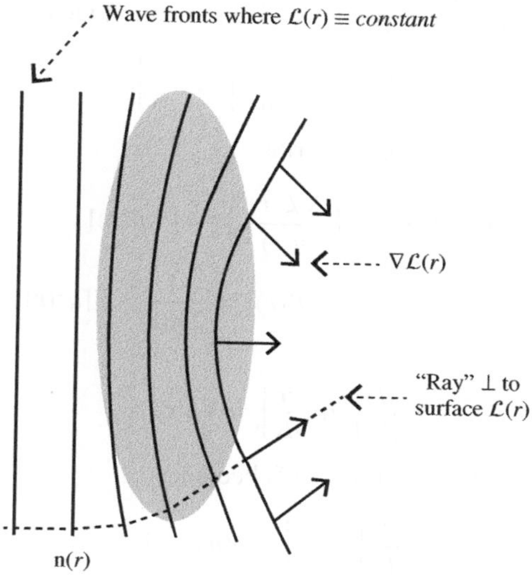

FIGURE 3.1 Illustration of eikonal equation scenario.

The result is that equation of the ray can be written as

$$n\frac{d\boldsymbol{r}}{ds} = \nabla\mathcal{L}(\boldsymbol{r}), \tag{3.5}$$

which implies that the electric and magnetic vectors at every point are perpendicular to the ray. The meaning of Eq. (3.5) is that

$$\frac{d\mathcal{L}}{ds} = \frac{d\boldsymbol{r}}{ds} \cdot \nabla\mathcal{L}(\boldsymbol{r}) = n. \tag{3.6}$$

Integrating Eq. (3.6) over a curve C yields the following relationship:

$$\mathcal{L}(\boldsymbol{r}) \equiv \text{optical path length} = \int_C n(\boldsymbol{r})\, ds. \tag{3.7}$$

Fermat's Principle states that the optical path

$$\mathcal{L}(\boldsymbol{r}) = \int_{v_1}^{v_2} n(\boldsymbol{r})\, ds \tag{3.8}$$

of an actual ray between any two points v_1 and v_2 is shorter than the optical path of any other curve that joins these points and that lies in a certain regular neighborhood of it [4, pp. 128–130]. In short, light travels the optical path that takes the minimum time to traverse. This is important to know.

Example 3.1 How do ray travel in homogeneous mediums? Taking the square root of Eq. (3.3), we obtain

$$\nabla\mathcal{L}(\boldsymbol{r}) = n(\boldsymbol{r})\hat{s} = n(\boldsymbol{r})\frac{d\boldsymbol{r}}{ds}. \tag{3.9}$$

Taking the derivative with respect to s of this equation yields

$$\frac{d}{ds}[\nabla\mathcal{L}(\boldsymbol{r})] = \frac{d}{ds}\left[n(\boldsymbol{r})\frac{d\boldsymbol{r}}{ds}\right]. \tag{3.10}$$

Using the chain rule for ∇, we can write

$$\frac{d}{ds}[\nabla\mathcal{L}(\boldsymbol{r})] = \frac{d\boldsymbol{r}}{ds}\cdot\nabla[\nabla\mathcal{L}(\boldsymbol{r})] = \left(\frac{\nabla\mathcal{L}(\boldsymbol{r})}{n(\boldsymbol{r})}\right)\cdot\nabla[\nabla\mathcal{L}(\boldsymbol{r})]$$
$$= \frac{1}{2\,n(\boldsymbol{r})}\cdot\nabla[\nabla\mathcal{L}(\boldsymbol{r})\cdot\nabla\mathcal{L}(\boldsymbol{r})] = \frac{1}{2\,n(\boldsymbol{r})}\cdot\nabla[n^2(\boldsymbol{r})] = \nabla n(\boldsymbol{r}) \tag{3.11}$$

or

$$\frac{d}{ds}\left[n(\boldsymbol{r})\frac{d\boldsymbol{r}}{ds}\right] = \nabla n(\boldsymbol{r}). \tag{3.12}$$

Now if $n(\boldsymbol{r}) \equiv \text{constant} = n_0$, Eq. (3.12) becomes

$$\frac{d}{ds}\left[n_0\frac{d\boldsymbol{r}}{ds}\right] = \nabla n_0 = 0. \tag{3.13}$$

This implies that $\boldsymbol{r}(s) = s\hat{c}$, with $\hat{c}$ being a constant as well, or in specific terms, light rays travel in straight lines while in a homogeneous medium, following Fermat's principle.

3.3 REFRACTION AND REFLECTION OF LIGHT RAYS

Assume that we have a monochromatic plane wave of light propagating in a medium of refractive index n and is incident at angle θ to the interface with another medium of refractive index $n_2 > n_1$. The wave fronts can be considered surfaces of equal phase. Figure 3.2 illustrates the principles of refraction. Part of the incident plane wave is refracted at a new angle θ' into the new medium. Upon entering the new medium, the velocity of the wave front reduces by the new refractive index. The result is that the wave front spacing compresses by the ratio $\frac{n_1}{n_2}$. It also is clear that the new refraction angle is closer to the interface normal.[1] It is quantified by Snell's Law, which is given by

$$\theta' = \sin^{-1}\left(\frac{n_2}{n_1}\sin\theta\right). \tag{3.14}$$

The remainder of the light that is not refracted is reflected at the same angle as the original plane wave, but in the opposite direction. No change in the wave front spacing occurs. Figure 3.3 summarizes the laws of refraction and reflection in terms of wave front rays.

Unfortunately, the dielectric interface affects the light amplitude going from one medium to the other. Some of the light is reflected in the opposite direction of its

[1]Note: In all optical media, the index of refraction varies with wavelength of the light. The refractive index is generally higher at the shorter wavelengths than at the longer wavelengths. In other words, blue light will be refracted at a greater angle than red light, placing it closer to the interface normal. This means that a polychromatic plane wave will experience angular dispersions upon entering the second medium, that is, rainbow effect.

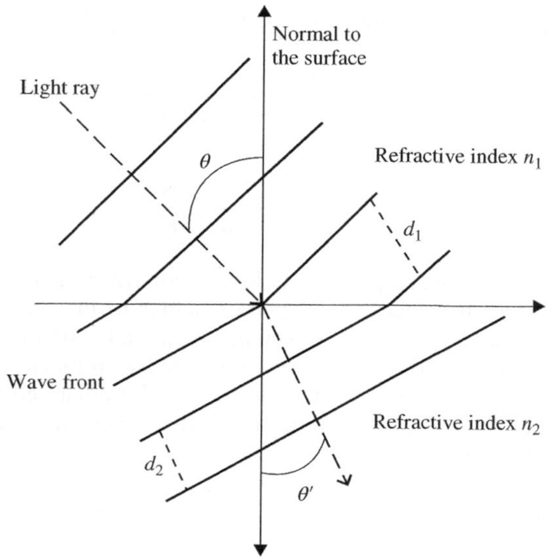

FIGURE 3.2 Principle of refraction.

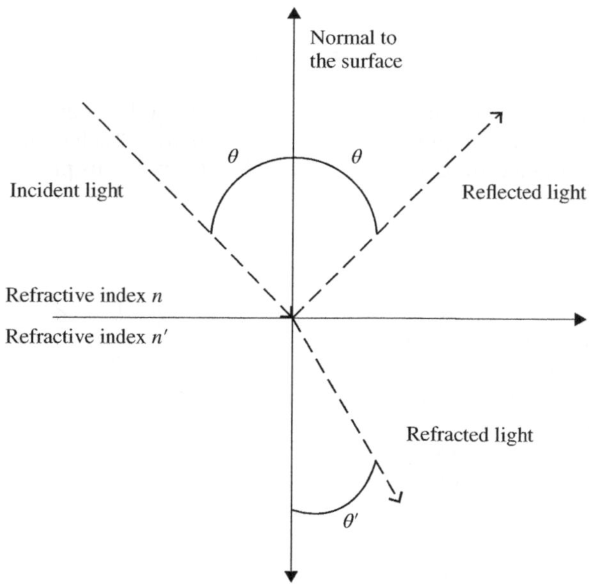

FIGURE 3.3 Principle of refraction and reflections.

incidence angle, reducing the amount of light transmitted into the second medium. It is polarization dependent. The reflection coefficient for s-polarized light is given by

$$r_s = \frac{n_1 \cos\theta - n_2 \cos\theta'}{n_1 \cos\theta + n_2 \cos\theta'} \tag{3.15}$$

and

$$r_p = \frac{n_1 \cos\theta' - n_2 \cos\theta}{n_1 \cos\theta' + n_2 \cos\theta} \tag{3.16}$$

for s- and p-polarized fields, respectively. Here, s-polarized means that the incident light is polarized with its electric field perpendicular to the plane containing the incident, reflected, and refracted rays. This plane is called the *plane of incidence*. The term p-polarized means the incident light is polarized with its electric field parallel to the plane of incidence. The reflectance for the s- and p-polarized intensities is just the square of Eqs. (3.15) and (3.16), respectively. If the incident light is unpolarized, the reflectance of the intensities then is given by

$$R_{\text{unpol}} = \frac{1}{2}\left[R_s + R_p\right] = \frac{1}{2}\left[r_s^2 + r_p^2\right]. \tag{3.17}$$

By the conservation of energy, the transmittances for the s- and p-polarized intensities equal

$$T_s = \left[1 - R_s\right] \tag{3.18}$$

and

$$T_p = \left[1 - R_p\right]. \tag{3.19}$$

Let us now turn to refraction of spherical waves on a flat aperture, so we can see what happens in this case. Figure 3.4a depicts our situation. Each point on the aperture interface surface has a different incident ray, unlike the previous plane wave example.

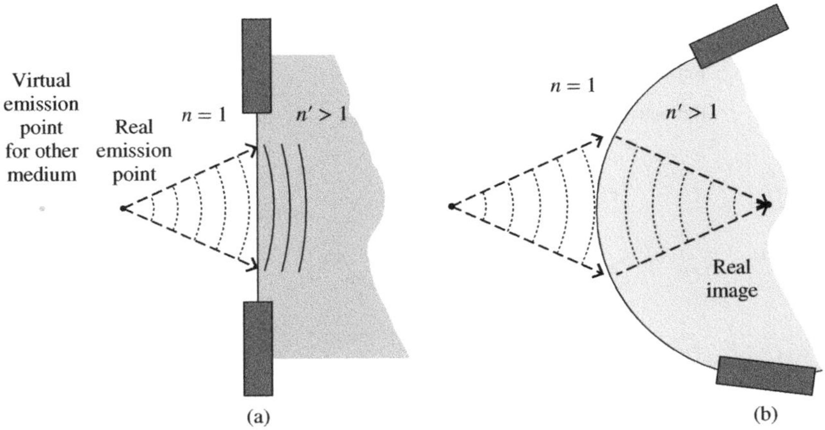

FIGURE 3.4 Spherical wave fronts incident on (a) a flat aperture and (b) curved aperture; both separating two mediums of different refractive indices.

The result is the same wave front compression we saw before, but the wave front has flattened out as well. In effect, this newest set of wave fronts effectively has a different emission point, which we denote as the virtual emission point in Figure 3.4a. This suggests that if the surface is curved, then one can get the wave fronts to be concave, as depicted in Figure 3.4b. This essentially created a "real image" of the original source. That is, a real image is formed by a diverging beam parallel to the direction of propagation (as seen in Figure 3.4b). Conversely, a "virtual" image is created by a diverging beam opposite the direction of propagation (as seen in Figure 3.4a).

The above discussion assumes ideal conditions. In reality, when a spherical wave front enters a medium of differing refractive index, the resulting wave front is no longer spherical. That is, if you broke up the wave fronts, each section would have a different virtual mage point. This is because the sine function has the form

$$\sin\theta = \theta - \frac{\theta^3}{3!} + \frac{\theta^5}{5!} - \cdots \tag{3.20}$$

If $\sin\theta \approx \theta$, then we are dealing with spherical or plane waves, and everything said above is totally valid. This is known as the *paraxial approximation* and is true when $\theta < 5°$. When the other terms are involved, we are not dealing with a spherical and the waves become distorted.

The amount deals with how many terms are involved. The distortion effects are known as *aberrations*.

3.4 GEOMETRICAL OPTICS NOMENCLATURE

To understand first-order optical design, or image formation, one needs to understand the terminology of lens theory. Figure 3.5 illustrates a simplified optical system and the key parameters used to characterize it. The optical axis of the systems is the dashed light running through the middle of the optical systems from left to right. This figure assumes we have a perfect optical system, obeying the paraxial approximation. Let us begin with the cardinal points of an optical system, which are the first and second *focal points*, *principal points*, and *nodal points*.

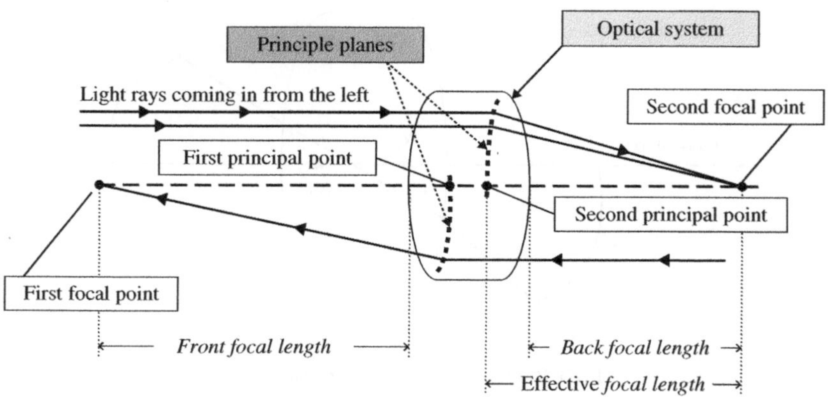

FIGURE 3.5 Locations of the key parameters characterizing an optical system.

The first focal point is the point to the left of the optical system that will create a collimated output from the optical system from emitted spherical light from that point. On the other hand, the second focal point is the point where collimated or plane wave input from the left side of the systems will focus to a point on the right side of the optics. Conversely, if light was being emitted by the second focal point on the right side of the system, to the left, the output of the optics propagating to the left would be a collimated beam. In all cases, the collimated input and output parallel the optical axis.

If the converging rays from the focal points and the incoming collimated beams are traced back to their intersection, the points of intersection would create a "virtual" surface, which are referred to the principal planes. They are depicted in Figure 3.4 within the optical system. In a perfect system, these surfaces would be a spherical and centered on the object and image points on the optical axis. However, in the paraxial regime close to the optical axis, these surfaces would be planar (this is shown in Figure 3.5 as well). This situation is the origin of their name, "Principal Planes." The intersections of these surfaces with the optical axis are the principal points, which also are shown in Figure 3.5. Thus, the second focal and principal points are those defined by light rays approaching from the left. The first focal and principal points are defined by rays from the right.

Referring to the same figure, the front and back focal lengths (ffl and bfl) are the distances from the first focal point to the first surface of the optical system and from the second focal point to the last surface of the optical systems, respectively. The effective focal length (efl) of the system is the distance from the second principal point to the second focal point.

The nodal points are two axial points where the ray hitting the first nodal point appears to be emerging out of the second nodal point paralleling it in propagation direction. This situation is depicted in Figure 3.6. When the optical system has air at both the entrance and exit of the optical systems of interest, which is generally true in most applications, the nodal points coincide with the principal points.

The power of a lens, or an optical system, is the reciprocal of the effective focal length and is denoted by the symbol ϕ. If the focal length is in meters, then the power has units of m^{-1} or *diopters*.

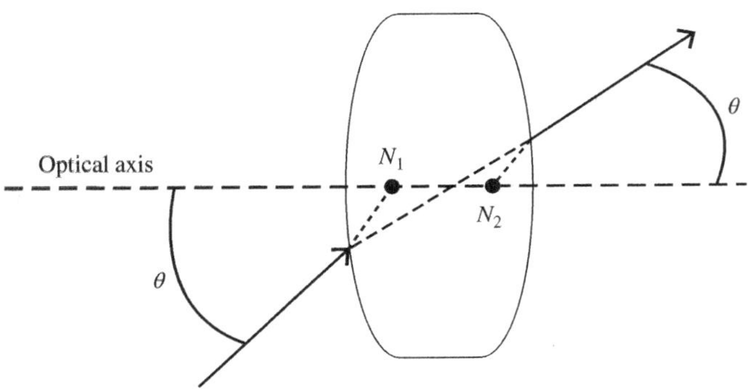

FIGURE 3.6 Locations of the nodal point.

The ratio of the focal length to the clear aperture of the lens, or optical system, is called the relative aperture, f − number, or speed. That is, we have

$$f - \text{number} = \frac{\text{efl}}{\text{clear aperture}}. \tag{3.21}$$

3.5 IMAGING SYSTEM DESIGN BASICS

Figure 3.7 provides the lens design model with key parameters located. Object space ends when one enters the optical system and is contained in a medium of refractive index n. Image space begins when light exits the optical system and is contained in a medium of refractive index n. Let us now define the conventions we use in this section.

Modern optics books use the Cartesian Sign Convention, which is given as follows:

- All figures are drawn with light traveling from left to right.
- All distances are measured from a reference surface, such as a wave front or a refracting surface. Distances to the left of the surface are negative.
- The refractive power of a surface that makes light rays more convergent is positive. The focal length of such a surface is positive.
- The distance of a real object is negative.
- The distance of a real image is positive.
- Heights above the optic axis are positive.
- Angles measured clockwise from the optic axis are negative.

Assume that we want to image an object located on the left of the optical system. Let y be the height of said object, and it is positive if the object is above the optical axis and negative if below. Let A and A' be the principal planes; here, A' is the conjugate plane of A. F and F' are the front and back focal points of the optical system, respectively. Assume that we also have the following distance delegations:

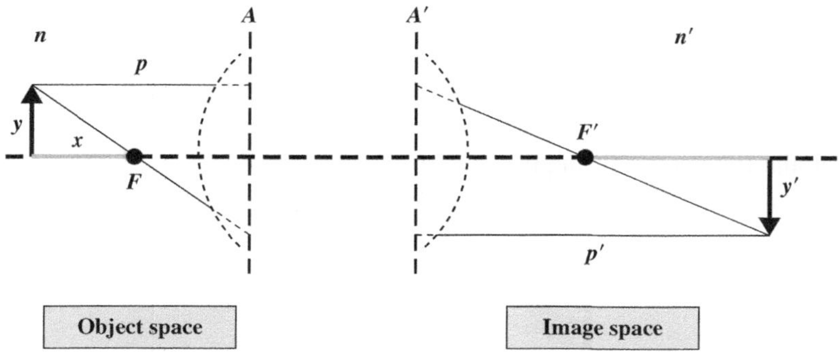

FIGURE 3.7 Locations of the key parameters in imaging system model.

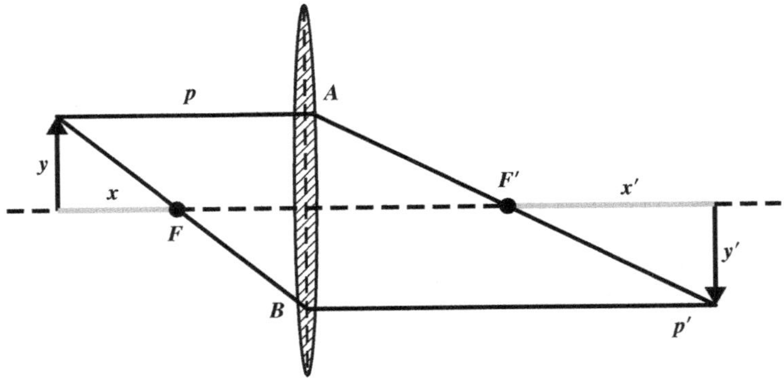

FIGURE 3.8 Locations of the key parameters in a thin lens.

- $p \equiv$ distance from the object to the plane A (positive if the object is to the left of A)
- $p' \equiv$ distance from the object to the plane A' (positive if the object is to the right of A')
- $x \equiv$ distance from the object to the front focal point
- $x' \equiv$ distance from the image to the back focal point
- $f \equiv$ distance from the focal point F to the plane A
- $f' \equiv$ distance from the focal point F' to the plane A' (= efl).

Let us now focus on the case of the thin lens. In this case, the first and second principal planes of said lens are in the same place internal to the lens, as well as the first and second principal points being in the same location, which is the center of the lens. This results in the focal length of a thin lens as just the distance from the center of the lens to either focal point. Figure 3.8 redraws Figure 3.7 under the thin lens assumptions.

Referring to Figure 3.8, we see that

$$\frac{y}{-x} = \frac{-y'}{f} \tag{3.22}$$

and

$$\frac{-y'}{x'} = \frac{y}{f'} \tag{3.23}$$

from similar triangles. This implies that

$$\frac{-x}{f} = \frac{y}{-y'} = \frac{f'}{x'} \tag{3.24}$$

or

$$ff' = -xx'. \tag{3.25}$$

If $n = n'$, then $f = f' =$ efl and we can write

$$x' = -\frac{f^2}{x}. \tag{3.26}$$

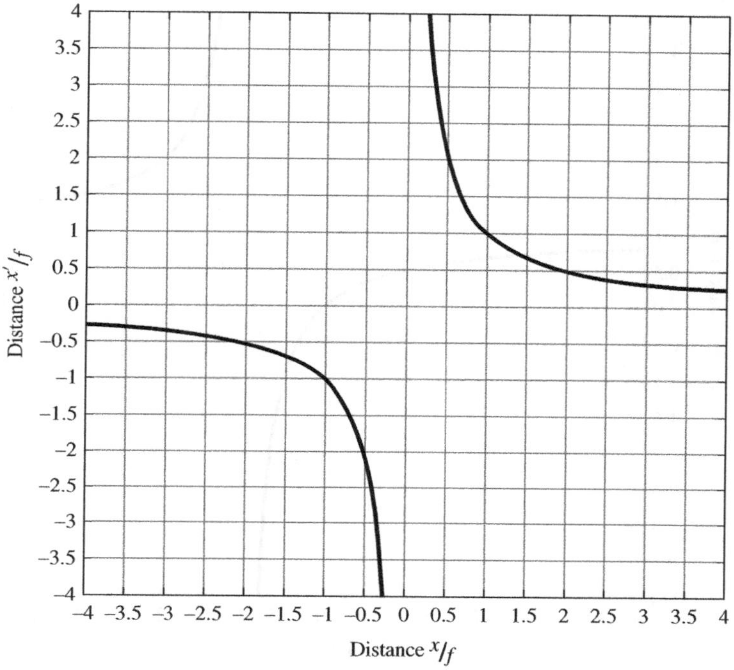

FIGURE 3.9 Graph of $\frac{x'}{f}$ as a function of $\frac{x}{f}$.

Equation (3.26) is known as the "Newtonian" form of the image equation. It was first derived by Sir Isaac Newton. Figure 3.9 shows a plot of x' as a function of x, both parameters normalized to the focal length of the optical system.

From similar triangles in Figure 3.8, we see that

$$\frac{-y'}{f} = \frac{y - y'}{p} \tag{3.27}$$

and

$$\frac{y}{f} = \frac{y - y'}{p'}. \tag{3.28}$$

Adding Eqs. (3.27) and (3.28) together, we find that

$$\frac{y - y'}{p'} + \frac{y - y'}{p} = \frac{y}{f} - \frac{y'}{f} = \frac{y - y'}{f}$$

or

$$\frac{1}{p'} + \frac{1}{p} = \frac{1}{f}. \tag{3.29}$$

Equation (3.29) is called the Gaussian form of the lens equation and equals the lens law defined in Eq. (2.73b). It was developed by the Mathematician Karl F. Gauss. Besides in diffraction theory, it is found in many introductory textbooks. Figure 3.9

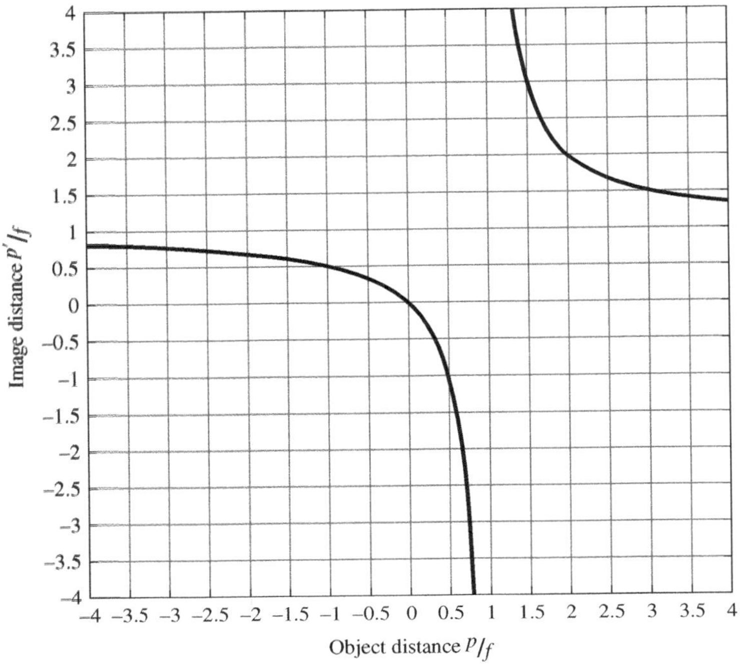

FIGURE 3.10 Graph of the image distance $\frac{p'}{f}$ as a function of object distance $\frac{p}{f}$.

shows a graph of the normalized image distance $\frac{p'}{f}$ as a function of the normalized object distance $\frac{p}{f}$. If the above lens equation yields a negative image distance, then the image is a virtual image on the same side of the lens as the object. If it yields a negative focal length, then the lens is a diverging lens rather than the converging lens in the illustration. This equation has been used to calculate the image distance for either real or virtual images and for either positive or negative lenses. However, it is not applicable to thick lens or complex optics.[2]

Example 3.2 Figure 3.11 illustrates seven possible image formation examples for a thin lens developed by Sears [5, p. 96]. Cases (1)–(5) follow the expected image formation performance one would expect from previous discussions. However, case (6) places the image inside the lens. Although seemly odd, it could be the result of the preceding lens (or lens system) placing the image in that position, or a real object placed adjacent the lens. In this case, we really cannot differentiate between the object and the image as in the previous cases. The object/image is just input light to what comes next to the right of that lens; it is just the object/image with no magnification or image

[2]Smith [3] developed an alternate form for the Gaussian lens equation for multiple-lens systems and more complex optical instruments. Namely, he showed that

$$\frac{1}{p'} = \frac{1}{f} + \frac{1}{p}.$$

However, for your purposes, Eq. (3.29) is sufficient.

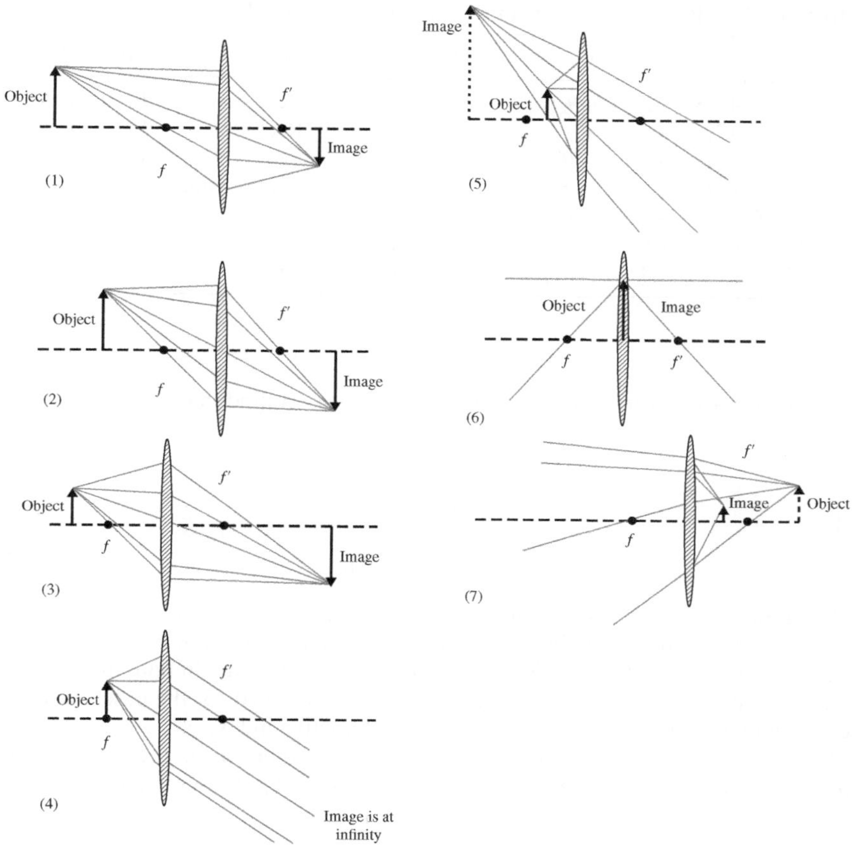

FIGURE 3.11 Seven examples of thin lens image formation.

inversion. Finally, case (7) depicts a converging cone of rays incident on the lens forming the real image, which by our theory implies it comes from a virtual image located to the right of the lens. By necessity, there must be a lens to the left of the lens for this to happen. Table 3.1 provides the key parameters for each of these seven examples [5, p. 97]. The reader should study Figures 3.8–3.10 and this table to understand how the results in this example related to the various developed equations [5, p. 96].

The *lateral or transverse magnification*, denoted by the symbol m, for the above optical system is given by the ratio of the image to object sizes $\frac{y'}{y}$. This implies that

$$m = -\frac{y'}{y} = -\frac{f}{x} = -\frac{-x'}{f}. \tag{3.30}$$

Substituting $x = p + f$ into Eq. (3.25) yields

$$m = -\frac{f}{p+f} = -\frac{p'}{p} \tag{3.31}$$

TABLE 3.1 Key Parameters for Thin Lens Images in Figure 3.11

Case	p	p'	x	x'	m	Object	Image	Erect or Inverted
1	$+3f$	$+1.5f$	$+2f$	$+\frac{1}{2}f$	$-\frac{1}{2}$	Real	Real	Inverted
2	$+2f$	$+2f$	$+f$	$+f$	-1	Real	Real	Inverted
3	$+1.5f$	$+3f$	$+\frac{1}{2}f$	$+2f$	-2	Real	Real	Inverted
4	$+f$	$\pm\infty$	0	$\pm\infty$	$\mp\infty$	Real	Real or virtual	Erect or inverted
5	$+2/3f$	$-2f$	$-1/3f$	$-3f$	$+3$	Real	Real	Erect
6	0	0	$-f$	$-f$	$+1$	Real or virtual	Real or virtual	Erect
7	$-2f$	$2/3f$	$-3f$	$-1/3f$	$+1/3$	Virtual	Virtual	Erect

by comparing similar triangles. Rewriting Eq. (3.31), we find that

$$p = f\left(1 - \frac{1}{m}\right) \tag{3.32}$$

and

$$p' = -f(m - 1). \tag{3.33}$$

Equations (3.30)–(3.33) show negative magnification. In general, a positive sign of the magnification means the image is noninverted; a negative sign implies the image is inverted. The absolute value of the magnification is the same in both instances. As a final comment, all of the above equations were developed for the case when both the object and image spaces are air with unity refractive index.

The *longitudinal magnification* denoted by the symbol $\overline{m}$, is the magnification along the optical axis. It can be shown that for all practical purposes that [3, pp. 24–25]

$$\overline{m} \approx m^2. \tag{3.34}$$

This indicates that this magnification is always positive.

Example 3.3 Assume that we have a thin-lens system with a positive focal length of 25 cm. What is the position and size of an image formed of an object 13.5 cm high that is located 100 cm to the left of the first focal point of the system?

Recall that we have

$$x' = -\frac{f^2}{x} = \frac{-(25\text{ cm})^2}{-100\text{ cm}} = +6.26\text{ cm}.$$

This answer implies that the position of the image is 6.26 cm to the right of the second focal point. The image height is found by first calculating the magnification, which equals

$$m = \frac{y'}{y} = \frac{-x'}{f} = \frac{-(6.26\text{ cm})}{26\text{ cm}} = -0.25$$

and then solving the following equation:

$$y' = my = (-0.25)(13.5\text{ cm}) = -3.13\text{ cm}.$$

Thus, we find from the magnification we have an inverted image a quarter of the original size, which is reflected in the y' calculation.

Example 3.4 Again, assume that we have a thin-lens system with a positive focal length of 25 cm. What is the position and size of an image formed of an object 13.5 cm high that is located 5 cm to the right of the first focal point of the system?

Recall that we have

$$x' = -\frac{f^2}{x} = \frac{-(25\ \text{cm})^2}{5\ \text{cm}} = -125\ \text{cm}.$$

This answer implies that the position of the image is 125 cm to the left of the second focal point, which for a thin-lens system, will be to the left of both the optical system and the object. The image height is found by first calculating the magnification, which equals

$$m = \frac{y'}{y} = \frac{-x'}{f} = \frac{+(125\ \text{cm})}{25\ \text{cm}} = +5$$

and then calculating the image location:

$$y' = my = (+5)(13.5\ \text{cm}) = +63.5\ \text{cm}.$$

These results say that both the magnification and image height are positive. In this case, a virtual image is formed. The implication is that the virtual image can only be seen by viewing though the lens from the right; it cannot be displayed on a screen and viewed standing elsewhere.

Example 3.5 If the object in Example 3.3 is 0.25 cm thick, what is the apparent thickness of the image?

The longitudinal magnification is

$$\overline{m} \approx m^2 = (5)^2 = 25.$$

This implies that the thickness of the image is approximately equal to (25)(0.25) = 6.25 cm.

3.6 OPTICAL INVARIANT

Let us now look at what happens to an arbitrary ray that passes through the optical system. Figure 3.12 shows such a ray, which is depicted by the dotted line running from the base of the object on the optical axis to the base of the image on the optical axis. Let s_1 be the distance from the lens to the object and s_2 is the distance from the lens to the image. This figure shows we have chosen a maximal ray, that is, the ray that makes the maximal angle with the optical axis as it leaves the object, passing through the lens at its maximum clear aperture. This choice makes it easier to visualize what is happening in the system, but the following argument can be made for any ray that is actually chosen.

Assume that this arbitrary ray goes through the lens at a height h above the optical axis. Since the thin lens theory assumes that paraxial approximation is valid, we have

$$n_1\theta_1 \approx \frac{h}{s_1} \tag{3.35}$$

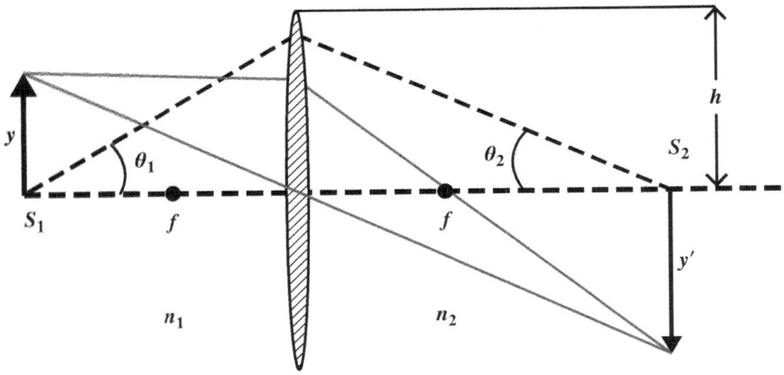

FIGURE 3.12 Example of an imaging thin-lens system showing a maximal ray.

and

$$n_2\theta_2 \approx \frac{h}{s_2}. \tag{3.36}$$

Assume that this arbitrary ray goes through the lens at a height h above the optical axis. Since the thin lens theory assumes that paraxial approximation is valid, we have

$$n_1\theta_1 \approx \frac{h}{s_1} \tag{3.37}$$

and

$$n_2\theta_2 \approx \frac{h}{s_2}. \tag{3.38}$$

Using the definition of magnification in Eq. (3.32), Eq. (3.38) can be rewritten as

$$n_2\theta_2 \approx \frac{h}{m\,s_1} = \frac{h}{s_1}\left(\frac{y_1}{y_2}\right) = n_1\theta_1\left(\frac{y_1}{y_2}\right)$$

or

$$n_2\theta_2 y_2 \approx \frac{h}{m\,s_1} = \frac{h}{s_1}\left(\frac{y_1}{y_2}\right) = n_1\theta_1 y_1. \tag{3.39}$$

Equation (3.39) is a fundamental law of optics, known as the Lagrange, Optical, or Smith–Helmholtz Invariant. It says that in any optical system that comprises lenses only, the product of the image size and ray angle is a constant, of the system. The above result is valid for the paraxial approximation and for any number of lenses, as can be verified by tracing the ray through a series of lenses. This equation also assumes perfect, aberration-free lenses. The addition of aberrations changes the equation; specifically, the equal sign in Eq. (3.39) would be replaced by a greater-than-or-equal sign. This means that aberrations always increase the product and nothing can be done to decrease it.

3.7 ANOTHER VIEW OF LENS THEORY

Again, we will assume that the paraxial approximation is valid. Let R_1 be the radius of the lens curvature on the left and R_2 be the radius of the lens curvature on the right. The lens is refractive index n, and the mediums on the left and right of the lens is assumed air. Figure 3.13 illustrates the imaging geometry of a conventional lens. s_1 represents the distance from an object pq to the left-hand side of the lens system.

Referring to the above figure, we see that a set of rays are emitted from a point q of an object pq and some of them intersect the portion of the lens of curvature R_1. These incident rays on the first surface of the lens form a virtual image of q at point q'. This virtual image is the "object emitting rays" intersecting the second surface of radius R_2 and results in the real image of q' at point q''. Here, s_1' depicts the distance from the virtual object $p'q'$ to the left-hand side of the lens system. It can be shown that this distance is given by

$$\frac{1}{s_1} + \frac{n}{s_1'} = \frac{n-1}{R_1} \tag{3.40}$$

using Snell's Law and the paraxial approximation [5, Chapter 3, Eq. (3.5), p. 63]. Assume that the thickness of the thin lens is t. Then, the distance from the virtual image $p'q'$ to the lens surface on the right-hand side on the optical axis is given by

$$s_2 = t - s_1' \tag{3.41}$$

using the distance convention listed in Section 3.8, that is, image distance s_1' must be a negative quantity.

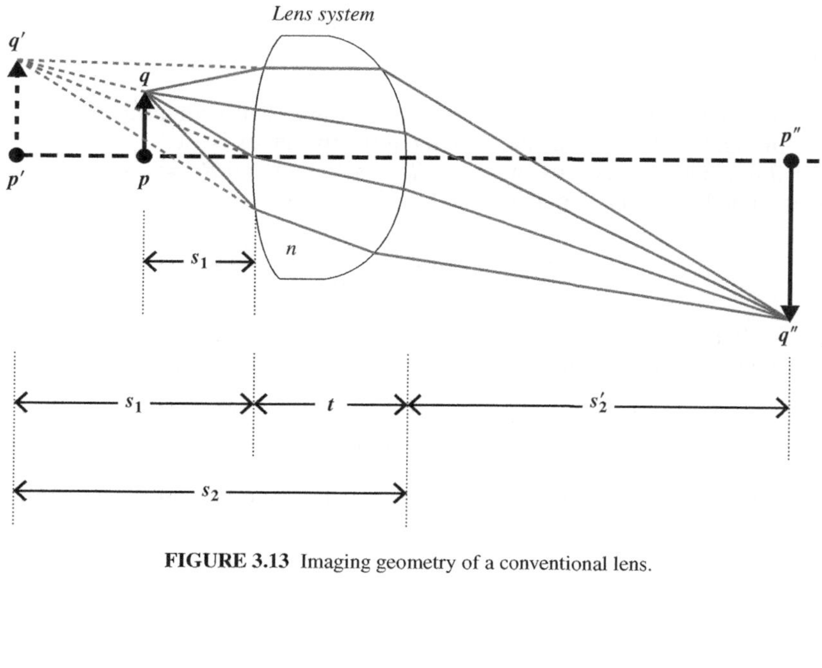

FIGURE 3.13 Imaging geometry of a conventional lens.

Looking at the second surface, we find in this case that

$$\frac{n}{s_2} + \frac{1}{s'_2} = \frac{1-n}{R_2} \tag{3.42}$$

from the same development used to create Eq. (3.35) [5].

Let us now focus on the case of the thin lens. Recall that the first focal point of any lens is the point on the optical axis for an object located at infinity. To find that focal point, we set s'_2 equal to infinity, which yields

$$\frac{n}{s_2} = \frac{1-n}{R_2}. \tag{3.43}$$

Since the thickness of the lens is negligible, $s_2 = -s'_1$ and we can rewrite Eq. (3.43) as

$$\frac{n}{s_2} = -\frac{n}{s'_1}. \tag{3.44}$$

This implies that

$$\frac{1}{s_1} = (n-1)\left(\frac{1}{R_1} - \frac{1}{R_2}\right). \tag{3.45}$$

Since the first and second principal planes of said lens are in the same place internal to the lens, and the first and second principal points being in the same location, Eq. (3.40) can be rewritten as

$$\frac{1}{f} = (n-1)\left(\frac{1}{R_1} - \frac{1}{R_2}\right) \tag{3.46}$$

as $s_1 = f, f$ being the focal length of the thin lens in air. The above equation is known as the *lensmaker's equation.*

Example 3.6 Find the focal length of the plano-convex lens with refractive index 1.5 shown in Figure 3.14. Outside the lens, the refractive index is 1.0.

For this plano-convex lens, we have $R_1 = +30$ mm and $R_2 = +\infty$. Substituting these radii and our refractive index into Eq. (3.41) gives

$$\frac{1}{f} = (1.5-1)\left(\frac{1}{30\ \text{mm}} - \frac{1}{\infty}\right) = \frac{0.5}{30\ \text{mm}},$$

or, $f = +60$ mm.

What happens if the planar surface faces the illuminating light? Then, we have

$$\frac{1}{f} = (1.5-1)\left(\frac{1}{\infty} - \frac{1}{-30\ \text{mm}}\right) = \frac{0.5}{30\ \text{mm}},$$

or, $f = +60$ mm. Therefore, for a thin lens, the first and second focal lengths are the same.

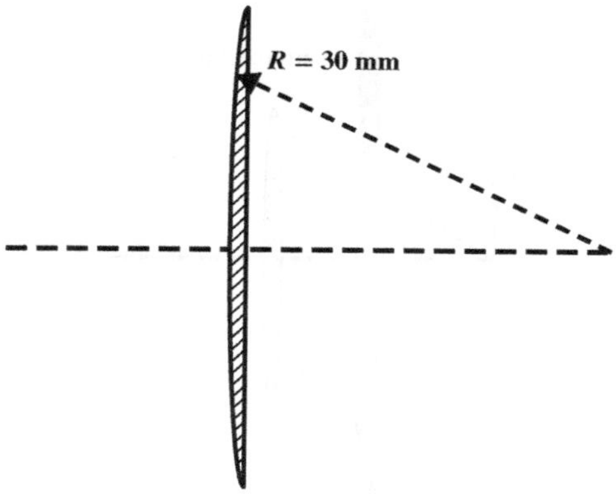

FIGURE 3.14 Depiction of a plano-convex lens.

3.8 APERTURES AND FIELD STOPS

Two important aspects of any imaging system are the amount of light transmitted by the system and the extent of an object that is seen by the system. Apertures and Field Stops limit the brightness (intensity) of an image and the field of view (FOV) of such an optical system. They each have a unique role and location within any system. This section describes the various stops and apertures that characterize the imaging performance of an optical system.

3.8.1 Aperture Stop

The Aperture Stop (AS) is a real element like a planar sheet perpendicular to the optical axis with a hole in the center, which physically limits the solid angle of light rays passing through the optical system from an on-axis object point. The aperture stop limits the optical energy that is admitted to the system. Figure 3.15 illustrates an example of AS before a lens.

It should be noted that even though aperture stops generally are smaller than the lens comprising the optical system, some of the rays will miss the lens entirely and/or some of the lens will have no light impinging on it. These effects are what is called *vignetting*. If present, the image off-axis appears dimmer than the image on-axis. It usually does not affect systems with small FOVs, but can be of significant importance to other optical instruments.

The Entrance Pupil is the opening an observer would identify as the limitation on the solid angle of rays diverging from an on-axis object point; that is, it is the image of the aperture stop as seen through all the optics before the aperture stop. Again, it can be a real or virtual image, depending on the location of the aperture stop.

On the other hand, the Exit Pupil is the image of the aperture stop formed by the light rays after they have passed through the optical system, that is, it is the image of the aperture stop as seen through all the optics beyond the aperture stop. It also can be a real or virtual image, depending on the location of the aperture stop.

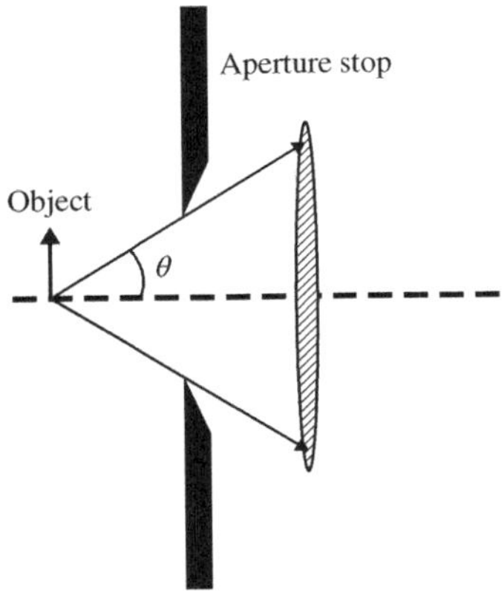

FIGURE 3.15 Depiction of an aperture stop.

3.8.2 Entrance and Exit Pupils

A pupil may, or may not, be a physical object. Figure 3.16 illustrates the entrance and exit apertures for a notional multi-element optical system. The *Entrance and Exit Pupils* are the images of the aperture stop in the object space and image space, respectively. Let's be more specific.

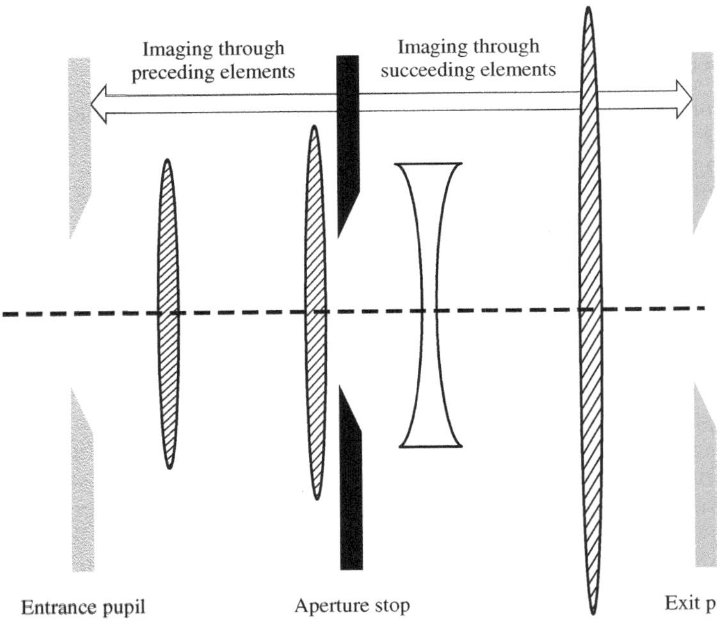

FIGURE 3.16 Entrance and Exit Pupil locations in a notional multielement optical system.

The entrance pupil is the opening an observer would identify as the limitation on the solid angle of rays diverging from an on-axis object point, i.e., it is the image of the aperture stop as seen through all the optics before the aperture stop. Again, it can be a real or virtual image, depending on the location of the aperture stop.

On the other hand, the exit pupil is the image of the aperture stop formed by the light rays after they have passed through the optical system, i.e. it is the image of the aperture stop as seen through all the optics beyond the aperture stop. It also can be a real or virtual image, depending on the location of the aperture stop.

3.8.3 Field Stop and Chief and Marginal Rays

Figure 3.17 illustrates a different notional multielement optical imaging systems with aperture and field stops included. As before, the Entrance and Exit Pupils are shown, which are images of the aperture stop in the object space and image space, respectively. Unlike Figure 3.16, the Entrance Pupil is located to the right of the aperture stop, not to the left. This figure introduces the concept of a Field Stop (FS), which defines the FOV of the optical system.

The Marginal Ray (MR) is a ray that runs from an object point on optical axis that passes at edge of the Entrance Pupil. It also passes at the edge of the Exit Pupil. Conventionally, this ray is in the y–z-plane, usually called the meridian plane. This ray is useful, because it crosses the optical axis again at the locations where an image will be formed. The distance of the marginal ray from the optical axis at the locations

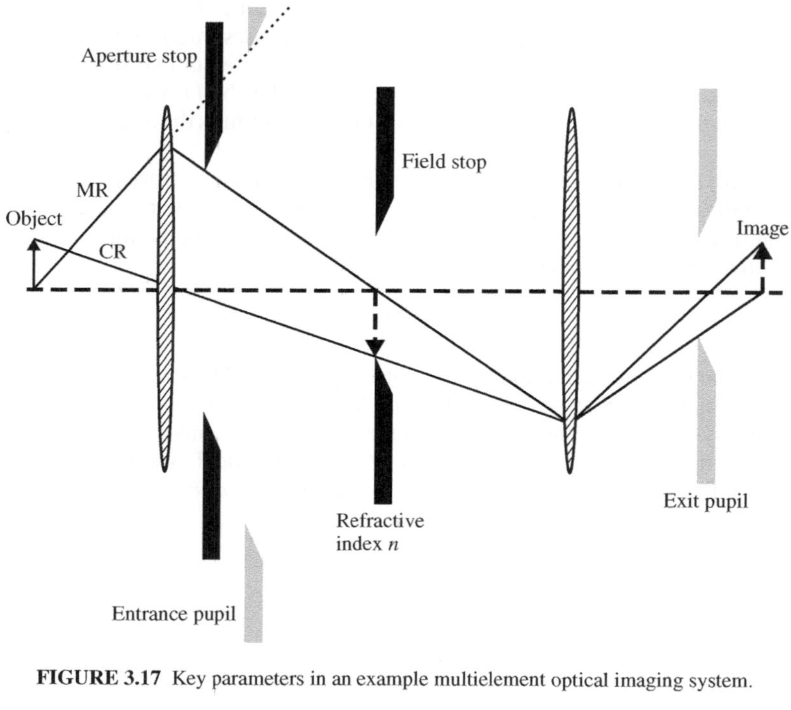

FIGURE 3.17 Key parameters in an example multielement optical imaging system.

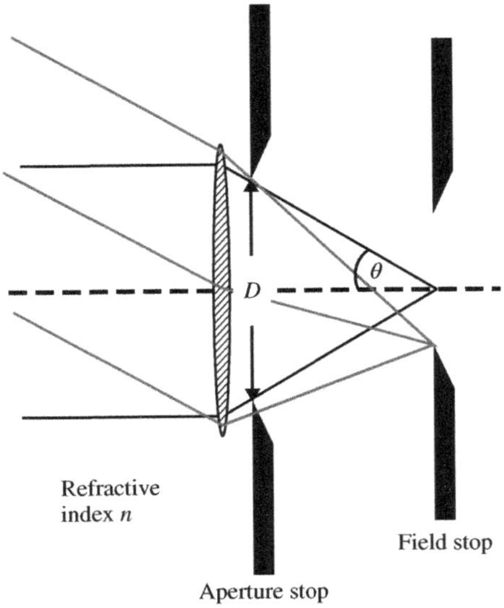

FIGURE 3.18 Key parameters in an example multielement optical imaging system.

of the Entrance and Exit Pupils defines the sizes of each pupil (since the pupils are images of the aperture stop).

The Chief Ray (CR) is a ray that runs from an object point at the edge of field, passing through the center of the aperture stop. It also goes through the center of the Entrance and Exit Pupils. Although there can be an infinite number of such rays, we can usually assume, at least for centered systems, that the chief ray also is restricted to the meridian plane. From Figure 3.17, we see that the FS limits the angular acceptance of the CRs.

Figure 3.18 shows the left portion of Figure 3.17, which comprises the first lens, and the aperture and field stops. In this figure, θ is the half-angle of the cone of light rays entering the system from the axial point of an object. The Numerical Aperture (NA) of the optical system equals

$$\mathrm{NA} = n \sin\theta \tag{3.47}$$

where n is the index of refraction for the object space. The NA specifies which angles of the Marginal Rays that just make it through the edges of the aperture stop.

When a system is designed to work at large distances, a convenient measure of its light-gathering ability is the f − number $(f/\#)$ of the optical system, defined as the ratio of the focal length to the diameter of system's *clear aperture*; mathematically, we write

$$f - \text{number} = \frac{f}{D}, \tag{3.48}$$

where f is the focal length of the optical system and D the clear aperture diameter [4, p. 186].

For aplanatic systems (where coma and spherical aberrations are corrected) with an infinite object distance, we find that the f − number is related to NA via the expression [3, p. 144]

$$f - \text{number} = \frac{1}{2\,\text{NA}}. \tag{3.49}$$

The f − number is associated with the *speed* of the optical system [3, p. 144]. A lens with a large clear aperture has a small f − number and the associated optical system has a "fast" or "high" speed. On the other hand, a lens with a small clear aperture has a small f − number and is said to be a "slow" or "low" speed. These terms come from photography where a large aperture allows a short (or "fast") exposure time to obtain enough energy to expose the photographic film. A small aperture requires long (or "slow") exposure time for the same amount of optical energy. In general, fast optics (large NA) are compact, but possess tight tolerances and are hard to manufacture. Slow optics (small NA) are just the opposite.

3.8.4 Entrance and Exit Windows

A *window* is an image of the field stop (the image area), viewed from a point on the optical axis at either end of the optical imaging systems.

The Entrance Window is the virtual image of the system's field stop, illuminated from the image space and viewed from the object end of the imaging system. Figure 3.19 shows the notional relationship among the Entrance and Exit Pupils and Entrance and Exit Windows. This window is the image that subtends the smallest angle φ at the center of the Entrance Pupil, which defines the system's FOV.

The Exit Window is the virtual image of the system's field stop, illuminated from the object space and viewed from the observer end of the imaging system. It defines the image field angle φ', that is, the acceptance angle of the ranges on the image space side of the optical system.

Example 3.7 Find the locations of the Entrance and Exit Pupils and Windows for the single-lens system shown in Figure 3.20.

From the above, we see that $z < f$ and $s' > f$. We have one physical aperture to the right of lens of diameter a at a distance z from said lens, and one physical aperture of diameter w' to the left at a distance s' from the lens. The angular acceptance with regard to an axial point object is limited by the left aperture. It is clear that there are no optical elements to the left of the left aperture, so that aperture is the Entrance Pupil.

The location of the Exit Pupil z' can be found by solving the following equation:

$$\frac{1}{z'} + \frac{1}{z} = \frac{1}{f},$$

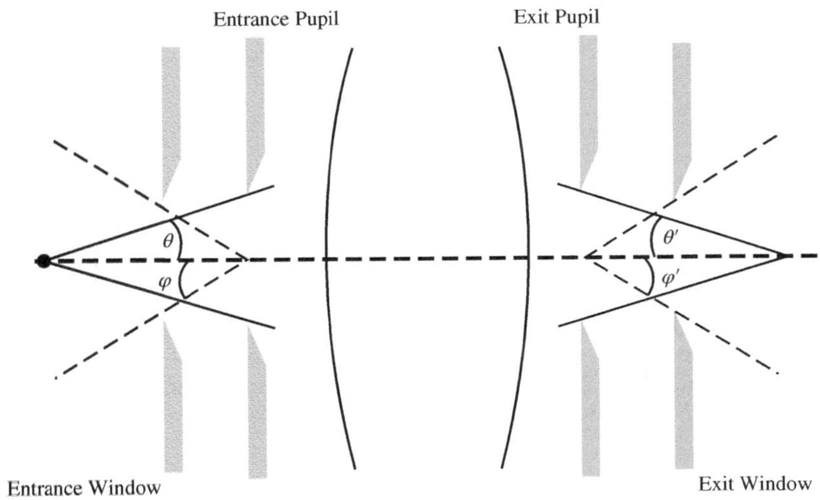

FIGURE 3.19 Entrance and Exit Window locations in a notional multielement optical system.

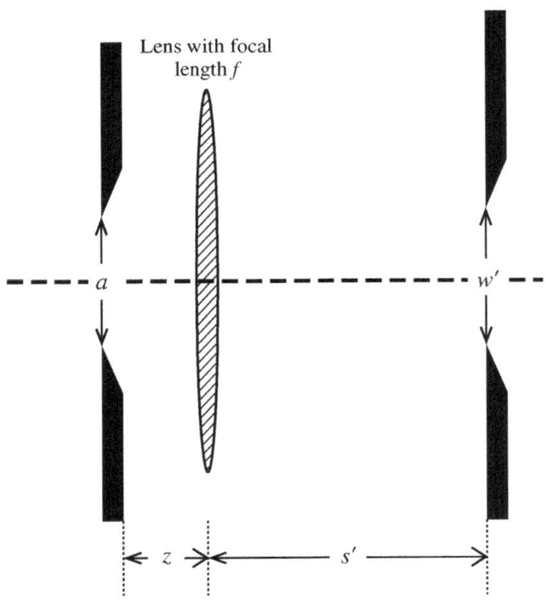

FIGURE 3.20 Layout of a single-lens imaging system.

which implies that

$$z' = \frac{zf}{z-f}\ (< 0).$$

The Exit Pupil size is given by

$$a' = -\frac{z'}{z}\, a = \frac{af}{z-f}.$$

The MR cutoff with regard to an off-axis point object is limited by w', which means that w' is the field stop. Referring to Figure 3.20, we see that there are no optical elements to the right of the right aperture, so the right aperture also is the Exit Window. Following a similar procedure as above, we find that the Entrance Window location equals

$$s = \frac{s'f}{s'-f}(> 0).$$

(Note: The above location also is the location of the in-focus object.) The Entrance Window size is equal to

$$w = -\frac{s}{s'}w' = -\frac{w'f}{s'-f}.$$

3.8.5 Baffles

A *baffle* is an interior obstruction, constructed as part of a diaphragm or tube, that blocks stray light and internal reflections from entering the image area and introducing glare or reducing contrast in the image. Baffles do not affect the aperture, image dimensions, or appearance of aberrations. Because they are not stops, baffle edges are always just outside the circumference of the light cone concentrated by the objective in other words, the key to baffle layout is to arrange them so that no part of the detector, eye or image screen can "see" any part of it.

3.9 PROBLEMS

Problem 3.1. One end of a cylindrical glass rod of refractive index 1.5 is ground and polished to a hemispherical shape of radius $R = 20$ mm. A 1 mm high, arrow-shaped object is located in air, 80 mm to the left of the spherical end of the rod on the optical axis, that is, its base is on said axis and the arrowhead is 1 mm up from that axis. Find the position and magnification of the image.

Problem 3.2. Assume the details of Problem 3.1, except now that the arrow and rod are immersed in water, where the refractive index is 1.5 and not unity. Find the position and magnification of the image.

Problem 3.3. What type of mirror is required to form an image on a wall 3 m, from the mirror, of a filament of a headlight lamp 10 cm in front of the mirror? What is the height of the image if the object height is 5 mm?

HINT: The equation for a mirror is

$$\frac{1}{s_1} - \frac{1}{s_1'} = -\frac{2}{R}.$$

Problem 3.4. Find the focal length and the positions of the focal points and principle points of a single thick lens shown in the following figure. The index of the lens is 1.5, its axial thickness 25 mm, the radius of the first surface 22 mm, and that of the second surface 16 mm.

HINT: The equation for the focal length for a thick lens is given by

$$f = s_1'\left(-\frac{s_2'}{s_2}\right),$$

not Eq. (3.46); the former reduces to the latter in limit of a thin lens.

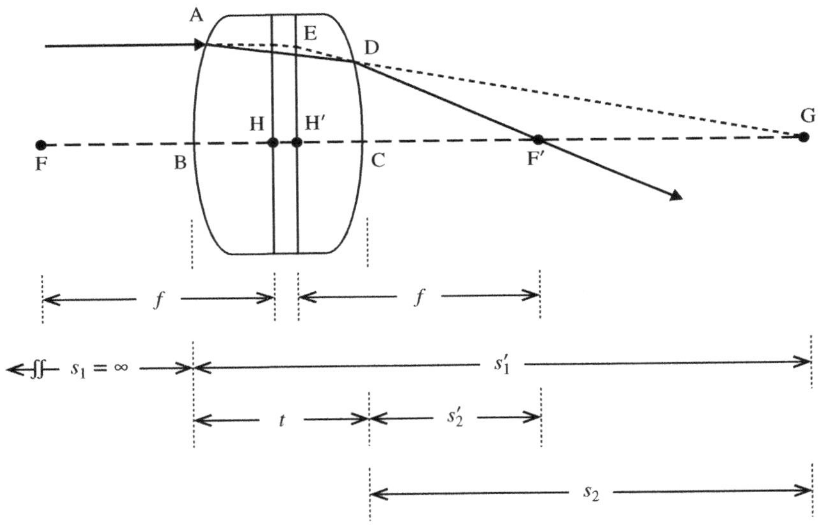

Problem 3.5. Ten-inch focal length lens forms an image of a telephone pole that is 200 ft away (from its first principal point). Where is the image located (a) with respect to the focal point of the lens and (b) with respect to the second principal point?

Problem 3.6. (a) Referring to Problem 3.5, how big is the image if the telephone pole is 50 ft high? (b) What is the magnification?

Problem 3.7. A 1-in. cube is 20 in. away from the first principal point of a negative lens with a negative 5-in. focal length. Where is the image and what are its height, width, and thickness?

Problem 3.8. Find the position and diameter of the Entrance and Exit Pupils of a 100-mm focal length lens with an aperture 20 mm to the right of the lens. Assume that the lens and aperture diameters are 15 and 10 mm, respectively.

Problem 3.9. What is the f – number of lens in Problem 3.5 with light (a) from the left and (b) from the right?

REFERENCES

1. Kasunic, K.J. (2011) *Optical Systems Engineering*, McGraw Hill Professional, Technology & Engineering, 448 pages.
2. Fischer, R., Tadic-Galeb, B., and Yoder, P.R. (2008) *Optical System Design*, 2nd edn, McGraw-Hill Companies, New York, NY.
3. Smith, W.J. (2008) *Modern Optical Engineering; The Design of Optical Systems*, 4th edn, SPIE Press, Bellingham, WA.
4. Born, M. and Wolf, E. (1999) *Principles of Optics: Electromagnetic Theory of Propagation, Interference and Diffraction of Light*, Cambridge University Press, London.
5. Sears, F.W. (1958) *Optics*, 3rd, Principles of Physics Series edn, Addison-Wesley Publishing Company, Reading, MA.

4

RADIOMETRY

4.1 INTRODUCTION

Radiometry characterizes the distribution of the optical power or energy in space, as opposed to *photometry*, which describes the optical radiation's interaction with the human eye [1–6]. In practice, the term usually refers to the measurement of infrared, visible, and ultraviolet light using optical instruments, generally known as *radiometers*.[1] The optical portion of the spectrum covers the five-decade frequency range from 3×10^{11} to 3×10^{16} Hz, corresponding to the wavelength range from 10 nm to 1000 μm, as shown in Figure 4.1 [7]. It is distinct from quantum processes such as photon counting because the theory uses "ray tracing" as its means for depicting optical radiation transfer one point to another, ignoring the dual nature of light [8]. Finally, radiometry assumes incoherent radiation and is sometimes supplemented with diffraction analysis at apertures, which we covered in Chapter 2, and polarization of light (typically with Stokes parameters along each ray path) [9]. Although an important area once, photometry is not discussed here because most optical instruments are not limited to just the visible response of the eye. The reader can consult books by McCluney [1] and Smith [3], for example, to obtain information on this area.

There are two aspects of radiometry: theory and practice. The practice involves the scientific instruments and materials used in measuring light, including bolometers, photodiodes, photomultiplier tubes, focal plane arrays, and a plethora of others. This area is ever changing and is tough for an introductory textbook to stay current.

[1]The use of radiometers to determine the temperature of objects and gases by measuring radiation flux is called *pyrometry*. Handheld pyrometer devices are often marketed as infrared thermometers.

Free Space Optical Systems Engineering: Design and Analysis, First Edition. Larry B. Stotts.

Companion website: www.wiley.com\go\stotts\free_space_optical_systems_engineering

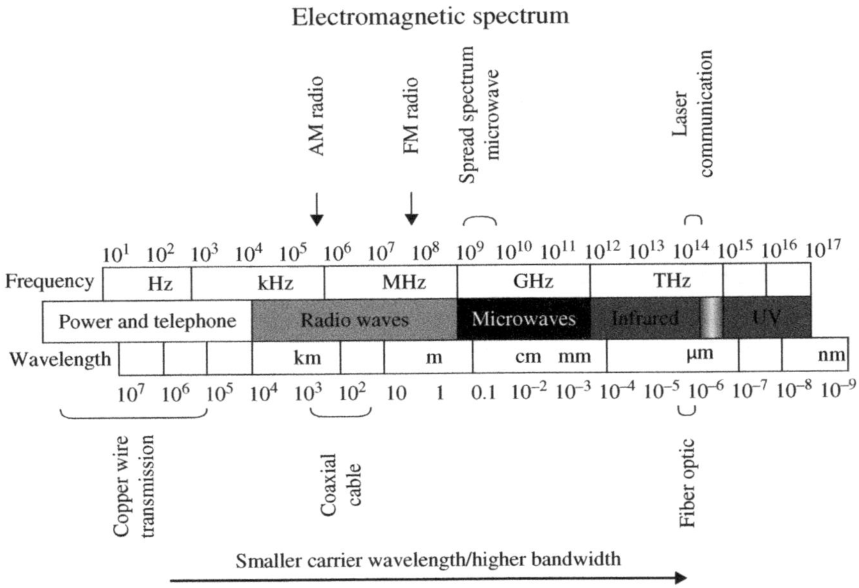

FIGURE 4.1 The electromagnetic frequency spectrum. Source: Alkholidi, 2014 [7]. Used under http://www.intechopen.com/books/contemporary-issues-in-wireless-communications/free-space-optical-communications-theory-and-practices.

The result is that we only focus on the underlining theory and leave the practice to the reader to research independently. The definition of the key characteristics such as radiant energy and reflectance in this chapter follows Forsyth [4]. We begin with the basic geometrical definitions necessary to better understand radiometry and the other material to come.

4.2 BASIC GEOMETRICAL DEFINITIONS

Let us begin with the definition of solid angle. *Solid angle* (symbol: Ω) is the two-dimensional angle in three-dimensional space that an object subtends at a point. It is a measure of how large the object appears to an observer looking from that point. In the International System of Units (SI), a solid angle is expressed in a dimensionless unit called a *steradian* (symbol: sr). Figure 4.2a and b depicts "solid angle" in two-dimensional (planar) and three-dimensional (spherical) geometries, respectively. In Figure 4.2a, we see that the angle θ in radians equals the ratio of arc length s to radius R. In Figure 4.2b, the solid angle Ω equals the ratio of surface area on a circle A to the radius-squared, R^2. Mathematically, we have

$$\theta = \frac{s}{R} \tag{4.1}$$

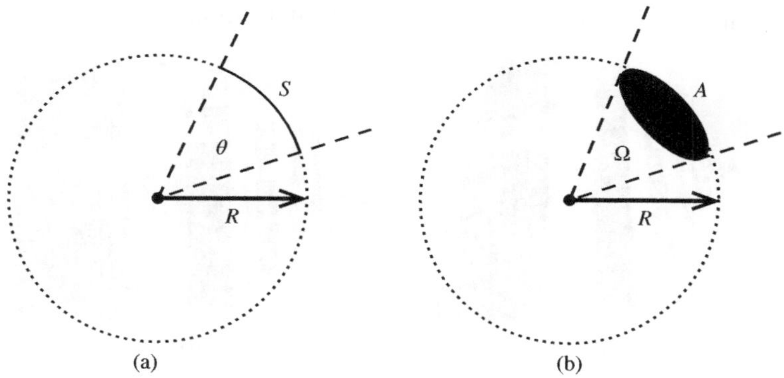

FIGURE 4.2 (a) Planar and (b) spherical geometrical view of solid angle.

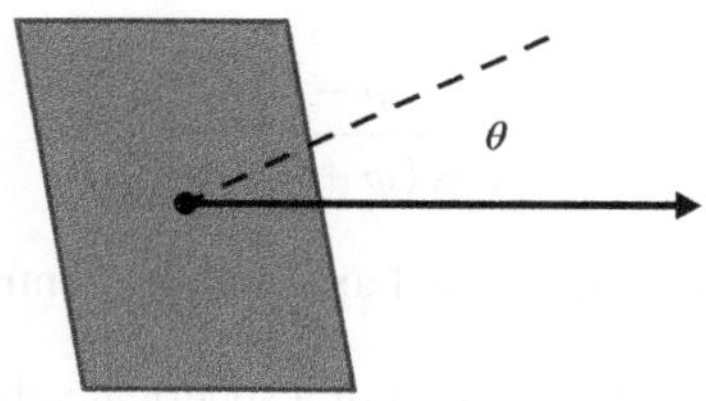

FIGURE 4.3 Depiction of projected area.

and

$$\Omega = \frac{A}{R^2}, \tag{4.2}$$

respectively, for the two definitions of solid angle.

The geometrical definition of *projected area* is "the rectilinear parallel projection of a surface of any shape onto a plane." Figure 4.3 illustrates this concept visually. This translates into the equation

$$A_{\text{proj}} = \int_A \cos\theta \, dA \tag{4.3}$$

with A_{proj} being the projected area, A the original area, and θ the angle between the normal to the surface A and the normal to the arbitrary plane onto which we project. In most practical applications, we have $A_{\text{proj}} \cong A\cos\theta$.

Example 4.1 An example of projected area is the keystone effect, also known as the tombstone effect. Figure 4.4 illustrates this effect and its correction. It is caused by attempting to image a scene onto focal plane pixels at an angle relative to the camera's optical axis; it is a distortion of the image dimensions, making it look like a trapezoid. The distortion (on a two-dimensional model and for small focus angles)

FIGURE 4.4 Example of the (a) keystone effect and (b) its correction. Source: https://en.wikipedia.org/wiki/File:Vertical-keystone.jpg.

is best approximated by

$$\frac{\cos\left(\varphi - \frac{\vartheta}{2}\right)}{\cos\left(\varphi + \frac{\vartheta}{2}\right)}, \tag{4.4}$$

where φ is the angle between the optical axis and the central ray from the projector and ϑ the width of the focus.

We now need to look at the projection of a spherical solid angle onto flat surface, that is, the flat projection of a spherical surface area divided by distance squared. Using the definitions given above, we can write this projection as

$$d\Omega = \frac{dA_{\text{proj}}}{R^2} \cong \frac{dA\cos\theta}{R^2} = d\omega\cos\theta, \tag{4.5}$$

where

$$d\omega = \frac{dA}{R^2}. \tag{4.6}$$

Example 4.2 Let us calculate the solid angle for a circular cone, hemisphere, and a sphere. Let us assume that θ_c is the half angle for the cone and φ is the azimuthal angle about that point shown in Figure 4.5. The solid angle for a circular cone is then

$$\omega_{\text{cone}} = \int_0^{2\pi}\int_0^{\theta_c} \sin\theta\, d\theta\, d\varphi = 2\pi\,(-\cos\theta)|_0^{\theta_c} = 2\pi\left[1 - \cos\theta_c\right]. \tag{4.7}$$

For a hemisphere, $\theta_c = \frac{\pi}{2}$ and $\omega_{\text{hemisphere}} = 2\pi$. For a sphere, $\theta_c = \pi$ and $\omega_{\text{sphere}} = 4\pi$.

To find the projected solid angle of a circular cone of half angle θ_c, we integrate Eq. (4.5) over the integration limits as above, while projecting each elemental area with the weighting of $\cos\theta$:

$$\omega_{\text{proj}} = \int_0^{2\pi}\int_0^{\theta_c} \cos\theta\sin\theta\, d\theta\, d\varphi = 2\pi\left(\frac{\sin^2\theta}{2}\right)\Bigg|_0^{\theta_c} = \pi\sin^2\theta_c. \tag{4.8}$$

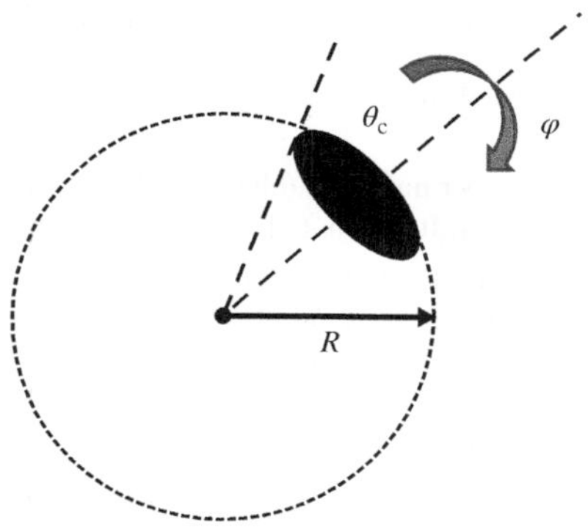

FIGURE 4.5 Integration geometry for a circular cone.

Again for hemisphere, $\theta_c = \frac{\pi}{2}$ and the projected solid angle is given by $\omega_{\text{proj hemisphere}} = \pi$, which is half of the solid angle of circular cone. For small areas of radius r_{cone}, we have

$$\omega_{\text{proj}} = \pi \sin^2\theta_c \approx \pi\theta_c{}^2 = \pi \left[\frac{r_{\text{cone}}{}^2}{R^2}\right] = \frac{A_{\text{cone}}}{R^2}. \tag{4.9}$$

4.3 RADIOMETRIC PARAMETERS

Radiometry is composed of the language, mathematics, and instrumentation for characterizing the propagation and effects of energy interacting with various media. It includes the refraction, reflection, absorption, scattering, and transmission by various substances in any of their possible phases: gaseous, liquid, or solid. To understand, one needs to know the language of radiometry, which means understand the definition of the key entities used to describe the above phenomena. This and the following section will try to provide that education to the reader. Table 4.1 depicts the fundamental radiometric quantities used in optics. All of these entities have a spectral and time dependence. Example 4.3 helps the reader get oriented to these quantities and how they fit in the hierarchy of radiometry. We define these entities in more detail in the sections to come.

Example 4.3 For a point source, the optical power appears to obviously emulate from a single point in space. Its *radiant energy* E_e is radiating uniformly in all angular directions. If the rate at which the radiant energy is emitted equals the *radiant power*, P_e, then the source has a radiant intensity of J_e equating to $\frac{P_e}{4\pi}$ watts per steradian because the volume the optical energy is being emitted into is a sphere.[2]

[2] Although no point sources truly exist, any optical source considered very small in extent compared to the distance between it and the observation plane can be treated a "quasi-point source" for all practical purposes. In this case, the radiant intensity emitted in the directions the source does radiate can be expressed approximately in watts per steradian.

TABLE 4.1 List of Fundamental Radiometric Quantities and Their Standard Notation

- *Radiant Energy* is *energy* carried from any electromagnetic field. It is denoted by E_e. Its SI unit is *joule* (J).
- *Radiant Flux* is radiant energy per unit time (also called *Radiant Power*); it is considered the fundamental radiometric unit. It is denoted by P_e. Its SI unit is *watt* (W).
- *Radiant Intensity* is radiant flux emitted from a point source per unit *solid angle*. It is denoted by J_e. Its SI unit is *watt per steradian* (W/sr).
- *Irradiance* is radiant flux emitted from an extended source per unit *projected source area*. It is denoted by H_e. Its SI unit is *watt per square meter* (W/m^2).
- *Radiance* is radiant flux emitted from an extended source per unit *solid angle* and per unit *projected source area*. It is denoted by N_e. Its SI unit is *watt per steradian per square meter* (W/sr m^2).
- *Spectral* in front of any of these quantities implies the same SI unit per *unit wavelength*. Typically, the spectral quantities will be proportional to inverse microns (μm).
  - For example, spectral radiances would be in watt per steradian per square meter per micron (W/sr m^2 μm) and spectral radiant energy is the energy radiated per unit wavelength micron (J/μm).

Let us now turn the problem around. Let us assume that we have a surface located at a distance R from the point source. An elemental area dA on this surface will subtend a solid angle from the source equal to Eq. (4.6):

$$d\omega = \frac{dA}{R^2}$$

oriented at the point where the normal to the surface from the source intersects the surface if R is very large. The *irradiance* H_e on the surface is the radiant power per unit area, which is the product of the radiant intensity for the source and its solid angle, divided by the unit area. Mathematically, we write

$$H_e = J_e \frac{d\omega}{dA} = \left(\frac{P_e}{4\pi}\right)\frac{1}{R^2} = \frac{P_e}{4\pi R^2}. \tag{4.10}$$

Equation (4.10) is known as the *inverse square law*. It says that the irradiance on a surface is inversely proportional to the square of the distance from the surface to the point source.

Example 4.4 Let us assume that we have a uniformly radiating source emitting a radiant power equal to 10 W. The radiant intensity then is given by

$$J_e = \frac{P_e}{4\pi} = \frac{10\ \mathrm{W}}{12.57\ \mathrm{sr}} = 0.796\ \mathrm{W/sr}.$$

The irradiance hitting the normal to a surface 100 cm away is then

$$H_e = \frac{P_e}{4\pi R^2} = \frac{0.796\ \mathrm{W/sr}}{(100\ \mathrm{cm})^2} = 0.796 \times 10^{-4}\ \mathrm{W}/\left(\mathrm{cm}^2\ \mathrm{sr}\right) \approx 80\ \mu\mathrm{W}/\left(\mathrm{cm}^2\ \mathrm{sr}\right).$$

For points off this normal, this irradiance will be reduced by a factor of $\cos^3\theta$, where θ is the angle between this point makes with the surface normal. This point is discussed later in this chapter.

Example 4.5 Let us look at the amount of power generated by the Sun. The Sun is located at a distance, $R_{\text{sun}} \approx 1.5 \times 10^{11}$ m (1 a.u.) from the Earth. The irradiance of the Sun outside the Earth's atmosphere, $H_{\text{sun}} \approx 1367\ \text{W/m}^2$ (Solar Constant). The Earth has a cross-sectional area, $A_{\text{earth}} \approx \pi(6350\ \text{km})^2 \approx 1.27 \times 10^{14}\ \text{m}^2$. This implies that the solar power hitting the Earth by the Sun equals

$$H_{\text{sun}}A_{\text{sun}} \approx \left(1367\ \text{W/m}^2\right)\left(1.27 \times 10^{14}\ \text{m}^2\right) = 1.783 \times 10^{17}\ \text{W}.$$

The solid angle subtended by the Earth relative to the Sun is given by

$$\Omega_{\text{earth–sun}} = \frac{A_{\text{earth}}}{R_{\text{sun}}{}^2} = \frac{\left(1.27 \times 10^{14}\ \text{m}^2\right)}{\left(1.5 \times 10^{11}\ \text{m}\right)^2} \approx 5.6 \times 10^{-9}\ \text{sr}.$$

If we assume that the Sun is the "point source," then the fractional solid angle the Earth subtends relative to the 4π sr emission pattern of the Sun is equal to

$$\frac{\Omega_{\text{earth–sun}}}{4\pi\ \text{sr}} \approx \frac{5.6 \times 10^{-9}\ \text{sr}}{4\pi\ \text{sr}} = 4.46 \times 10^{-10}$$

or around 4.45×10^{-8}% of the energy radiated by the Sun. This means that the total radiant power of the Sun equals

$$P_{\text{sun}} \approx \frac{4\pi\ H_{\text{sun}}A_{\text{earth}}}{\Omega_{\text{earth–sun}}} \approx 4.0 \times 10^{26}\ \text{W}.$$

In reality, light is radiant energy. When light is absorbed by a physical object, its energy is converted into some other form. For example, visible light causes an electric current to flow in a photographic light meter when its radiant energy is transformed into the kinetic energy by the emission of a photoelectron. There are two broad classes of radiant energy:

1. A broadband optical source such as the sun emits most of its radiant energy within the visible portion of the spectrum. Radiant energy (denoted as E_e) in this case is measured in joules.
2. A single-wavelength laser is a monochromatic source; all of its radiant energy is emitted at one specific wavelength. Spectral radiant energy (denoted as $E_e(\lambda)$) is measured in joules per unit wavelength, for example, joules per nanometer (J/nm) or joules per micron (J/μm).

4.3.1 Radiant Flux (Radiant Power)

Energy per unit time is power, which we measure in joules per second, or watts. A laser beam, for example, has so many milliwatts or watts of radiant power. As we saw in Chapter 2, light propagates through space, which can be considered as "flowing" through space. This concept leads to radiant power being more commonly referred to as the "time rate of flow of radiant energy" or radiant flux. In general terms, radiant flux or radiant power is the radiant energy emitted, reflected, transmitted, or received per unit time and spectral flux or spectral power is the radiant flux per unit wavelength.

In terms of a photographic light meter measuring visible light, the instantaneous magnitude of the electric photocurrent is directly proportional to the radiant flux. The total amount of current measured over a period of time is directly proportional to the radiant energy absorbed by the light meter during that time. This is how a photographic flash meter works – it measures the total amount of radiant energy received from a camera flash.

4.3.2 Radiant Intensity

Radiant intensity is the radiant flux emitted, reflected, transmitted, or received per unit solid angle and spectral intensity is the radiant intensity per unit wavelength. These are directional quantities. These entities also are distinct from irradiance and radiant exitance. This is because an infinitesimally small point source of light emits radiant flux in every direction, but has no spatial extent associated with that emission. Since we now are only talking about radiant per unit solid angle only, we cannot use irradiance and must have a special term to describe this situation. The amount of radiant flux emitted in a given direction can be represented by a ray of light contained in an elemental cone. This gives us the definition of radiant intensity.

4.3.3 Radiance

Radiance and irradiance/radiant exitance are related, so we start with the former first. *Radiance* is defined as "the amount of energy traveling at some point in a specified direction, per unit time, per unit area perpendicular to the direction of travel, per unit solid angle [10]." That is, it comes from an extended (nonpoint) source and is a function of wavelength, position, and direction. It can come from an elemental surface area, or from an elemental volume of hot gases or particulates illuminated by some optical source. Mathematically, the radiance at a point in space is usually denoted $N_e(\boldsymbol{r}, t; \theta, \varphi; \lambda)$, where $\boldsymbol{r} = (x, y, z)$ is a position coordinate in three-space or a point on surface, t represents time, (θ, φ) is a direction specified by a certain coordinate standard, and λ is the wavelength of light. In regard to formalizing the direction vector, one can use (θ, φ) coordinates that established the three coordinate axes for free space or by the surface normal, as illustrated in Figure 4.6, or by writing $\boldsymbol{r}_1 \rightarrow \boldsymbol{r}_2$, meaning in the direction from point $\boldsymbol{r}_1$ to $\boldsymbol{r}_2$.

Radiance has the highly desirable property that, for two points $\boldsymbol{p}_1$ and $\boldsymbol{p}_2$ that have an unoccluded line of sight between them, the radiance leaving $\boldsymbol{p}_1$ in the direction of $\boldsymbol{p}_2$ is the same as the radiance arriving at $\boldsymbol{p}_2$ from the direction of $\boldsymbol{p}_1$. This is a very important point and deserves a detailed look into why this is true.

Figure 4.7 shows a (lower) patch of surface radiating in a particular direction. If the radiance at the lower patch is $N_1\left(\boldsymbol{p}_1, t; \theta, \varphi; \lambda\right)$, then the energy transmitted by the patch into an infinitesimal region of solid angle $d\omega$ around the direction (θ, φ) in time dt is

$$N_1\left(\boldsymbol{p}_1, t; \theta, \varphi; \lambda\right) \cos\theta_1 dA_1 d\omega\, dt. \tag{4.11}$$

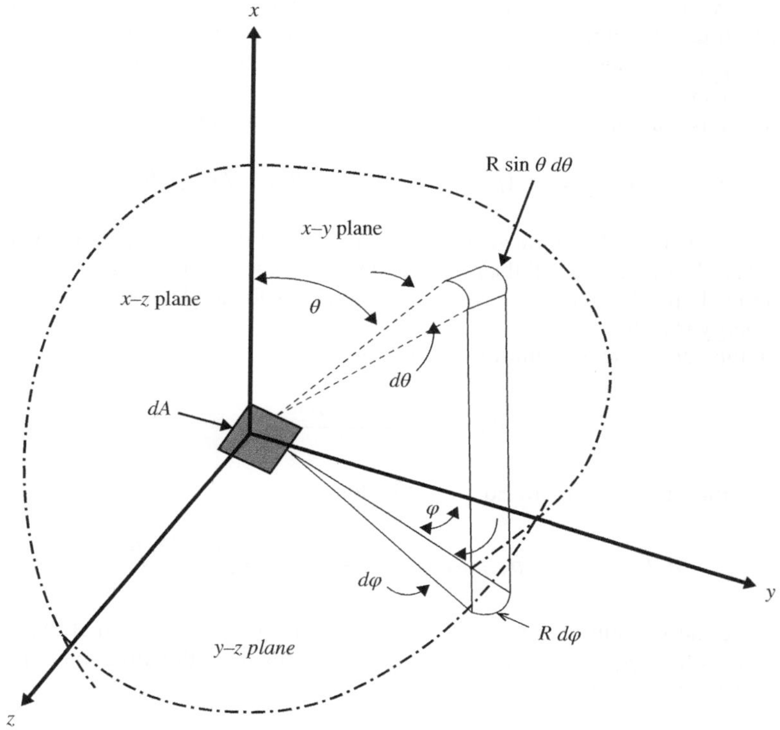

FIGURE 4.6 Coordinate systems for setting the direction vector (θ, φ).

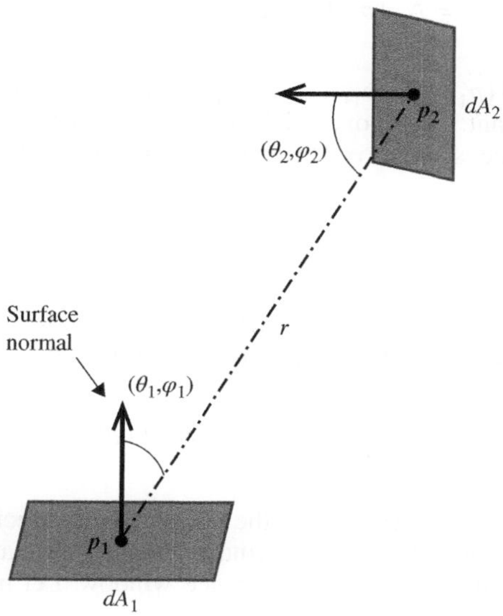

FIGURE 4.7 Energy transfer from point $\boldsymbol{p}_1$ to point $\boldsymbol{p}_2$.

Let us now look at the energy transfer from the lower patch to the upper patch illustrated in Figure 4.6. The radiance leaving $\boldsymbol{p}_1$ in the direction of $\boldsymbol{p}_2$ is $N_1\left(\boldsymbol{p}_1, t; \theta_1, \varphi_1; \lambda\right)$ and the radiance arriving at $\boldsymbol{p}_2$ from the direction of $\boldsymbol{p}_1$ is $N_2\left(\boldsymbol{p}_2, t; \theta_2, \varphi_2; \lambda\right)$.

This means that, in time dt, the energy leaving $\boldsymbol{p}_1$ toward $\boldsymbol{p}_2$ is

$$d^3E_{1\to 2} = N_1\left(\boldsymbol{p}_1, t; \theta_1, \varphi_1; \lambda\right) \cos\theta_1 dA_1 d\omega_{2\to 1}\, dt, \tag{4.12}$$

where $d\omega_{2\to 1}$ is the solid angle subtended by patch 2 at patch 1 (energy emitted into this solid angle arrives at 2; all the rest disappears into the void). The notation $d^3E_{1\to 2}$ implies that there are three infinitesimal terms – area, solid angle, and time – involved in this energy transfer.

From the geometrical definitions section, we know that

$$d\omega_{2\to 1} = \frac{\cos\theta_2 dA_2}{r^2}. \tag{4.13}$$

Substituting Eq. (4.12) into Eq. (4.11) yields

$$d^3E_{1\to 2} = N_1\left(\boldsymbol{p}_1, t; \theta_1, \varphi_1; \lambda\right) \frac{\cos\theta_1 \cos\theta_2 dA_1 dA_2\, dt}{r^2}. \tag{4.14}$$

Because the medium is a vacuum, it does not absorb energy, so that the energy arriving at $\boldsymbol{p}_2$ from $\boldsymbol{p}_1$ is the same as the energy leaving $\boldsymbol{p}_1$ in the direction of $\boldsymbol{p}_2$. The energy arriving at $\boldsymbol{p}_2$ from $\boldsymbol{p}_2$ is

$$d^3E_{2\to 1} = d^3E_{1\to 2} = N_2\left(\boldsymbol{p}_2, t; \theta_2, \varphi_2; \lambda\right) \cos\theta_2 dA_2 d\omega_{1\to 2}\, dt \tag{4.15}$$

$$= N_2\left(\boldsymbol{p}_2, t; \theta_2, \varphi_2; \lambda\right) \frac{\cos\theta_1 \cos\theta_2 dA_1 dA_2\, dt}{r^2}, \tag{4.16}$$

which means that $N_1\left(\boldsymbol{p}_1, t; \theta_1, \varphi_1; \lambda\right) = N_2\left(\boldsymbol{p}_2, t; \theta_2, \varphi_2; \lambda\right)$ and the radiance along straight lines is constant. The above is known as the *Radiance Theorem* and it basically says that radiance is *conserved* through a lossless optical system.

4.3.4 Étendue

Let us now look at the total power transmitted by a perfectly transmitting optical system (i.e., no vignetting, absorption, etc.).

The received power by an optical system is given by

$$P = \iint N\left(\boldsymbol{r}, \hat{\boldsymbol{n}}\right) \cos\theta dA\, d\Omega, \tag{4.17}$$

where $N\left(\boldsymbol{r}, \hat{\boldsymbol{n}}\right)$ is the source radiance of the point $\boldsymbol{r}$ in the direction of unit vector $\hat{\boldsymbol{n}}$. The surface integral is over the entrance window, and the solid angle integral extends over the solid angle subtended by the entrance window. Let us assume a constant source independent of view angle, that is, $N(\boldsymbol{r}, \hat{\boldsymbol{n}}) = N_0$. This implies that Eq. (4.17) equals

$$\begin{aligned} P &= \iint N_0 \cos\theta dA\, d\Omega \\ &= N_0 \iint \cos\theta dA\, d\Omega = \frac{N_0}{n_0^2}\mathcal{E}, \end{aligned} \tag{4.18}$$

where n_0 is the index of refraction of the medium in the Object Space and

$$\begin{aligned} \mathcal{E} &\equiv \text{étendue} \\ &= n_0^2 \iint \cos\theta dA\, d\Omega. \end{aligned} \tag{4.19}$$

Equation (4.18) says that the collected power is the product of $\mathcal{E}$ and the basic radiance of the source or the geometric *étendue* $\mathcal{E}$ is the product of the area of the source times the solid angle Ω of the system's entrance pupil, as seen from the source. It is a purely geometric quantity that is a measure of the flux gathering capability of the optical system. In other words, it represents the maximum beam size that optical imaging can accept. The étendue never increases in any optical system. At best, a perfect imaging system produces an image with the same étendue as the source.

Consider the optical system depicted in Figure 4.8. Suppose that A_1 represents the source area and is so small that the solid angle subtended by the Entrance Pupil of the optical system does not change over the source. We can write the étendue as

$$\mathcal{E} = n_1^2 \iint \cos\theta dA\, d\Omega = n_1^2 A_1 \int \cos\theta d\Omega \tag{4.20}$$

$$= n_1^2 A_1 \Omega_{\text{proj},1}, \tag{4.21}$$

where $\Omega_{\text{proj},1}$ is the projected solid angle from the source.

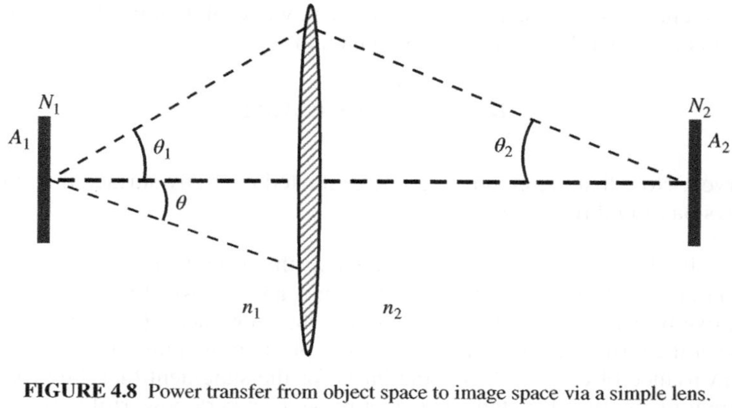

FIGURE 4.8 Power transfer from object space to image space via a simple lens.

Since the entrance pupil subtends the half angle θ_1, we find that

$$\Omega_{\text{proj},10} = \int_0^{2\pi} d\varphi \int_0^{\theta_1} \cos\theta \sin\theta d\theta \tag{4.22}$$

$$= \pi \sin^2\theta_1 \tag{4.23}$$

and the étendue is equal to

$$\mathcal{E} = \pi n_1^2 A_1 \sin^2\theta_1. \tag{4.24}$$

Now Abbe's Sine Condition (Problem 4.11) states that

$$N_1 h_1 \sin\theta_1 = N_2 h_2 \sin\theta_2, \tag{4.25}$$

where h_1 and h_2 are the heights of the object and image, respectively. Equation (4.25) says that $N\,h \sin\theta$ is conserved between the object and the image in a well-corrected imaging system. This implies that the étendue is invariant between the object and the image planes.

Example 4.6 Let us consider an element of the étendue given by

$$d^2\mathcal{E} = n^2 \cos\theta \, dA \, d\theta. \tag{4.26}$$

The amount of element power for the same situation equals

$$d^2P = N \cos\theta \, dA \, d\theta. \tag{4.27}$$

Taking the ratios of Eqs. (4.27) and (4.26) allows us to write

$$d^2P = \frac{N}{n^2} d^2\mathcal{E}. \tag{4.28}$$

Now in any lossless optical system, the elemental power is conserved because of conservation of energy. From the Radiance Theorem, we recall that the N/n^2 is invariant. These facts mean that the étendue is invariant and

$$\mathcal{E}\mathcal{E} = n^2 \iint \cos\theta dA \, d\Omega \tag{4.29}$$

is conserved. Note that the étendue can be evaluated over any surface that intersects all the rays passing through the system.

In some books and papers, it is also known as the *acceptance*, *throughput*, or *AΩ product*. For example, throughput and AΩ product are extensively used in radiometry and radiative transfer where they are related to the view factor (or shape factor). No matter what the term, étendue is a central concept in nonimaging optics.

The invariance of étendue forms the basis for the statement that the product $A\Omega$ is a constant in an optical system. This fact provides a very powerful tool for optical lens system design.

Example 4.7 Let us assume that a source of diameter d_1 is creating an image diameter d_2 by a simple lens with focal length f. Since $A\Omega \equiv$ constant, we have

$$A_1\Omega_1 = A_2\Omega_2$$

or

$$d_1\theta_1 \approx d_2\theta_2$$

under the paraxial approximation. From the relationship between numerical aperture the f − number found in Chapter 3, the above can be rewritten as

$$\frac{d_1}{f - \text{number}_1} \approx \frac{d_2}{f - \text{number}_2}.$$

4.3.5 Radiant Flux Density (Irradiance and Radiant Exitance)

Radiant flux density is the radiant flux per unit area at a point on a surface. There are two possible conditions. The flux can be arriving at the surface, in which case the radiant flux density is referred to as *irradiance*, or can be leaving, The flux can also be leaving the surface due to emission and/or reflection. The radiant flux density then is referred to as *radiant exitance*.

The irradiance is the radiant power received by a surface per unit area, and spectral irradiance is the irradiance of a surface per unit wavelength. A surface illuminated by radiance $N_i\left(\boldsymbol{r}, t; \theta_i, \varphi_i; \lambda\right)$ coming in from a differential region of solid angle $d\omega$ at angles $\left(\theta_i, \varphi_i\right)$ creates the following incremental irradiance:

$$dH_i\left(\boldsymbol{r}, t; \theta_i, \varphi_i; \lambda\right) = N_i\left(\boldsymbol{r}, t; \theta_i, \varphi_i; \lambda\right) \cos\theta_i \, d\omega, \tag{4.30}$$

where we have multiplied the incoming radiance by the projected area factor and by the solid angle to get irradiance. One would compute the entire irradiance at that point $H_i\left(\boldsymbol{r}, t; \theta_i, \varphi_i; \lambda\right)$ by integrating the above incremental irradiance over the whole input hemisphere.

Let us now turn to radiant exitance, denoted by the symbol W. In terms of emission, it refers to optical blackbody sources and we discuss them later in this chapter. In terms of reflective radiant exitance, we now need to ask: "What is the relationship between incoming illuminations and reflected light?" As one might expect, it will be a function of both the direction in which light arrives at a surface and the direction in which it leaves. For example, a mirror will reflect the light in only one direction while a piece of frost glass will disperse the light in many directions. To describe these phenomena, we need to introduce some additional parameters.

4.3.6 Bidirectional Reflectance Distribution Function

The most general entity describing local reflection is the bidirectional reflectance distribution function (BRDF). The BRDF is defined as "the ratio of the radiance in the outgoing direction to the incident irradiance [10]." This implies that if the surface cited above is to emit radiance $N_e\left(\boldsymbol{r}, t; \theta_o, \varphi_o; \lambda\right)$, its BRDF would be given by

$$\rho_{dH}\left(\boldsymbol{r}, t; \theta_o, \varphi_o; \theta_i, \varphi_i; \lambda\right) = \frac{N_o\left(\boldsymbol{r}, t; \theta_o, \varphi_o; \lambda\right)}{dH_i\left(\boldsymbol{r}, t; \theta_i, \varphi_i; \lambda\right)}. \tag{4.31}$$

The BRDF has units of inverse steradians (1/sr). The BRDF is symmetric in the incoming and outgoing direction, a fact known as the Helmholtz reciprocity principle.

We now use Eq. (4.17) to calculate the outgoing surface radiance that comes the total irradiance. Manipulating the equation and then integrating over the entire incoming solid angle, we find that

$$N_e\left(\boldsymbol{r},t;\theta_o,\varphi_o;\lambda\right) = \int_{\Omega} \rho_{dH}\left(\boldsymbol{r},t;\theta_o,\varphi_o;\theta_i,\varphi_i\,;\lambda\right) N_i\left(\boldsymbol{r},t;\theta_i,\varphi_i;\lambda\right)\cos\theta_i\,d\omega. \tag{4.32}$$

This implies that the BRDF is not an arbitrary symmetric function in four variables.

4.3.7 Directional Hemispheric Reflectance

The light reflected by many surfaces is largely independent of the exit angle. A natural measure of a surface's reflective properties in this situation is the directional hemispheric reflectance (DHR), which is defined as "the fraction of the incident irradiance in a given direction that is reflected by the surface, whatever the direction of reflection [6]." The DHR of a surface is obtained by integrating the radiance leaving the surface over all directions and dividing by the irradiance in the direction of illumination. Mathematically, we have

$$\begin{aligned}
\rho_{\mathrm{DHR}}\left(\boldsymbol{r},t;\theta_i,\varphi_i;\lambda\right) &= \frac{\iint_{\Omega} N_o\left(\boldsymbol{r},t;\theta_o,\varphi_o;\lambda\right)\cos\theta_o\,d\omega_o}{N_i\left(\boldsymbol{r},t;\theta_i,\varphi_i;\lambda\right)\cos\theta_i\,d\omega_i} \\
&= \iint_{\Omega}\left[\frac{N_o\left(\boldsymbol{r},t;\theta_o,\varphi_o;\lambda\right)\cos\theta_o}{N_i\left(\boldsymbol{r},t;\theta_i,\varphi_i;\lambda\right)\cos\theta_i\,d\omega_i}\right] d\omega_o \\
&= \iint_{\Omega} \rho_{dH}\left(\boldsymbol{r},t;\theta_o,\varphi_o;\theta_i,\varphi_i\,;\lambda\right)\cos\theta_o d\omega_o.
\end{aligned} \tag{4.33}$$

This parameter is dimensionless and goes between 0 and 1.

4.3.8 Specular Surfaces

Another important class of surfaces is the glossy or mirror-like surfaces, often known as specular surfaces. An ideal specular reflector behaves like an ideal mirror. What that means is once a normal to the surface at the point of incidence is set, the angle of reflection for any light ray equals its angle of incidence and there will be no energy loss upon reflection. However, one does not work with ideal specular reflectors. A nonideal specular reflector will have some fraction of incoming radiation absorbed, but the law of reflection still will apply. This means that the nonideal BRDF can be written as

$$\rho_{\mathrm{spec}}\left(\boldsymbol{r},t;\theta_o,\varphi_o;\theta_i,\varphi_i\,;\lambda\right) = \rho_s\left(\theta_i\right)\ \delta\left(\theta_o-\theta_i\right)\ \delta\left(\varphi_o-\varphi_i-\pi\right), \tag{4.34}$$

where $\rho_s\left(\theta_i\right)$ is the fraction of radiation that leaves the surface, or *surface albedo*. Dirt, surface roughness, and other surface effects can cause angular spreading off the surface, that is, specular lobes. Fortunately, this spreading is small compared to diffuse reflector and can be ignored for first-order analytic situation.

4.4 LAMBERTIAN SURFACES AND ALBEDO

For some surfaces the DHR does not depend on illumination direction. Examples of such surfaces include frosted glass, matte paper, and matte paints. This implication is that the BRDF is independent of the outgoing direction (and, by the reciprocity principle, of the incoming direction as well). This means the radiance leaving the surface is independent of angle. Such surfaces are known as ideal diffuse surfaces or Lambertian surfaces. Their DHR is often referred to as having *diffuse reflectance* or *albedo*. The DHR in this case is denoted by ρ_d. Since we know that $\rho_{dH}\left(\boldsymbol{r}, t; \theta_o, \varphi_o; \theta_i, \varphi_i\,; \lambda\right) = \rho \equiv$ constant, we can write

$$
\begin{aligned}
\rho_d &= \iint_\Omega \rho_{dH}\left(\boldsymbol{r}, t; \theta_o, \varphi_o; \theta_i, \varphi_i\,; \lambda\right) \cos\theta_o d\omega_o \\
&= \iint_\Omega \rho \cos\theta_o d\omega_o \\
&= \int_0^{2\pi} \int_0^{\pi/2} \rho \cos\theta_o \sin\theta_o d\theta_o d\varphi \\
&= \pi\rho.
\end{aligned} \tag{4.35}
$$

The BRDF therefore can be rewritten as

$$
\rho_{dH}\left(\boldsymbol{r}, t; \theta_o, \varphi_o; \theta_i, \varphi_i\,; \lambda\right) = \frac{\rho_d}{\pi}. \tag{4.36}
$$

Because our sensations of brightness essentially correspond to measurements of radiance, a Lambertian surface will look equally bright from any direction, whatever the direction along which it is illuminated. However, this source will exhibit a radiant intensity given by

$$
J_e^{\mathrm{Ex}}(\theta) = J_0 \cos\theta. \tag{4.37}
$$

Equation (4.37) is known as *Lambert's law*. The parameter $J_e^{\mathrm{Ex}}(\theta)$ represents radiant intensity of a small incremental area at an angle θ from the normal to the surface, and the term J_0 is the radiant intensity of a small incremental area in the direction of the said normal.

Example 4.8 Assume that we have a radiating disk of area $1\ \mathrm{cm}^2$ emitting $1\ \mathrm{W/(cm^{-2}\ sr)}$. The radiant intensity from this disk is equal to

$$
J_e = N_e A = \left(1\ \mathrm{W/(cm^{-2}\ sr)}\right)\left(1\ \mathrm{cm}^2\right) = 1\ \mathrm{W/sr}
$$

in a direction normal to the radiating surface. In the direction 45° to this normal, the radiance is reduced by $\cos 45 = 0.707$, which implies that $J_e\left(45^\circ\right) = 0.707\ \mathrm{W/sr}$.

4.5 SPECTRAL RADIANT EMITTANCE AND POWER

Let us calculate the *spectral radiant exitance* and total power radiated into a hemisphere from a Lambertian source of area A and radiance $N_{\text{Lambertian}}$. This is another important parameter in radiometry and is denoted by the symbol, W. Its units are in watts per meter-squared.

Figure 4.9 depicts the propagation geometry for light being emitted by said source. The spectral radiance, N_e, is related to the spectral radiant *exitance* by the relationship

$$W_e = \iint_{\Omega_{\text{hemisphere}}} N_e \cos\theta d\Omega, \tag{4.38}$$

where $d\Omega = \sin\theta\, d\theta\, d\varphi$. This implies that

$$W_{\text{Lambertian}} = \int_0^{2\pi} \int_0^{\frac{\pi}{2}} N_{\text{Lambertian}} \cos\theta \ \sin\theta \ d\theta d\varphi \tag{4.39}$$

$$= 2\pi \int_0^{\frac{\pi}{2}} N_{\text{Lambertian}} \cos\theta \ \sin\theta d\theta$$

$$= 2\pi\, N_{\text{Lambertian}} \left. \frac{\sin^2\theta}{2} \right|_0^{\frac{\pi}{2}} = \pi\, N_{\text{Lambertian}}. \tag{4.40}$$

Thus, the spectral radiant exitance and spectral radiance are related by a constant for Lambertian sources.

Let us now turn to the total power projected into a hemisphere. The intensity for this source equals

$$J_{\text{Lambertian}} = J_0 \cos\theta = N_{\text{Lambertian}} A \cos\theta.$$

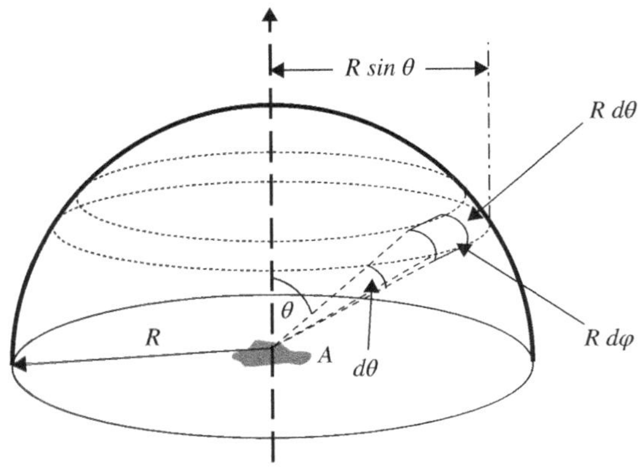

FIGURE 4.9 Geometry of a Lambertian source emitting into a hemisphere.

The incremental ring area between the two dotted rings can easily be shown to be equal to

$$2\pi\, R^2 \sin\theta\, d\theta \tag{4.41}$$

and subtends a solid angle given by

$$\frac{2\pi\, R^2 \sin\theta\, d\theta\, d\varphi}{R^2} = 2\pi \sin\theta\, d\theta. \tag{4.42}$$

The light captured by this ring equals

$$dP_{\text{Lambertian}} = 2\pi\, J_0\, R^2 \sin\theta\, d\theta = 2\pi\, N_{\text{Lambertian}} A \cos\theta \sin\theta d\theta. \tag{4.43}$$

Integrating Eq. (4.43) over the angular range 0 to $\frac{\pi}{2}$ yields the total received power in the hemisphere, or

$$P_{\text{Lambertian}} = 2\pi A \int_0^{\frac{\pi}{2}} N_{\text{Lambertian}} \cos\theta\ \sin\theta\, d\theta \tag{4.44}$$

$$= 2\pi N_{\text{Lambertian}} A\ \frac{\sin^2\theta}{2}\bigg|_0^{\frac{\pi}{2}} = \pi\, N_{\text{Lambertian}} A$$

$$= \pi\, N_{\text{Lambertian}} A \tag{4.45a}$$

$$= W_{\text{Lambertian}} A. \tag{4.45b}$$

One might expect that since the solid angle is 2π, the received power should be twice the above value; however, the cosine-weight of the intensity decreases the effective radiated power into the hemisphere, resulting in the above values.

4.6 IRRADIANCE FROM A LAMBERTIAN SOURCE

Let us now look at the irradiance created by a Lambertian source at a point on the optical axis.

Let the Lambertian source be positioned normal to the z axis with radiance $N_{\text{Lambertian}}$ and area A. Define $\boldsymbol{x}_0 = (0,\ 0,\ S)$ be a point located a distance S from the source on the normal axis. Figure 4.9 shows this situation graphically. The radiant intensity of a small element of area dA in the direction of $\boldsymbol{x}_0$ is equal to

$$dJ_{\text{Lambertian}} = N_{\text{Lambertian}} dA \cos\theta.$$

Since the distance from dA to $\boldsymbol{x}_0$ is $S \sec\theta$, the incremental irradiance at $\boldsymbol{x}_0$ is

$$dH_{\text{Lambertian}} = N_{\text{Lambertian}}\, dA \cos\theta\ \left[\frac{\cos^3\theta}{S^2}\right] = \frac{N_{\text{Lambertian}}\, dA \cos^4\theta}{S^2} \tag{4.46}$$

$$= \frac{N_{\text{Lambertian}}\ 2\pi r\, dr \cos^4\theta}{S^2} \tag{4.47}$$

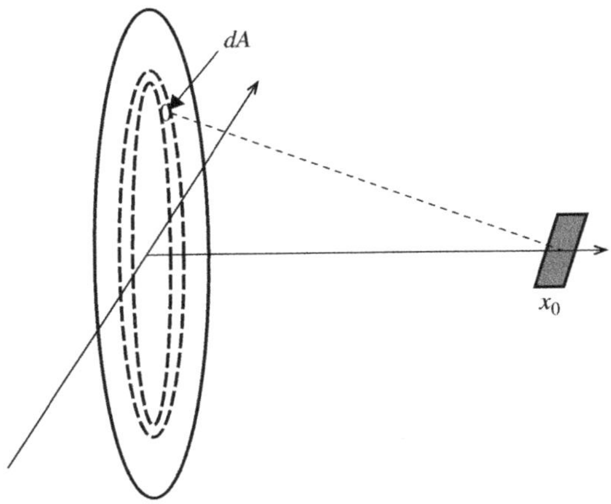

FIGURE 4.10 Elemental are illuminating a point $\boldsymbol{x}_0$.

because we obtain the same radiance from each incremental area of the ring depicted in Figure 4.10, which allows us to substitute $2\pi r\, dr$ for dA. To simplify the integration of the above, we note that

$$r = S \tan\theta$$

and

$$dr = S \sec^2\theta\, d\theta.$$

This implies that

$$\begin{aligned} dH_{\text{Lambertian}} &= \frac{N_{\text{Lambertian}}\, 2\pi\, S \tan\theta S \sec^2\theta\, d\theta \cos^4\theta}{S^2} \\ &= 2\pi\, N_{\text{Lambertian}} \tan\theta \cos^2\theta\, d\theta \\ &= 2\pi N_{\text{Lambertian}} \sin\theta \cos\theta\, d\theta. \end{aligned} \tag{4.48}$$

Integrating this elemental irradiance over the entire source area yields

$$H_{\text{Lambertian}} = \pi\, N_{\text{Lambertian}} \sin^2\theta_{\text{source}},$$

where θ_{source} is the source half-angle relative to the point $\boldsymbol{x}_0$.

If the point $\boldsymbol{x}_0$ does not lie on the optical axis like Figure 4.9, then the irradiance is subject to the cosine to the fourth law. Specifically, if the off-axis point $\boldsymbol{x}_1$ in the observation plane makes an angle from the optical axis equal to, say $\theta_{\text{off-axis}}$, the irradiance at that new point is given by

$$H_{\text{off-axis}} = H_{\text{Lambertian}} \cos^4\theta_{\text{off-axis}}. \tag{4.49}$$

Example 4.9 The *cosine to the fourth law* is a very important effect in the design of any optical system. It is more than the keystone effect. The cosine to the fourth law includes the keystone effect, plus three other angular effects, which account for the

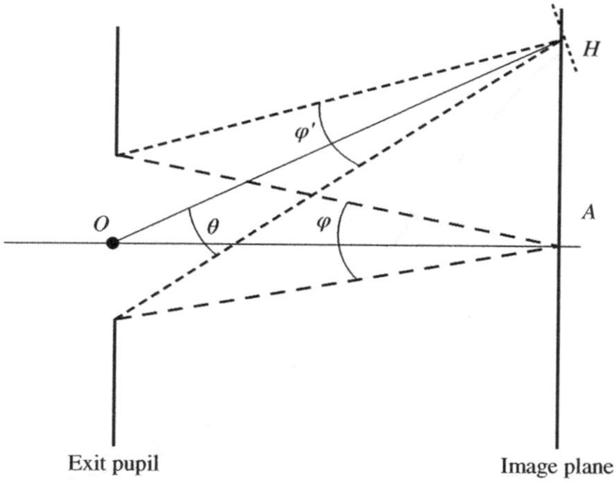

FIGURE 4.11 Schematic of the relationship between the exit pupil and the image plane.

irradiance reduction at an axis point, located an angle θ off the optical center line. In this example, we provide the derivation of the "cosine to the fourth law."

Figure 4.11 shows the relationship between the exit pupil and the image plane for a point A on the optical axis and an off-axis point H. The irradiance received at the latter point is proportional to the solid angle that the exit pupil subtends from the point. Let us now look at the details using Smith [3, p. 144].

The solid angle subtended from point A is the area of the exit pupil, divided by the square of the distance OA, the slant distance from the center of the exit pupil to the point H. From point H, the solid angle is the projected area of the pupil, divided by the square of the distance OH. Since OH is greater than OA, the distance from the exit pupil to the image plane by a factor of $\cos^{-1}\theta$, this increased distance reduces the irradiance by a factor of $\cos^2\theta$. Here, θ is the angle between point A and point H relative to the exit pupil. In addition, the exit pupil is viewed obliquely from the point H, and its projected area is reduced by a factor proportional to $\cos\theta$ when OH is large compared to the size of the exit pupil. The result is that the irradiance at H is reduced by $\cos^3\theta$.

Finally, this last result only is true for illumination on a plane normal to the line OH, indicated by the small dashed line at H in Figure 4.10. However, we want the irradiance in the plane AH, which reduces the irradiance by another factor of cos θ. This implies that the irradiance at H in the plane AH is equal to the irradiance at point A times $\cos^4\theta$. Figure 4.12 depicts cosine to the fourth as a function of angle. The importance of this effect comes clear for wide-angle imaging systems since the irradiance reduction at 30° (0.524 rad), 45° (0.785 rad), and 60° (1.047 rad) are 0.56, 0.25, and 0.06, respectively, which are quite significant. Figure 4.13 shows a comparison between the angular response of a retroreflector and the cosine to the fourth law. Although not a perfect comparison, it shows that the law can be useful to the working engineer when doing a first-order analysis.

As a final note, the above is good approximation when OH is large compared to the size of the exit pupil. For large angles where this is not true, *P*. Foote gave a more

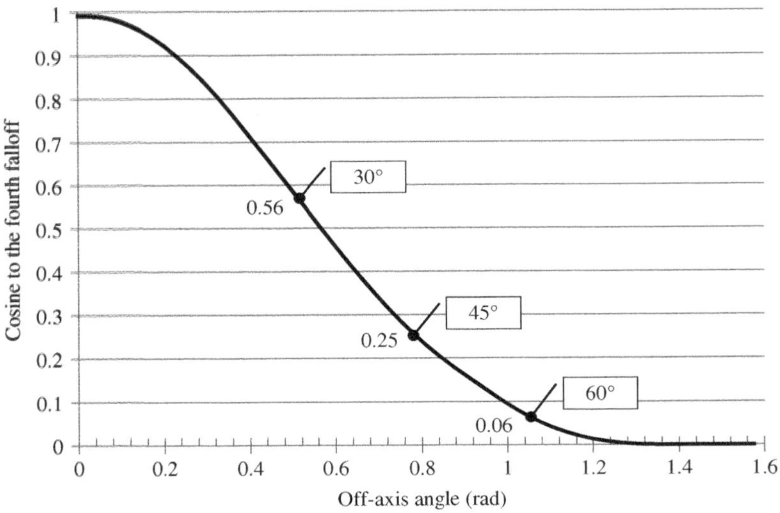

FIGURE 4.12 Cosine to the fourth as a function of angle.

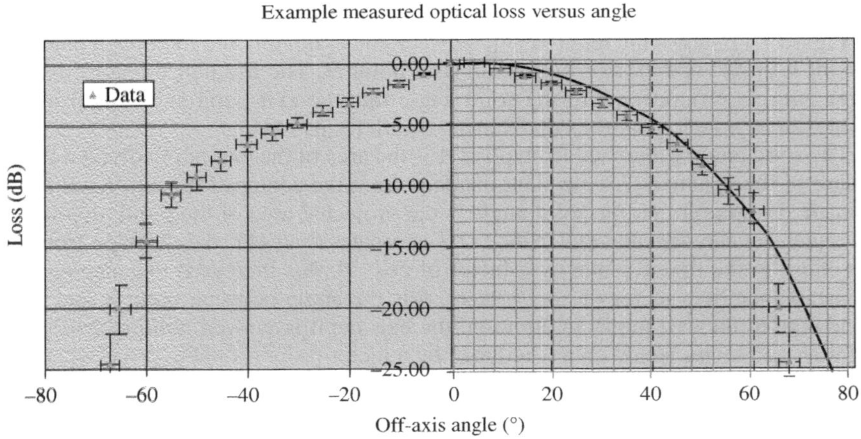

FIGURE 4.13 Comparison of reflector's angular response and the cosine to the fourth law.

accurate equation for the irradiance at point H [11], namely,

$$H = \frac{\pi N}{2}\left[1 - \frac{\left(1 + \tan^2\varphi - \tan^2\theta\right)}{\sqrt{\tan^4\varphi + 2\tan^2\varphi\left(1 - \tan^2\theta\right) + \frac{1}{\cos^4\theta}}}\right]. \tag{4.50}$$

For extreme cases, Smith showed that Eq. (4.37) can provide a 42% larger irradiance that the cosine to the fourth [3]. However, for most practical applications, cosine to the fourth is sufficient.

4.7 THE RADIOMETRY OF IMAGES

When an optical source is imaged by an optical system, the resulting image has a special form of radiance [3]. Specifically, the observed image only has radiance within the solid angle subtended by the image captured by the clear aperture of said optical system and zero, elsewhere. This fact affects the quantity of optical energy from the original image that can be captured. Let us see how we can model this effect.

Figure 4.14 depicts an aplanatic optical system imaging setup.[3] Assume that object A is a Lambertian source. What is its image in the observation plane? The radiance there is hypothesized to be formed through a generalized incremental area p in the first principal surface of the optical system. The radiance of the source is given by N_e and the projected area A in the direction of θ is $A\cos\theta$. From previous work, we know that the solid angle subtended by the incremental area p from A is p/S^2, where the distance from the object to the first principal surface is S. This results in a radiant power of the form

$$P_e = \frac{N_e\, p\, A\cos\theta}{S^2}. \tag{4.51}$$

The above radiant power is imaged into a projected area for A' in the observation plane, that is, $A'\cos\theta'$, through the solid angle $\frac{p'}{S'^2}$. This implies that the radiance in the observation plane is given by

$$N_e' = \gamma\,\frac{N_e\, p\, A\cos\theta}{S^2}\left[\frac{S'^2}{p'A'\cos\theta'}\right], \tag{4.52}$$

where γ is the transmission of the optical system. The above two incremental areas are related to each other by the laws of first-order optics and if we are media of the same refractive index, then we have

$$AS'^2 = A'S^2. \tag{4.53}$$

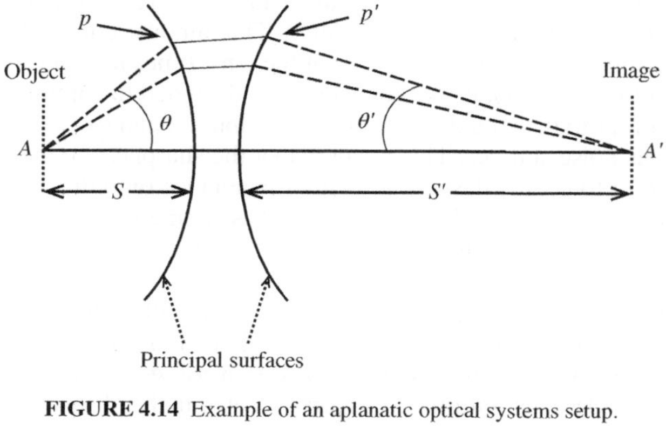

FIGURE 4.14 Example of an aplanatic optical systems setup.

[3] An aplanatic optical system is one where there is no coma or spherical aberrations; the principal "planes" are spherical surfaces and are center on the object and image [3].

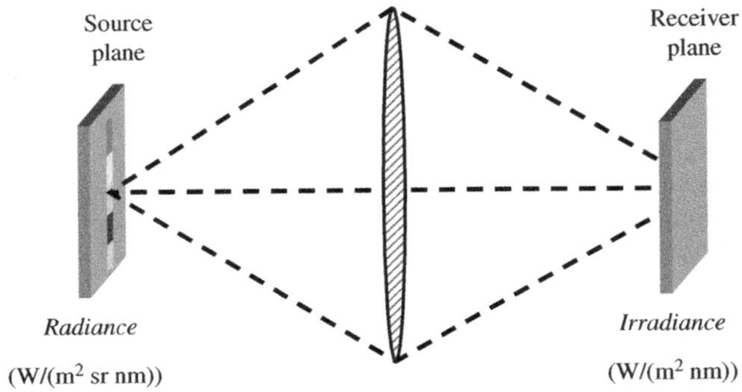

FIGURE 4.15 Scene radiance mapping into image irradiance.

In addition, the principal surfaces can be assumed to be unit images of each other, which implies that

$$p \cos\theta = p' \cos\theta'. \tag{4.54}$$

Inserting of Eqs. (4.51) and (4.52) into Eq. (4.50), we find that

$$N_e' = \gamma\, N_e. \tag{4.55}$$

The irradiance produced in the observation plane equals

$$H_e' = \gamma\, \pi N_e \sin^2\theta' = \gamma\, N_e \omega, \tag{4.56}$$

where θ' is the half-angle subtended by the exit pupil of the lens.

It is clear from the form of Eq. (4.56) that the source radiance distribution uniquely is mapped into the receiver plane as an irradiance distribution, not the true radiance distribution. This is illustrated in Figure 4.15 for a simple lens setup. Each point in the image plane contains the radiance in the equivalent position as it is in the object plane, but only captures a portion of the light emitted. That amount equals the source radiance times the solid angle of the exit pupil of the lens, times the transmittance of the lens. As we will find out, optical systems must be designed to capture as much light as possible in order to detect a desired signal distribution and overcome background light and various noise sources. This result is that the sharpness of the image will be sacrificed in order to get the desired amount of irradiance onto the detector element, fiber amplifier, or pixel plane. This can be viewed as geometrical optics interpretation of the effect of diffraction on an image's light distribution.

Example 4.10 Assume that we have a circular source with radiance 2 W/ $\left(\text{cm}^2\ \text{sr}\right)$ radiating light into an 1 in.-diameter imaging lens. See Figure 4.16. The distance from the lens to the observation plane is 100 in. The transmission of the lens equals 0.8. What is the received power within a 1 cm^2 area located in the observation plane about the optical axis?

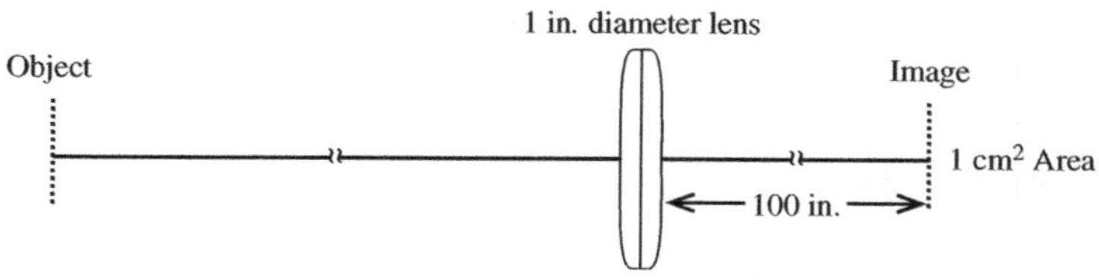

FIGURE 4.16 Example optical systems setup.

Using Eq. (4.56), we find that the power in the observation plane captured by the 1 cm^2 area in the observation plane is equal to

$$P_e = H'_e \left(1 \text{ cm}^2\right) = \gamma N_e \omega \left(1 \text{ cm}^2\right)$$
$$= (0.8)\left(2 \text{ W}/\left(\text{cm}^2 \text{ sr}\right)\right)\left[\frac{1 \text{ in.}^2}{(100 \text{ in.})^2}\right]\left(1 \text{ cm}^2\right)$$
$$= (1.6 \text{ W/sr})\left(10^{-4} \text{ sr}\right) = 160 \text{ μW}.$$

Example 4.11 Let us derive a special form for the received power within an A' area located in the observation plane about the optical axis.

Again, using Eq. (4.56), we write for the total received power,

$$P_e = H'_e A' = \gamma\, N_e \omega A' = \gamma\, N_e \left(\frac{A_{\text{rec}}}{S'^2}\right) A'. \tag{4.57}$$

where A_{rec} is the area of the clear aperture of the lens. Rewriting Eq. (4.57), we obtain

$$P_e = \gamma\, N_e A_{\text{rec}} \left(\frac{A'}{S'^2}\right) = \gamma\, N_e A_{\text{rec}} \Omega_{\text{FOV}}. \tag{4.58}$$

with Ω_{FOV} being the field of view of the receiving area A', for example, like that of a photodetector or focal plane array detector.

4.8 BLACKBODY RADIATION SOURCES

A *blackbody* in thermal equilibrium emits a broad spectrum of electromagnetic radiation. This emission is called blackbody radiation. This radiation is emitted according to Planck's law, meaning that it has a spectrum that is determined by the temperature alone. The rule of thumb is that a blackbody must be at a temperature greater than 700 K to create useful levels of visible radiation; otherwise, it will appear "black" to visible sensors and/or the naked eye. Its importance is its ability to realistically emulate light emission or reflection from various sources useful optical engineering, such

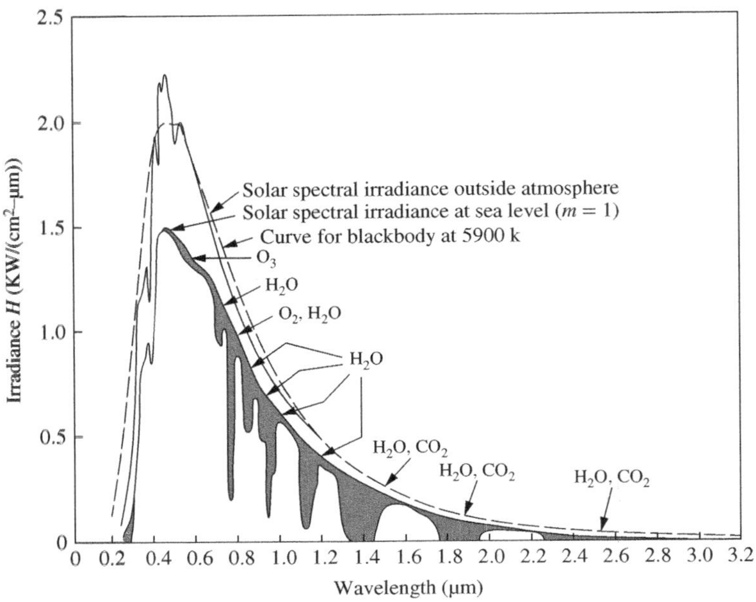

FIGURE 4.17 Comparison of exo- and endo-atmospheric solar spectral irradiance and 5900 K blackbody (the shaded areas indicate atmospheric absorption by the molecular constituents noted) [12].

as the Sun, Moon, the Sky, the Earth, the Stars, and the Planets [12, 13]. Figure 4.17 is an example, showing a comparison among exo- and endo-atmospheric Solar spectral irradiance, and a 5900 K blackbody source. It is clear that the blackbody does a very good job in mimicking the exo-atmospheric spectral irradiance. Blackbody sources in thermal equilibrium have two notable properties: (a) they are ideal emitters and (b) they are diffuse emitter. The first property (a) implies that they emit as much energy at every frequency as any other body at the same temperature; the second (b) implies that the energy is radiated isotopically, independent of direction. Real materials emit energy at a fraction – called the emissivity – of blackbody energy levels. By definition, a blackbody in thermal equilibrium has an emissivity of $\varepsilon = 1.0$. A source with lower emissivity independent of frequency often is referred to as a *Greybody* [12, 13].

A perfectly insulated enclosure that is in thermal equilibrium internally will house blackbody radiation and will radiate it through a hole made in its wall, provided the hole is small enough to have negligible effect upon the equilibrium. The *Stefan–Boltzmann law* describes the radiant exitance of a blackbody; mathematically, it can be written as

$$W_{BB} = \sigma T^4. \tag{4.59}$$

where T is the temperature of the blackbody in degree kelvin (K). The constant σ in the above equation is called the *Stefan–Boltzmann constant* or *Stefan's constant*. It comes from other known constants of nature; specifically, it is given by the equation

$$\sigma = \frac{2\pi^5 k^4}{15c^2h^3} = 5.670373 \times 10^{-8}\,\mathrm{W/m^2\,K^4}, \tag{4.60}$$

where k is the Boltzmann constant, h Planck's constant, and c the speed of light in vacuum. Using Eq. (4.59), we find that the blackbody radiance is given by

$$N_{BB} = \frac{W_{BB}}{\pi} = \frac{\sigma}{\pi} T^4. \tag{4.61}$$

As noted earlier, we may not have a perfect radiator and the true exitance for the blackbody will be given by

$$W_{BB} = \varepsilon \sigma T^4, \tag{4.62}$$

where the above ϵ represents the emissivity of the Greybody and is less than 1. In reality, the emissivity will be wavelength dependent, that is, $\epsilon = \varepsilon(\lambda)$.

Shortly after this development, Wien found that Blackbodies of different temperature emit spectra that peak at different wavelengths [14]. Specifically, he found that the spectral radiance of blackbody radiation peaks at the wavelength λ_{max} given by

$$\lambda_{max} = \frac{2.8977729\,(17) \times 10^{-3}\ \text{m K}}{T}. \tag{4.63a}$$

$$\approx \frac{2900\ \mu\text{m K}}{T}. \tag{4.63b}$$

and the radiant exitance at that wavelength is given by

$$W_{\lambda_{max}} = 1.288 \times 10^{-15}\, T^5 \ \left(\text{W/cm}^2\ \mu\text{m}\right). \tag{4.64}$$

Equation (4.63) is called the *Wien's displacement law*. What it tells us is that

- Hotter objects emit most of their radiation at *shorter* wavelengths; hence they will appear to be *bluer*.
- Cooler objects emit most of their radiation at *longer* wavelengths; hence they will appear to be *redder*.

Implicit with law is that for any wavelength, a hotter object radiates more (is more luminous) than a cooler one.

Example 4.12 The sunlight spectrum is peaked at $\lambda_{max} \approx 0.475\ \mu\text{m}$ (Green). What is the surface temperature of the Sun? Rewriting Eq. (4.63b), we have

$$T \approx \left.\frac{2900\ \mu\text{m K}}{\lambda_{max}}\right|_{\lambda_{max}=500} = \frac{2900\ \mu\text{m K}}{0.475\ \mu\text{m}} = 5894\ \text{K},$$

which is close to 5900 K temperature we used to compare blackbody irradiance to that of measured solar irradiance in Figure 4.17.

Planck's law describes the electromagnetic radiation emitted by a blackbody in thermal equilibrium at a definite temperature, which we noted earlier [15]. It is a pioneering result of modern physics and quantum theory.

Let dN_{BB} be the spectral radiance of a blackbody within the frequency range ν and $\nu + d\nu$ be given by

$$dN_{BB}(\nu) = \left\{ \left(\frac{2h\nu^3}{c^2} \right) \frac{1}{e^{\frac{h\nu}{kT}} - 1} \right\} d\nu, \tag{4.65}$$

where ν represents the frequency of the light. The function in $\{\cdots\}$ is known as *Planck's law* and has SI units equal to $\mathrm{W}/\left(\mathrm{sr\ m^{-2}\ Hz}\right)$. Planck's law gives the intensity radiated by a blackbody as a function of frequency or wavelength; to obtain the latter, we note that

$$\nu = \frac{c}{\lambda}. \tag{4.66}$$

Differentiating Eq. (4.66) with respect to wavelength, we have

$$d\nu = -\frac{c}{\lambda^2}\, d\lambda, \tag{4.67}$$

which implies that

$$\begin{aligned} dN_{BB}(\lambda) &= \left\{ \left(\frac{2hc}{\lambda^3} \right) \frac{1}{e^{\frac{hc}{\lambda kT}} - 1} \right\} \frac{c}{\lambda^2}\, d\lambda \\ &= \left\{ \left(\frac{2hc^2}{\lambda^5} \right) \frac{1}{e^{\frac{hc}{\lambda kT}} - 1} \right\} d\lambda. \end{aligned} \tag{4.68}$$

The function in the function inside $\{\cdots\}$ has SI units equal to $\mathrm{W/(sr\ m^{-3})}$. Figure 4.18 illustrated Eq. (4.68) for a temperature of 5900 K.

In the limit of low frequencies (i.e., long wavelengths), Planck's law tends to the Rayleigh–Jeans law

$$N_{BB}^{\nu}(T) = \left(\frac{2\nu^2 kT}{c^2} \right) \tag{4.69}$$

or

$$N_{BB}^{\lambda}(\lambda, T) = \left(\frac{2ckT}{\lambda^4} \right), \tag{4.70}$$

while in the limit of high frequencies (i.e., small wavelengths), it tends to the Wien's law. Figure 4.18 shows the comparison among these three irradiance distributions as a function of wavelength.

One also can analyze the output of the blackbody in terms of a number n that is the density of photons within a frequency range ν to $\nu + d\nu$ [16]. In general, the photon density can be written as

$$n\left(E_{\mathrm{Photon}}\right) \Delta E_{\mathrm{Photon}} = g\left(E_{\mathrm{Photon}}\right) f\left(E_{\mathrm{Photon}}\right) \Delta E_{\mathrm{Photon}}, \tag{4.71}$$

where E_{Photon} is the energy per photon, $n\left(E_{\mathrm{Photon}}\right)$ the number of photons per unit volume in the interval $\Delta E_{\mathrm{Photon}}$, $g\left(E_{\mathrm{Photon}}\right)$ the number of energy states per unit volume in the interval $\Delta E_{\mathrm{Photon}}$, $f\left(E_{\mathrm{Photon}}\right)$ the distribution function (probability) that

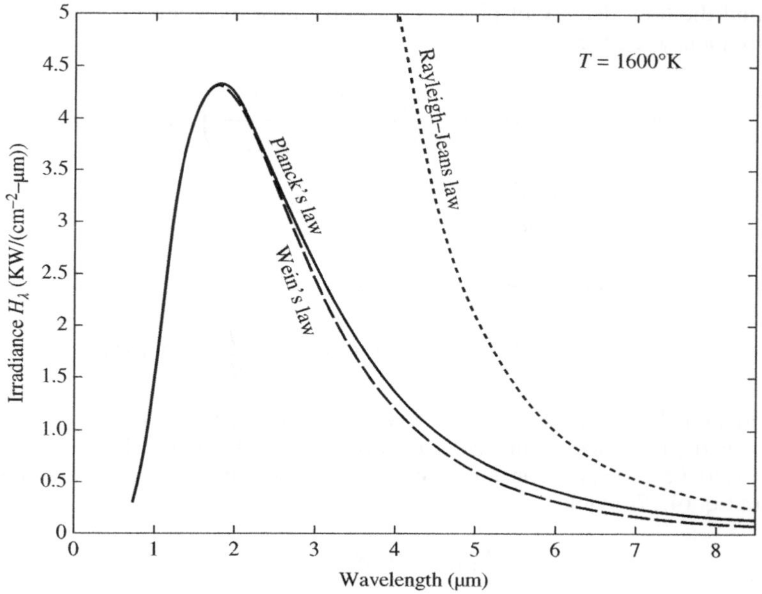

FIGURE 4.18 Comparison among Planck's law, Wien's law, and the Rayleigh–Jeans law irradiance distributions as a function of wavelengths.

a photon is in energy state E_{Photon}, and ΔE_{Photon} the energy interval. Since photons are bosons (particles with zero mass) rather than fermions (particles with mass), the above distribution function is the *Bose–Einstein distribution*, which is given by the equation

$$f\left(E_{\text{Photon}}\right) = \frac{1}{e^{\frac{E_{\text{Photon}}}{kT}} - 1}. \tag{4.72}$$

This distribution describes the probability that a given energy state will be occupied, but must be multiplied by the density of states function to weight the probability by the number of states available at a given energy.

Since we know that $E_{\text{Photon}} = h\nu$, we can rewrite Eq. (4.71) in terms of light frequency ν and interval $d\nu$. Specifically, we have

$$n(\nu)\,d\nu = M(\nu)\,f(\nu)\,d\nu. \tag{4.73}$$

In the above, $M(\nu)$ represent the modal density at frequency ν in a unit volume V, which equals [17]

$$M(\nu) = \frac{8\pi\nu^2}{c^3} \equiv n_0\nu^2. \tag{4.74}$$

The photon density expressed in Eq. (4.73) then is given by

$$n(\nu)\,d\nu = n_0\left(\frac{\nu^2}{e^{\frac{h\nu}{kT}} - 1}\right)d\nu \tag{4.75}$$

To find the total density of photons N_{Ph}, one must integrate Eq. (4.68) over the entire frequency regime:

$$N_{\text{Ph}} = \int_0^\infty n(\nu)\,d\nu = \frac{8\pi}{c^3}\int_0^\infty \left(\frac{\nu^2}{e^{\frac{h\nu}{kT}} - 1}\right) d\nu \tag{4.76}$$

$$= 8\pi\left(\frac{kT}{hc}\right)^3 \int_0^\infty \left(\frac{x^2}{e^x - 1}\right) dx \tag{4.77a}$$

$$= 8\pi\left(\frac{kT}{hc}\right)^3 \Gamma(3)\ \zeta(3) \tag{4.77b}$$

$$= 8\pi\left(\frac{kT}{hc}\right)^3 2!\,(1.20205) = 8\pi\left(\frac{kT}{ch}\right)^3 (2.404) \tag{4.77c}$$

using Eq. (4.411.1) in Gradshteyn and Ryzhik [18]. In Eq. (4.77b), $\Gamma(x)$ is the gamma function and $\zeta(x)$ is Riemann's zeta function [19]. In the above, $\Gamma(n+1) = n!$ and $\zeta(3) = 1.20205$ (Apéry's constant). Substituting the numerical values for the constants in Eq. (4.77c) yields

$$N_{\text{Ph}} = 2.02 \times 10^7\ T^3, \tag{4.78}$$

where N_{Ph} has units of m^{-3}. This equation shows that the photon density goes as the third power of the temperature.

The photon energy density of a photon gas follows from its number density. Specifically, we can write this density as

$$D_{Ph} = \int_0^\infty E(\nu)\ n(\nu)\,d\nu = \frac{8\pi}{(hc)^3}\int_0^\infty \left(\frac{E^3}{e^{\frac{E}{kT}} - 1}\right) d\nu \tag{4.79}$$

$$= \frac{8\pi(kT)^4}{(hc)^3}\int_0^\infty \left(\frac{x^3}{e^x - 1}\right) dx \tag{4.80a}$$

$$= \frac{8\pi(kT)^4}{(hc)^3}\,\Gamma(4)\ \zeta(4) \tag{4.80b}$$

$$= \frac{8\pi(kT)^4}{(hc)^3}\left(\frac{\pi^4}{15}\right) = \frac{8\pi(kT)^4}{(hc)^3}\ (6.49)\,. \tag{4.80c}$$

as $\zeta(4) = \frac{\pi^4}{90} = 1.0824$. Substituting the numerical values for the constants in Eq. (4.80c), we obtain

$$D_{\text{Ph}} = 7.565 \times 10^{-16}\ T^4, \tag{4.81}$$

where D_{Ph} has units of J/m^3. This equation shows that the energy density goes as the fourth power of the temperature. Dividing these densities gives the average energy per photon, or

$$\frac{\text{Energy}}{\text{Photon}} = \left[\frac{\frac{8\pi(kT)^4}{(hc)^3}\ (6.49)}{8\pi\left(\frac{kT}{hc}\right)^3\ (2.402)}\right] = 2.7\ kT. \tag{4.82}$$

4.9 PROBLEMS

Problem 4.1. What is irradiance 2 m from a 100 W light bulb? What is radiance of a typical 100 W light bulb? Here assume that the filament area is 1 mm^2.

Problem 4.2. Your night light has a radiant flux of 10 W. What is the irradiance on your book you were reading that fell 2 m from the light when you fell asleep (assuming the book is perpendicular to the night light)?

Problem 4.3. A point source radiates 10 W/sr toward a 10 cm -diameter lens. How much power is collected by the lens when its distance from the source is (a) 3 m and (b) 100 m?

Problem 4.4. Using the Stefan–Boltzmann law, compute the radiant exitance of a blackbody at $T = 27\ ^\circ\text{C}\ \left(80.6\ ^\circ\text{F}\right)$. What are the maximum wavelength at this temperature and the radiant exitance at that wavelength?

Problem 4.5. Using Planck's law, derive the Wien's law irradiance distribution used to create the curve in Figure 4.18.

Problem 4.6. What is the spectral radiant exitance of a 5900 K blackbody in the region of 0.5 μm? What is its radiance?

Problem 4.7. If you have a source at a temperature of 1000 K, and an emissivity of 0.93, what is the radiant exitance of the source?

Problem 4.8. (a) What is the spectral radiant emittance of a 1000 K blackbody in the region of 2 μm wavelength? (b) What is the radiance? (c) If an idealized band-pass filter only transmitting between 1.95 and 2.05 μm is used, what is the total power falling on a 1 cm^2 detector placed 1 m from a 1 cm^2 blackbody?

Problem 4.9. What is the approximate wavelength that humans radiate at?

Problem 4.10. Given a lamp with a radiant intensity of 0.1 W/sr illuminating a Lambertian diffuser 10 cm away with a 1 cm diameter aperture located just beyond the diffuser; this "new" 1 cm diameter source then illuminates a detector 100 cm from the lamp. What is the irradiance on the detector if the transmission of the diffuser is 0.60?

Problem 4.11. Suppose that we have a source of height h_1 in medium of index n_2, creating an image by a simple lens that has the height h_2 in medium of index n_2. Prove Abbe's Sine Condition:

$$N_1 h_1 \sin\theta_1 = N_2 h_2 \sin\theta_2.$$

REFERENCES

1. McCluney, W.R. (2014) *Introduction to Radiometry and Photometry*, Artech House Photonics Series, 2nd edn, Artech House, Norwood, MA.
2. Palmer, J.M. and Grant, B.G. (2009) *The Art of Radiometry*, Volume: PM184, ISBN: 9780819472458, SPIE Press, Bellingham, WA, pp 384.

3. Smith, W.J. (2008) *Modern Optical Engineering; The Design of Optical Systems*, 4th edn, SPIE Press, Bellingham, WA.
4. JM Forsyth, *Contemporary Optics*, Chapter 2, reprinted by Czech Technical University in Prague University, http://cmp.felk.cvut.cz/cmp/courses/dzo/resources/chapter_radiometry_forsyth.pdf.
5. Wolfe, W.L. (1996) *Introduction to Infrared System Design*, SPIE Press, Bellingham, WA.
6. Boreman, G.D. (1998) *Basic Electro-Optics for Electrical Engineers*, SPIE Press, Bellingham, WA.
7. Alkholidi, A.G. and Altowij, K.S., Chapter 5 (2014) in *Contemporary Issues in Wireless Communications* (ed. M. Khatib), InTech, Rijeka, Croatia, EUROPEAN UNION, 252 pages, ISBN 978-954-51-1732-2
8. Lohmann, A.W. (2006) in *Optical Information Processing* (ed. S. Sinzinger), Universitätsverlag, Ilmenau, Germany, ISBN 4-939474-00-6
9. Born, M. and Wolf, E. (1999, pp. 33–34 and pp. 554–555) *Principles of Optics: Electromagnetic Theory of Propagation, Interference and Diffraction of Light*, Cambridge University Press, London.
10. Sillion, F.X. and Puech, C. (1994) *Radiosity and Global Illumination*, Eurographics technical report, Morgan Kaufmann Publishers, ISBN 1558602771; 9781558602779, 251 pages.
11. Foote, P. (1915) The Bulletin of the Bureau of Standards 12, p. 583.
12. Wolfe, W.L. and Zissis, G.J. (1978) *The Infrared Handbook*, The Infrared Information and Analysis Center of the Environmental Research Institute of Michigan for the Office of Naval Research, Washington, DC.
13. Spiro, I.J. and Schlessinger, M. (1989) *Infrared Technology Fundamentals*, Marcel Dekker, Inc., New York, pp. 75–77.
14. Boltzmann, L. (1884) *Annalen der Physik*, **xxii**, 31, 291.
15. Wien, W. (1896) *Annalen der Physik*, **lviii**, 662.
16. Boal, D. (2001) PHYS 390 Lecture 14 – Photons, Simon Fraser University.
17. C.R. Nave, *Photon Energy Density*, HyperPhysics (Contact Dr. Carl R. (Rod) Nave, Department of Physics and Astronomy, Georgia State University, Email: RodNave@gsu.edu), Atlanta, Georgia, http://hyperphysics.phy-astr.gsu.edu/hbase/quantum/phodens.html.
18. Gradshteyn, I.S. and Ryzhik, I.M. (1965) in *Table of Integrals, Series, and Products*, 4th edn (eds Y.V. Geronimus and M. Yu Tseytlin), Translated by Alan Jeffery, Academic Press, New York.
19. Planck, M. (1900) *Verhandlungen der Deutschen physikalischen Gesellschaft*, **Ii**, 237.

5

CHARACTERIZING OPTICAL IMAGING PERFORMANCE

5.1 INTRODUCTION

In Chapters 2 and 3, we developed the theories of physical and geometrical optics, respectively, under the assumption of the perfect, lossless imaging of an object or source. In reality, this is never the case. We noted that optical aberrations such as coma and astigmatism can cause image blurring and distortion effects that degrade the quality of the image. However, this is just the "tip of the iceberg." Most optical systems are immersed in some sort of channel such as the atmosphere or the ocean or clouds, which can further distort and degrade the optical signal in both space and time. The optical engineer must deal with these problems by either correcting or compensating for all of them as best as he/she can. The question now arising is how one characterizes all these effects in such diverse media? Fortunately, the last examples in Chapter 2 suggest a first-order means for this characterization – convolution. Specifically, we find that convolution can describe the effects of an imperfect optical system or an optical channel such as the optical scatter channel or turbulent channel on an input distribution, as well as the result from coherent and incoherent signal processing. That is great news! One theory for all!

This chapter deals with the *convolutional theory of image formation*. Although we use the word image here, one can consider laser radar (Ladar) and free-space optical communications receivers as "one pixel" imagers and the theory thus covers a large gambit of applications. The material here covers coherent and incoherent situations, providing the terminology of each unique area. Chapter 6 discusses partially coherent analysis. However, the theory for all is the same to first order. Authors like Karp and

Free Space Optical Systems Engineering: Design and Analysis, First Edition. Larry B. Stotts.

Companion website: www.wiley.com\go\stotts\free_space_optical_systems_engineering

Stotts [1], Lohmann [2], Goodman [3], Born and Wolf [4], and Papoulis [5] elaborated on the general theory in their respective textbooks and Andrews and Phillips [6] and Gagliardi *et al.* [7] discuss the characterization of the effects of the turbulent and particulate channels on incoming signals as well. We shall leverage these works in the sections to come.

5.2 LINEARITY AND SPACE VARIANCE OF THE OPTICAL SYSTEM OR OPTICAL CHANNEL

Example 2.3 and the section on imperfect imaging suggest a model for optical degradation of the form

$$v(x,y) = u(x',y') \circledast F(x,y) = \int\int_{-\infty}^{\infty} u(x',y')\, F(x-x',y-y')\, dx'dy', \tag{5.1}$$

where $v(x,y)$ is the output from the optical system or channel, $u(x)$ the initial source distribution, and $F(x,y)$ an impulse response function, which was the complex phase term in Eq. (2.70) in the cited example. The symbol $\circledast$ denotes a convolution operation. The function $F(x,y)$ comes from the input of a delta function point source $\delta(x,y)$ into the optical system or channel. What are the desired properties of the "system or channel?" They are linearity (a continued theme) and space invariance. In general, a system can be as follows:

- Linear and space invariant
- Linear and space variant
- Nonlinear and space invariant
- Nonlinear and space variant.

Let us now discuss what these characteristics mean. We temporarily limit our mathematics to one dimension for simplicity.

Focusing on the first bullet, when a system is called space invariant, it means that a source shifted off the optical axis creates the same image function $F(x)$ in the image plane, but at same shifted location; that is,

$$\delta(x-x') \to F(x-x'). \tag{5.2}$$

We now introduce the property of *homogeneity*, which is an underlining assumption to the first bullet. It has the connotation of invariance as all components of an equation have the same degree of value whether or not each of these components is scaled to different values; for example, by multiplication or addition. This means that if we multiply the left side of Eq. (5.2) by $u(x)$, then we have

$$u(x')\delta(x-x') \to u(x')F(x-x') \tag{5.3}$$

as the result. Another underlining assumption is *additivity*. That is, if we have two different shifted versions of $u(x)\delta(x)$, then we have

$$u(x')\delta(x-x') + u(x)\delta(x-x'') \to u(x'')F(x-x') + u(x)F(x-x''). \tag{5.4}$$

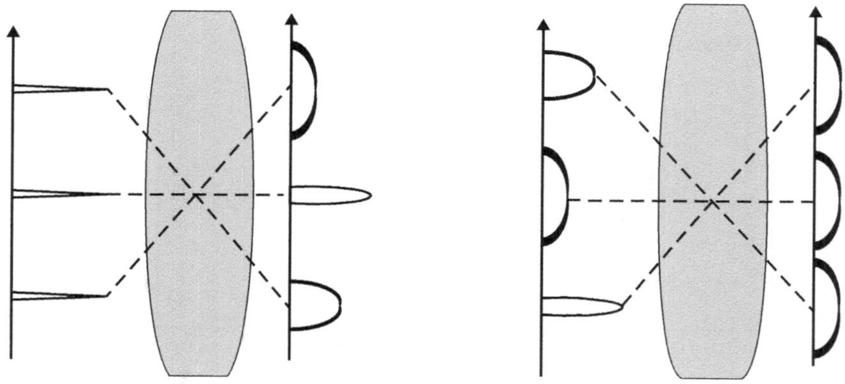

FIGURE 5.1 Example output images created by a linear, space-variant system using input functions that comprised (a) shifted delta functions and (b) shifted arbitrary functions.

If there are an infinite set of summand in the object plane, then Eq. (5.4) becomes

$$\int u(x')\delta(x - x')\, dx' = u(x) \rightarrow \int u(x')F(x - x')dx' = v(x). \tag{5.5}$$

The properties of homogeneity and additivity define the property of *linearity*. It is clear why space invariance and linearity are so desirable. These two properties make things considerably easier to understand the imaging results, as well as the characterization of the "system." We can see that is true in what follows.

Now if the system is linear, but space variant, we have the example situation shown in Figure 5.1. This shows that space variance complicates our ability to understand the "system." Figure 5.1a shows that the shift invariant property is no longer true. One gets a different output for the three shifted delta functions, which implies the impulse response is different for each shift input. However, Figure 5.1b illustrates that three different input functions create the same output function for the impulse responses given in Figure 5.1a. The conclusion is that although this makes the "system" appear space invariant, in reality, one could not guarantee that every input function would create the same output from result at every location. This would require one to characterize every foreseeable input function for any possible shift to truly understand if we are space invariant or not. This really complicates things. Unfortunately, this is exacerbated when we add a second (y-direction) or possibly a third dimension (time) depending on the system we are interested in, which would happen if the "system" was the turbulent channel for instance.

Now a nonlinear system can be useful at times when space variants are not useful. This can be seen in Figure 5.2. These figures show the effects of high-end clipping [saturation → (a)], low-end clipping [saturation → (b)], and high/low clipping [saturation and thresholding → (c)]. Looking at the image (a) in Figure 5.2, we see that most of the pictures are washed out, with finer details barely visible. Image (b) shows larger object profiles very nicely, but the finer details are mostly gone from the building and cars. Both effects are typical for which is about typical for photography (except for

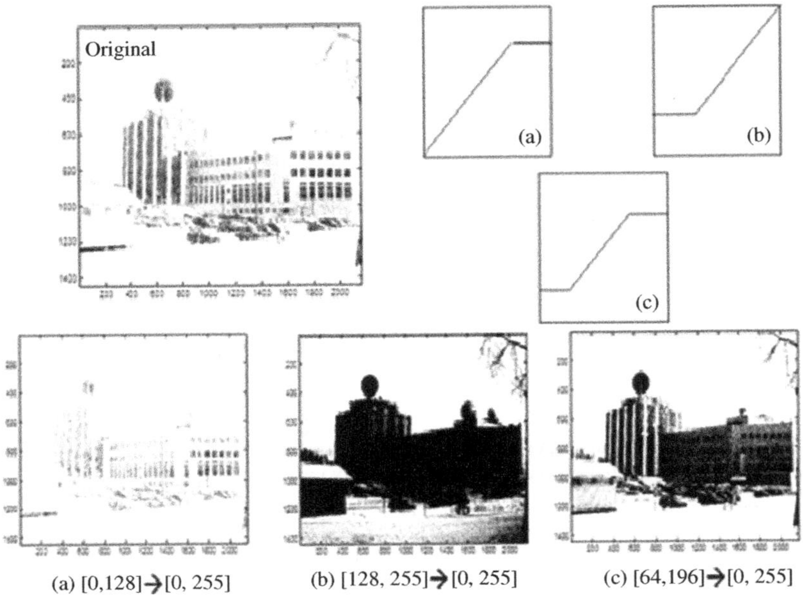

FIGURE 5.2 Example output images created by a nonlinear processing an input functions where (a) high, (b) low, and (c) high and low intensities are zeroed.

the sharp edges). However, image (c) shows a reasonable compromise between the other two and in fact, the original as well, nicely showing the finer details and profiles with better visibility. By the way, image (c) illustrates where nonlinearity and space variance were desirable features, the situation where the objects of interest are restricted. In other words, we had some *a priori* information about the objects. We are interested in the building and car details, and the nonlinear banding highlights these areas very well. As a final comment, this type of nonlinear effect belongs to the simplest class of nonlinearities, the so-called point-to-point nonlinearities, because each image point depends only on one object point. For time signal, the corresponding class of nonlinearities is called "memoryless."

5.3 SPATIAL FILTER THEORY OF IMAGE FORMATION

The conclusion from the last section is that the convolution integral in Eq. (5.1) can describe the effects of an imperfect optical system or an optical channel, ideally if the entity of interest is linear and space invariant. In other words, this integral suggests a straightforward way to model these phenomena. Specifically, it comes from the Fourier transform of Eq. (5.1), which is equal to

$$\hat{v}(k_x, k_y) = \int\int_{-\infty}^{\infty} v(x, y) e^{-2\pi i(k_x\, x + k_y\, y)}\, dxdy = \hat{u}(k_x, k_y)\, \hat{F}(k_x, k_y), \tag{5.6}$$

where

$$\hat{u}(k_x, k_y) = \iint_{-\infty}^{\infty} u(x, y) e^{-2\pi i(k_x x + k_y y)} \, dxdy \tag{5.7}$$

and

$$\hat{F}(k_x, k_y) = \iint_{-\infty}^{\infty} F(x, y) e^{-2\pi i(k_x x + k_y y)} \, dxdy \tag{5.8}$$

in two dimensions. Equation (5.6) is the basis for the theory of coherent image formulation that was developed by Ernst Abbe almost 150 years ago.[1] His theory says that one can determine the output image of an optical system by modeling it as a simple linear filtering process in the Fourier domain, reducing the complexity of the mathematical developments as alluded to in the last section. The function $\hat{F}(k_x, k_y)$ is called the *Spatial Filter Function*, or sometimes, the *Complex Pupil Function*. The function $F(x, y)$ is known as the *Coherent Point Spread Function* or more commonly, the *Amplitude Spread Function* (ASF).

Figure 5.3a and b illustrates example configurations for incoherent and coherent filtering, respectively. (They are shown in only one dimension for simplicity.) In Figure 5.3, a self-emitting object in Figure 5.3a, or illuminated transparent object in Figure 5.3b, has its light distribution Fourier transformed into an image plane (filter) a focal length away from the transforming perfect lens. A filter is placed in this filter plane that creates the light distribution given in the right side of Eq. (5.6). This composite distribution is then inverse Fourier transformed by the next lens and the result is imaged into the output plane one focal length away from this lens, creating $v(x, y)$. Let us see if that is true. We begin with a development we performed previously, which covered the light propagation of the object to $z = 2f$, which corresponds to the filter plane in Figure 5.3b.

Recall that Section 2.6 of Chapter 2 showed that the two-dimensional output electric field in the filter plane ($z = 2f$) is given by

$$u(x, y, 2f - 0) = \hat{u}_0 \left(\frac{x}{\lambda f}, \frac{y}{\lambda f} \right) e^{i\frac{\pi}{2}}.$$

If there now is a filter at $z = 2f$, then using Eq. (5.6) we have

$$u(x, y, 2f + 0) = \hat{v} \left(\frac{x}{\lambda f}, \frac{y}{\lambda f} \right) e^{i\frac{\pi}{2}}. \tag{5.9}$$

For simplicity, we again employ an approximate form of the parabolic approximation of the Huygens–Fresnel–Kirchhoff integral. In this case, we obtain the following field amplitude from the propagation form $z = 2f + 0$ to $z = 3f - 0$:

$$u(x'', y'', 3f - 0) \approx \iint_{-\infty}^{\infty} u(x', y', 2f) \, e^{i\frac{\pi}{\lambda f}\left[(x''-x')^2 + (y''-y')^2\right]} dx'dy'. \tag{5.10}$$

[1]Coherent light is orderly in its phase distributions, while incoherent light is completely random in its phase distribution. Laser light is almost completely coherent, while light from thermal sources (the sun, bulbs, etc.) is almost completely incoherent. In addition, in-between situations exist if laser light or thermal light is properly manipulated by moving diffusers, pinholes, spectral filters, and the like. However, this in-between situation, which is called "partial coherence," occurs mainly in connection with interference.

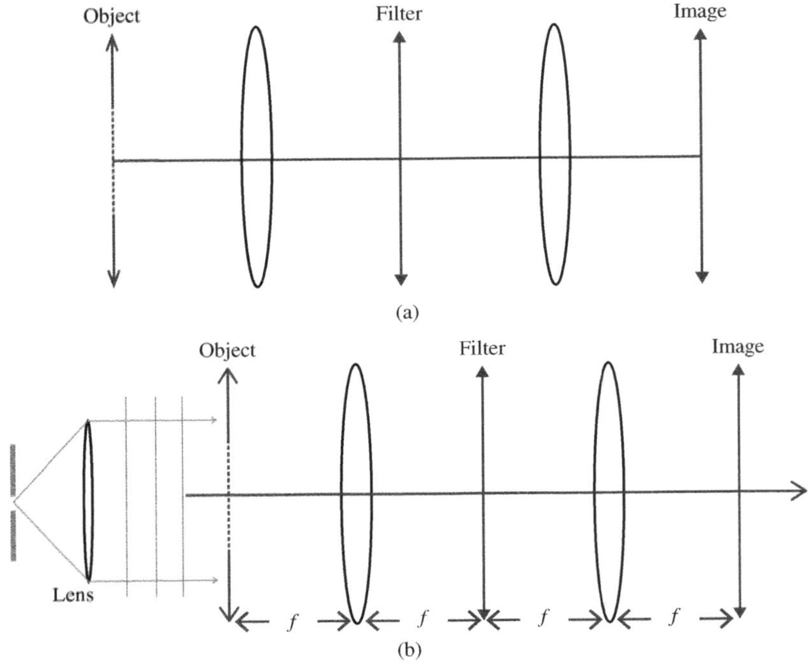

FIGURE 5.3 (a) Lens layout and (b) telecentric lens layout for filtering an incoherent and coherent image.

Propagating through the simple lens, we have

$$u(x'', y'', 3f + 0) = u(x'', y'', 3f - 0)e^{-i\pi \frac{\left[x''^2 + y''^2\right]}{\lambda f}}, \tag{5.11}$$

where the exponential in the above equation is the phase factor induced by the simple lens. Finally, the received field amplitude in the observation plane $z = 2f$ is given by

$$u(x, y, 4f) \approx \int\int_{-\infty}^{\infty} u(x'', y'', 3f + 0)\, e^{i\frac{\pi}{\lambda f}\left[(x-x'')^2 + (y-y'')^2\right]} dx''dy'' \tag{5.12}$$

$$= \int\int_{-\infty}^{\infty}\int\int_{-\infty}^{\infty} u(x', y', 2f + 0)e^{i\frac{\pi}{\lambda f} g(x,y;\, x',y';\, x'',y'')}\, dx'dy'dx''dy'' \tag{5.13}$$

with

$$\begin{aligned} &g(x, y;\ x', y';\ x'', y'') \\ &= \left[(x'' - x')^2 + (y'' - y')^2\right] + \left[(x - x'')^2 + (y - y'')^2\right] - \left[x''^2 + y''^2\right] \\ &= \left[x''^2 + y''^2 + x'^2 + y'^2 + x^2 + y^2 - 2x'x'' - 2xx'' - 2y'y'' - 2yy''\right]. \end{aligned} \tag{5.14}$$

Substituting Eq. (5.9) into Eq. (5.13) gives

$$u\,(x,y,4f) = e^{i\frac{\pi}{2}} \int\int_{-\infty}^{\infty}\int\int_{-\infty}^{\infty} \hat{v}\left(\frac{x'}{\lambda f},\frac{y'}{\lambda f}\right) e^{-i\,\frac{\pi}{\lambda f}\,g(x,y;\,x',y';\,x'',y'')}\,dx''dy''dx'dy', \tag{5.15}$$

where

$$g\left(x,y;\,x',y';\,x'',y''\right) = \left[(x''-(x'+x))^2-(x'+x)^2+(y''-(y'+y))^2 \right.$$
$$\left. -(y'+y)^2+x'^2+y'^2+x^2+y^2\right]. \tag{5.16}$$

From previous work, integrating over x'' and y'' yields constants. This implies that we have

$$u\,(x,y,2f) = (\lambda f)^2 e^{i\pi} \int\int_{-\infty}^{\infty} \hat{v}\left(\frac{x'}{\lambda f},\frac{y'}{\lambda f}\right) e^{-i\,\frac{\pi}{\lambda f}\,g(x,y;\,x',y')}dx'dy'\,, \tag{5.17}$$

where

$$\begin{aligned} g(x,y;\,x',y') &= \left[-(x'+x)^2-(y'+y)^2+x'^2+y'^2+x^2+y^2\right] \\ &= \left[-(x'^2+2xx'+x^2)-(y'^2+2yy'+y^2)+x'^2+y'^2+x^2+y^2\right] \\ &= \left[-2xx'-2yy'\right]. \end{aligned} \tag{5.18}$$

This means that we now have

$$\begin{aligned} u\,(x,y,4f) &= (\lambda f)^2 e^{i\pi} \int\int_{-\infty}^{\infty} \hat{v}\left(\frac{x'}{\lambda f},\frac{y'}{\lambda f}\right) e^{-2\pi i\left[\left(\frac{x}{\lambda f}\right)x'+\left(\frac{x}{\lambda f}\right)y'\right]}dx'dy' \\ &= (\lambda f)^4 e^{i\pi} \int\int_{-\infty}^{\infty} \hat{v}(\mu,\nu)e^{-2\pi i\,[\mu x+\nu y]}d\mu d\nu \\ &= (\lambda f)^4 e^{i\pi}\,v(x,y) \propto \int\int_{-\infty}^{\infty} u(x',y')\,F(x-x',y-y')\,dx'dy', \end{aligned} \tag{5.19}$$

since

$$v(x,y) = \int\int_{-\infty}^{\infty} \hat{v}(\mu,\nu)e^{-2\pi i\,[\mu x+\nu y]}d\mu d\nu = \int\int_{-\infty}^{\infty} \hat{u}(\mu,\nu)\,\hat{F}(\mu,\nu)e^{-2\pi i\,[\mu x+\nu y]}d\mu d\nu. \tag{5.20}$$

In general, the spatial filter function $\hat{F}(\mu,\nu)$ cannot hang in space without some sort of nontransparent holder in this filter plane. The result is that the high frequencies will be truncated whether or not energy is present. Assume the latter for the moment. This type of blocking filter function is known as a "low-pass filter." The circular aperture analyzed in Chapter 2 is an example of a low-pass filter; it degraded a point source input. In particular, the high-frequency cutoff of a low-pass filter is the initial reason for a limit of resolution in any optical system. We will discuss resolution more in this chapter, but the takeaway of the analysis thus far is that coherent/incoherent image formation can be described as a linear filter with spatial frequency transmittance properties that are characterized by the filter function $\hat{F}(\mu, \nu)$. As a final comment, optical low-pass filters typically have sharp edges, which is unique in signal processing. This cannot happen in time-frequency domain processing. If their filters had sharp cutoff frequencies, these filters would violate the causality axiom.

5.4 LINEAR FILTER THEORY OF INCOHERENT IMAGE FORMATION

Assume we have the self-luminous object in the setup illustrated in Figure 5.3a. It is emitting amplitude $A(x)$ and time-varying phase $\varphi(x, t)$. The initial source field is thenwritten as

$$u_o(x, t) = A(x)\, e^{i\ \varphi(x,t)}, \tag{5.21}$$

where

$$\langle u_o(x, t)\rangle = 0 \tag{5.22}$$

and

$$\langle |u_o(x, t)|^2\rangle = |A(x)|^2 = I_0(x). \tag{5.23}$$

In the above equations, $\langle \cdots \rangle$ means either time-averaging, or time integration, or mathematically,

$$\langle \cdots \rangle = \lim_{T \to Big} \int_{-\frac{T}{2}}^{\frac{T}{2}} \cdots dt. \tag{5.24}$$

From the previous discussions, it is clear that

$$\delta(x - x') \overset{\text{yields}}{\to} F(x - x') \tag{5.25}$$

$$u_o(x, t)\, \delta(x - x') \overset{\text{yields}}{\to} u_o(x, t)\, F(x - x') \tag{5.26}$$

$$\int u_o(x, t)\, \delta(x - x') dx' \overset{\text{yields}}{\to} \int u_o(x, t)\, F(x - x')\, dx' = v(x, t). \tag{5.27}$$

The complex image amplitude $v(x, t)$ has a zero-time-average if the phase $\varphi(x, t)$ varies wildly during the integration time T. That is, we have

$$\langle v(x, t)\rangle = \int \langle e^{i\varphi(x',t)} \rangle A(x)\, F(x - x')\, dx' = 0. \tag{5.28}$$

because $\langle e^{i\varphi(x',t)}\rangle = 0$. Fortunately for us, optical detectors respond to optical intensity $|v(x, t)|^2$, not optical amplitude $v(x, t)$, and this entity is nonzero. To see that, we write

$$\langle |v(x, t)|^2\rangle = \iint \left\langle e^{i\ [\varphi(x',t) - \varphi(x'',t)]} \right\rangle A(x')\, A(x'')\, F(x - x')\, F^*(x - x'') dx'\, dx'' \tag{5.29}$$

$$= \int A^2(x')\, |F(x - x')|^2\, dx', \tag{5.30}$$

since

$$\iint \langle e^{i\,[\varphi(x',t)-\varphi(x'',t)]}\rangle = \delta(x-x') \tag{5.31}$$

because again, the phase $\varphi(x,t)$ varies wildly during the integration time.

Equation (5.30) can be rewritten in terms of intensity; namely, we can write

$$I_b(x) = \int I_0(x)\,P(x-x')\,dx', \tag{5.32}$$

where

$$I_b(x) = \langle |v(x,t)|\rangle^2, \tag{5.33}$$

$$I_0(x) = A^2(x'), \tag{5.34}$$

and

$$P(x) = F(x)\,F^*(x) = |F(x)|^2. \tag{5.35}$$

Equation (5.32) states that the received image intensity I_b is the convolution of the object intensity I_0 and the incoherent *Point Spread Function* (*PSF*) *P*. Fourier transforming Eq. (5.32), we have

$$\hat{I}_b(k_x) = \hat{I}_0(k_x)\,\hat{\Theta}(k_x), \tag{5.36}$$

where

$$\hat{I}_b(k_x) = \int_{-\infty}^{\infty} I_b(x)e^{-2\pi i k_x\,x}\,dx, \tag{5.37}$$

$$\hat{I}_0(k_x) = \int_{-\infty}^{\infty} I_0(x)e^{-2\pi i k_x\,x}\,dx, \tag{5.38}$$

and

$$\hat{\Theta}(k_x) = \int_{-\infty}^{\infty} P(x)e^{-2\pi i k_x\,x}\,dx \tag{5.39a}$$

$$= \int_{-\infty}^{\infty} |F(x)|^2 e^{-2\pi i k_x\,x}\,dx. \tag{5.39b}$$

$\hat{\Theta}(k_x)$ is called the *Optical Transfer Function* (*OTF*) or *Contrast Transfer Function* (*CTF*). The OTF also can be written as

$$\hat{\Theta}(k) = |\hat{\Theta}(k)|\,e^{i\Phi(k_0)}, \tag{5.40}$$

where $|\hat{\Theta}(k)|$ is the *Modulation Transfer Function* (*MTF*) and $\Phi(k)$ is the *Phase Transfer Function* (*PTF*).

In general, the $\hat{\Theta}(k_x)$, and its two-dimensional version $\hat{\Theta}(k_x, k_y)$, obeys the following reality symmetry properties:

$$\hat{\Theta}(k_x) = \hat{\Theta}^*(-k_x), \tag{5.41}$$

$$|\widehat{\Theta}(k_x)| = |\widehat{\Theta}(-k_x)|, \tag{5.42}$$

and

$$\Phi(k_x) = -\Phi(-k_x) \tag{5.43}$$

for any arbitrary phase shift, $\Phi(k_x)$.

5.5 THE MODULATION TRANSFER FUNCTION

The modulus of the OTF is commonly known as the MTF because the normal target used to characterize that function is comprised a series of alternating light and dark bars of equal width. Figure 5.4 shows the most common target set of this type, the 1951 USAF Resolution Test Chart. Figure 5.5 shows an example bar series set in terms of its (a) spatial and (b) radiance distributions. If the line pair (black and white bar set) period is $\frac{1}{N}$ in millimeters, then the number of lines per millimeter is N; this is normally the metric used in any testing with this chart. That is, when this chart has

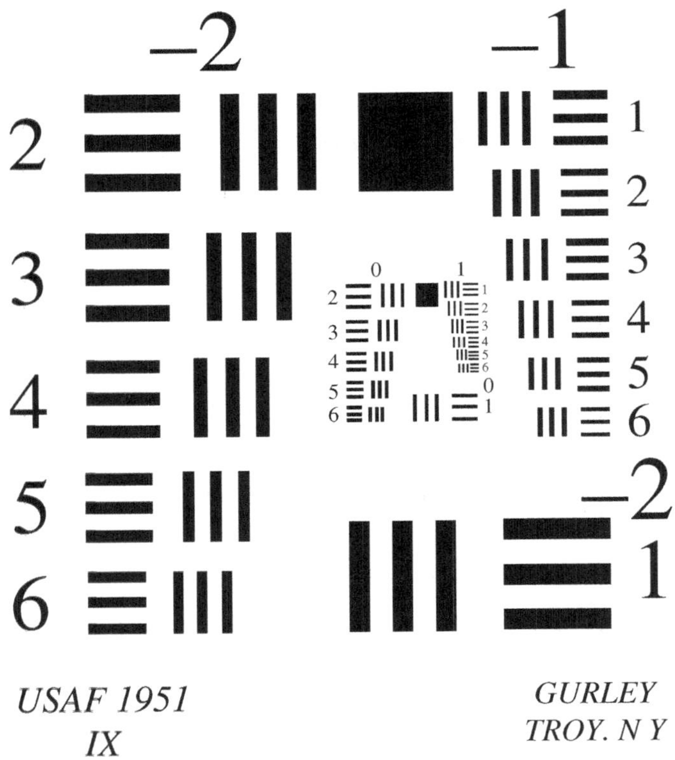

FIGURE 5.4 Image of the USAF 1951 Resolution Test Chart.

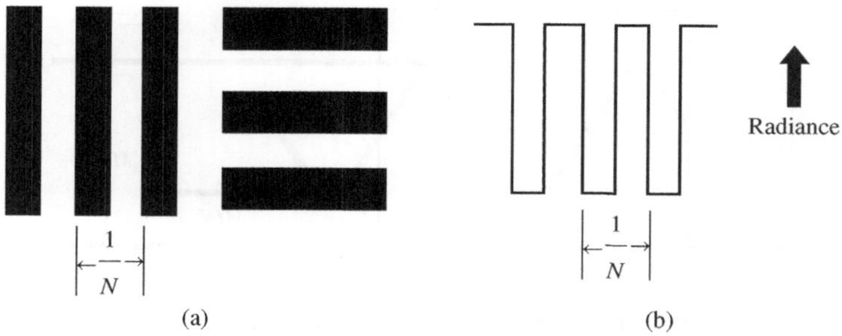

FIGURE 5.5 (a) Spatial and (b) radiance distribution of an arbitrary bar series.

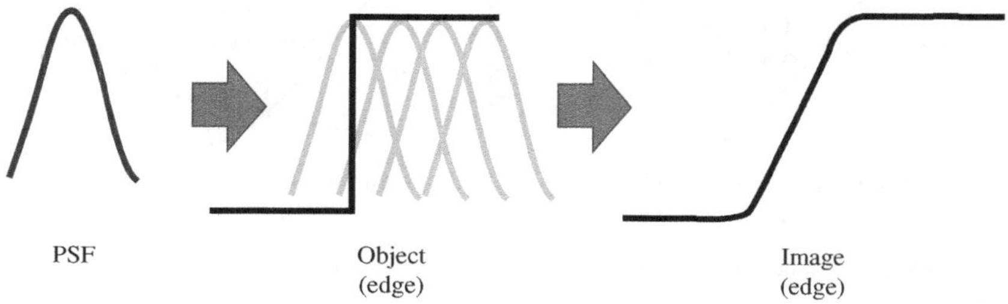

FIGURE 5.6 Effect on imaged edge by the point spread function of the optics.

been imaged by an optical system under test, the finest bar series that can be resolved, that is, line structure that can be discerned, is called the *limit of resolution* for the system. This limit is expressed as a certain number of "line pairs per millimeter."

The reason that there is a limit in resolution is that the contrast between light and dark bars gets reduced; that is there comes a point in this chart where the eye cannot perceive the contrast between the line pairs. The simple way to begin to look at this is shown in Figure 5.6. When a line pair is imaged by an optical system, each point of the object radiance is blurred by the PSF, which degrades the edge into more of a "steep ramp" than the "cliff" inherent to the original radiance distribution. The maximum radiance value reduces as well because of the smearing. In other words, the effect of the PSF is to round off the sharp edge (corner) of the "square wave" radiance pattern, causing it to look more like a "sine wave." Another result of this blurring is that the bounds of the image radiance levels shrink, as illustrated in Figure 5.7. In other words, the image "modulation," defined as

$$C = \frac{N_{max} - N_{min}}{N_{max} + N_{min}}, \tag{5.44}$$

decreases as the number of line pairs per millimeter N increases. In Eq. (5.44), N_{max} and N_{min} are maximum and minimum values of the imaged sine-wave, as illustrated in Figure 5.7. The parameter C defined above is called the *Michelson contrast or visibility*.

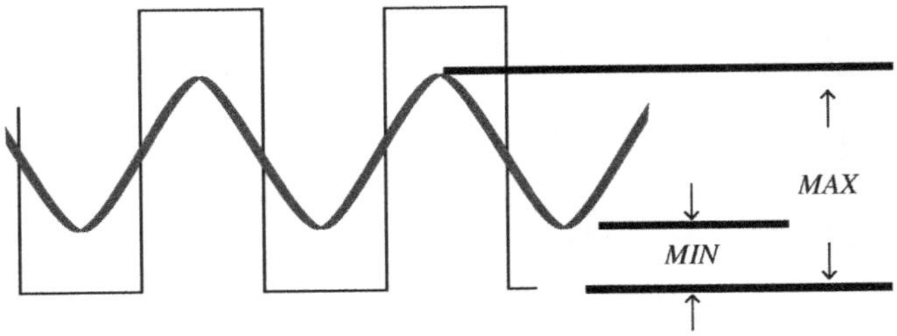

FIGURE 5.7 Blurring effect on an imaged square wave.

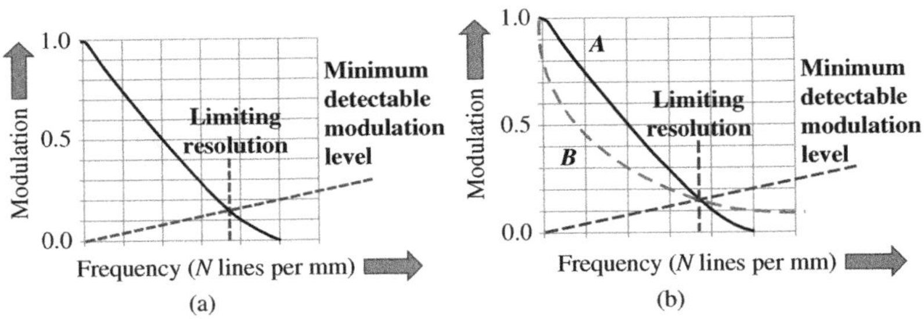

FIGURE 5.8 Plot of modulation as function of line pairs per millimeter for (a) one system and (b) two systems.

Referring back to Figure 5.4, let us now assume an infinite set of bar series in the chart. This allows us to plot the image contrast as a continuous function of line pairs per millimeter, recognizing there is a minimum detectable modulation level that defines the limit of resolution. This is illustrated in Figure 5.8a. What happens if we have two systems with different OTFs but the same limit of resolution? Figure 5.8b depicts an example of that situation. One might conclude that the system (A) with the higher modulation at lower frequencies is the better choice because the coarser images will be sharper and have better contrast. However, if one system exhibits higher limiting resolution while the system that has higher contrast at low frequencies, the choice of superiority is not so clear. The designer will have to consult the customer on what are the more important characteristics desired for the intended applications.

The following example shows the relationship between the referenced contrast reduction and the OTF.

Example 5.1 Let us assume the periodic intensity distribution given in Figure 5.9a:

$$I_0(x) = 1 + A_0 \cos(2\pi f_0 x). \tag{5.45}$$

Fourier transforming Eq. (5.45) yields its associated spectrum:

$$\hat{I}_0(f) = \delta(f) + \tfrac{A_0}{2} \left[\delta(f + f_0) + \delta(f + f_0)\right]. \tag{5.46}$$

Let us assume the OTF can be written as

$$\hat{\Theta}(f) = |\hat{\Theta}(f)| \, e^{i\Phi(f)}. \tag{5.47}$$

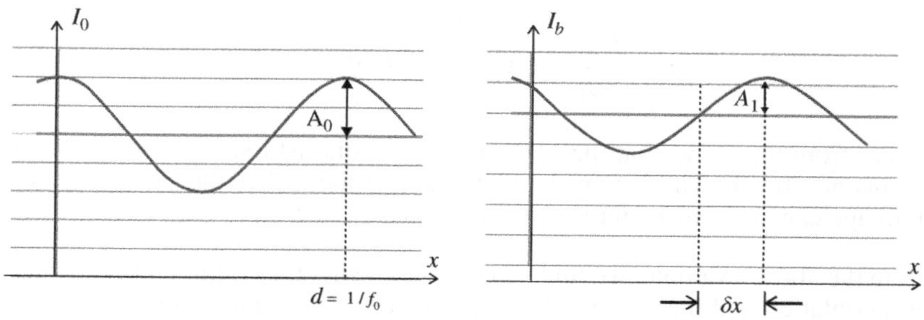

FIGURE 5.9 Plot of (a) $1 + A_0 \cos(2\pi f_0 x)$ and (b) $1 + A_1 \cos(2\pi f_0 x + \varphi(f_0))$.

Multiplying $\hat{I}_0(f)$ and $\Theta(f)$ together, and inverse Fourier transforming the result, we obtain the intensity distribution:

$$I_b(x) = 1 + A_b \cos(2\pi f_0 x + \Phi(f_0)), \tag{5.48}$$

where

$$A_b = A_0 |\hat{\Theta}(f_0)| \tag{5.49}$$

and

$$\delta x = \tfrac{-\Phi(f_0)}{2\pi f_0}. \tag{5.50}$$

The intensity distribution given in Eq. (5.48) is depicted in Figure 5.9b and shows a cosine function modulation like the original object modulation but with a lateral phase shift. This example shows that a cosine intensity distribution is always imaged as a cosine intensity distribution. It also shows that if the PTR is linear with frequency, then it creates a lateral shifting of the pattern. However, if the PTR is nonlinear, there will be an effect on image quality. For example, a phase shift of 180° creates a phase reversal. We will see an example of this in a later section.

It is customary to normalize the OTF at zero frequency, that is, $\hat{\Theta}(0) = 1$, which translates into

$$\int I_0(x)\, dx = \hat{I}_0(0) = \hat{I}_b(0) = \int I_b(x)\, dx. \tag{5.51}$$

In reality, not all of the object energy per time $\int I_0(x)\, dx$ will arrive in the image plane since the object radiates in much wider angle than the optical systems can collect; this is another degradation that the finite apertures occurring at the entrance, within, and exit of any optical system will create, besides image blurring. Consequently, most analyses one does deal only with relative brightness, not the absolute brightness.

Using the information in this example, we see for the object and image Michelson contrasts that

$$C_{\text{obj}} = \frac{(1 + A_0) - (1 - A_0)}{(1 + A_0) + (1 - A_0)} = \frac{2A_0}{2} = A_0, \tag{5.52}$$

$$C_{\text{img}} = \frac{(1 + A_b) - (1 - A_b)}{(1 + A_b) + (1 - A_b)} = \frac{2A_b}{2} = A_b = A_0|\hat{\Theta}(f_0)|, \tag{5.53}$$

and

$$\frac{C_{\text{img}}}{C_{\text{obj}}} = |\widehat{\Theta}(f_0)| \leq |\widehat{\Theta}(0)|. \tag{5.54}$$

It is clear from the above that the MTF denotes a contrast reduction in the image at any frequency, but dc. This is why the OTF is sometimes called the CTF. The contrast of an image can never be better than the original.

From the above example, we find that we can predict the output contrast using the input modulation and the system OTF. Specifically, for an input signal of the form, we saw that an input of the form

$$I_{\text{input}}(x) = 1 + A_0 \cos(2\pi f_0 x)$$

yields the system output image

$$I_{\text{output}}(x) = 1 + A_0 |\widehat{\Theta}(f_0)| \cos(2\pi f_0 x + \Phi(f_0)) . \tag{5.55}$$

The above output image was created by Fourier transforming the product of the input modulation and OTF. It is clear that magnitude of the cosine modulation term in the output image will be dictated by the MTF, which typically decreases with frequency, and the phase shift by the PTF.

The following example shows the relationship between the referenced contrast reduction and the OTF.

Example 5.2 Assume that the MTF of the optical systems is depicted at the top of Figure 5.10 and $\Phi(f_0) = 0$. Figure 5.10a–c illustrates the output function given by Eq. (5.55) created by those MTF and PTF for input sine wave frequencies (a) $f_0 = 0.2$, (b) $f_0 = 0.4$, and (c) $f_0 = 0.8, f_0 = 0.2$, respectively. It is clear that magnitude of the output is diminished by the MTF at that frequency, but no shift in the sine wave results.

When the object image really is a "sine wave" and not a "square wave," we still get a sine wave image from the optical system, as seen in the above examples. This has led to the community into adopting the MTF as the means for characterizing the performance of any optical system using this type of input function. In this case, the MTF is the ratio of the modulation in the image to that of the object as a function of frequency of the sine wave pattern. Figure 5.11 shows the image contrast for a chirped intensity waveform (These results are from a digital simulation of an analog (optical) system. Please note the effect of image sampling that is apparent at high frequencies in the second graph.). This is a more concise way of displaying effect on the MTF, and OTF in general, on input function comprised on multiple spatial frequencies. The plot of the MTF against frequency is the metric of choice for characterizing any optical instrument or mechanism, for example, film, CCDs, the eye, telescopes, and night vision goggles.

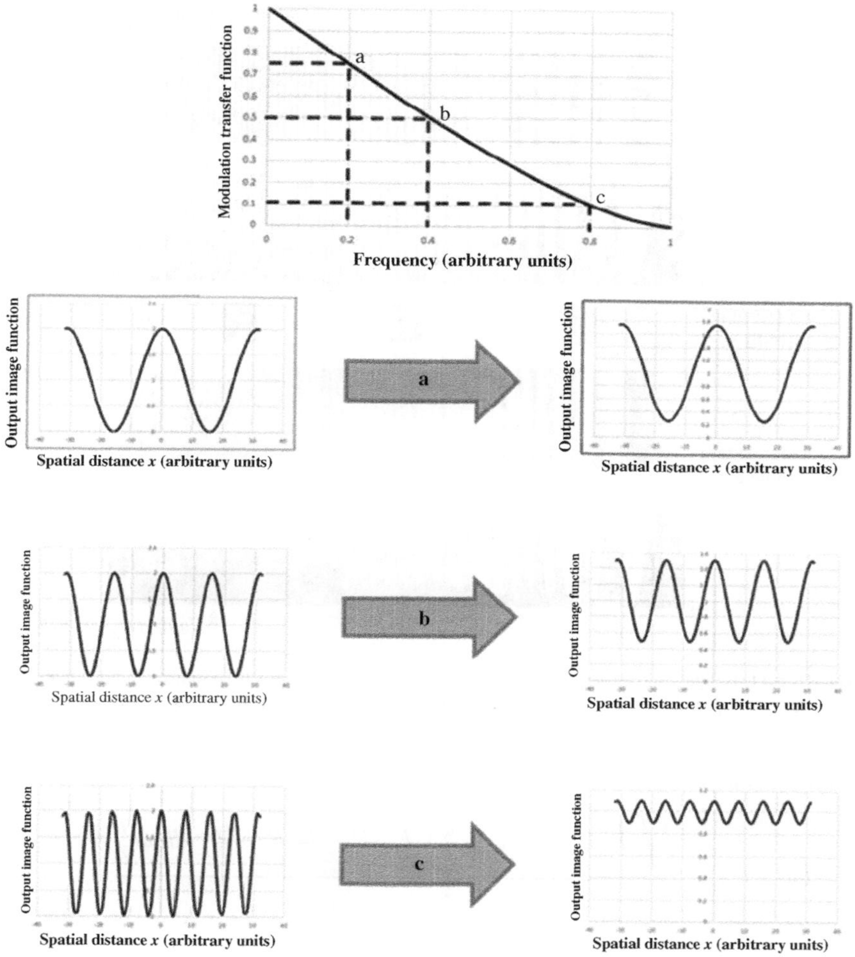

FIGURE 5.10 Plot of $1 + A_0\, |\widehat{\Theta}(f_0)|\, \cos(2\pi f_0\, x + \Phi(f_0))$ for a specific MTF and $\Phi(f_0) = 0$.

5.6 THE DUFFIEUX FORMULA

Equation (5.39a) stated that the OTF as the (normalized) Fourier transform of the incoherent PSF. In terms of the ASF in two-dimensions, we have

$$\widehat{\Theta}(k_x, k_y) = \int\int_{-\infty}^{\infty} |F(x, y)|^2 e^{-2\pi i(k_x\, x + k_x\, y)} dxdy, \tag{5.56}$$

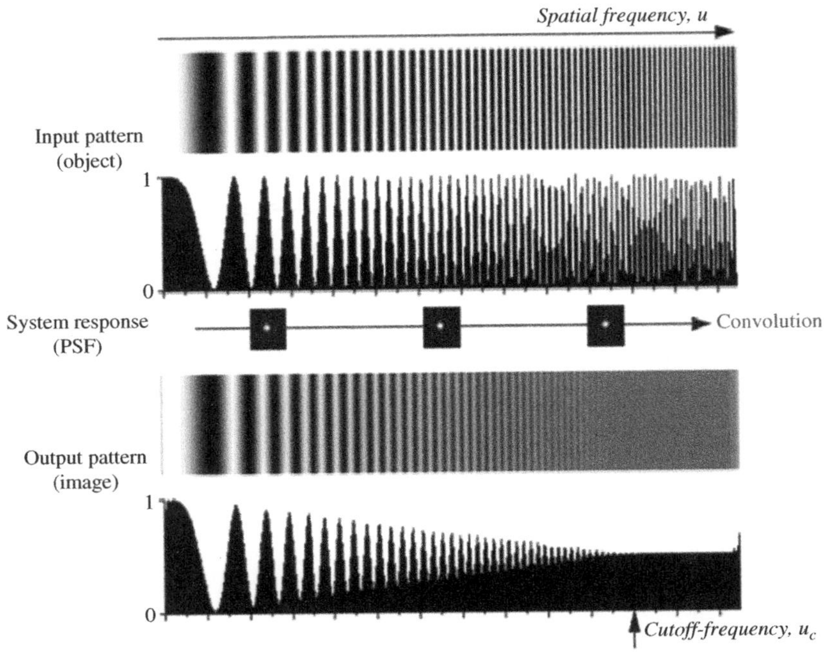

FIGURE 5.11 Image contrast for a chirped spatial intensity waveform [8]. Source: Schowengerdt [8]. Reproduced with permission from Professor Emeritus Robert A. Schowengerdt, University of Arizona.

which leads to

$$\hat{\Theta}(k_x,k_y) = \frac{\iint \hat{F}\left(k_x' + \frac{k_x}{2}, k_y' + \frac{k_y}{2}\right) \hat{F}^*\left(k_x' - \frac{k_x}{2}, k_y' - \frac{k_y}{2}\right) dk_x' dk_y'}{\iint |\hat{F}(k_x',k_y')|^2 dk_x' dk_y'}. \tag{5.57}$$

Writing Eq. (5.56) in terms of true coordinates, we have

$$\hat{\Theta}(k_x,k_y) = \int\int_{-\infty}^{\infty} |F(x,y)|^2 e^{-\frac{2\pi i}{\lambda f}(xx'+yy')}\, dxdy. \tag{5.58}$$

Making use of the well-known theorem for the Fourier transform of a product we used earlier, Eq. (5.57) becomes

$$\hat{\Theta}(k_x,k_y) = \frac{\iint \hat{F}\left(x + \frac{\lambda f k_x}{2}, y + \frac{\lambda f k_y}{2}\right) \hat{F}^*\left(x - \frac{\lambda f k_x}{2}, y - \frac{\lambda f k_y}{2}\right) dxdy}{\iint |\hat{F}(x,y)|^2 dxdy}. \tag{5.59}$$

Equations (5.57) and (5.59) are known as the *Duffieux Formula*. The numerator of formula is the autocorrelation function of the Complex Pupil Function, $\hat{F}(k_x,k_y)$. The denominator is the area under the Complex Pupil Function, which is the normalizing quantity. Thus, the OTF for a diffraction-limited optical system can be computed either by the Fourier transform of the PSF, or by the normalized complex convolution of the ASF and its conjugate.[2] The importance of the latter approach is that it does not require one to deal with the details of a complicated diffraction pattern from a point source. In fact, for a uniform transmission over an aperture, one may write

$$\hat{F}(k_x,k_y) = \begin{cases} e^{ik\,\Delta(k_x,k_y)}; & \text{for } k_x,k_y \subset A \\ 0; & \text{for } k_x,k_y \not\subset A \end{cases}, \tag{5.60}$$

where $\Delta(k_x,k_y)$ represents the deviations of the actual wavefront (the aberrations) from a reference spherical wavefront [10].

Example 5.3 The pupil function for a perfect rectangular lens with no aberrations is given by

$$\hat{F}(x,y) = \text{rect}\left(\frac{x}{2a}\right)\text{rect}\left(\frac{y}{2b}\right). \tag{5.61}$$

The numerator in Eq. (5.59) is given by the overlapping area of two shifted rectangles with x and y lengths $2a$ and $2b$, respectively. Figure 5.12 depicts this situation graphically.

Substituting Eq. (5.61) into Eq. (5.59), it is clear that the two integrations are independent of the other parameter. Let us first look at the case where the centers of the rectangles are centered on the x-axis. This is shown in Figure 5.13. This means

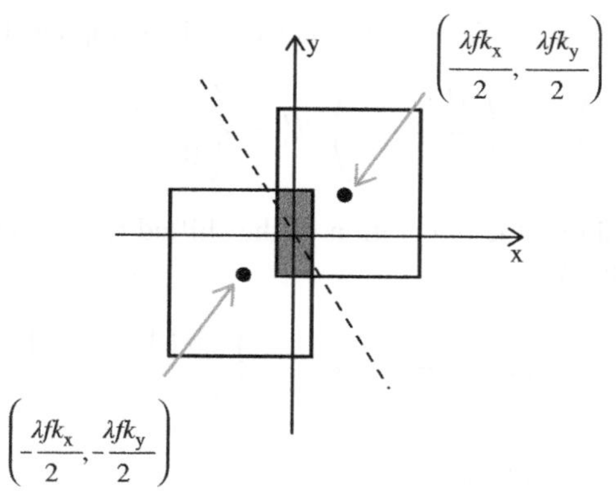

FIGURE 5.12 Graphical depiction of the overlap between two circles of diameter D and center positions $\left(\frac{\lambda f k_x}{2}, \frac{\lambda f k_y}{2}\right)$ and $\left(-\frac{\lambda f k_x}{2}, -\frac{\lambda f k_y}{2}\right)$.

[2]This analysis in this part ignores the phase term in the ASF, which is important in coherent imaging. The effect of this is that the results are qualitatively correct, but we have lost the speckle effects that occur in coherent light. The effects of the phase term are considered in the Goodman's book on speckle [9], but a detailed description of speckle effects are beyond this introductory textbook. The author recommends Ref. [9] to interested readers.

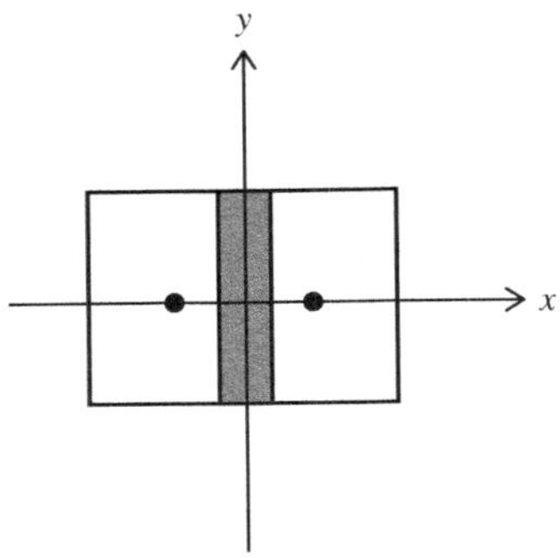

FIGURE 5.13 Picture showing the area of overlap of two displaced rectangles centered on the x-axis.

that the integrations of the shifted rectangles in the x-direction on the x-axis are given by

$$\int_{-\infty}^{\infty} \mathrm{rect}\left(\frac{x' + \frac{\lambda f k_x}{2}}{2a}\right) \mathrm{rect}\left(\frac{x' - \frac{\lambda f k_x}{2}}{2a}\right) dx' = \int_{-a+\frac{\lambda f k_x}{2}}^{a-\frac{\lambda f k_x}{2}} dw = 2a\left(1 - \frac{\lambda f k_x}{2a}\right) \tag{5.62a}$$

and

$$\int_{-\infty}^{\infty} \mathrm{rect}\left(\frac{x' + \frac{\lambda f k_x}{2}}{2a}\right) \mathrm{rect}\left(\frac{x' - \frac{\lambda f k_x}{2}}{2a}\right) dx' = 2a\left(1 + \frac{\lambda f k_x}{2a}\right) \tag{5.62b}$$

for $0 \le \lambda f k_x < a$ and $-a \le \lambda f k_x < 0$, respectively. This implies that we can write

$$\int_{-\infty}^{\infty} \mathrm{rect}\left(\frac{x' + \frac{\lambda f k_x}{2}}{2a}\right) \mathrm{rect}\left(\frac{x' - \frac{\lambda f k_x}{2}}{2a}\right) dk_x' = \begin{cases} 2a\left(1 - \frac{|\lambda f k_x|}{2a}\right); & \text{for } |\lambda f k_x| \le a \\ 0; & \text{otherwise} \end{cases}. \tag{5.63}$$

Similarly, we find that the integration of the shifted rectangles in the y-direction on the x-axis is given by

$$\int_{-\infty}^{\infty} \mathrm{rect}\left(\frac{y' + \frac{\lambda f k_y}{2}}{2b}\right) \mathrm{rect}\left(\frac{y' - \frac{\lambda f k_y}{2}}{2b}\right) dy' = \begin{cases} 2b\left(1 - \frac{|\lambda f k_y|}{2b}\right); & \text{for } \ |\lambda f k_y| \le b \\ 0; & \text{otherwise} \end{cases}. \tag{5.64}$$

This implies that Eq. (5.52) equals

$$\begin{aligned} \hat{\Theta}(k_x, k_y) &= \frac{(2a)(2b)\left(1 - \frac{|\lambda f k_x|}{2a}\right)\left(1 - \frac{|\lambda f k_y|}{2b}\right)}{(2a)(2b)} \\ &= \left(1 - \frac{|\lambda f k_x|}{2a}\right)\left(1 - \frac{|\lambda f k_y|}{2b}\right) = \Lambda\left(\frac{|\lambda f k_x|}{2a}\right)\Lambda\left(\frac{|\lambda f k_y|}{2b}\right) && (5.65a) \\ &= \Lambda\left(\frac{|k_x|}{2\gamma_{xc}}\right)\Lambda\left(\frac{|k_y|}{2\gamma_{yc}}\right), && (5.65b) \end{aligned}$$

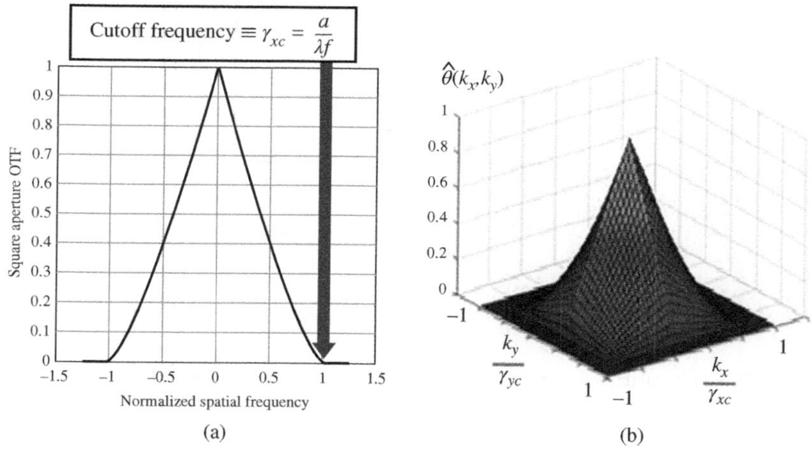

FIGURE 5.14 Graphs of the OTF for a square aperture in (a) one dimension and (b) two dimensions.

where

$$\Lambda(x) = \begin{cases} 1 - |x| ; & \text{for } |x| \le 1 \\ 0; & \text{otherwise} \end{cases}, \tag{5.66}$$

$$\gamma_{xc} = \frac{a}{\lambda f}, \tag{5.67a}$$

and

$$\gamma_{yc} = \frac{b}{\lambda f}. \tag{5.67b}$$

Here the denominator of Eq. (5.59) is just the area of the rectangle. Figure 5.14 shows Eq. (5.65b) in (a) the x-dimension and (b) two dimensions.

Example 5.4 The pupil function for a perfect lens with no aberrations is given by

$$\hat{F}(x, y) = 1, \quad if\,(x^2 + y^2) \le \frac{D}{2}; 0, \quad \text{otherwise}. \tag{5.68}$$

The numerator in Eq. (5.59) is given by the overlapping area of two shifted circles with diameter D and center positions $\left(\frac{\lambda f k_x}{2}, \frac{\lambda f k_y}{2}\right)$ and $\left(\frac{-\lambda f k_x}{2}, \frac{-\lambda f k_y}{2}\right)$. Figure 5.15 depicts this situation graphically. Let us now calculate its OTF.

The overlap area can be calculated using simple geometry. Let us assume the overlap orientation shown in Figure 5.16a. Figure 5.16b depicts a quarter sector of the overlap area in Figure 5.16a, which we designate "B." The area of an "ice cream cone" sector created by one of the circles is equal to

$$A_{\text{ice cream cone}} \approx \frac{2\theta}{2\pi}\left[\pi\left(\frac{D}{2}\right)^2\right].$$

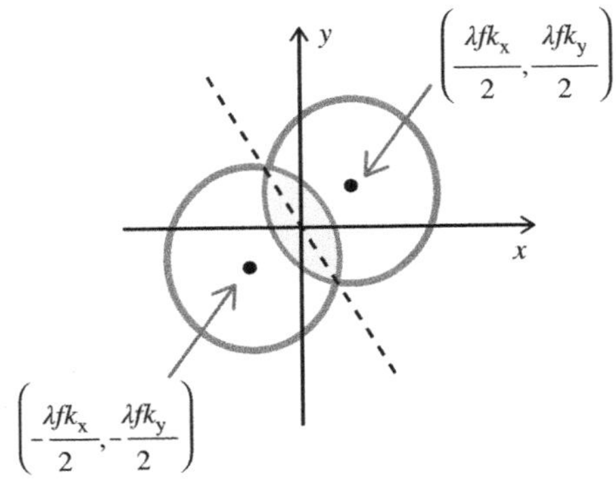

FIGURE 5.15 Graphical depiction of the overlap between two circles of diameter D and center positions $\left(\frac{\lambda f k_x}{2}, \frac{\lambda f k_y}{2}\right)$ and $\left(-\frac{\lambda f k_x}{2}, -\frac{\lambda f k_y}{2}\right)$.

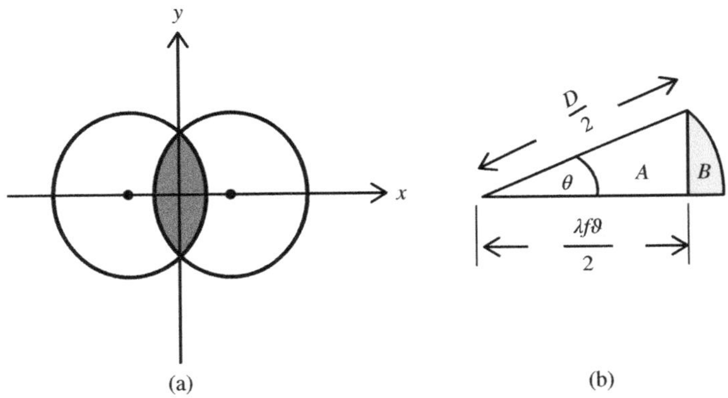

FIGURE 5.16 (a) Pictures showing the area of overlap of two displaced circles and (b) the geometry for calculations specific fraction of that area.

The area of the triangle shown in Figure 5.5b is equal to

$$A_{\text{Triangle}} \approx \frac{1}{2}\left(\frac{(\lambda f \vartheta)}{2}\right)\left(\frac{D}{2}\right)\sin\theta.$$

From that same triangle, we see that

$$\cos\theta = \frac{\frac{(\lambda f \vartheta)}{2}}{\left(\frac{D}{2}\right)} = \frac{\lambda f \vartheta}{D}$$

or

$$\theta = \cos^{-1}\left(\frac{\lambda f \vartheta}{D}\right).$$

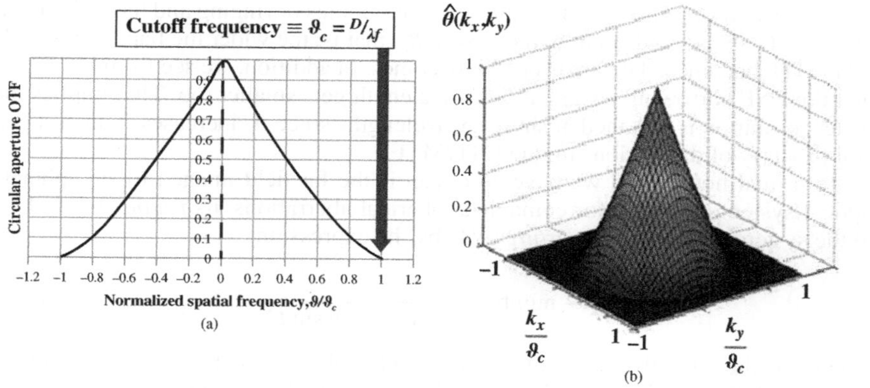

FIGURE 5.17 Graphs of the OTF for a circular aperture in (a) one dimension and (b) two dimensions.

This implies that

$$\begin{aligned}4\,(\text{Area } B) &= \frac{4\theta}{2\pi}\left[\pi\left(\frac{D}{2}\right)^2\right] - 2\left(\frac{(\lambda f\vartheta)D}{4}\right)\sin\theta \\ &= 2\left(\frac{D}{2}\right)^2\left[\theta - \left(\frac{\lambda f\vartheta}{D}\right)\sqrt{1-\cos^2\theta}\right] \\ &= 2\left(\frac{D}{2}\right)^2\left[\cos^{-1}\left(\frac{\lambda f\vartheta}{D}\right) - \left(\frac{\lambda f\vartheta}{D}\right)\sqrt{1-\left(\frac{\lambda f\vartheta}{D}\right)^2}\right].\end{aligned}$$

The above implies that the OTF for a circular aperture lens equals

$$\hat{\Theta}\left(\frac{\vartheta}{\vartheta_c}\right) = \begin{cases}\frac{2}{\pi}\left[\cos^{-1}\left(\frac{\vartheta}{\vartheta_c}\right) - \left(\frac{\vartheta}{\vartheta_c}\right)\sqrt{1-\left(\frac{\vartheta}{\vartheta_c}\right)^2}\right]; & \text{for } \vartheta \le \vartheta_c = \frac{D}{\lambda f} \\ 0; & \text{for } \vartheta > \vartheta_c\end{cases}, \qquad (5.69)$$

where

$$\vartheta_c = \frac{D}{\lambda f} = \frac{1}{\lambda\left(\frac{f}{D}\right)} = \frac{1}{\lambda(f - \text{number})} \qquad (5.70)$$

after normalization. Thus, we have a system with diameter $D = 1$ cm and a focal length $f = 50$ mm, the f − number $= \frac{f}{D}$ is equal to 5, and the cutoff frequency is equal to 363 cycles/mm at $\lambda = 0.55$ μm. This is a very high cutoff. As a comparison, the human vision has a cutoff frequency of 10 cycles/mm (object scale) at normal viewing distances.

Figure 5.17 depicts Eq. (5.69). Figure 5.17a is a plot of Eq. (5.69) versus normalized spatial frequency in one dimension and Figure 5.17b is a two-dimensional image of the OTF.

Unlike some OTFs, this OTF exhibits no harmonics or ringing and has a sharp cut-off at ϑ_c. One also finds that by increasing ϑ_c, that is, the wider the bandwidth of the OTF/MTF becomes, the narrower PSF becomes. In addition, we see from Eq. (5.70), that this PSF narrowing occurs as the wavelength gets smaller too. That implies that better resolution is obtained at shorter wavelengths since ϑ_c increases. This is a general property of diffraction-limited OTF/MTF.

As a final note, when we have an object is the far field imaged by an aplanatic optical system corrected for coma and spherical aberrations, its *f-number* is related to the system's *numerical aperture* (*NA*) by the expression:

$$f - \text{number} = \frac{1}{2NA} = \frac{1}{2n \sin U}, \tag{5.71}$$

where n is the refractive index of the medium the object is and U is the half-angle of the cone of illumination. One finds that NA is used to characterize systems that work at finite conjugates like microscope objectives, and *f-number* that image objects at long distances away.

Let us now provide some additional terminology definitions related to the *f-number*. The terms *fast* and *slow* often are applied to the *f*-number of the optical system to describe its *speed* [1, p. 144]. A system with a large entrance aperture captures more light and has a small *f-number*, but is said to be "fast" or to have high "speed." Conversely, an optical system with a small entrance aperture, which has a large *f-number*, will be characterized as "slow" or of low "speed."

5.7 OBSCURED APERTURE OTF

The diffraction-limited OTF/MTF is the upper limit in performance of any optical system. That is, no optical systems can perform better than its diffraction-limited OTF/MTF because any aberrations in the systems will only make things worse, dragging down the curve, thus essentially lowering its useable spatial frequency band. However, an exception occurs when one puts an obstruction in the middle of the entrance aperture, like what is found in reflecting telescopes. In this case, the "donut" aperture lowers the OTF at low spatial frequencies and also increases the OTF at higher spatial frequencies. Computer calculations confirm similar tendencies for both circular apertures and circular obscurations [9, p. 198].

Example 5.5 Let us see what happens to the circular aperture OTF with a central circular obscuration, following O'Neil [10]. Figure 5.18 shows the geometry of this problem.

Assume the radius of the outer aperture equals a and the radius of the obscuration is εa. Recalling Eq. (2.107), we can write the obscured aperture field amplitude in the observation plane as

$$\hat{u}(r) = 2\pi \int_{\varepsilon a}^{a} J_0(r\rho)\, \rho \, d\rho, \tag{5.72}$$

and its normalized results is given by

$$\hat{u}(r) = \frac{1}{(1-\varepsilon^2)} \left[\frac{2J_1(ar)}{ar} - \epsilon^2 \frac{2J_1(\varepsilon a\ r)}{\varepsilon a\ r} \right]. \tag{5.73}$$

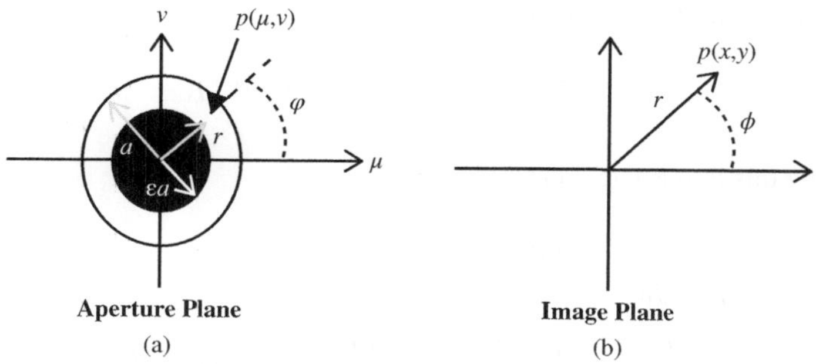

FIGURE 5.18 Geometry for (a) a centrally obscured circular lens and (b) the image plane.

The resulting intensity therefore is

$$I(r) = |\hat{u}(r)|^2 = \frac{1}{(1-\varepsilon^2)^2}\left\{\left[\frac{2J_1(S)}{S}\right]^2 + \int^4 \left[\frac{2J_1(\varepsilon S)}{\varepsilon S}\right]^2 - 8\varepsilon\frac{J_1(S)J_1(\varepsilon)}{S^2}\right\}$$

$$= I(S), \tag{5.74}$$

where

$$S = ar. \tag{5.75}$$

The OTF is equal to

$$\hat{\Theta}(\omega) = \frac{1}{a^2}\int_0^\infty \int_0^{2\pi} I(S) e^{i\left(\frac{w}{a}\right)S\cos\theta}\, S\, dS\, d\theta \tag{5.76}$$

$$= \frac{1}{(a\,(1-\varepsilon^2))^2}\{4B_1 + 4\varepsilon^4 B_2 - 8\varepsilon B_3\}. \tag{5.77}$$

In Eq. (5.77), we have

$$B_1 = \int_0^\infty \int_0^{2\pi} \left[\frac{J_1(S)\ \cos}{\sqrt{S}}\right]^2 e^{i\Omega S\cos\theta}\, dS\, d\theta, \tag{5.78}$$

$$B_2 = \frac{1}{\varepsilon^2}\int_0^\infty \int_0^{2\pi} \left[\frac{J_1(\varepsilon S)}{\sqrt{\varepsilon S}}\right]^2 e^{i\left(\frac{\Omega}{\varepsilon}\right)\varepsilon S\cos\theta}\, d(\varepsilon S)\, d\theta, \tag{5.79}$$

and

$$B_3 = \int_0^\infty \int_0^{2\pi} \frac{J_1(S)J_1(\varepsilon S)}{S}\, e^{i\Omega S\cos\theta}\, dS\, d\theta. \tag{5.80}$$

The above three equations all have the form

$$J = \int_0^\infty \int_0^{2\pi} \frac{J_1(a\rho)J_1(b\rho)}{\rho}\, e^{iz\rho\cos\theta}\, d\rho\, d\theta. \tag{5.81}$$

To solve Eq. (5.78), O'Neil noted that

$$e^{iz\rho\,\cos\theta} = 1 + i\rho\cos\theta \int_0^z e^{ix\rho\,\cos\theta} dx. \tag{5.82}$$

Substituting this into Eq. (5.81) yields

$$\begin{aligned} J &= \int_0^\infty \int_0^{2\pi} \frac{J_1(a\rho)J_1(b\rho)}{\rho}\, d\rho\, d\theta \\ &\quad + i \int_0^z \int_0^\infty \int_0^{2\pi} \frac{J_1(a\rho)J_1(b\rho)}{\rho}\, e^{ix\rho\,\cos\theta} \rho\cos\theta\, d\rho\, d\theta\, dx \\ &= \int_0^\infty \int_0^{2\pi} \frac{J_1(a\rho)J_1(b\rho)}{\rho}\, d\rho\, d\theta \\ &\quad + i \int_0^z \int_0^\infty \int_0^{2\pi} J_1(a\rho)J_1(b\rho)\, e^{ix\rho\,\cos\theta} \cos\theta\, d\rho\, d\theta dx. \end{aligned} \tag{5.83}$$

Integrating Eq. (5.83) over θ allows us to write

$$J = 2\pi \int_0^\infty \frac{J_1(a\rho)J_1(b\rho)}{\rho}\, d\rho - \int_0^z \int_0^\infty J_1(a\rho)J_1(b\rho)J_1(x\rho)\, d\rho\, dx \tag{5.84a}$$

$$= 2\pi \int_0^\infty \frac{J_1(a\rho)J_1(b\rho)}{\rho}\, d\rho - \int_0^z G(x)\, dx \tag{5.84b}$$

with

$$G(x) = \int_0^\infty J_1(a\rho)J_1(b\rho)J_1(x\rho) d\rho. \tag{5.85}$$

From Ref. [11, Eq. (1), p. 405], we know that the first integral in Eq. (5.84b) follow the following equation:

$$2\pi \int_0^\infty \frac{J_\alpha(a\rho)J_\alpha(b\rho)}{\rho}\, d\rho = \begin{cases} \frac{1}{2\alpha}\left(\frac{b}{a}\right)^\alpha; & for\ b \le a \\ \frac{1}{2\alpha}\left(\frac{a}{b}\right)^\alpha; & for\ b \ge a \end{cases}, \tag{5.86}$$

which in our case equals

$$2\pi \int_0^\infty \frac{J_1(a\rho)J_1(b\rho)}{\rho}\, d\rho = \frac{1}{2}\left(\frac{b}{a}\right) \tag{5.87}$$

because of the size of $\in < 1$.Now the integral in Eq. (5.85) also is a form found in Ref. [11]; specifically, we find that

$$\int_0^\infty J_\mu(a\rho)J_\mu(b\rho)J_\mu(x\rho) d\rho = \frac{2^{\mu-1}\Delta^{2\mu-1}}{(abx)^\mu\, \Gamma\left(\mu + \frac{1}{2}\right)\Gamma\left(\frac{1}{2}\right)}, \tag{5.88}$$

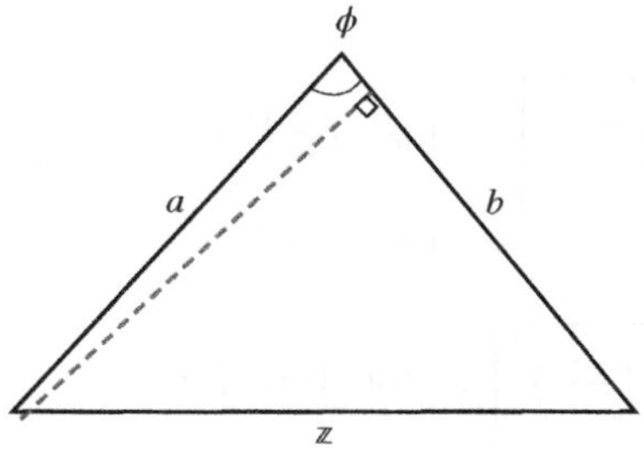

$$\Delta = \frac{1}{2}\, ab \sin \phi;\ \text{for } a - b \le z \le a + b$$

$$z^2 = a^2 + b^2 - 2ab \cos \phi$$

$$\frac{dz}{z^2} = \frac{ab \sin \phi \, d\phi}{z^2} = \frac{ab \sin \phi \, d\phi}{a^2 + b^2 - 2ab \cos \phi}$$

FIGURE 5.19 Geometrical interpretation of Δ given in Eq. (5.92).

which becomes in our case

$$\int_0^{\infty} J_1(a\rho) J_1(b\rho) J_1(x\rho) d\rho = \frac{\Delta}{(abx)\,\Gamma\left(\frac{3}{2}\right)\Gamma\left(\frac{1}{2}\right)} = \frac{\Delta}{(abx)\left(\frac{\sqrt{\pi}}{2}\right)(\sqrt{\pi})} = \frac{2\Delta}{\pi(abx)}, \tag{5.89}$$

where a, b, and x are the sides of a triangle whose area is Δ [11, Eq. (3), p. 411]. If a, b, and x are not the sides of a triangle, then the integral is zero. The interpretation of the Δ is shown in Figure 5.19. In the above two equations, $\Gamma(x)$ again is the Gamma function.

Using the above, we can rewrite Eq. (5.84b) as

$$J = \begin{cases} \dfrac{\pi b}{a}; & \text{for } z < a - b \\ \dfrac{\pi b}{a} - 2ab \displaystyle\int_0^{\pi} \frac{\sin^2 \phi}{a^2 + b^2 - 2ab \cos \phi} d\phi; & \text{for } a - b < z < a + b \\ \dfrac{\pi b}{a} - 2ab \displaystyle\int_0^{\pi} \frac{\sin^2 \phi}{a^2 + b^2 - 2ab \cos \phi} d\phi; & \text{for } z > a + b \end{cases}, \tag{5.90}$$

where

$$\phi = \cos^{-1}\left[\frac{a^2 + b^2 - z^2}{2ab}\right]. \tag{5.91}$$

Performing the ϕ-integration in Eq. (5.90) yields

$$J = \begin{cases} \dfrac{\pi b}{a}; & \text{for } z < a - b \\ \dfrac{\pi b}{a} - \sin \phi - \left[\dfrac{a^2 + b^2}{2ab}\right]\phi & \\ \quad + \left[\dfrac{a^2 - b^2}{ab}\right]\tan^{-1}\left[\left(\dfrac{a + b}{a - b}\right)\tan \dfrac{\phi}{2}\right] & \text{for } a - b < z < a + b \\ 0; & \text{for } z > a + b \end{cases}. \tag{5.92}$$

Now returning to our original equation given in Eq. (5.76), we have

$$\widehat{\Theta}(\omega) = \frac{1}{(1 - \varepsilon^2)}[B_4 + B_5 + B_6], \tag{5.93}$$

where

$$B_4 = \begin{cases} \frac{2}{\pi}\left[\cos^{-1}\left(\frac{\Omega}{2}\right) - \left(\frac{\Omega}{2}\right)\sqrt{1-\left(\frac{\Omega}{2}\right)^2}\right]; & \text{for } 0 \le \frac{\Omega}{2} \le 1 \\ 0; & \text{for } \frac{\Omega}{2} > 1 \end{cases}, \tag{5.94}$$

$$B_5 = \begin{cases} \frac{2\varepsilon^2}{\pi}\left[\cos^{-1}\left(\frac{\Omega}{2\varepsilon}\right) - \left(\frac{\Omega}{2\varepsilon}\right)\sqrt{1-\left(\frac{\Omega}{2\varepsilon}\right)^2}\right]; & \text{for } 0 \le \frac{\Omega}{2\varepsilon} \le 1 \\ 0; & \text{for } \frac{\Omega}{2\varepsilon} > 1 \end{cases}, \tag{5.95}$$

and

$$B_6 = \begin{cases} -2\varepsilon^2; & \text{for } 0 \le \frac{\Omega}{2} \le \frac{1-\varepsilon}{2} \\ -2\varepsilon^2 + \frac{2\varepsilon}{\pi}\sin\phi + \left[\frac{1+\varepsilon^2}{\pi}\right]\phi & \\ \quad - \left[\frac{2(1-\varepsilon^2)}{\pi}\right]\tan^{-1}\left[\left(\frac{1+\varepsilon}{1-\varepsilon}\right)\tan\frac{\phi}{2}\right]; & \text{for } \frac{1-\varepsilon}{2} < \frac{\Omega}{2} < \frac{1+\varepsilon}{2} \\ 0; & \text{for } \frac{\Omega}{2} > \frac{1+\varepsilon}{2} \end{cases}. \tag{5.96}$$

In the above equations, we have

$$\phi = \cos^{-1}\left[\frac{1+\varepsilon^2-\Omega^2}{2\varepsilon}\right]. \tag{5.97}$$

Figure 5.20 depicts the a centrally obscured circular lens OTF for $\varepsilon = 0$, 0.25, 0.50, and 0.75. This figure shows the high spatial frequency response of the system is improved at the expense of low-frequency detail (contrast). This implies a decrease in the width of the central maximum with more light filling up the outer rings of the point image diffraction pattern. O'Neil pointed out that since the curves shown are normalized, it must be remembered that any gain achieved is accompanied by a serious loss in the incoming irradiance as a result of the obstruction. Finally, for a thin annular ring, he also noted that from the shape of the curve a sharp peak near the resolution limit of the system appears to be obtainable as ε approaches unity. This is true, and is nothing more than the two-dimensional analog of a Young's double slit aperture, the peak occurring at the spatial frequency of the fringes [10].

As a final note, one also can use variable transmission filter or coating at the aperture to modify the OTF to modify the diffraction pattern [12, p. 360]. This is called *apodization*. Typically, the filter provides minimal attenuation at the center of the pupil, but increases it as one moves radially away from the center. This amounts to a "softening" of the edges of the aperture through the introduction of an attenuating mask. We saw in Chapter 2 that diffraction by hard apertures is created by edge waves originating around the rim of the said aperture. The "softening" of the edge by apodization spreads the origin of these diffracted waves over a broader area around the edges of the pupil, thereby suppressing any ringing effects that would originate by edge waves with a highly localized origin. This approach provides a partial

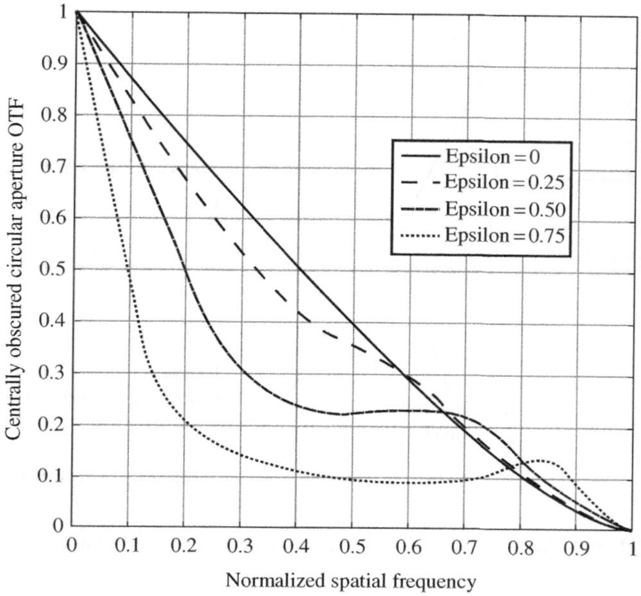

FIGURE 5.20 Plot of the centrally obscured circular Lens OTF for various obscuration levels.

suppression of the secondary maxima by an appropriate modification of the pupil function [4, p. 417]. In spectroscopic analysis, it facilitates the detection of spectral lines; in astronomical applications, it helps resolve double stars with very different apparent brightness.[3]

5.7.1 Aberrations

Recall that the ideal effect of a lens is to impart a quadratic phase shift, multiplied by an aperture function. Aberrations distort this quadratic phase shift. The result is that their effects can be modeled as a complex phase shifting plate at the exit pupil. In particular, the model is given by

$$F_{\mathrm{ab}}(x, y) = F(x, y)e^{ikW(x,y)}, \tag{5.98}$$

[3]The term *brightness* is the same as radiance, but is weighted by the light response of the human eye. It is a term used in photometry, which deals with the radiation that human eye can see. For example, the luminous flux emitted by a source with spectral power $P(\lambda)$ is given by

$$F = 680 \int V(\lambda)\, P(\lambda)\, d\lambda$$

where $V(\lambda)$ is the response of the eye [12, Figure 5.10]. The brightness can be calculated paralleling the radiometry definitions given in Chapter 3. As noted earlier, a more detailed discussion of photometry is not covered in this textbook, but can be found in other books [3, 12, 13].

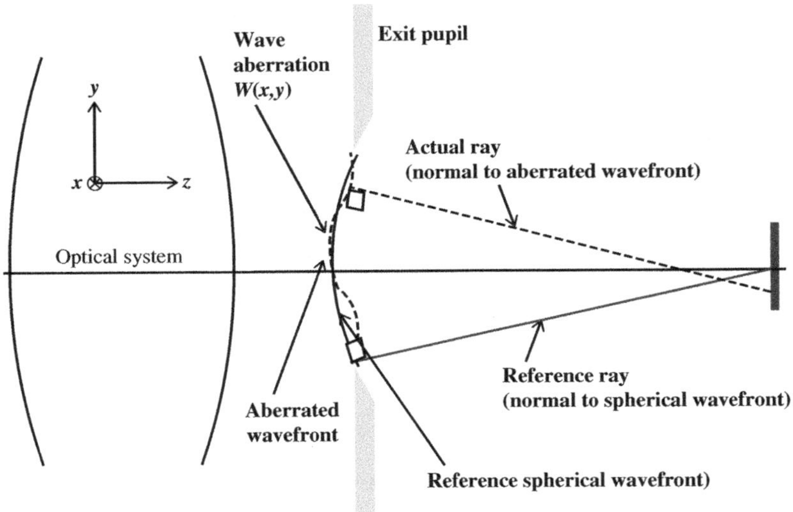

FIGURE 5.21 Wave aberration function at the exit pupil of an optical imaging system.

where $W(x, y)$ is the effective path length error [14, 15]. The wave aberration function, $W(x, y)$, is defined as the distance, in optical path length (product of the refractive index and path length), from the reference sphere to the wavefront in the exit pupil measured along the ray as a function of the transverse coordinates (x, y) of the ray intersection with a reference sphere centered on the ideal image point. It is not the wavefront itself, but it is the departure of the wavefront from the reference spherical wavefront (OPD) as indicated in Figure 5.21. It is generally characterized as the peak-to-valley or root-mean-squared (RMS) Optical Path Difference (OPD) [12, pp. 327–340], defined as

$$\text{RMS-OPD} = \frac{1}{A}\iint W(x, y)^2 dA. \tag{5.99}$$

The convention for describing the OTF with aberrations is through the *CTF*, which is related to the Fourier transform of Eq. (5.59). Specifically, the CTF or Generalized Pupil Function (GPF), is given by the above model with a change in parameters, or

$$\hat{F}_{\text{ab}}(k_x,k_y) = \hat{F}(-\lambda z_i k_x, -\lambda z_i k_y) e^{ikW(-\lambda z_i k_x, -\lambda z_i k_y)}. \tag{5.100}$$

The integral of the square of Eq. (5.100) is the PSF we discussed previously. Again, $W(x, y)$ is the cumulative optical path length error across the pupil for the optical system, $k = \frac{2\pi}{\lambda}$ and λ is the wavelength of light.

It is clear from the above that aberrations introduce a pure phase distortion within an unaffected passband defined by $\hat{F}(-\lambda z_i k_x, -\lambda z_i k_y)$. For coherent illumination, the effects are described by Eq. (5.100). For incoherent illumination, we must deal with the OTF as stated before. Let us focus on circular apertures for the moment.

Recall from Example 5.4 that the OTF for diffraction-limited circular aperture is the normalized overlap area of displaced pupils. Its OTF is given by

$$\widehat{\Theta}_{\mathrm{DL}}(k_x,k_y) = \frac{\displaystyle\int\int_{A(k_x,k_y)} dx\,dy}{\displaystyle\int\int_{A(0,0)} dx\,dy}. \tag{5.101}$$

With aberrations, the OTF is written as

$$\widehat{\Theta}_{\mathrm{ab}}(k_x,k_y) = \frac{\displaystyle\int\int_{A(k_x,k_y)} e^{ik\left[W\left(x+\frac{\lambda f k_x}{2},y+\frac{\lambda f k_y}{2}\right)-W\left(x-\frac{\lambda f k_x}{2},y-\frac{\lambda f k_y}{2}\right)\right]} dx\,dy}{\displaystyle\int\int_{A(0,0)} dx\,dy}. \tag{5.102}$$

The aberration can only decrease the MTF, that is,

$$\left|\widehat{\Theta}_{\mathrm{ab}}(k_x,k_y)\right|^2 \le \left|\widehat{\Theta}_{\mathrm{DL}}(k_x,k_y)\right|^2. \tag{5.103}$$

They do not reduce the absolute cutoff frequency but do reduce the contrast at some frequencies as we showed previously. For a diffraction-limited system, the OTF is positive definite. With aberrations, the OTF can be negative at certain frequencies, resulting in a contrast reversal.

Example 5.6 Let us see what happens to the OTF of a perfect square aperture lens system if he now has a focus error.

The ideal phase distribution across the exit pupil to the focus point in the ideal image plane is given by

$$\Phi_i(x,y) = \frac{\pi}{\lambda f}\left(x^2+y^2\right). \tag{5.104}$$

With a slight blur from a focus error, this phase distribution becomes

$$\Phi_{\mathrm{err}}(x,y) = \frac{\pi}{\lambda(f+\varepsilon)}\left(x^2+y^2\right) \tag{5.105a}$$

$$= \frac{\pi}{\lambda f\left(1+\frac{\varepsilon}{f}\right)}(x^2+y^2) \approx \frac{\pi}{\lambda f}\left(1-\frac{\varepsilon}{f}\right)\left(x^2+y^2\right) \tag{5.105b}$$

$$\approx \frac{\pi}{\lambda f}\left(x^2+y^2\right) - \frac{\pi\varepsilon}{\lambda f^2}\left(x^2+y^2\right). \tag{5.105c}$$

The path length error then equals to

$$W(x,y) = \frac{1}{k}[\Phi_i(x,y) - \Phi_{err}(x,y)] = \frac{\pi\varepsilon}{k\lambda f^2}(x^2+y^2) = \frac{\varepsilon}{2f^2}(x^2+y^2), \tag{5.106}$$

which is seen to depend quadratically on the space variables in the exit pupil.

Let us assume the linear dimension of the square aperture is $2a$. The maximum path errors occur at the edge of the aperture along the x- and y-axes. Thus, the maximum path error can be written as

$$W_{\max} \equiv W(a,0) = \frac{\mathcal{E}a^2}{2f^2}, \tag{5.107}$$

where $\gamma_c = \frac{2a}{\lambda f}$ is the cutoff frequency we previously found for a square aperture. This implies that Eq. (5.106) can be rewritten as

$$W(x,y) = \frac{W_{\max}}{a^2}\left(x^2+y^2\right). \tag{5.108}$$

The integration area $A(k_x,k_y)$ is the area of overlap of the two pupil functions. Let us now calculate the slightly defocused OTF using Eq. (5.59); specifically, we can write

$$\hat{\Theta}_{\mathrm{ab}}(k_x,k_y) = \frac{\iint_{A(k_x,k_y)} e^{ik\left[W\left(x+\frac{\lambda f k_x}{2},y+\frac{\lambda f k_y}{2}\right)-W\left(x-\frac{\lambda f k_x}{2},y-\frac{\lambda f k_y}{2}\right)\right]}\,dxdy}{\iint_{A(0,0)} dxdy} \tag{5.109}$$

$$= \frac{\left[\frac{\sin\left[\frac{2kW_{\max}}{2a^2}[\lambda f k_x](\lambda f|k_x|-2a)\right]}{\frac{kW_{\max}}{a^2}[\lambda f k_x]}\right]\left[\frac{\sin\left[\frac{2kW_{\max}}{2a^2}[\lambda f k_y](\lambda f|k_y|-2a)\right]}{\frac{kW_{\max}}{a^2}[\lambda f k_y]}\right]}{\iint_{A(0,0)} dxdy}$$

$$= \frac{a^2}{\iint_{A(0,0)} dxdy}\left[\frac{\sin\left[\frac{2kW_{\max}}{a}[\lambda f k_x]\left(1-\frac{\lambda f|k_x|}{2a}\right)\right]}{\frac{kW_{\max}}{a}[\lambda f k_x]}\right]$$

$$\times\left[\frac{\sin\left[\frac{2kW_{\max}}{a}[\lambda f k_y]\left(1-\frac{\lambda f|k_y|}{2a}\right)\right]}{\frac{kW_{\max}}{a}[\lambda f k_y]}\right]$$

$$= \left[\frac{\sin\left[\frac{2kW_{\max}}{a}[\lambda f k_x]\left(1-\frac{\lambda f|k_x|}{2a}\right)\right]}{\frac{2kW_{\max}}{a}[\lambda f k_x]}\right]\left[\frac{\sin\left[\frac{2kW_{\max}}{a}[\lambda f k_y]\left(1-\frac{\lambda f|k_y|}{2a}\right)\right]}{\frac{2kW_{\max}}{a}[\lambda f k_y]}\right].$$

$$= \left(1-\frac{|k_x|}{\gamma_c}\right)\left(1-\frac{|k_y|}{\gamma_c}\right)\left[\frac{\sin\left[4\pi\frac{W_{\max}}{\lambda}\left[\frac{k_x}{\gamma_c}\right]\left(1-\frac{|k_x|}{2\gamma_c}\right)\right]}{4\pi\frac{W_{\max}}{\lambda}\left[\frac{k_x}{\gamma_c}\right]\left(1-\frac{|k_x|}{2\gamma_c}\right)}\right]$$

$$\times\left[\frac{\sin\left[4\pi\frac{W_{\max}}{\lambda}\left[\frac{k_x}{\gamma_c}\right]\left(1-\frac{|k_y|}{2\gamma_c}\right)\right]}{2\pi i\frac{W_{\max}}{\lambda}\left[\frac{k_x}{\gamma_c}\right]\left(1-\frac{|k_x|}{2\gamma_c}\right)}\right] \tag{5.110}$$

$$= \Lambda\left(\frac{|k_x|}{\gamma_c}\right)\Lambda\left(\frac{|k_y|}{\gamma_c}\right)\sin c\left[\frac{8W_{\max}}{\lambda}\left[\frac{k_x}{2\gamma_c}\right]\left(1-\frac{|k_x|}{2\gamma_c}\right)\right]$$
$$\times \sin c\left[\frac{8W_{\max}}{\lambda}\left[\frac{k_y}{2\gamma_c}\right]\left(1-\frac{|k_y|}{2\gamma_c}\right)\right]. \quad (5.111)$$

Figure 5.22 depicts Eq. (5.111) for several fraction values of defocusing. In this figure, $W_{\max} = 0$ equates to diffraction-limited imaging, which is given by Eq. (5.65b). For nonzero values of $W_{\max}$, the OTF degrades and even experiences contrast reversals for values of w greater than half a wavelength. An example is seen in Figure 5.23 for an image created slightly out of focus circular aperture optical system. This situation is known as *spurious resolution* since a line pattern is seen, but it is the inverse of the real object's bar series. This phenomenon is usually seen in defocused, well-corrected lenses, or in lenses whose image of a point is nearly uniform.

Since our analysis is true solely at a single plane, there will be a slight defocusing in real systems. The concept of depth of focus rests on the assumption that in any given imaging system; there exists a blur circle small enough that it will not significantly affect the performance of the system. The amount of tolerable defocus is characterized by the *depth of focus*. The depth of focus is the amount by which an image may be longitudinally shifted with respect to some reference plane and produce no more than the maximum acceptable blur. (The amount that the object may

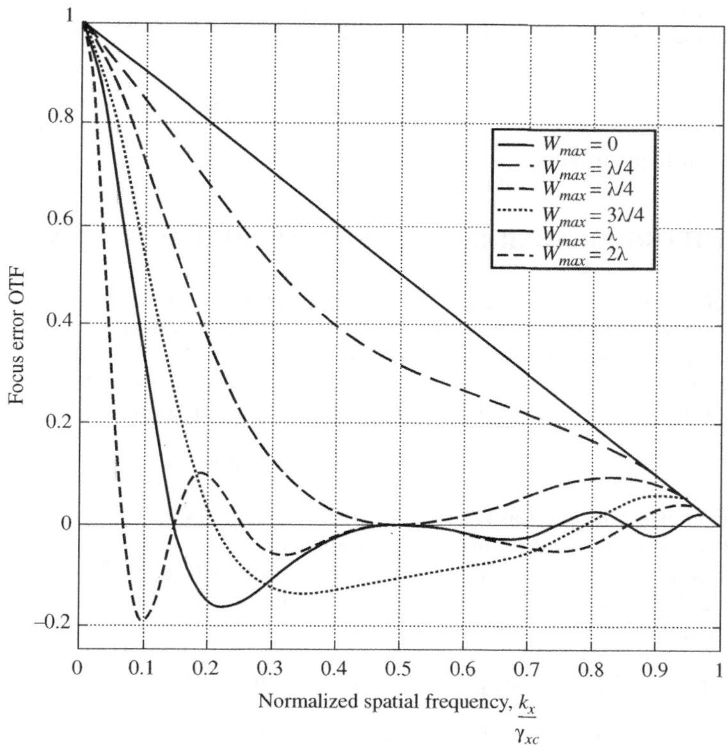

FIGURE 5.22 Cross section of the square aperture OTF with a focusing error.

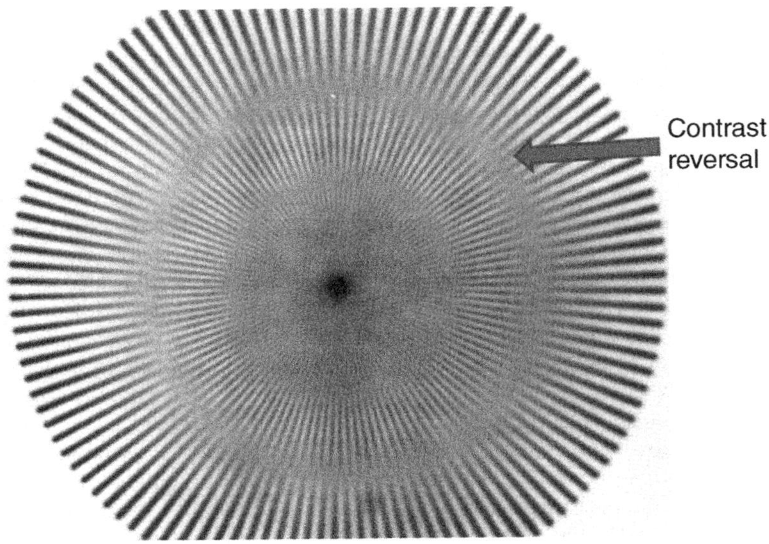

FIGURE 5.23 Defocused image of a starburst pattern [3, p. 151].

longitudinally shift to yield the maximum acceptable blur is called the *depth of field.*) Mathematically, the depth of focus is written as

$$\delta = 2\lambda\,(f - \text{number})^2 = \frac{\lambda}{2NA^2}. \tag{5.112}$$

The depth of focus usually refers to an OPD of $\frac{\lambda}{4}$.

5.8 HIGH-ORDER ABERRATION EFFECTS CHARACTERIZATION

Fortunately, most optical systems have circular apertures, so we focus our discussion of higher order aberration under this fact. Circular apertures imply rotational symmetry. This further simplifies the analysis to come.

Any system with rotational symmetry suggests that any path length error it is subject to can be characterized in polar rather than rectangular coordinates, that is,

$$W(x, y) \to W(r\cos\theta, r\sin\theta) = W(r, \theta), \tag{5.113}$$

where the aberration function W depends on the object height, y (or image height y') and (r, θ) is the pupil location in polar coordinates.

A standard way of describing the wave aberration is to use a Taylor expansion polynomial in field (object height) and pupil coordinates [16]. Specifically, the series expands as follows:

$$\begin{aligned} W(r,\theta) = {} & W_{020}\, r^2 + W_{040}\, r^4 + W_{131} h\, r^3 \, \cos\theta + W_{222} h^2\, r^2\, (\cos\theta)^2 \\ & + W_{220} h^2\, r^2 + W_{331} h^3\, r^3 \, \cos\theta + \text{higher order terms}, \end{aligned} \tag{5.114}$$

where W_{ijk} is the wave aberration coefficient for the particular term or mode. In the above, the specifically cited modes represent the following aberrations:

- $r^2 \leftrightarrow$ Defocus
- $r^4 \leftrightarrow$ Sphereical aberration
- $h\, r^3 \cos\theta \leftrightarrow$ Coma
- $h^2\, r^2\, (\cos\theta)^2 \leftrightarrow$ Astigmatism
- $h^2\, r^2 \leftrightarrow$ Field curvature
- $h^3\, r^3 \cos\theta \leftrightarrow$ Distortion.

These modes, with the exception of Defocus, are the Primary, or Seidel, Aberrations. These modes are monochromatic aberrations, which all do not depend on wavelength; other modes that depend on wavelength are called Chromatic Aberrations. The higher order modes cited in Eq. (5.114) are Secondary, Tertiary Aberrations, and so on.

The wavefront variance is derived from the integration of Eq. (5.114) over the pupil function [14, 15]. This suggests we can absorb image height terms into their respective wave aberration coefficients. If we now define a normalized pupil radius as

$$\rho = \frac{r}{a}, \tag{5.115}$$

and the power series in Eq. (5.114) becomes

$$W(\rho,\theta) = a_{11}\,\rho\cos\theta + a_{20}\,\rho^2 + a_{22}\,\rho^2(\cos\theta)^2 + a_{31}\,\rho^3\cos\theta + a_{33}\,\rho^3(\cos\theta)^3 + a_{40}\,\rho^4 + \cdots. \tag{5.116}$$

Although this series looks nice, it is not composed of an orthogonal set of basis functions [15]. This implies that it would not be very useful in calculating wavefront variance, data fitting, and for describing any experimental measurements of wavefront aberrations. Fortunately, there is balanced, complete set of basis functions orthonormal over a unit circle. They are known as the Zernike polynomials [14–21]. Zernike polynomials are a unique basis set for characterizing optical aberrations as they are

- polynomials characterized in only two variables, ρ and θ;
- orthonormal over a unit circle;
- efficient representation of common lens errors;
- invariant with respect to rotation of the coordinate axes about the origin; and
- polynomials characterized by permissible pairs on m and n coefficient indices [15].

Figure 5.24 shows the propagation geometry.[4] Specifically, a phase (or phase error) distribution of an optical imaging systems can be expressed in the following finite

[4]The formulation shown is traditional notation, measures the clockwise from the x-axis. Other different Zernike polynomial definitions exist. For example, the convention adopted by the Optical Society of America specifies the right-hand coordinate system, with the angle θ being measured counterclockwise from the x-axis.

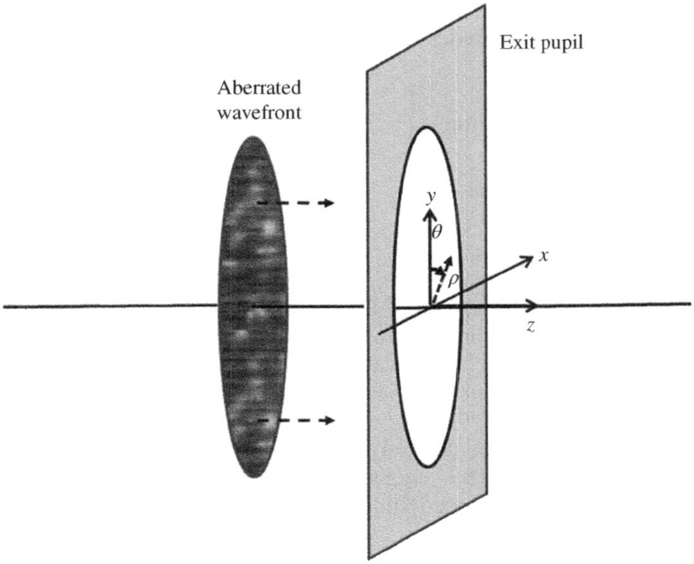

FIGURE 5.24 Propagation geometry for Zernike polynomials.

element series:

$$\Phi(\rho,\theta) = k\ W(\rho,\theta) = A_{00} + \frac{1}{\sqrt{2}}\sum_{n=2}^{\infty} A_{n0}\ \mathcal{R}_n^0(\rho)$$

$$+ \sum_{n=1}^{\infty}\sum_{m=1}^{n} \mathcal{R}_n^m(\rho)[A_{nm}\ \cos m\theta + B_{nm}\ \sin m\theta], \tag{5.117}$$

where n and m are positive integers, with $n - m$ always even and greater than zero (n−m) rule. The radial function in the above equation equals

$$\mathcal{R}_n^m(\rho) = \sum_{s=0}^{\frac{(n-m)}{2}} \frac{(-1)^s (n-s)!}{s!\ \left(\frac{(n+m)}{2} - s\right)!\ \left(\frac{(n-m)}{2} - s\right)!}\ \rho^{n-2s}. \tag{5.118}$$

From Eq. (5.117), we see that radial (index n) and azimuthal (index m) polynomials are preceded by Zernike coefficients A_{nm} and B_{nm}, which completely describe the wavefront up to the order specified by the largest n or m [16, p. 8]. Figure 5.25 depicts the first 10 Zernike polynomial modes.

Besides the $(n - m)$ rule, there is another important rule the reader should be aware of. The first is that each Zernike term of order n is composed of the highest power that that n takes, plus all the other powers of n of the same even- or oddness [17, p. 27]. For example, the Zernike focus term where $(n, m) = (2, 0)$ equals

$$\mathcal{R}_2^0 = 2\rho^2 - 1 \tag{5.119}$$

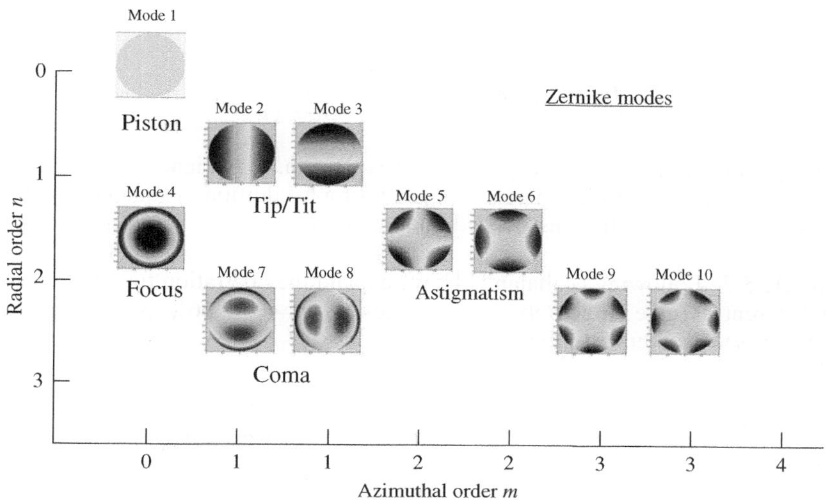

FIGURE 5.25 Depictions of the first 10 Zernike polynomial modes.

and contains the constant 1, the piston term where $(n, m) = (0, 0)$. In a similar way, the Zernike spherical aberration term where $(n, m) = (4, 0)$ contains the focus and piston terms. For odd azimuthal terms, the same rule applies [17, p. 27]. For example, the Zernike coma term where $(n, m) = (3, 1)$ contains the Zernike tilt term $(n, m) = (1, 1)$.

It also should be pointed out at this point that ordering of the Zernike polynomials could be different from the above. In fact, different lens design and optical systems analysis software sometimes use their own ordering convention. The takeaway is that the reader should be careful and pay attention to the ordering when reading papers and textbooks, and in using optical SW applications.

Following Nijboer [22–24], we expand the aberration function in terms of the Zernike circle polynomials:

$$W(\rho, \theta) = \sum_{n=0}^{\infty} \sum_{m=0}^{n} \varepsilon_{nm} C_{nm} \, Z_n^m(\rho, \theta), \tag{5.120}$$

where

$$\varepsilon_{nm} = \begin{cases} \frac{1}{\sqrt{2}}; & m = 0, n \neq 0 \\ 1; & \text{otherwise} \end{cases} \tag{5.121}$$

and

$$Z_n^m(\rho, \theta) = \mathcal{R}_n^m(\rho) \, \cos m\theta. \tag{5.122}$$

The inclusion of the ε_{nm} term helps the form of possible analytical results [4, p. 466].

From elementary trigonometry, we know that

$$\int_0^{2\pi} \cos m\theta \cos m'\theta \, d\theta = \pi(1 + \delta_{m0}) \, \delta_{mm'} \tag{5.123}$$

and from Born and Wolf,

$$\int_0^1 \mathcal{R}_n^m(\rho)\, \mathcal{R}_{n\prime}^m(\rho)\, \rho\, d\rho = \frac{1}{2(n+1)}\, \delta_{nn'} . \tag{5.124}$$

Thus, by using the above Zernike polynomial expansion, the variance of the phase distribution or aberration function becomes just a simple addition of squared expansion coefficients divided by their associated $2(n + 1)$ term.

Example 5.7 Following Mahajan [14], we expand the aberration function in terms of a different Zernike circle polynomials series [4, pp. 464–466]; specifically, it can be shown (see Problem 5.6) that

$$\sigma_W^2 = \frac{1}{\pi} \int_0^{2\pi} \int_0^1 W^2(\rho,\theta)\, \rho\, d\rho\, d\theta - W_{\text{AVG}}^2 = \sum_{n=1}^{\infty} \sum_{m=0}^{n} \frac{C_{nm}^2}{2(n+1)}, \tag{5.125}$$

where

$$W_{\text{AVG}} = C_{00}, \tag{5.126}$$

$$W(\rho,\theta) = \sum_{n=0}^{\infty} \sum_{m=0}^{n} \varepsilon_{nm} D_{nm}\, Z_n^m(\rho,\theta), \tag{5.127}$$

and

$$Z_n^m(\rho,\theta) = \sqrt{\frac{2\,(n+1)}{1+\delta_{m0}}}\, \mathcal{R}_n^m(\rho)\, \cos m\theta . \tag{5.128}$$

Clearly, the functions $Z_n^m(\rho,\theta)$ are orthonormal in this case, that is,

$$\frac{1}{\pi} \int_0^1 \int_0^{2\pi} Z_n^m(\rho,\theta)\, Z_{n'}^{m'}(\rho,\theta)\, \rho\, d\rho\, d\theta = \delta_{nn'}\, \delta_{mm'} . \tag{5.129}$$

In this formalism, the Zernike expansion coefficients equal

$$C_{nm} = \frac{1}{\pi} \sqrt{2(n+1)(1+\delta_{m0})} \int_0^1 \int_0^{2\pi} W(\rho,\theta)\, \mathcal{R}_n^m(\rho)\, \cos m\theta\, \rho\, d\rho\, d\theta . \tag{5.130}$$

The orthonormal Zernike polynomials and the names associated with some of them when identified with aberrations are given in Table 5.1 for $n \leq 8$ [14]. The number of Zernike (or orthogonal) aberration terms in the expansion through a certain order n is written as

$$N_n = \begin{cases} \left(\frac{n}{2}+1\right)^2; & \text{for } n \equiv \text{even} \\ \frac{(n+1)(n+3)}{4}; & \text{for } n \equiv \text{odd} \end{cases} . \tag{5.131}$$

TABLE 5.1 Selected orthonormal Zernike circle polynomials and associated aberrations

n	m	Orthognal Zernike polynomial $Z_n^m(\rho,\theta) = \sqrt{\frac{2\ (n+1)}{1+\delta_{m\ 0}}}\Re_n^m(\rho)\ \cos m\theta.$	Aberration name
0	0	1	Piston
1	1	$\rho\ \cos\ \theta$	Distortion (Tilt)
2	0	$\sqrt{3}\ (2\rho^2 - 1)$	Field curvature (defocus)
2	2	$\sqrt{6}\ \rho^2\ \cos 2\theta$	Primary astigmatism
3	1	$\sqrt{8}(3\rho^3 - 2\rho)\ \cos\theta$	Primary coma
3	3	$\sqrt{8}\ \rho^3\ \cos 3\theta$	
4	0	$\sqrt{5}(6\rho^4 - 6\rho^2 + 1)$	Primary spherical
4	2	$\sqrt{10}\ (4\rho^4 - 3\rho^2)\ \cos 2\theta$	Secondary astigmatism
4	4	$10\ \rho^4\ \cos 4\theta$	
5	1	$\sqrt{12}(10\ \rho^5 - 12\rho^3 + 3\rho)\ \cos\theta$	Secondary coma
5	3	$\sqrt{12}(5\ \rho^5 - 4\rho^3)\ \cos 3\theta$	
5	5	$\sqrt{12}\rho^5\ \cos 5\theta$	
6	0	$\sqrt{7}(20\rho^6 - 30\rho^4 + 12\rho^2 - 1)$	Secondary spherical
6	2	$\sqrt{14}(15\rho^6 - 20\rho^4 + 6\rho^2)\ \cos 2\theta$	Tertiary astigmatism
6	4	$\sqrt{14}(6\rho^6 - 5\rho^4)\ \cos 4\theta$	
6	6	$\sqrt{14}\rho^6\ \cos 6\theta$	
7	1	$4\ (35\rho^7 - 60\rho^5 + 30\rho^3 - 4\rho)$	Tertiary coma
7	3	$4\ (21\rho^7 - 30\rho^5 + 10\rho^3)\cos 3\theta$	
7	5	$4\ (7\rho^7 - 6\rho^5)\cos 5\theta$	
7	7	$4\rho^7\ \cos 7\theta$	
8	0	$3\ (70\rho^8 - 140\rho^6 + 90\rho^4 - 20\rho^2 + 1)$	Tertiary spherical

More generally, the Zernike circle polynomials can be expressed in complex notation, namely, as

$$V_n^m(\rho\ \cos\theta, \rho\sin\theta) = \mathcal{R}_n^m(\rho)\ e^{im\theta}, \tag{5.132}$$

where $m \gtreqless 0$ and $n \geq 0$ are integers, $n \geq |m|$ and $n - |m|$ is even [4, pp. 464–466]. As expected, the orthogonality and normalization properties essentially are the same, that is,

$$\int\int_{X^2+Y^2\leq 1} V_n^{*\ m}(X,Y)V_{n'}^{m'}(X,Y)dX\ dY = \frac{\pi}{n+1}\ \delta_{nn'}\ \delta_{mm'}. \tag{5.133}$$

The radial function $\mathcal{R}_n^m(\rho)$ obeys Eq. (5.124), but now is given by

$$\mathcal{R}_n^{\pm k}(\rho) = \sum_{s=0}^{\frac{(n-k)}{2}} \frac{(-1)^s(n-s)!}{s!\left(\frac{(n+k)}{2} - s\right)!\left(\frac{(n-k)}{2} - s\right)!}\ \rho^{n-2s}, \tag{5.134}$$

where $k = |m|$. The normalization is such that for all values of n and k, we have

$$\mathcal{R}_n^{\pm k}(1) = 1.0. \tag{5.135}$$

The following relation is of great importance to the Nijboer–Zernike theory:

$$\int_0^1 \mathcal{R}_n^m(\rho)\, J_m(\mu\rho)\, \rho\, d\rho = (-1)^{\frac{(n-m)}{2}} \frac{J_{n+1}(\mu)}{\mu} \tag{5.136}$$

with $J_n(x)$ being a Bessel function of the first kind.

Example 5.8 Let us calculate the PSF of an aberrated imaging system. Assume the aberrated phase function is given by

$$\Phi(\rho,\theta) = k\, W(\rho,\theta) = \sum_{n=0}^{\infty}\sum_{m=0}^{n} \alpha_{nm} V_n^m(\rho,\theta). \tag{5.137}$$

Now

$$F(w,\varphi) = \frac{1}{\pi}\int_0^1\int_0^{2\pi} e^{i\Phi(\rho,\theta)}\, e^{2\pi i w\rho\,\cos(\theta-\varphi)} \rho\, d\rho\, d\theta\,. \tag{5.138}$$

For small aberrations where $\theta \ll 2\pi$, we have

$$e^{i\Phi(\rho,\theta)} \approx 1 - i\Phi(\rho,\theta) + \cdots, \tag{5.139}$$

which implies that

$$F(w,\varphi) \approx \int_0^1\int_0^{2\pi} [1 + i\Phi(\rho,\theta)]\, e^{2\pi i w\rho\,\cos(\theta-\varphi)} \rho\, d\rho\, d\theta \tag{5.140}$$

$$= \int_0^1\int_0^{2\pi} \left[1 + i\sum_{n=0}^{\infty}\sum_{m=0}^{n} \alpha_{nm} V_n^m(\rho,\theta)\right] e^{2\pi i w\rho\,\cos(\theta-\varphi)} \rho\, d\rho\, d\theta$$

$$= \int_0^1\int_0^{2\pi} \left[\sum_{n=0}^{\infty}\sum_{m=0}^{n} \alpha'_{nm} V_n^m(\rho,\theta)\right] e^{2\pi i w\rho\,\cos(\theta-\varphi)} \rho\, d\rho\, d\theta\,, \tag{5.141}$$

where

$$\alpha'_{nm} = \begin{cases} i\alpha_{nm}; & \text{for } n \neq 0 \\ 1 + i\alpha_{00}; & \text{for } n = 0 \end{cases}. \tag{5.142}$$

Substituting V_n^m into Eq. (5.141), we obtain

$$F(w,\varphi) \approx \sum_{n=0}^{\infty}\sum_{m=0}^{n} \alpha'_{nm} \int_0^1\int_0^{2\pi} \mathcal{R}_n^{|m|}(\rho)\, e^{im\theta}\, e^{2\pi i w\rho\,\cos(\theta-\varphi)} \rho\, d\rho\, d\theta\,. \tag{5.143}$$

To evaluate Eq. (5.143), we note that Richards and Wolf found that

$$\int_0^{2\pi} e^{im\theta}\, e^{2\pi i r\rho\,\cos(\theta-\varphi)} \rho\, d\rho\, d\theta = 2\pi i^m\, J_m(2\pi r)\, e^{im\theta} \tag{5.144}$$

with $J_m(x)$ the Bessel function of the first kind and of order m [25, Eq. (2.9)]. This means that

$$F(w,\varphi) \cong \pi \sum_{n=0}^{\infty}\sum_{m=0}^{n} \alpha'_{nm}\, i^{\frac{(n-m)}{2}} \left(\frac{2J_{n+1}(2\pi w)}{2\pi w}\right) e^{im\varphi} \tag{5.145}$$

using Eq. (5.136) [25, Eq. (13)]. Here,

$$\int_0^1 \mathcal{R}_n^{|m|}(\rho)\, J_m(\mu\rho)\, \rho\, d\rho = (-1)^{\frac{(n-m)}{2}} \frac{J_{n+1}(\mu)}{\mu}. \tag{5.146}$$

Equation (5.145) means that the PSFs for Zernike aberrations are related to higher order Bessel functions.

5.9 THE STREHL RATIO

When aberrations affect an incoherent optical imaging system, the transmitted irradiance reduction can be characterized by a factor commonly called the *Strehl Ratio* (*SR*). The modern definition of the SR is the ratio of the observed peak intensity at the detection plane of a telescope or other imaging system from a point source compared to the theoretical maximum peak intensity of a perfect imaging system working at the diffraction limit. In other words, it is defined as the ratio of the light intensity at the maximum of the PSF of the optical system with aberrations to the maximum of the PSF of the same system without aberrations. Alternatively, when aberrations are severe, the SR is equal to the normalized volume under the OTF of an optical system with aberrations, or more specifically,

$$\mathrm{SR} = \frac{\displaystyle\int_{-\infty}^{\infty}\int_{-\infty}^{\infty} \widehat{\Theta}(k_x,k_y)|_{\text{with Aberrations}} dk_x\, dk_y}{\displaystyle\int_{-\infty}^{\infty}\int_{-\infty}^{\infty} \widehat{\Theta}(k_x,k_y)|_{\text{without Aberrations}} dk_x\, dk_y}. \tag{5.147}$$

Let us look at the SR for aberrated lens systems in more detail.

For our unity circle pupil function, the SR in polar coordinates can be written as

$$\mathrm{SR} = \frac{1}{\pi^2}\left|\int_0^{2\pi}\int_0^1 e^{i\Phi(\rho,\theta)}\rho\, d\rho\, d\theta\right|^2 \tag{5.148}$$

$$\begin{aligned} &= |\langle e^{i(\Phi(\rho,\theta)-\langle\Phi\rangle)}\rangle|^2 \\ &= |\langle\cos(\Phi-\langle\Phi\rangle)\rangle|^2 + |\langle\sin(\Phi-\langle\Phi\rangle)\rangle|^2 \\ &\geq |\langle\cos(\Phi-\langle\Phi\rangle)\rangle|^2, \end{aligned} \tag{5.149}$$

where the angle brackets indicate an average across the pupil [15, p. 84]. Expanding the cosine term in Eq. (5.149) and retaining only the first two terms under the assumption of small aberrations, we see that

$$\mathrm{SR} \geq \left(1 - \frac{\sigma_\Phi^2}{2}\right)^2, \tag{5.150}$$

where

$$\sigma_\Phi^2 = \langle(\Phi - \langle\Phi\rangle)^2\rangle \tag{5.151a}$$

$$= \langle\Phi^2\rangle - \langle\Phi\rangle^2 \tag{5.151b}$$

is the phase distribution variance across the pupil and

$$\langle\Phi^k\rangle = \frac{1}{\pi^2}\int_0^{2\pi}\int_0^1 \Phi^k(\rho,\theta)\,\rho\,d\rho\,d\theta. \tag{5.152}$$

For small aberration, Eq. (5.150) can be written in three ways:

$$\mathrm{SR}_1 \geq \left(1 - \frac{\sigma_\Phi^2}{2}\right)^2, \tag{5.153a}$$

$$\mathrm{SR}_2 \geq 1 - \sigma_\Phi^2, \tag{5.153b}$$

and

$$\mathrm{SR}_3 \geq e^{-\sigma_\Phi^2}. \tag{5.153c}$$

The first is the *Maréchal formula*, the second is the commonly used expression obtained when the σ_Φ^4 term is neglected, and the third is an empirical expression giving a better fit to the actual numerical results for various aberration [14, p. 84]. It is clear from the above that the SR depends solely on the phase distribution variance, and not on the specific details of either the phase distribution or the path length error.

Example 5.9 Alternatively, for small aberration, we can expand the exponential in Eq. (5.148) as

$$e^{-i\Phi(\rho,\theta)} \approx 1 - i\Phi(\rho,\theta) + \frac{1}{2}\Phi^2(\rho,\theta). \tag{5.154}$$

Substituting Eq. (5.154) into Eq. (5.148) yields

$$\begin{aligned}
\mathrm{SR} &\approx \frac{1}{\pi^2}\left|\int_0^{2\pi}\int_0^1 \left(1 - i\Phi(\rho,\theta) + \frac{1}{2}\Phi^2(\rho,\theta)\right)\rho\,d\rho\,d\theta\right|^2 \\
&= \left[1 - \frac{1}{2\pi}\int_0^{2\pi}\int_0^1 \Phi^2(\rho,\theta)\rho\,d\rho\,d\theta\right]^2 + \frac{1}{\pi^2}\left|\int_0^{2\pi}\int_0^1 \Phi(\rho,\theta)\,\rho\,d\rho\,d\theta\right|^2 \\
&\approx 1 - \frac{1}{\pi}\int_0^{2\pi}\int_0^1 \Phi^2(\rho,\theta)\rho\,d\rho\,d\theta + \frac{1}{\pi^2}\left|\int_0^{2\pi}\int_0^1 \Phi(\rho,\theta)\,\rho\,d\rho\,d\theta\right|^2
\end{aligned} \tag{5.155}$$

ignoring higher order terms.

Now, the wavefront variance σ_W^2 is given by

$$\begin{aligned}
\sigma_W^2 &= \frac{1}{\pi}\int_0^{2\pi}\int_0^1 (W(\rho,\theta) - W_{\mathrm{AVG}})^2\,\rho\,d\rho\,d\theta \\
&= \frac{1}{\pi}\int_0^{2\pi}\int_0^1 W^2(\rho,\theta)\,\rho\,d\rho\,d\theta - W_{\mathrm{AVG}}^2
\end{aligned} \tag{5.156}$$

with

$$W_{\mathrm{AVG}} = \frac{1}{\pi}\int_0^{2\pi}\int_0^1 W(\rho,\theta)\,\rho\,d\rho\,d\theta. \tag{5.157}$$

This implies that

$$\begin{aligned} \mathrm{SR} &\approx 1 - \frac{1}{\pi^2}\int_0^{2\pi}\int_0^1 \Phi^2(\rho,\theta)\rho\,d\rho\,d\theta + \frac{1}{\pi^2}\left|\int_0^{2\pi}\int_0^1 \Phi(\rho,\theta)\,\rho\,d\rho\,d\theta\right|^2 \\ &= 1 - \frac{k^2}{\pi}\int_0^{2\pi}\int_0^1 W^2(\rho,\theta)\rho\,d\rho\,d\theta + \frac{k^2}{\pi^2}\left|\int_0^{2\pi}\int_0^1 W(\rho,\theta)\,\rho\,d\rho\,d\theta\right|^2 \\ &= 1 - k^2\sigma_W^2 \approx e^{-k^2\sigma_W^2}. \end{aligned} \tag{5.158}$$

All of the above equations, derived for small aberrations, suggest that the SR can be maximized by minimizing the phase distribution or wavefront error variances. The following example illustrates the idea of balanced aberration as an example [15, p. 86].

Example 5.10 For small lens errors, the SR is greatest when the aberration variance is least. One way to obtain this reduced variance is to balance the various effects against one another when it makes sense. The simplest is to balance spherical aberrations against defocus [14, p. 86]. Specifically, we have

$$\Phi(\rho,\theta) = A_s\,\rho^4 + B_d\rho^2. \tag{5.159}$$

with B_d being the adjustable parameter. To determine its specific value, we calculate the phase distribution variance and differentiate with respect to B_d. It can be shown that the optimum value for B_d is $-A_s$. The standard deviation for this optimally balanced aberration is $\frac{A_s}{6\sqrt{5}}$, which is a factor 4 less than the standard deviation when $B_d = 0$. Since the standard deviation has been reduced to a factor of 4 by balancing spherical aberrations against defocus, the optical tolerance has been increased by this same factor.

5.10 MULTIPLE SYSTEMS TRANSFER FUNCTION

When one has multiple systems operating on an input object distribution, the resulting transfer function is just the multiplication of the various OTFs/MTFs involved [26, p. 8]. That is, the image distribution is given by

$$g(x,y) = f(x,y) \circledast P(x,y), \tag{5.160}$$

where $f(x,y)$ is the object radiance distribution and $P(x,y)$ the inverse Fourier transform of the following transfer function:

$$\hat{\theta}(k_x,k_y) = \hat{\theta}_1(k_x,k_y) \times \hat{\theta}_2(k_x,k_y) \times \hat{\theta}_3(k_x,k_y) \times \cdots \times \hat{\theta}_M(k_x,k_y)\,. \tag{5.161}$$

Alternately, $g(x,y)$ can be calculated by taking the Inverse Fourier transform of

$$\hat{G}(k_x,k_y) = \hat{F}(k_x,k_y) \times \hat{\theta}(k_x,k_y)\,, \tag{5.162}$$

where $\widehat{G}(k_x, k_y)$ and $\widehat{F}(k_x, k_y)$ are the Fourier transforms of $g(x, y)$ and $f(x, y)$, respectively. Boreman has summarized a number of the system and environmental MTF in his 2001 book [27]. The following example provides some environmental MTF expressions to illustrate the above procedure.

Example 5.11 Let us look at the MTF for turbulent and particulate media that can be used to reconstruct the original image intensity.

In free space, we found that the MTF of an imaging system can be shown to be given by

$$\mathrm{MTF}_0(\nu) = \begin{cases} \frac{2}{\pi}\left[\cos^{-1}\left(\frac{\nu}{\vartheta_c}\right) - \left({}^{\nu}/_{\vartheta_c}\right)\sqrt{1-\left(\frac{\nu}{\vartheta_c}\right)^2}\right]; & \text{for } \nu \le \vartheta_c = \frac{D}{\lambda f} \\ 0; & \text{for } \nu > \vartheta_c \end{cases},$$

where ν represents spatial frequency and ϑ_c is the system's cutoff frequency.

In a particulate atmosphere that comprised aerosols, the MTF is given by

$$\mathrm{MTF}_{\mathrm{aero}}(\nu) = \begin{cases} \exp\left\{-a_{\mathrm{aero}}z - b_{\mathrm{aero}}z\left(\frac{\nu}{\nu_{\text{aero-cutoff}}}\right)^2;\right\} & \nu \le \nu_{\text{aero-cutoff}} \\ \exp\{-a_{\mathrm{aero}}z - b_{\mathrm{aero}}z\}; & \nu > \nu_{\text{aero-cutoff}} \end{cases}. \tag{5.163}$$

In the above, a and b are the volume absorption and scattering coefficients, respectively [1, Chapter 10] and $\nu_{\text{aero-cutoff}}$ is the aerosol cutoff spatial frequency. Note that for frequencies above $\nu_{\text{aero-cutoff}}$, the MTF equals the normal atmospheric transmittance expression from Beer's law, that is, atmospheric transmittance is given by $e^{-\tau}$, where $\tau = cz$. The parameter τ is the optical thickness or depth of the propagation path and c $(= a + b)$ is the volume extinction coefficient [1, Chapter 10]. The cutoff frequency is proportional to the averaged radius of the aerosols and inverse wavelength [26, pp. 616–617]. The average aerosol radius is typically on the order of the wavelength of light. Thus, the MTF of the propagation path between the source and detector is given by

$$\mathrm{MTF}_{\mathrm{total}}(\nu) = \mathrm{MTF}_0(\nu) \times \mathrm{MTF}_{\mathrm{aero}}(\nu). \tag{5.164}$$

In a turbulent atmosphere, its MTF is given by

$$\mathrm{MTF}_{\mathrm{turb}}(\nu) = \exp\left\{-3.44\left(\frac{\lambda f \nu}{r_0}\right)\right\}, \tag{5.165}$$

where $\mathrm{MTF}_{\mathrm{turb}}(\nu)$ is the long-term atmospheric MTF, ν the spatial frequency of image plane, f the focal length of receiving system, and r_0 is approximately the atmospheric coherence diameter, generally known as the *Fried parameter* [1, 26, 28]. Equation (5.164) neglects particulate scattering [26, p. 617]. Mathematically, the Fried parameter is equal to

$$r_0 = \left[0.42\, k^2 \int_0^z C_n^2(z')\, dz'\right]^{-\frac{3}{5}} \tag{5.166}$$

with $C_n^2(z)$ being the refractive index structure parameter [1, Chapter 10]. The total MTF for an imaging system in a turbulent atmosphere then is

$$\text{MTF}_{\text{total}}(\nu) = \text{MTF}_0(\nu) \times \text{MTF}_{\text{aero}}(\nu) \times \text{MTF}_{\text{turb}}(\nu). \tag{5.167}$$

Hou et al. stated that the channel MTF in particulate, turbulent media like seawater is given by the equation

$$\text{MTF}\,(\nu, r)_{\text{channel}} = \text{MTF}\,(\nu, z)_{\text{path}} \times \text{MTF}\,(\nu, z)_{\text{par}} \times \text{MTF}\,(\nu, z)_{\text{turb}} \tag{5.168}$$

$$= \left(\frac{1}{1 + N_{\text{path}}}\right) \exp\left\{- cz + bz \left(\frac{1 - e^{-2\,\pi\,\theta_0\,\nu}}{2\,\pi\,\theta_0\nu}\right)\right\} \times \exp\left\{- S_n\, \nu^{\frac{5}{3}}\, z\right\}, \tag{5.169}$$

where θ_0 is the root-mean-square scattering angle of the particulate media and N_{path} relates to the path radiance [29]. S_n contains parameters that are dependent on the refractive index structure function and can be expressed in terms of the turbulence dissipation rate of temperature, salinity, and kinetic energy, assuming a Kolmogorov power spectrum type [28]. The total MTF equals to Eq. (5.168) times $\text{MTF}_0(\nu)$.

5.11 LINEAR SYSTEMS SUMMARY

A summary of the key concepts found in this chapter is shown in Figure 5.26, with their associated image quality criteria noted.

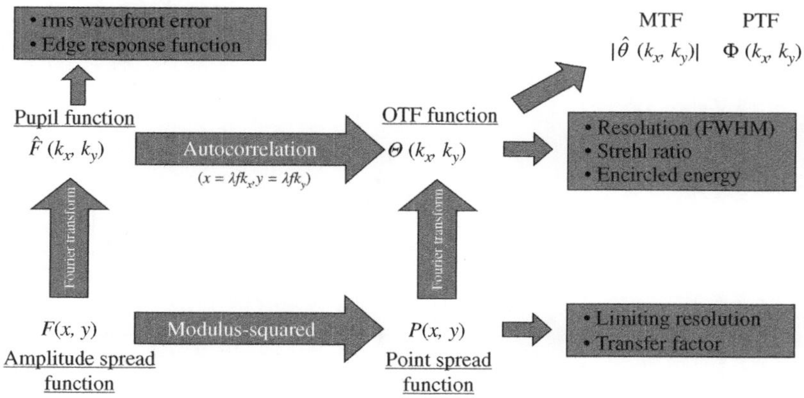

FIGURE 5.26 Linear system summary.

Problem 5.1. Calculate the OTF for a square annular aperture where the square central obstruction is half the size of the aperture. Graphically compare its performance against its diffraction-limited form (i.e., no central obscuration), showing that the former increases image quality at the higher spatial frequencies.

Problem 5.2. The line spread function of a two-dimensional imaging system is defined to be the response of that system to a one-dimensional delta function passing through the origin of the input plane.

(a) In the case of a line excitation lying along the x-axis, show that the line spread function l and the point spread function p are related by

$$l(y) = \int_{-\infty}^{\infty} p(x, y)\, dx,$$

where l and p are to be interpreted as amplitudes or intensities, depending on whether the system is coherent or incoherent, respectively.

(b) Show that for a line source oriented along the x-axis, the (1D) Fourier transform of the line spread function is equal to a slice through the (2D) Fourier transform of the point-spread function, the slice being along the f_y-axis. In other words, if the Fourier transform of l is $\hat{L}$ and the Fourier transform of p is $\hat{P}$, then $\hat{L}(f) = \hat{P}(0,f)$.

(c) Find the relationship between the line spread function and the step response of the system, that is, the response to a unit step excitation oriented parallel to the x-axis.

Problem 5.3. Two circular apertures are placed in the exit pupil of an imaging system as shown in Figure P5.1. Light from these two small openings create a fringe pattern in the image plane.

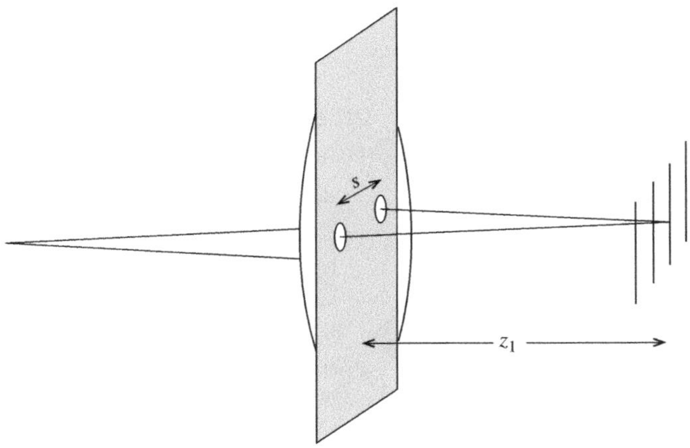

FIGURE P5.1

(a) Find the spatial frequency of this fringe in terms of the center-to-center spacings of the two openings, the wavelength A, and the image distance z_i.

(b) The openings are circular and have diameter d. Specify the envelope of the fringe pattern caused by the finite openings in the pupil plane.

Problem 5.4. Consider a pinhole camera shown in the below figure. Assume that the object is incoherent and nearly monochromatic, the distance R, from the object is so large that it can be treated as infinite, and the pinhole is circular with diameter $2d$.

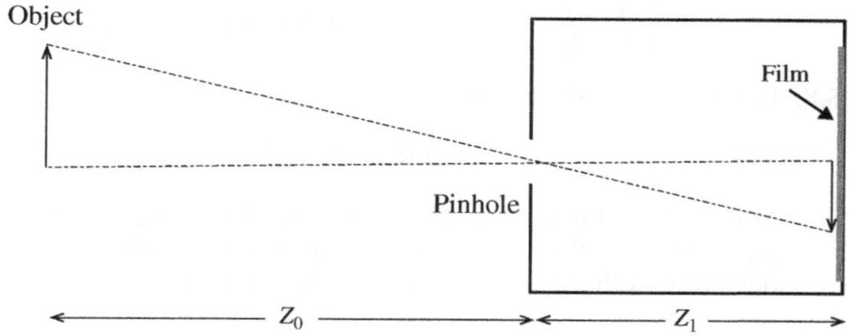

(a) Under the assumption that the pinhole is large enough to allow a purely geometrical optics estimation of the point-spread function, find the optical transfer function of this camera. If we define the "cutoff frequency" of the camera to be the frequency where the first zero of the OTF occurs, what is the cutoff frequency under the above geometrical-optics approximation? (Hint: First find the intensity point-spread function, then Fourier transform it. Remember the second approximation above.)

(b) Again calculate the cutoff frequency, but this time assuming that the pinhole is so small that Fraunhofer diffraction by the pinhole governs the shape of the point-spread function.

(c) Considering the two expressions for the cutoff frequency that you have found, can you estimate the "optimum" size of the pinhole in terms of the various parameters of the system? Optimum in this case means the size that produces the highest possible cutoff frequency.

Problem 5.5. A 100 mm focal length circular lens with a F-number $=4$ is used to image a 100 line/mm cosine grating with a contrast of 0.5 in a one-to-one imaging system illuminated with incoherent light of approximately 500 nm. Calculate the contrast of the image. If this same lens is used to image this grating with a magnification of (a) 0.25, (b) 0.5, (c) 2, (d) 4. What is the contrast of the image in each case?

Problem 5.6. Equation (5.120) can be rewritten as

$$W(\rho,\theta) = C_{00} + \frac{1}{\sqrt{2}}\sum_{n=2}^{\infty} C_{n0}\mathcal{R}_n^0(\rho) + \sum_{n=1}^{\infty}\sum_{m=1}^{n} C_{nm}\mathcal{R}_n^m(\rho)\cos m\theta.$$

Using the above equation, show that the variance of the aberration function is given by

$$\sigma_W^2 = \overline{W^2(\rho,\theta)} - \overline{W(\rho,\theta)}^2 = \sum_{n=1}^{\infty}\sum_{m=0}^{n} C_{nm}^2.$$

Problem 5.7. Using the Zernike polynomial definition in Eq. (5.128), show that

$$\frac{1}{\pi}\int_0^1\int_0^{2\pi} Z_n^m(\rho,\theta)\ Z_{n'}^{m'}(\rho,\theta)\ \rho\ d\rho\ d\theta = \delta_{nn'}\ \delta_{mm'}.$$

Problem 5.8. Equation (5.153b) states that

$$\mathrm{SR}_2 \geq 1 - \sigma_\Phi^2.$$

Maréchal postulated that a well-corrected imaging system translates into a SR of 0.8 or higher. Using this equation, what is the standard deviation of the wavefront error for that condition?

REFERENCES

1. Karp, S. and Stotts, L.B. (2013) *Fundamentals of Electro-Optic Systems Design: Communications, Lidar, and Imaging*, Cambridge Press, New York.
2. Lohmann, A.W. (2006) in *Optical Information Processing* (ed. S. Sinzinger), Universitätsverlag, Ilmenau, Germany, ISBN 2-939472-00-6.
3. Goodman, J.W. (2004) *Introduction to Fourier Optics*, 3rd edn, Roberts and Company, Englewood, CO.
4. Born, M. and Wolf, E. (1999) *Principles of Optics: Electromagnetic Theory of Propagation, Interference and Diffraction of Light*, London, Cambridge University Press.
5. Papoulis, A. (1968) *Systems and Transforms with Applications in Optics*, McGraw-Hill Series in Systems Science, Edition Reprint, Robert Krieger Publishing Company, Malabar, Florida, 474 pp., ISBN 0898743583, 9780898743586.
6. Andrews, L.C., Phillips, R.L., Bagley, Z.C. *et al.* (September, 2012) in *Advanced Free Space Optics (FSO) A Systems Approach*, vol. **186** Editor-in-Chief, Chapter 9, Springer Series in Optical Sciences (ed. W.T. Rhodes), Springer, New York.
7. Karp, S., Gagliardi, R.M., Moran, S.E., and Stotts, L.B. (1988) *Optical Channels: Fiber, Atmosphere, Water and Clouds*, Plenum Publishing Corporation, New York.
8. R. Schowengerdt, University of Arizona ECE 425 Course Notes, 2000.
9. Goodman, J. (2007) *Speckle Phenomena in Optics: Theory and Applications*, Roberts & Company, Edgewood, CO.
10. O'Neil, E.L. (1956. Also, E.L. O'Neil (1956) Errata: transfer function for an annular aperture, **46**(12), 1096) Transfer function for an annular aperture. *J. Opt. Soc. Am.*, **46** (**4**), 285–288.
11. Watson, G.N. (1996) *A Treatise on the Theory of Bessel Functions*, 2nd edn, Cambridge University Press, London, England.
12. Warren, J.S. (2008) *Modern Optical Engineering; The Design of Optical Systems*, 4th edn, SPIE Press, Bellingham, Washington.

13. Kopeika, N.S. (1998) *A System Engineering Approach to Imaging*, SPIE Optical Engineering Press, Bellingham, WA.
14. Mahajan, V.N. (2011) *Aberration Theory Made Simple*, 2nd edn, SPIE Press, Tutorial Press, Bellingham, WA.
15. S.-Y. Yuan, "Tutorial on Strehl ratio, wavefront power series expansion, Zernike polynomials expansion in small aberrated optical systems," Source: http://fp.optics.arizona.edu/optomech/student%20reports/tutorials/YuanTutorial1.pdf
16. Tyson, R.K. (2004) in *Field Guide to Adaptive Optics, SPIE Field Guides*, vol. **FG03** (ed. J.E. Greivenkamp), SPIE Press, Bellingham, WA.
17. Tyson, R.K. (2000) in *Introduction to Adaptive Optics*, vol. **TT41**, Tutorial Texts in Optical Engineering (ed. A.R. Jr Weeks), SPIE Press, Bellingham, WA.
18. Roggemann, M.C. and Welsh, B.M. (1996) *Imaging Through Turbulence*, CRC Press, Boca Raton, FL.
19. Hardy, J.W. (June 1994) Adaptive optics. *Sci. Am.*, **260** (**6**), 60–65.
20. Tyson, R.K. (1991) *Principles of Adaptive Optics*, 2nd edn, Academic Press, New York.
21. Hardy, J.W. (1978) Adaptive optics – a new technology for the control of light. *Proc. IEEE*, **66**, 651–697.
22. B. R. A. Nijboer, The Diffraction Theory of Aberrations, Ph.D. thesis, University of Groningen, The Netherlands, 1942.
23. Zernike, F. and Nijboer, B. R. A. (1949) Contributions to *La théorie des images optiques. Réunions organisées en octobre 1946 avec l'appui de la Fondation Rockefeller*, Fluery, P. (eds), Centre national de la recherche scientifique (France), Colloque international, La Revue d'optique, Paris, France, p. 127.
24. Richards, B. and Wolf, E. (1959) Electromagnetic diffraction in optical systems II. Structure of the image field in an aplanatic system. *Proc. R. Soc. London Ser. A*, **253**, 358–379.
25. Van Haver, S. and Janssen, A.J.E.M. (2013) Advanced analytic treatment and efficient computation of the diffraction integrals in the extended Nijboer–Zernike theory. *J. Eur. Opt. Soc.*, Rapid Publications, Europe, **8**, 13044-1–13044-29.
26. Andrews, L.C. and Phillips, R.L. (2005) *Laser Propagation Through Random Media*, 2nd edn, SPIE Press, Bellingham, WA.
27. Boreman, G.D. (2001) *Modulation Transfer Functions in Optical and Electro-Optical Systems*, Tutorial Text Series, SPIE Press, Bellingham, WA.
28. Walters, D.L. (April 1981) Atmospheric modulation transfer function for desert and mountain locations: r_0 measurements. *J. Opt. Soc. Am*, **71** (**4**), 406–409.
29. Hou, W., Woods, S., Goode, W. *et al.* (May 2011) Impact of optical turbulence on underwater imaging. *Proc. SPIE*, **8030**, 803009-1–803008-7.

6

PARTIAL COHERENCE THEORY

6.1 INTRODUCTION

In Chapter 2, we established that there are impulse responses and associated Fourier transforms for both coherent and incoherent light traversing free-space and optical imaging systems. The differences between the two were highlighted in our discussion on image resolution, which depended on the *interference* created by a coherence state between the two sources. Specifically, we assumed that the light waves from the two sources, which overlapped in the plane of observation, could be coherent in three specific states, "in phase," "in quadrature," and "out of phase." We then compared the resulting diffraction patterns for the three states to that of two incoherent lights under the Rayleigh criterion. The former created interference patterns that caused our resolving power to be eliminated, remained the same, or improved. In general, the interference, or fringes, sometimes is be barely visible as well, exhibiting only low contrast effects. This situation is called *partial coherence*. Partial coherence theory is considered the most difficult subject in optics and much is written on this subject to clarify what is going on; sometimes successfully, sometimes not.

This chapter focuses on describing partial coherence theory based on previously described results, hoping to provide a more intuitive insight into the theory's development rather than in the abstract. It leverages the discussions in Lohmann [1, Section 30], Saleh and Teich [2, Chapter 11], Ross [3], and Reynolds *et al.* [4].

Free Space Optical Systems Engineering: Design and Analysis, First Edition. Larry B. Stotts.

Companion website: www.wiley.com\go\stotts\free_space_optical_systems_engineering

6.2 RADIATION FLUCTUATION

Lohmann once said that light was like Voltaire [1]:

> *Voltaire was born a Catholic, lived a Protestant and died a Catholic. Light is born a particle, lives as a wave and dies as a particle.*

What he is referring to is the duality of light where it exhibits both the properties of a particle and a wave. In this section, we look at the statistical fluctuations in incoming and received light in order to introduce the quantum nature of light and its coherence properties.

Let us assume a single atom with a number of electrons about it. Electron excitation is the transfer of one or more electron(s) to a more energetic, but still bound, state(s). This can be done by photoexcitation, where the electron absorbs a photon and gains all its energy or by electrical excitation, where the electron receives energy from another, energetic electron. Within a semiconductor crystal lattice, thermal excitation is a process where lattice vibrations provide enough energy to transfer electrons to a higher energy band. If the excited electron falls back to a state of lower energy, then the energy loss, e.g., photon, can occur at some known time in the future without external influence. This type of event is called spontaneous emission. This is typical for thermal light sources.

Besides spontaneous emission, it was predicted early last century that the presence of a light wave of a particular energy near the above elevated electron may cause the said electron to drop to a lower energy state and emit light. That energy drop will create a light wave whose energy equals the energy state difference, a value that is the same as the passing light's energy. It also is in-phase with the original light wave. In this case, the light is acting like a particle of energy E_{Photon}. This is the so-called stimulated emission, postulated by Einstein in 1905. The energy level involved is given by

$$E_{\text{Photon}} = h\nu = \frac{hc}{\lambda}, \tag{6.1}$$

where ν is the frequency of light emitted, λ its wavelength, c the speed of light, and h Planck's constant.

From Eq. (4.72) in Chapter 4, we know that light particles called *photons*, follows the Bose–Einstein distribution formula. This implies that the mean number of photons in a single quantum state is given by

$$\overline{n}_s = \frac{1}{e^{\frac{E_{\text{Photon}}}{kT}} - 1}. \tag{6.2}$$

To calculate the amount of fluctuations the emission and detection of light is subject to, we must go back to a more general Bose–Einstein distribution formula that is derived for gas particles, namely,

$$\overline{n}_g = \frac{1}{e^{\frac{(E_s-\mu)}{kT}} - 1}, \tag{6.3}$$

where E_s is the kinetic energy (equal to E_{Photon} for photons) and μ the chemical potential (equal to zero for photons). Using this equation, the mean-squared fluctuations may be obtained from the equation

$$\overline{\Delta n_g^2} = kT\,\frac{\partial \overline{n}_g}{\partial \mu} = \frac{e^{\frac{(E_s-\mu)}{kT}}}{\left(e^{\frac{(E_s-\mu)}{kT}} - 1\right)^2}, \tag{6.4}$$

which represents the mean-squared fluctuations of a portion of the gas at a constant temperature and volume. Therefore, in the case of light Eq. (6.4) becomes

$$\overline{\Delta n_s^2} = \frac{e^{\frac{E_{\text{Photon}}}{kT}}}{\left(e^{\frac{E_{\text{Photon}}}{kT}} - 1\right)^2} = \overline{n}_s \frac{e^{\frac{E_{\text{Photon}}}{kT}}}{e^{\frac{E_{\text{Photon}}}{kT}} - 1} = \overline{n}_s(1 + \overline{n}_s). \tag{6.5}$$

Equation (6.5) is the mean-squared fluctuations of photons for each quantum state and gives the resulting noise in terms of photons that exist in blackbody radiation.

It can be shown that given a source with bandwidth $\Delta\nu$, a receiver will detect $\overline{n}$ photoelectrons in time Δt [3, p. 16]. Mathematically, this is written as

$$\overline{n} = \frac{\Omega}{\lambda^2} A_d \overline{n}_s\, \Delta t\, \Delta\nu, \tag{6.6}$$

where $\overline{n}_s$ is the number of photoelectrons received per quantum state, Ω the solid angle of the received light, λ the "monochromatic" wavelength of received light and A_d is the receiver's detection area. By putting this equation in terms of photoelectrons, we take into account the *quantum efficiency* η of the detection process, which says that not every photon gets detected and contributes to the detected energy. The output current, the average number of photo-electrons per unit time, from a detector is given by

$$i = \frac{\eta q}{h\nu} P_s, \tag{6.7}$$

where η is the quantum efficiency of the detector, q is the charge of an electron, h Planck's constant, ν the frequency of light, and P_s the received power before the detector.

The term $\frac{\lambda^2}{\Omega}$ is called the *coherence area* A_c [3, p. 19], which defines the receiver area in which incoherent sources can beat or interfere with each other [5]. For example, a beam of filtered sunlight can create interference fringes within an area of $5 \times 10^4\ \mu\text{m}^2 = 5 \times 10^{-2}\ \text{mm}^2$, which was derived from direct measurement [6].[1] (The size of this coherence area shows why interference effects are not observed in everyday experience: for sunlight, interference only can occur within a spot roughly

[1] It should be noted that this value is different from what would be calculated by the coherence area given above. This calculation assumes that the Earth is in the far zone of the Sun, an assumption that is incorrect. Agarwal *et al.* showed that fact and provided guidance how to correctly calculate the coherence area when this assumption is not valid [7].

a tenth of a millimeter across!) On the other hand, filtered starlight has coherence area of $\sim 6\,\mathrm{m}^2$. The difference between the two is essentially that stars subtend a much smaller solid angle than that of the sun. The spatial coherence factor, here we denote by M, is the ratio of the above coherence area to the detector area; specifically, for a detector of surface area A_d, we have

$$M = \frac{A_d}{\left(\frac{\lambda^2}{\Omega}\right)} = \frac{A_d}{A_c}. \tag{6.8}$$

Equation (6.8) represents the number of independent coherences areas within the detector area.

Not too surprising, there also is a time interval that incoherent sources can create fringes. It is called the *coherence time* τ_c and is the inverse of the source's bandwidth $\Delta\nu$. Recall that Eq. (6.5) is valid for a single quantum state. Therefore, within any time interval, there can be many of these quantum states existing and the span of this interval will dictate how much phase averaging occurs. The temporal coherence factor, here we denote by M', depends on the ratio of source's coherence time to the time measurement interval Δt, or we write

$$M' = \frac{\Delta t}{\tau_c}. \tag{6.9}$$

The above implies that within any measurement interval greater than the coherence time, a receiver with a detection area greater than the coherence area has a composite coherence factor given by

$$L = \left(\frac{\Delta t}{\tau_c}\right)\left(\frac{A_d}{A_c}\right), \tag{6.10}$$

which reflects the number of independent quantum states. Since these states are statistically independent, there is no correlation among them and the fluctuations must add in a random fashion. Therefore, if we detect n useful photons in time interval Δt and detector area A_d, we have for each independent quantum state,

$$\overline{\Delta n_L^2} = \frac{\overline{n}}{L} + \left(\frac{\overline{n}}{L}\right)^2 \tag{6.11}$$

[3, p. 17]. Now adding up all the noise fluctuations due to all states, the total fluctuations detected by a receiver in the measurement period Δt is given by

$$\overline{\Delta n^2} = L\overline{\Delta n_L^2} = \overline{n} + \frac{\overline{n}^2}{L} = \overline{n}\left(1 + \overline{n}\left(\frac{\tau_c}{\Delta t}\right)\left(\frac{A_c}{A_d}\right)\right). \tag{6.12}$$

The first term on the right-hand-side (RHS) represents the quantum fluctuations from the source and the second term is the quantum interference fluctuations. The greater the measurement time and detector area compared to the coherence time and area, respectively, the less interference fluctuations will be realized. Alternately, as the ratio of $\frac{\Delta t}{\tau_c}$ $\left(\text{or } \frac{A_d}{A_c}\right)$ approaches 1, the number of independent, uncorrelated waves become less until the limit of $\Delta t < \tau_c$, where within each measurement time Δt, only one "optical wave" is received. This means that full coherence or correlation exists and maximum fluctuation is possible. Specifically, when $\Delta t \ll \tau_c$ and $A_d \ll A_c$, $L = 1$, Eq. (6.12) reduces to

$$\overline{\Delta n^2} = \overline{n}(1 + \overline{n}), \tag{6.13}$$

where $\overline{n}$ is the average number of detected photoelectrons measured in the time interval Δt and area A_d.

Assume for the moment that $A_d \ll A_c$. Since the rate that photoelectrons are received can be written as $\overline{n}_r = \frac{\overline{n}}{\Delta t}$, Eq. (6.12) can be rewritten as

$$\overline{\Delta n^2} = \overline{n}_r \Delta t \left(1 + \frac{\overline{n}_r}{\Delta \nu}\right) = \frac{\overline{n}_r}{2B_e}\left(1 + \frac{\overline{n}_r}{B_o}\right) \tag{6.14}$$

with

$$B_e = 1/2\Delta t \tag{6.15}$$

being the receiver's electrical bandwidth and $B_o = \Delta \nu$ the optical bandwidth of the source. In Eq. (6.14), the ratio $\frac{\overline{n}_r}{B_o}$ is the number of photoelectrons per unit spectral interval. If, on the average, there is at least one effective photon in each spectral interval, then the wave beating happens. Otherwise, we have

$$\overline{\Delta n^2} = \overline{n}_r \Delta t = \frac{\overline{n}_r}{2B_e}. \tag{6.16}$$

What this implies is that incoherent sources cannot generate the needed spectral density to create interference.

6.3 INTERFERENCE AND TEMPORAL COHERENCE

Chapter 2 dealt with monochromatic waves with a definite frequency and wave number, an idealized situation that does not arise in practice. Even with the most monochromatic light source, it will have a finite spread of frequencies or wavelengths. The finite duration of the source, inherent broadening, or a number of other mechanisms can cause this spread. One now needs to include time as well as space in the solution of Maxwell equations.

Recall from Eq. (2.18) that

$$\nabla^2 \boldsymbol{E} - \varepsilon\mu_0 \frac{\partial^2 \boldsymbol{E}}{\partial t^2} = 0.$$

A similar equation can be derived for $\boldsymbol{H}$. Equation (2.19) was the general solution to these equations and was a decomposition of the electric field into time frequencies. Specifically, we have the following scalar solution:

$$u(\boldsymbol{r}, t) = \frac{1}{\sqrt{2\pi}} \int_{-\infty}^{\infty} \int_{-\infty}^{\infty} \hat{A}(\boldsymbol{k}) e^{i\boldsymbol{k}\cdot\boldsymbol{r} - i\omega(\boldsymbol{k})t} d^2 k, \tag{6.17}$$

where

$$\hat{A}(\boldsymbol{k}) = \frac{1}{\sqrt{2\pi}} \int_{-\infty}^{\infty} \int_{-\infty}^{\infty} u(\boldsymbol{r}, 0) e^{-i\boldsymbol{k}\cdot\boldsymbol{r}} d^2 r \tag{6.18}$$

and

$$|\boldsymbol{k}| = \frac{\omega}{v} = \sqrt{\mu\varepsilon}\, \frac{\omega}{c}. \tag{6.19}$$

The parameter $v = \frac{c}{\sqrt{\mu\varepsilon}}$ is a constant with dimensions of velocity, a function of the characteristics of the medium, its permittivity ε and permeability μ, and the speed of light in vacuum c. The parameter $\hat{A}(\boldsymbol{k})$ characterizes the properties of the linear superposition of the different waves, e.g.,

$$\hat{A}(\boldsymbol{k}) = \frac{1}{\sqrt{2\pi}}\, \delta(\boldsymbol{k} - \boldsymbol{k}_0). \tag{6.20}$$

For simplicity, we restrict our attention to one dimension only for this discussion without loss of generality.

If $u(x, 0)$ represents a harmonic wave $e^{ik_0 x}$ for all values of x, the Fourier transform relationship given by Eq. (2.75) implies that Eq. (6.20) corresponds to a monochromatic traveling wave of the form

$$u(x, t) = e^{ik_0 x - i\omega(k_0)t}. \tag{6.21}$$

On the other hand, if at $t = 0$, $u(x, 0)$ is a finite wave packet comprised a continuous light wave of a certain frequency, constrained within an envelope of finite extent, and $\hat{A}(k)$ is not given by Eq. (6.20). Rather, $\hat{A}(k)$ is peaked function of width Δk, centered about $k = k_0$. Figure 6.1 shows an example of (a) this type of wave amplitude $\hat{A}(k)$ and (b) its associated wave train $|u(x, 0)|$ with their wave number spread Δk and spatial spread Δx shown, respectively.

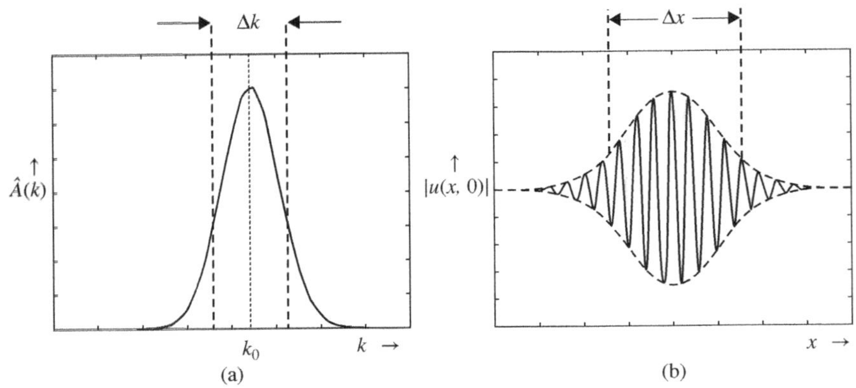

FIGURE 6.1 Spatial example of (a) an amplitude spectrum $\hat{A}(k)$ and (b) its wave train.

If Δk and Δx are the root-mean-square deviations from the average values of k and x derived from the intensities $|\hat{A}(k)|^2$ and $|u(x,0)|^2$, then we have

$$\Delta x\,\Delta k \geq \frac{1}{2} \tag{6.22}$$

[8, pp. 323–324]. This equation implies that short wave trains containing only a few wavelengths are comprised of very wide distribution of wave numbers of monochromatic waves, and large wave trains tend toward monochromaticity [8, pp. 323–324]. Let us now look at the behavior of the wave train in time.

Now, if $\hat{A}(k)$ is sharply peaked about $k = k_0$, then we can expand the frequency $\omega(k)$ in a Taylor series:

$$\omega(k) = \omega_0 + \left.\frac{d\omega(k)}{dk}\right|_{k=k_0}(k-k_0) + \cdots. \tag{6.23}$$

Substituting Eq. (6.23) into a one-dimensional version of Eq. (6.16), keeping only the first two terms in the Taylor series because of $\hat{A}(k)$ peak nature, we write

$$u(x,t) = \frac{1}{\sqrt{2\pi}}\int_{-\infty}^{\infty}\hat{A}(k)e^{ikx-i\omega(k)t}\,dk \tag{6.24}$$

$$\approx \frac{1}{\sqrt{2\pi}}\int_{-\infty}^{\infty}\hat{A}(k)e^{ikx-i\left[\omega_0+\left.\frac{d\omega(k)}{dk}\right|_{k=k_0}(k-k_0)\right]t}\,dk \tag{6.25}$$

$$\approx \frac{e^{-i\left[\omega_0-k_0\left.\frac{d\omega(k)}{dk}\right|_{k=k_0}\right]t}}{\sqrt{2\pi}}\int_{-\infty}^{\infty}\hat{A}(k)e^{ik\left[x-t\left.\frac{d\omega(k)}{dk}\right|_{k=k_0}\right]}\,dk \tag{6.26}$$

$$\approx u\left(x-t\left.\frac{d\omega(k)}{dk}\right|_{k=k_0},\,0\right)e^{-i\left[\omega_0-k_0\left.\frac{d\omega(k)}{dk}\right|_{k=k_0}\right]t}. \tag{6.27}$$

Ignoring the phase term in Eq. (6.27), this equation implies that the pulse travels with a fixed shape with a velocity given by

$$v_g = \left.\frac{d\omega(k)}{dk}\right|_{k=k_0}. \tag{6.28}$$

The velocity v_g is called the *group velocity*. It represents the speed of the energy transport for the pulse. For light waves, we have

$$\omega(k) = \frac{ck}{n(k)}, \tag{6.29}$$

where $n(k)$ is the index of refraction of the medium [7, p. 325]. The phase velocity is equal to

$$v_p = \frac{\omega(k)}{k} = \frac{c}{n(k)}. \tag{6.30}$$

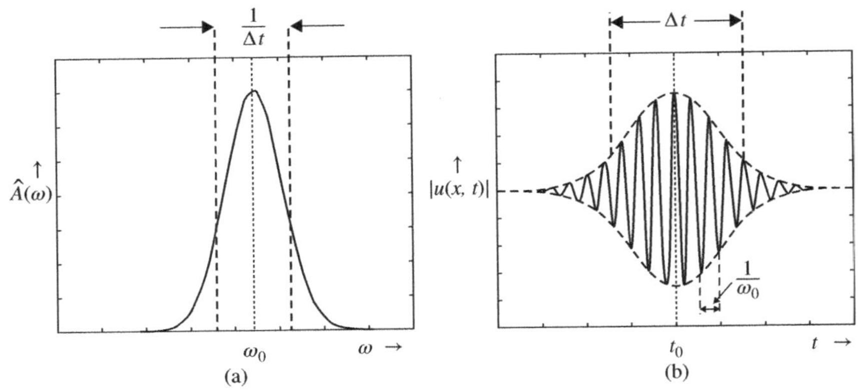

FIGURE 6.2 Temporal example of (a) an amplitude spectrum $\hat{A}(k)$ and (b) its wave train.

Clearly, the phase velocity can be greater than or less than the speed of light depending on whether $n(k)$ is greater than or less than one, respectively. In general, $n(k) > 1$. On the basis of the above, it is not surprising that we have temporal wave packet and a spectrum result similar to what we found in the spatial discussion. Their basic characteristics are shown in Figure 6.2. $\hat{A}(\omega)$ again is peaked function of width f, centered about $\omega = \omega_0$. Here, Δt is the temporal pulse duration. The sine wave peak separation length is equal to $\frac{1}{\omega_0}$. We now need to define some important quantities and concepts. Let us now look at an example interference experiment with a common source to do so.

Figure 6.3 shows the setup for creating interference at an observation point O using an optical source and a mirror. The source located at point P sends direct illumination to the observation point. It also sends indirect illumination to the observation point via the source's refection in the mirror at point R. Assume the path length $\overline{PR}$ and $\overline{P'R}$ are of equal length. Let us now define

$$\overline{PO} = D \tag{6.31a}$$

and

$$\overline{P'O} = D'. \tag{6.31b}$$

The total field at the observation point O is equal to the sum of the direct and indirect illumination, or mathematically, we have

$$u_{\text{Total}}(O,t) = u(O,t) + u\left(O, t - \frac{(D' - D)}{c}\right). \tag{6.32}$$

where $t - \frac{(D'-D)}{c}$ is the time delay between the direct and indirect waves [1, p. 328].

The nonaveraged intensity $|u(\mathbf{r},t)|^2$ is called the *random intensity* or *instantaneous intensity*. If we let $B(O,t) = |u_{\text{Total}}(O,t)|^2$ be the instantaneous intensity at point O, the resulting intensity at that same point, at time t_0 is given by

$$I(t_0) = \frac{1}{T}\int_{t_0-\frac{T}{2}}^{t_0+\frac{T}{2}} B(O,t)\,dt \tag{6.33}$$

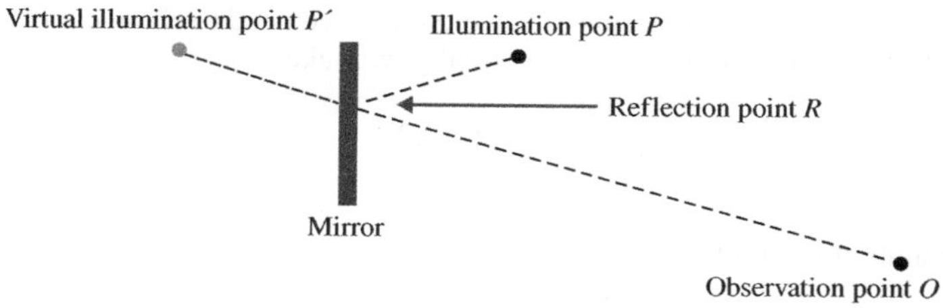

FIGURE 6.3 Interference experimental setup.

with T being the statistical averaging or integration time. Herein, T has to be large enough to enclose both wave trains, $u(O,t)$ and $u\left(O, t - \frac{(D'-D)}{c}\right)$. Let $\tau = t - \frac{(D'-D)}{c}$ be our shorthand for delay time. Then, we can write the intensity at $t = t_0$ as

$$I(t_0) = \frac{1}{T}\int_{t_0-\frac{T}{2}}^{t_0+\frac{T}{2}} |u(O,t)|^2\, dt + \frac{1}{T}\int_{t_0-\frac{T}{2}}^{t_0+\frac{T}{2}} |u(O,t+\tau)|^2\, dt$$
$$+ 2\,\mathrm{Re}\left\{\frac{1}{T}\int_{t_0-\frac{T}{2}}^{t_0+\frac{T}{2}} u(O,t)\, u^*(O,t+\tau)\, dt\right\} \tag{6.34}$$

with the asterisk again signifying the complex conjugate of the field. The function inside Re$\{\cdots\}$ is the autocorrelation of the complex random function, as introduced in Eq. (1.99) in Chapter 1.

$$2\,\mathrm{Re}\left\{\frac{1}{T}\int_{t_0-\frac{T}{2}}^{t_0+\frac{T}{2}} u(O,t)\, u^*(O,t+\tau)\, dt\right\}$$
$$= 2\,\mathrm{Re}\left\{\frac{e^{i[\omega_0 - k_0\, v_g]\tau}}{T}\int_{t_0-\frac{T}{2}}^{t_0+\frac{T}{2}} u(O - v_g t, 0)\, u^*(O - v_g\,(t+\tau), 0)\, dt\right\}. \tag{6.35}$$

Given that our observation point is fixed, the integral

$$\int_{\frac{T}{2}}^{\frac{T}{2}} u(O - v_g t, 0)\, u^*(O - v_g\,(t+\tau), 0)\, dt$$

will be nonzero if, and only if, $\tau < \Delta t$. The fact that we are dealing with energy transport in the above implies this criterion is true. However, it is not the most accurate one. The condition for interference comes from the power spectrum of the optical source. Specifically for interference to be observed, the width $\Delta\nu$ of the temporal spectral density be less than $\frac{1}{\tau}$. This can certainly be satisfied by a wave train of duration $\Delta t \approx \frac{1}{\Delta\nu}$, as previously pointed out, but completely different waves $u(\boldsymbol{r}, t)$, which are not zero over much longer intervals, can have the same power spectrum. Of course, these other wave trains would have different phase angles, but that really does not matter in this experiment. This means that the original criterion $\tau < \Delta t$ transforms into $\Delta\nu < \frac{1}{\tau}$ [1, pp. 329–330].

To get this new criterion in terms of wavelengths, we need to change the time delay τ into a path length difference ΔL. Specifically, we take

$$\tau = \frac{(D' - D)}{c}$$

and manipulate it to yield

$$c\tau = (D' - D) = \Delta L, \tag{6.36}$$

the desired path length difference. Recall that $\lambda\nu = c$. The spectral spread $\Delta\omega$ is related to wavelength via the relationship

$$\frac{\Delta\nu}{\Delta\lambda} \equiv \left|\frac{d\nu}{d\lambda}\right| = \frac{C}{\overline{\lambda}^2} \tag{6.37}$$

or

$$\Delta\nu = c\frac{\Delta\lambda}{\overline{\lambda}^2}, \tag{6.38}$$

where $\overline{\lambda}$ is the mean wavelength of the wave packet and $\Delta\lambda$ its wavelength spread. Thus, we find that

$$\Delta\nu = c\frac{\Delta\lambda}{\overline{\lambda}^2} < \frac{1}{\tau} = \frac{c}{\Delta L}$$

or

$$\frac{\Delta L}{\overline{\lambda}} < \frac{\overline{\lambda}}{\Delta\lambda}. \tag{6.39}$$

Equation (6.52) states that for interference to occur, the path difference, in terms of the number of mean wavelengths, must be less than the "spectral purity" of the optical source.

Without going into details, there are three additional conditions that negate interference creation besides $\frac{\Delta L}{\overline{\lambda}} < \frac{\overline{\lambda}}{\Delta\lambda}$ [1, p. 330]. They are as follows:

- Two waves that are perpendicularly polarized cannot interfere.
- Two waves coming from different sources that independently emit spontaneously cannot interfere.

Two waves, emitted from the same source, but with different temporal frequencies ν and $\nu + \delta\nu$ cannot interfere, if the duration of observation (time integral) T is such that $T \gg \frac{1}{\delta\nu}$.

In many problems, we are dealing with statistical processes that are *stationary*.[2] Under this condition, $u(O, t) = u(t)$ now is a *stationary complex random function* and

[2] A statistically stationary wave has an average intensity $I(r)$ that does not vary with time, that is, its instantaneous intensity varies with time, but its average is constant. Conversely, a statistically nonstationary wave has an average intensity $I(r)$ that does vary with time.

we can write

$$G(\tau) = \lim_{T\to\infty} \frac{1}{2T} \int_{-\frac{T}{2}}^{\frac{T}{2}} u(t)\, u^*(t-\tau)\, dt = \overline{u^*(t)u(t-\tau)}, \tag{6.40}$$

where $G(\tau)$ is the *temporal coherence function*. Furthermore, when the waves are stationary, the average intensity $\bar{I}(\boldsymbol{r})$ is given by

$$\bar{I}(\boldsymbol{r}) = \lim_{T\to\infty} \frac{1}{2T} \int_{-\frac{T}{2}}^{\frac{T}{2}} u(t)\, u^*(t)\, dt = \lim_{T\to\infty} \frac{1}{2T} \int_{-\frac{T}{2}}^{\frac{T}{2}} |u(t)|^2\, dt = G(0) \tag{6.41}$$

using Eqs. (5.23) and (5.24).[3] Besides the intensity, the temporal coherence function characterizes the coherence of the light. A useful measure is *complex degree of temporal coherence*, defined as

$$g(\tau) = \frac{G(\tau)}{G(0)} = \frac{\langle u^*(\boldsymbol{r},t)u(\boldsymbol{r},t+\tau)\rangle}{\langle u^*(\boldsymbol{r},t)\, u(\boldsymbol{r},t)\rangle}, \tag{6.42}$$

where $0 \le |g(\tau)| \le 1$. This parameter depicts the correlation between the wave train and a shifted version of itself. For most real sources, $|g(\tau)|$ decreases from its maximum value at $\tau = 0$ as τ increases.

If $|g(\tau)|$ is a monotonic function decreasing with τ, the convention is to define a specific parameter that characterizes the drop of the monotonic function to a specific number, say 0.5 or e^{-1}. It tells the user the useful range of utility for the said function relative to a specific property such as correlation. For the complex degree of temporal coherence, this parameter is the *coherence time*, τ_c, which we were introduced to earlier. It is given by

$$\tau_c = \int_{-\infty}^{\infty} |g(\tau)|^2\, d\tau \tag{6.43}$$

[2, p. 409]. One sometimes interested in the path length difference that good correlation is still achievable. The coherence path length is called the *coherence length* ℓ_c and is related to the coherence time via the expression,

$$\ell_c = c\, \tau_c. \tag{6.44}$$

A related function to this example is the *average spectrum* $U(\omega)$ of the light wave. It is defined as the Fourier transform of the field amplitude:

$$U(\omega) = \int_{-\infty}^{\infty} u(t)e^{-i\omega t}\, dt. \tag{6.45}$$

The quantity $\langle |U(\omega)|^2 \rangle$ represents the energy spectral density of the light. The above definition has $u(t)$ structured such that $U(\omega) = 0$ for negative frequencies.

[3]In this chapter, we have replaced "big" in Eq. (5.24) by infinity.

Since a truly stationary process means that $u(t)$ is infinite in duration and possesses infinite energy, the above definitions are not very useful in a practical system analysis. The result is that one uses a truncated spectral density $U_T(\omega)$ defined as

$$U_T(\omega) = \int_{-\frac{T}{2}}^{\frac{T}{2}} u(t)e^{-i\omega t}\, dt. \tag{6.46}$$

where T is the width of the observation window. The quantity $\left\langle |U_T(\omega)|^2 \right\rangle$ now represents the energy density of the light train. The *power (intensity) spectral density* is the energy per unit, so it is obtained by dividing the above energy density by T and taking the limit $T \to \infty$, or-

$$S(\omega) = \lim_{T\to\infty} \frac{\left\langle |U_T(\omega)|^2 \right\rangle}{T}. \tag{6.47}$$

Again, $S(\omega) = 0$ for negative frequencies. $S(\omega)$ is also known as the *spectral density* or *spectrum* [2, p. 410].

The total average intensity then is

$$I = \int_0^\infty S(\omega)e^{-i\omega t}\, d\omega. \tag{6.48}$$

The temporal coherence and spectral density functions are Fourier transform pairs:

$$S(\nu) = \int_{-\infty}^\infty G(t)e^{-2\pi i\nu t}\, dt. \tag{6.49}$$

Equation (6.49) is known as the *Wiener–Khinchin Theorem*, which was introduced in Eq. (1.129).

The spectral width, or linewidth, of the optical source is the width $\Delta\nu$ of the source's spectral density $S(\nu)$. Because of the Wiener–Khinchin Theorem, the widths of the temporal coherence and spectral density functions are inversely related to one another. There are several ways to define $\Delta\nu$ [2, p. 411]. If one uses the following expression for the spectral width

$$\Delta\nu_c = \frac{\left(\int_0^\infty S(\nu)\, d\omega\right)^2}{\int_0^\infty S^2(\nu)\, d\omega}, \tag{6.50}$$

then it can be shown that

$$\Delta\nu = \frac{1}{2\tau_c} \tag{6.51}$$

[2, p. 412].

Example 6.1 Assume that the observation point is in the far field of the source and its mirrored image and we have a stationary process. This situation thus creates two sets of plane waves interacting at the observation point with a time delay due to the path difference. If the light is deterministic and monochromatic, that is,

$$u(t) = A_0 e^{i\omega_0 t}, \tag{6.52}$$

where A_0 is a constant, then we can write

$$g_0(\tau) = \frac{1}{I_0} \lim_{T\to\infty} \frac{1}{T} \int_{-\frac{T}{2}}^{\frac{T}{2}} u(t)\, u^*(t+\tau)\, dt = e^{i\omega_0 \tau}. \tag{6.53}$$

Hence, $|g(\tau)| = 1$ for all values of τ. In this case, the intensity at point O can be written as

$$I = 2I_0 + 2\, I_0 \,\mathrm{Re}\{g_0(\tau)\} \tag{6.54}$$

$$= 2I_0 + 2\, I_0 \cos \omega_0 \tau, \tag{6.55}$$

where

$$I_0 = \lim_{T\to\infty} \frac{1}{T} \int_{-\frac{T}{2}}^{\frac{T}{2}} |u(t)|^2\, dt = \lim_{T\to\infty} \frac{1}{T} \int_{-\frac{T}{2}}^{\frac{T}{2}} |u(t+\tau)|^2\, dt = |A|^2. \tag{6.56}$$

Equation (6.55) creates a sinusoidal pattern for the normalized intensity versus the phase $\omega_0\tau$, as is depicted in Figure 6.4. As we noted in the last chapter, the visibility of the sinusoidal pattern is given by the Michelson contrast:

$$C_{\mathrm{Michelson}} = \frac{I_{\max} - I_{\min}}{I_{\max} + I_{\min}}.$$

In partial coherence theory, $C_{\mathrm{Michelson}}$ also is known as the *Visibility V* or the *modulation depth*, and can be written as

$$V = \frac{I_{\max} - I_{\min}}{I_{\max} + I_{\min}} = |g(\tau)|. \tag{6.57}$$

Example 6.2 Assume we have a partially coherent wave $U(t)$ of intensity I_0 and complex degree of temporal coherence $g(\tau)$. It can easily be shown that

$$I = \left\langle |U(t) + U(t+\tau)|^2 \right\rangle = 2I_0[1 + |g(\tau)| \cos(\varphi(\tau))], \tag{6.58}$$

where $\varphi(\tau) = \arg[g(\tau)]$ [2, pp. 419–421]. The relationship between I and τ is called an *interferogram* [2, p. 421].

If $g(\tau) = g_a(\tau)e^{i\omega_0\tau}$, then Eq. (6.58) becomes

$$I = 2I_0[1 + |g_a(\tau)| \cos(\omega_0\tau + \varphi_a(\tau))], \tag{6.59}$$

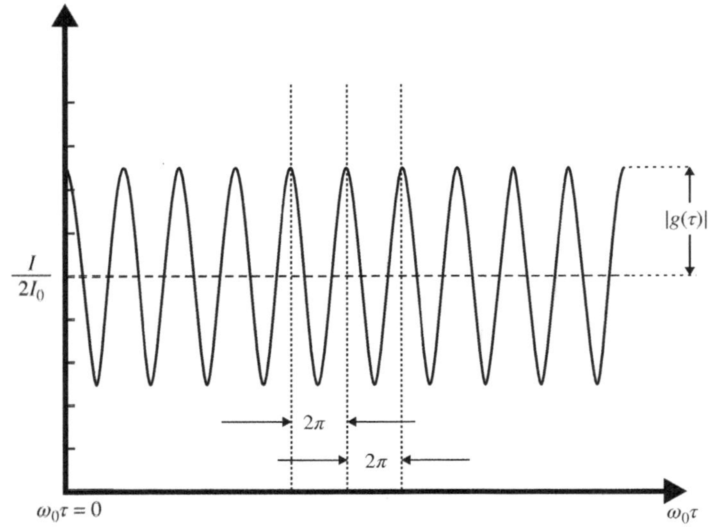

FIGURE 6.4 Normalized intensity as a function.

where $\varphi_a(\tau) = \arg[g_a(\tau)]$. Here, the spectral width Δv_c is inversely equal to the coherence time τ_c. If $\Delta v_c \ll v_0$, then both $|g_a(\tau)|$ and $\varphi_a(\tau)$ will vary slowly compared to the period $\frac{1}{v_0}$ and one will obtain an interference pattern like that shown in Figure 6.2b with $t_0 = 0$. The envelope follows $|g_a(\tau)|$.

6.4 INTERFERENCE AND SPATIAL COHERENCE

Let us extend the source to the coordinate space (x', y', z') as well as in temporal frequencies v. The resulting interference experiment is shown in Figure 6.5.

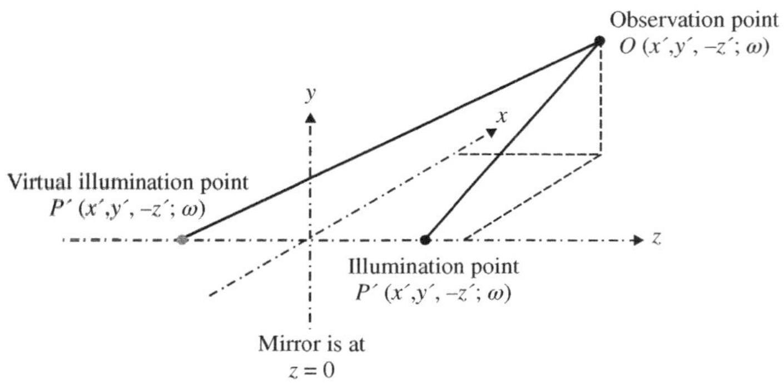

FIGURE 6.5 Interference from a source and a mirror.

Let the function S represent the temporal power spectrum density at point $(x', y', z'; \nu)$. There is a mirror image energy emission at point $(x', y', z'; \nu)$. Let $\eta(\nu)$ be the spectral sensitivity (quantum efficiency) of the detector. Using the results from Chapter 2, we can write the intensity at point O as

$$I(x, y, z; \nu) = \int \iiint \eta(\nu)\, S(x', y', z'; \nu) \left| \frac{e^{ikr_0}}{r_0} + \frac{e^{ikr_1}}{r_1} \right|^2 dx'dy'dz'd\nu, \tag{6.60}$$

where

$$r_0 = \sqrt{(x - x')^2 + (y - y')^2 + (z - z')^2}, \tag{6.61}$$

$$r_1 = \sqrt{(x - x')^2 + (y - y')^2 + (z + z')^2}, \tag{6.62}$$

and

$$k = \frac{2\pi}{\lambda} = \frac{2\pi\nu}{c}. \tag{6.63}$$

Equation (6.60) depicts the interaction of a pair of quasi-coherent spherical waves at frequency ω, coming from the point pair (x', y', z') and $(x', y', -z')$. From previous work, we know that

$$\left| \frac{e^{ikr_0}}{r_0} + \frac{e^{ikr_1}}{r_1} \right|^2 = \frac{1}{r_0^2} + \frac{1}{r_1^2} + \frac{2\,\cos(k(r_0 - r_1))}{r_0\, r_1}. \tag{6.64}$$

Example 6.3 For a monochromatic point source, we have

$$\eta(\nu)\, S(x', y', z'; \nu) = \delta(\nu - \nu_0)\, \delta(x', y', z' - z_0). \tag{6.65}$$

Equation (6.63) then becomes

$$I(x, y, z; \nu_0) = \frac{1}{r_0^2} + \frac{1}{r_1^2} + \frac{2\,\cos(k_0[r_0 - r_1])}{r_0\, r_1}, \tag{6.66}$$

where

$$r_0 = \sqrt{x^2 + y^2 + (z - z_0)^2}, \tag{6.67}$$

$$r_1 = \sqrt{x^2 + y^2 + (z + z_0)^2}, \tag{6.68}$$

and

$$k_0 = \frac{2\pi}{\lambda_0} = \frac{2\pi\nu_0}{c}. \tag{6.69}$$

The maxima of the interference fringes occur where the cosine is +1. This means when

$$k_0[r_0 - r_1] = 0, 2\pi, 4\pi, \ldots \tag{6.70a}$$

or

$$[r_0 - r_1] = 0, \lambda_0, 2\lambda_0, \ldots . \tag{6.70b}$$

If y is very large, then we have

$$r_0 \approx y\sqrt{\frac{(z-z_0)^2}{y^2}+1} \approx y\left(1+\frac{(z-z_0)^2}{2y^2}\right) \tag{6.71}$$

and

$$r_1 \approx y\sqrt{\frac{(z-z_0)^2}{y^2}+1} \approx y\left(1+\frac{(z-z_0)^2}{2y^2}\right). \tag{6.72}$$

Equation (6.66) becomes

$$I(x,y,z;\nu_0) \approx \frac{2}{z^2}\left(1+\cos\left(k_0 y\left[\left(1+\frac{(z-z_0)^2}{2y^2}\right)-\left(1+\frac{(z-z_0)^2}{2y^2}\right)\right]\right)\right) \tag{6.73}$$

$$\approx \frac{2}{y^2}\left(1+\cos\left(\frac{4\pi z z_0}{\lambda_0 y}\right)\right). \tag{6.74}$$

Figure 6.6 illustrates an example fringe pattern for Eq. (6.74). Included in this figure is a contour pattern to better show the fringe behavior. The fringe spacing equals $\frac{\lambda_0 y}{2z_0}$. We would find a similar equation for x being very large, with x replacing y in the above expression.

If z is very large, then we have

$$\begin{aligned} r_0 &\approx z\sqrt{\frac{x^2+y^2}{(z-z_0)^2}+1} \approx z\left(1+\left(\frac{x^2+y^2}{2z^2}\right)\left(1+\frac{z_0}{z}\right)\right) \\ &\approx z\left(1+\left(\frac{x^2+y^2}{2z^2}\right)\right)+z_0\left(\frac{x^2+y^2}{2z^2}\right) \end{aligned} \tag{6.75}$$

and

$$\begin{aligned} r_1 &\approx z\sqrt{\frac{x^2+y^2}{(z+z_0)^2}+1} \approx z\left(1+\left(\frac{x^2+y^2}{2z^2}\right)\left(1-\frac{z_0}{z}\right)\right) \\ &\approx z\left(1+\left(\frac{x^2+y^2}{2z^2}\right)\right)-z_0\left(\frac{x^2+y^2}{2z^2}\right). \end{aligned} \tag{6.76}$$

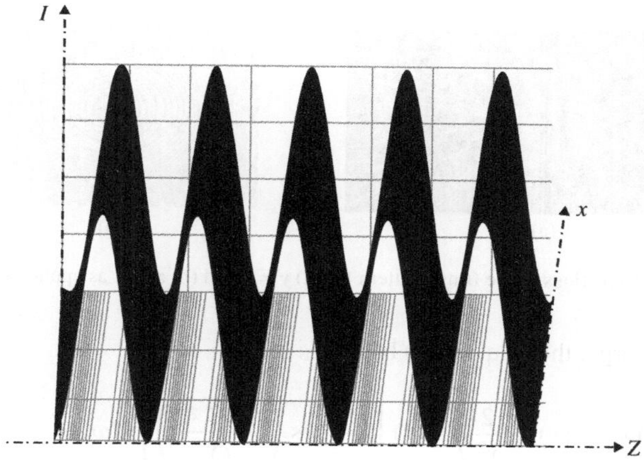

FIGURE 6.6 Depiction of the fringe pattern for large y applied to Eq. (6.74).

This implies that

$$r_0 - r_1 \approx z\left(1 + \left(\frac{x^2+y^2}{2z^2}\right)\right) + z_0\left(\frac{x^2+y^2}{2z^2}\right)$$
$$- z\left(1 + \left(\frac{x^2+y^2}{2z^2}\right)\right) + z_0\left(\frac{x^2+y^2}{2z^2}\right) \tag{6.77}$$
$$\approx +2z_0\left(\frac{x^2+y^2}{2z^2}\right). \tag{6.78}$$

Equation (6.66) becomes

$$I(x, y, z; \lambda_0) \approx \frac{2}{z^2}\left(1 + \cos\left(k_0\left[2z_0\left(\frac{x^2+y^2}{2z^2}\right)\right]\right)\right) \tag{6.79}$$
$$\approx \frac{2}{z^2}\left(1 + \cos\left(\frac{k_0 z_0(x^2+y^2)}{z^2}\right)\right) = \frac{2}{z^2}\left(1 + \cos\left(\frac{2\pi z_0(x^2+y^2)}{\lambda_0 z^2}\right)\right). \tag{6.80}$$

Figure 6.7 depicts contour plots for Eq. (6.80) when (a) $y = 0$ and (b) $z \gg 1$ as the fringe patterns are hard to discern in black-and-white in three dimensions [1, p. 332]. The hyperboloids in Figure 6.7a are located a distance of $2z_0$ apart. The pattern in Figure 6.7b essentially is Fresnel zone plate with radii proportional to $\frac{\lambda_0 z^2}{z_0}$ [2, p. 110].

Example 6.4 For a polychromatic point source, we have

$$\eta(\nu)\, S(x', y', z'; \nu) = S(\nu)\, \delta(x', y', z' - z_0) \tag{6.81}$$

with $S(\nu)$ being the power spectral density of the polychromatic point source.

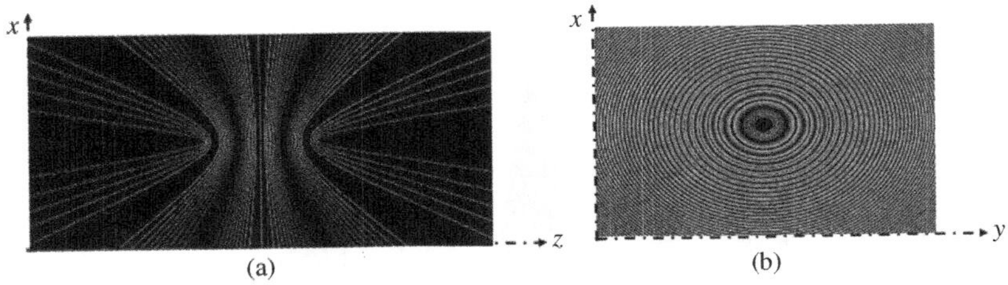

FIGURE 6.7 Depictions of the fringe pattern for (a) $y = 0$ and (b) $z \gg 1$ as applied to Eq. (6.66).

If y is very large, then Eq. (6.58) becomes

$$I(x, y, z; \nu) \approx \frac{2}{y^2} \int S(\nu) \left[1 + \cos\left(\frac{4\pi \nu z\, z_0}{cy} \right) \right] d\omega \tag{6.82}$$

$$= \frac{2}{y^2} \left[G(0) + \left| G\left(\frac{2z\, z_0}{cy} \right) \right| \cos\left(\frac{2z\, z_0}{cy} \sigma_{\tilde{R}} \right) \right], \tag{6.83}$$

since the temporal coherence function is related to the power spectrum via the relation

$$G(\rho) = |G(\rho)| e^{i\sigma_{\tilde{R}}(\rho)} = \int S(\omega)\, e^{-2\pi i \nu \rho}\, d\omega \tag{6.84}$$

and the assumption that $G = G^*$. We would find a similar equation for x being very large, with x replacing y in the above.

Since many sources have peaked spectrum around a mean frequency ω_0, we can write

$$S(\nu) = S_0(\nu - \nu_0), \tag{6.85}$$

which means that

$$|G(\rho)| = |G_0(\rho)| \tag{6.86}$$

and

$$\sigma_{\tilde{R}} = \sigma_{\tilde{R}0} + 2\pi \nu_0 \rho. \tag{6.87}$$

The intensity then is

$$I(x, y, z; \nu_0) = \frac{2}{y^2} \left[G(0) + \left| G\left(\frac{2z\, z_0}{cy} \right) \right| \cos\left(\frac{4\pi z\, z_0}{cy} \sigma_{\tilde{R}} \right) \right] \tag{6.88}$$

$$= \frac{2G(0)}{y^2} \left[1 + \left| \frac{G\left(\frac{2z\, z_0}{cy} \right)}{G(0)} \right| \cos\left(2\pi \omega_0 \left(\frac{2z\, z_0}{cy} \right) + \left(\frac{2z\, z_0}{cy} \right) \sigma_{\tilde{R}0} \right) \right] \tag{6.89a}$$

$$= \frac{2G(0)}{y^2}\left[1 + \left|g\left(\frac{2z\,z_0}{cy}\right)\right| \cos\left(2\pi\nu_0\left(\frac{2z\,z_0}{cy}\right) + \left(\frac{2z\,z_0}{cy}\right)\sigma_{\tilde{R}0}\right)\right], \tag{6.89b}$$

where $g(\rho)$ again is the complex degree of temporal coherence. Equations (6.89a) and (6.89b) have a visibility that goes down with z for a fixed value of y, with a possible fringe shift at $\frac{\sigma_{\tilde{R}0}}{2\pi\nu_0}$. Similarly, for z being very large, we obtain

$$I(x,y,z;\nu_0) = \frac{2G(0)}{z^2}\left[1 + \left|g\left(\frac{z_0(x^2+y^2)}{cz^2}\right)\right| \times \cos\left(2\pi\nu_0\left(\frac{z_0(x^2+y^2)}{cz^2}\right) + \sigma_{\tilde{R}0}\left(\frac{z_0(x^2+y^2)}{cz^2}\right)\right)\right]. \tag{6.90}$$

6.5 COHERENT LIGHT PROPAGATING THROUGH A SIMPLE LENS SYSTEM

In this section, we investigate how partially coherent light traverses an optical lens system in both space and time, introducing some other key characterizing parameters. This section follows the development and definitions of Reynolds *et al.* [4]. This development has slightly different definitions than above. One finds both views in the literature, so at the end we show the differences. As it turns out, it is just a notation difference, but it is important to reconcile.

Referring to Figure 6.8, we begin with the one-dimensional analysis. Let us assume that we have some coherent (complex) spatial source in the ($z = 0$)-plane, $u_0(x')$. This

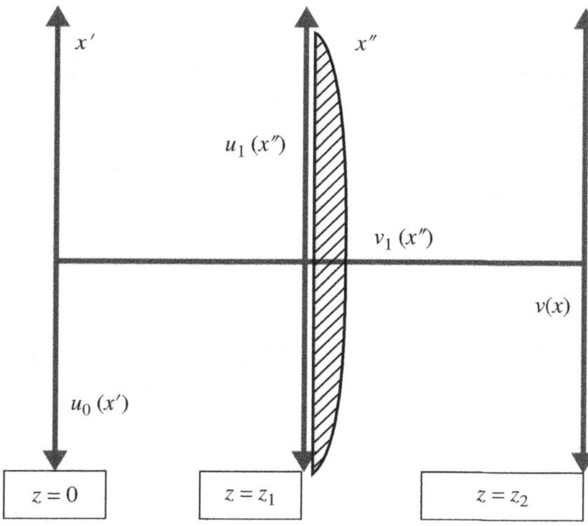

FIGURE 6.8 Geometry for image formation.

source will propagate to the $(z = z_1)$-plane, resulting in the distribution

$$u(x'', z) = u_1(x''). \tag{6.91}$$

From Chapter 2, we know that

$$u_1(x'') \propto \int_{-\infty}^{\infty} u_0(x')\, e^{\frac{i\pi\,(x'-x'')^2}{\lambda z_1}}\, dx'. \tag{6.92}$$

Let us assume a coherent point source located at (ξ) in the source plane. In this case, the above equation becomes

$$u_1(x'') = \frac{e^{i\left[kz+\frac{\pi}{2}\right]}}{\lambda z}\, e^{\frac{i\pi\,(x''-\xi)^2}{\lambda z_1}}. \tag{6.93}$$

At $(z = z_1)$-plane, we place a lens and the distribution passes through the lens to create the distribution $v_1(x'')$.

Following our convention, the incoming distribution depicted in Eq. (6.91) is multiplied by a location-dependent phase change and the a pupil function, $T(x)$, for the lens. Mathematically, we have

$$v_1(x') = u_1(x')\, \hat{T}(x')\, e^{\frac{-i\pi x'^2}{\lambda f}}, \tag{6.94}$$

where f is the focal length of the lens. This new source then will propagate to the $(z = z_2)$-plane, resulting in the distribution $v(x)$ given by

$$v(x) = \int_{-\infty}^{\infty} v_1(x'')\, e^{\frac{i\pi\,(x-x'')^2}{\lambda z_2}}\, dx''. \tag{6.95}$$

These again are the steps in the propagation from left to right for Figure 6.8. Substituting Eqs. (6.93) and (6.94) into the above yields

$$v(x) \propto \frac{1}{\lambda^2\, z_1 z_2} \int_{-\infty}^{\infty} \hat{T}(x')\, e^{\frac{i\pi\,(x-x'')^2}{\lambda z_2}}\, e^{\frac{-i\pi x''^2}{\lambda f}}\, e^{\frac{i\pi\,(x''-\xi)^2}{\lambda z_1}}\, dx''. \tag{6.96}$$

Now

$$\frac{i\pi\,(x-x'')^2}{\lambda z_2} + \frac{i\pi\, x''^2}{\lambda f} + \frac{i\pi\,(\xi - x'')^2}{\lambda z_1} = \frac{i\pi}{\lambda}\left[-2x''\left(\frac{x}{z_2} + \frac{\xi}{z_1}\right) + \frac{x^2}{z_2} + \frac{\xi^2}{z_1}\right]$$

using the Lens Law. This implies that

$$v(x) \propto \frac{1}{\lambda^2\, z_1 z_2} \int_{-\infty}^{\infty} \hat{T}(x')\, e^{\frac{i\pi}{\lambda}\left[-2x''\left(\frac{x}{z_2}+\frac{\xi}{z_1}\right)+\frac{x^2}{z_2}+\frac{\xi^2}{z_1}\right]}\, dx'' \tag{6.97}$$

or

$$F2\pi/z_1z_2(x+\xi/M) \propto \frac{1}{\lambda^2\, z_1z_2}\int_{-\infty}^{\infty}\hat{T}(x')\; e^{-\frac{i2\pi}{\lambda z_1}x''\left(x+\frac{\xi}{M}\right)}\, dx'' \tag{6.98}$$

since the Fourier transform of Complex Pupil Function $\hat{T}(x')$ is the Amplitude Spread Function $F(x)$. Here, $M = \frac{z_1}{z_2}$ is the magnification previously defined.

Equation (6.98) says that if the Lens Law is satisfied, $F(x)$, the impulse response of the coherent imaging systems, is just (within a factor of $\left[\frac{1}{\lambda z_1 z_2}\right]$) the Fourier transform of the pupil function $\hat{T}(x')$ centered at coordinate $x = \frac{\xi}{M}$.

Following the same procedure, it is easy to show the two-dimensional solution

$$v(x,y) \propto \frac{1}{\lambda^2\, z_1z_2}\iint_{-\infty}^{\infty}\hat{T}(x',y')\; e^{-\frac{i2\pi}{\lambda z_1}\left[x''\left(x+\frac{\xi}{M}\right)+y''\left(y+\frac{\eta}{M}\right)\right]}dx''dy'' \tag{6.99a}$$

or

$$v(x,y) \propto T(2\pi/z_1z_2(x+\xi/M), 2\pi/z_1z_2(y+\eta/M)), \tag{6.99b}$$

Let us now look at the linear systems approach to coherent light [4].

From previous work, we know that the resulting distribution created by a coherent image system is given by

$$v(x,y)^{\propto} = \iint_{-\infty}^{\infty} T(2\pi/z_1z_2(x+s), 2\pi/z_1z_2(y+p))u_0(-M_s, -M_p) \tag{6.100}$$

$$= \iint_{-\infty}^{\infty} T(2\pi/z_1z_2(x+s), 2\pi/z_1z_2(y+p))\tilde{u}_0(s,p)\, \mathrm{d}s\,\mathrm{d}p \tag{6.101}$$

using Eq. (6.97b). In the above, we have $s = \frac{\xi}{M}, p = \frac{\eta}{M}, \tilde{F} = |M|\; F$ and

$$\tilde{u}_0(s,p) = |M|\; u_0(-M_s, -M_p). \tag{6.102}$$

Using our list of Fourier transform properties, we have

$$\hat{F}(k_x,k_y) \propto \mathcal{F}\left\{\iint_{-\infty}^{\infty}\tilde{F}(\tilde{x},\tilde{y})\; e^{-\,2\pi/\lambda z_1[\tilde{x}x+\tilde{y}y]}d\tilde{x}\; d\tilde{y}\right\} \propto T(-\lambda z_1k_x, -\lambda z_1k_y). \tag{6.103}$$

Therefore, like in the incoherent case, the ideal coherent image produced by a diffraction-limited optical system is a scaled and inverted version of the coherent source image.

If we have coherent source of the form $u_0(x,y;t)\; e^{-2\pi i\bar{\nu}t}$, then the complex image coming out of a diffraction-limited optical system is given by

$$u(x,y;t) = \iint_{-\infty}^{\infty} F(x-\tilde{\xi}, y-\tilde{\eta})\; \tilde{u}_0(\tilde{\xi},\tilde{\eta}; t-\tau)\; d\tilde{\xi}\; d\tilde{\eta}, \tag{6.104}$$

where τ is a time delay associated with the propagation from $(\tilde{\xi},\tilde{\eta})$ to (x,y).

To calculate the image intensity, we must time-average the instantaneous intensity represented by $|u(x, y; t)|^2$. This is because the detector integration time usually is long compared with the reciprocal of the optical bandwidth. Mathematically, the intensity is written as

$$I_t(x, y) = |u(x, y; t)|^2 = \iint_{-\infty}^{\infty} d\tilde{\xi}_1 d\tilde{\eta}_1 \iint_{-\infty}^{\infty} d\tilde{\xi}_2 d\tilde{\eta}_2\, F(x - \tilde{\xi}_1, y - \tilde{\eta}_1)$$
$$F^*(x - \tilde{\xi}_2, y - \tilde{\eta}_2) \langle \tilde{u}_0(\tilde{\xi}_1, \tilde{\eta}_1; t - \tau_1)\, \tilde{u}_0^*(\tilde{\xi}_2, \tilde{\eta}_2; t - \tau_2) \rangle, \qquad (6.105)$$

where

$$\lim_{T\to\infty} \frac{1}{T} \int_{-\frac{T}{2}}^{\frac{T}{2}} \cdots \, dt.$$

For a fixed image point, the impulse response F is only nonzero over a small region about that point. That is, the integrand of Eq. (6.102) is nonzero only when $(\tilde{\xi}, \tilde{\eta})$ and (x, y) are very close to each other. This also implies that the time delays are negligible under the narrowband assumption; so the delays can be ignored. The result is that Eq. (6.102) can be rewritten as

$$I_t(x, y) = |u(x, y; t)|^2 = \iint_{-\infty}^{\infty} d\tilde{\xi}_1 d\tilde{\eta}_1 \iint_{-\infty}^{\infty} d\tilde{\xi}_2 d\tilde{\eta}_2\, F(x - \tilde{\xi}_1, y - \tilde{\eta}_1)$$
$$F^*(x = \tilde{\xi}_2, y - \tilde{\eta}_2)\, \langle \tilde{u}_0(\tilde{\xi}_1, \tilde{\eta}_1; t)\, \tilde{u}_0^*(\tilde{\xi}_2, \tilde{\eta}_2; t) \rangle$$
$$= \iint_{-\infty}^{\infty} d\tilde{\xi}_1 d\tilde{\eta}_1 \iint_{-\infty}^{\infty} d\tilde{\xi}_2 d\tilde{\eta}_2\, F(x - \tilde{\xi}_1, y - \tilde{\eta}_1)$$
$$F^*(x - \tilde{\xi}_2, y - \tilde{\eta}_2)\, J_0(\tilde{\xi}_1, \tilde{\eta}_1; \tilde{\xi}_2, \tilde{\eta}_2) \qquad (6.106)$$

with

$$J_0(\tilde{\xi}_1, \tilde{\eta}_1; \tilde{\xi}_2, \tilde{\eta}_2) = \langle \tilde{u}_0(\tilde{\xi}_1, \tilde{\eta}_1; t)\, \tilde{u}_0^*(\tilde{\xi}_2, \tilde{\eta}_2; t) \rangle. \qquad (6.107)$$

The function $J_0(\tilde{\xi}_1, \tilde{\eta}_1; \tilde{\xi}_2, \tilde{\eta}_2)$ is known as the *Mutual Intensity* and is a measure of the spatial coherence of the light at the two points.

If the light is completely coherent, the Mutual Intensity can be written as

$$J_0(\tilde{\xi}_1, \tilde{\eta}_1; \tilde{\xi}_2, \tilde{\eta}_2) = \tilde{u}_0(\tilde{\xi}_1, \tilde{\eta}_1)\, \tilde{u}_0^*(\tilde{\xi}_2, \tilde{\eta}_2), \qquad (6.108)$$

which gives

$$I_t(x, y) = \left| \iint_{-\infty}^{\infty} d\tilde{\xi}_2 d\tilde{\eta}_2\, F(x - \tilde{\xi}_1, y - \tilde{\eta}_1)\, \tilde{u}_0(\tilde{\xi}_1, \tilde{\eta}_1; t) \right|^2. \qquad (6.109)$$

If the light is incoherent, the mutual intensity can be written as

$$J_0(\tilde{\xi}_1, \tilde{\eta}_1; \tilde{\xi}_2, \tilde{\eta}_2) = \kappa\, I_0(\tilde{\xi}_1, \tilde{\eta}_1)\, \delta(\tilde{\xi}_1 - \tilde{\xi}_2, \tilde{\eta}_1 - \tilde{\eta}_2), \qquad (6.110)$$

which gives

$$I_t(x,y) = \kappa \iint_{-\infty}^{\infty} d\tilde{\xi}_2 d\tilde{\eta}_2 \, |F(x-\tilde{\xi}_1, y-\tilde{\eta}_1)|^2 I_0(\tilde{\xi}_1,\tilde{\eta}_1) \tag{6.111a}$$

$$= \kappa \iint_{-\infty}^{\infty} d\tilde{\xi}_2 d\tilde{\eta}_2 P\,(x-\tilde{\xi}_1, y-\tilde{\eta}_1) I_0(\tilde{\xi}_1,\tilde{\eta}_1). \tag{6.111b}$$

An observable quantity that can be measured is the *Mutual Coherence Function* (*MCF*) introduced by Wolf [This function (of a field assumed to be stationary in time) is given by

$$\Gamma_{12}(\tau) = \Gamma(\tilde{\xi}_1,\tilde{\eta}_1;\tilde{\xi}_2,\tilde{\eta}_2;\tau) = \langle u_1(\tilde{\xi}_1,\tilde{\eta}_1;t)\, u_2{}^*(\tilde{\xi}_2,\tilde{\eta}_2;t+\tau)\rangle, \tag{6.112}$$

where $\tau = t_2 - t_1$, given by $\frac{\Delta l}{c}$. This time is related to the path difference Δl between the two optical fields, divided by the speed of light c, as before.

When the two field point sets coincide, then the above cross-correlation becomes an autocorrelation and the MCF then reduces to the temporal coherence function $G(\tau)$. When the intensity at the point $(\tilde{\xi}_1,\tilde{\eta}_1)$ has a zero-time delay, that is,

$$\Gamma_{11}(0) = I(\tilde{\xi}_1,\tilde{\eta}_1), \tag{6.113}$$

then we obtain the average intensity $\overline{I}(\boldsymbol{r})$, which also equals $G(0)$.

The *complex degree of coherence* is defined as the normalized MCF, or

$$\gamma_{12}(\tau) = \frac{\Gamma_{12}(\tau)}{\sqrt{\Gamma_{11}(0)\,\Gamma_{22}(0)}}. \tag{6.114}$$

It should be pointed out that both the modulus and phase of this function are measureable quantities.

From the definition of $\Gamma_{12}(\tau)$ and the fact that $u(\tilde{\xi}_1,\tilde{\eta}_1;t)$ will satisfy the wave equation, it can be shown that the MCF in vacuum satisfies the following pair of coupled-wave equations:

$$\nabla_n^2 \Gamma(\tilde{\xi}_1,\tilde{\eta}_1;\tilde{\xi}_2,\tilde{\eta}_2;\tau) = \frac{1}{c^2}\frac{\partial^2}{\partial\tau^2}\Gamma(\tilde{\xi}_1,\tilde{\eta}_1;\tilde{\xi}_2,\tilde{\eta}_2;\tau), \quad n = 1,2 \tag{6.115}$$

[4, p. 76].

Example 6.5 Referring to Figure 6.9, let us assume that we have stationary, quasi-monochromatic source illuminating two slits, which creates light in an observation plane [4, pp. 76–79]. On the basis of our previous work, the light impinging on the point (ξ_p,η_p) can be represented as the superposition of the illumination leaving slits S_1 and S_2. Mathematically, we can write the field at (ξ_p,η_p) as equal to

$$u_p(t) = K_1\, u_1\left(t - \tfrac{r_1}{c}\right) + K_2\, u_2\left(t - \tfrac{r_2}{c}\right), \tag{6.116}$$

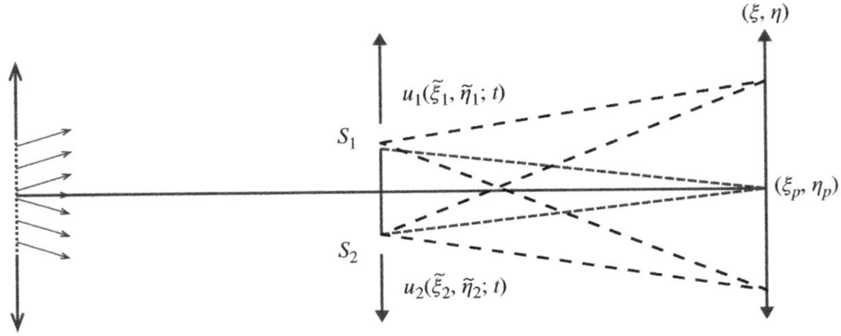

FIGURE 6.9 Geometry of a two slit experiment.

where K_n is the propagator for the light leaving slit S_n for point (ξ_p, η_p) and r_n the distance from slit S_n to the point (ξ_p, η_p). The intensity of light at the point (ξ_p, η_p) then can be written as

$$I_p(\xi_p, \eta_p) = \langle u_p(t)\, u_p{}^*(t)\rangle = \left\langle \left| K_1\, u_1 \left(t - \tfrac{r_1}{c}\right) + K_2\, u_2 \left(t - \tfrac{r_2}{c}\right)\right|^2 \right\rangle \tag{6.117}$$

or

$$I_p(\xi_p, \eta_p) = I_1 + I_2 + 2\,\mathrm{Re}\left\{K_1\, K_2{}^* \left\langle u_1 \left(t - \tfrac{r_1}{c}\right)\, u_2{}^* \left(t - \tfrac{r_2}{c}\right)\right\rangle\right\}, \tag{6.118}$$

where I_n is the intensity emanated from slit S_n to the point (ξ_p, η_p). Because of stationarity, the above intensity can be rewritten as

$$I_p(\xi_p, \eta_p) = I_1 + I_2 + 2\,\mathrm{Re}\{K_1\, K_2{}^*\Gamma_{12}(\tau)\} \tag{6.119a}$$

$$= I_1 + I_2 + 2\,|K_1\, K_2{}^*|\; |\Gamma_{12}(0)|\, \cos(\Phi_{12}), \tag{6.119b}$$

where $\Phi_{12} = 2\pi\omega_0\tau + \beta_{12}$. Here, $2\pi\omega_0\tau = \left(\frac{2\pi}{\lambda_0}\right)\, |r_2 - r_1|$ and $\beta_{12} = \arg[\Gamma_{12}]$. Equations (6.119a) and (6.119b) are sometimes referred to as *the generalized interference law for partially coherent light.* Since we have a quasi-monochromatic source, we can consider only small path differences where $\Delta l \ll \frac{c}{\Delta\omega}$, where $\Delta\omega$ is the spectral width of the radiation assumed very small compared to the mean frequency of the light source, that is, $\Delta\omega \ll \omega_0$.

Rewriting Eq. (6.119b) in terms of the normalized MCF, we obtain

$$I_p(\xi_p, \eta_p) = I_1 + I_2 + 2\,\sqrt{I_1\, I_2}\, |\gamma_{12}(\tau)|\, \cos(\Phi_{12}). \tag{6.120}$$

Equation (6.120) is referred to as the *generalized interference law for partially coherent quasi-monochromatic light.*

If the experimental source is incoherent, then $|\gamma_{12}(\tau)| = 0$ and Eq. (6.120) reduces to

$$I_p(\xi_p, \eta_p) = I_1 + I_2, \tag{6.121}$$

which is the incoherent limit. On the other hand, if the source is completely coherent, then $|\gamma_{12}(\tau)| = 1$ and Eq. (6.118) reduces to

$$I_p(\xi_p, \eta_p) = I_1 + I_2 + 2\sqrt{I_1 I_2}\cos(\Phi_{12}). \tag{6.122}$$

Equation (6.122) is the well-known *generalized interference law for completely coherent light.*

Although all of the effects of coherence are contained in the normalized MCF $\gamma_{12}(\tau)$, conventional has an artificial separation of coherence into two categories:

- *Spatial coherence*: This arises from the angular source size considerations.
- *Temporal coherence or coherence length*: This relates to the finite spectral *width of the radiation* $\Delta\nu$ (see the last section).

[4, p. 79]. In terms of measureable parameters, $\gamma_{12}(0)$ is the measure of spatial coherence effect and $\gamma_{11}(\tau)$ is the measure of temporal coherence [4, p. 79]. In particular, the visibility of the fringes in the interference pattern give us gives the value of $|\gamma_{12}(\tau)|$ (= $|g(\tau)|$) as we stated before.

Example 6.6 Referring to Figure 6.10, we now assume a stationary, monochromatic source passing through a single slit, creating the emission $S(x_0, y_0; \omega_0)$, and then illuminating two slits at a slight tilt. From the previous work, the wavefront out of the single slit is given by

$$e^{ik_0\left(\frac{xx_0+yy_0}{z}\right)} = e^{2\pi i\left(\frac{c}{\omega_0}\right)\left(\frac{xx_0+yy_0}{z}\right)}. \tag{6.123}$$

If we assume that our two slit objects have the same transmittance $u(x', y')$ at the center frequency, ω_0, then the wavefront after the object will be

$$u(x', y')\, e^{2\pi i\left(\frac{c}{\omega_0}\right)\left(\frac{x'x_0+y'y_0}{f}\right)}. \tag{6.124}$$

As we have seen before, the second lens shown in Figure 6.8 performs a Fourier transformation and we obtain in the observation plane the following output wavefront:

$$\iint u(x', y')\, e^{2\pi i\left(\frac{c}{\omega_0 f}\right)(x'x_0+y'y_0+x'x+y'y)} dx'dy'$$
$$= \iint\iint \tilde{u}(\nu, \mu)\, e^{2\pi i\left(\frac{c}{\omega_0 f}\right)(x'x_0+y'y_0+x'x+y'y)} e^{2\pi i(\nu x+\mu y)} d\nu\, d\mu\, dx'dy'. \tag{6.125}$$

The x'-integration and y'-integration in Eq. (6.123) yield

$$\delta\left(\nu - \left(\frac{c}{\omega_0 f}\right)[x_0 + x]\right) \quad \text{and} \quad \delta\left(\mu - \left(\frac{c}{\omega_0 f}\right)[y_0 + y]\right),$$

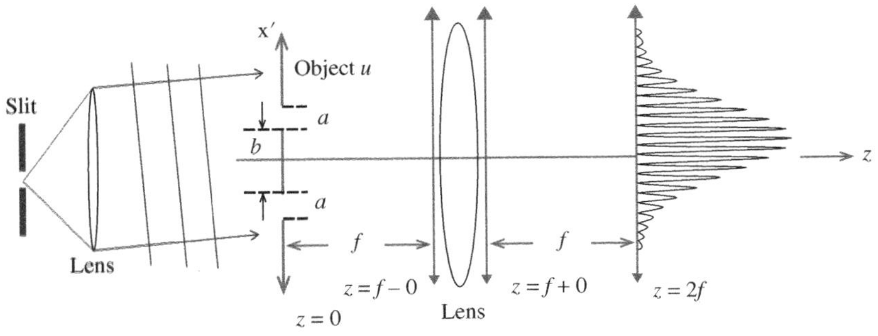

FIGURE 6.10 Another geometry for a two slit experiment.

respectively. Substituting into Eq. (6.125), the complex amplitude created in the observation plane is given by

$$\iint \tilde{u}(\nu,\mu)\,\delta\left(\nu - \left(\frac{c}{\omega_0 f}\right)[x_0 + x]\right)\delta\left(\nu - \left(\frac{c}{\omega_0 f}\right)[x_0 + x]\right)e^{2\pi i(\nu x+\mu y)}d\nu\, d\mu$$
$$= \tilde{u}\left(\left(\frac{c}{\omega_0 f}\right)[x_0 + x], \left(\frac{c}{\omega_0 f}\right)[y_0 + y]\right). \quad (6.126)$$

Based on our results of Chapter 2, the wavefront in the observation plane is unsurprisingly the Fourier transform of the object function. However, there are two new features:

- The transform is shifted sidewise by the source point position(x_0, y_0).
- A wavelength-proportional magnification $\left(M \propto \frac{1}{\frac{c}{\omega_0}} = \frac{\omega_0}{c} \propto \lambda^{-1}\right)$ is created.

[1, p. 339].

Example 6.7 As we noted earlier, monochromatic sources do have a small, but real, bandwidth. That means that the source emission is $S(x_0, y_0; \omega)$ rather than solely $S(x_0, y_0; \omega)$. Let us now take into account all the wavelengths in this bandwidth in addition to the center spectral component ω_0. Recall that we pointed out that different frequencies ω and light from different atoms does not interfere. This implies that we must add up intensities rather than complex amplitudes [1, p. 339]. Mathematically, this is given by

$$\int\iint \left|\tilde{u}\left(\left(\frac{c}{\omega f}\right)[x_0 + x], \left(\frac{c}{\omega f}\right)[y_0 + y]\right)\right|^2 S(x_0, y_0; \omega) dx_0\, dy_0\, d\omega. \quad (6.127)$$

The above represents an incoherent integration of the wavefront intensity; a coherent integration would have taken the modulus square of the integration of the complex amplitude times its spectral density, which is shown in the beginning of this section.

Let us begin with a point source. This means that

$$S(x_0, y_0; \omega) = \delta(x_0)\delta(y_0)S(\omega)$$

and Eq. (6.127) reduces to

$$I(x,y) = \int \left| \tilde{u}\left(\left(\frac{c}{\omega f}x\right), \left(\frac{c}{\omega f}y\right)\right) \right|^2 S(x_0, y_0; \omega)\, d\omega. \tag{6.128}$$

Let us see what happens if we assume two distinct slits of height h. In this case, the object function will be given by

$$u(x', y') = \left[\mathrm{rect}\left(\frac{x - \frac{b}{2}}{a}\right) + \mathrm{rect}\left(\frac{x + \frac{b}{2}}{a}\right)\right] \mathrm{rect}\left(\frac{y}{h}\right). \tag{6.129}$$

Its Fourier transform is equal to

$$\tilde{u}(\omega, \mu) = a\,h\ \sin c(\omega a) \sin c(\mu b) \cos(\pi \nu \omega), \tag{6.130}$$

which means that

$$|\tilde{u}(\omega, \mu)|^2 = (a\,h)^2\ \sin c^2(\omega a) \sin c^2(\mu b) \cos^2(\pi \omega b) \tag{6.131a}$$

$$= (a\,h)^2\ \sin c^2(\omega a)\ \sin c^2(\mu b)\ [1 + \cos\ 2(\pi \omega b)] \tag{6.131b}$$

$$\approx (a\,h)^2\ \sin c^2(\omega_0 a)\ \sin c^2(\omega_0 b)\ [1 + \cos\ 2(\pi \omega b)]. \tag{6.131c}$$

with the last equation assuming the spectral density function is very narrow, essentially approximating the center spectral component. This is justified because the sin c-function is a smooth function, almost constant, if ω does not vary much in the region of interest. This then allows us to write Eq. (6.131c) as

$$I(x,y) \approx (a\,h)^2\ \sin c^2(\omega_0 a)\ \sin c^2(\omega_0 b) \int \left[1 + \cos\left(\frac{2\pi\omega}{cf} xb\right)\right] S(\omega)\, d\omega. \tag{6.132}$$

Following a previous example, Eq. (6.132) can rewritten as

$$I(x,y) \approx I_0(x,y) \left[1 + V\left(\frac{x}{x_p \omega_0}\right) \cos\left(\frac{x}{x_p \omega_0} \sigma_{\tilde{R}}\right)\right], \tag{6.133}$$

since the cross-correlation between the two displaced wavefronts is again related to the power spectrum via the relation:

$$G(\rho) = |G(\rho)| e^{i\sigma_{\tilde{R}}(\rho)} = \int S(\omega)\, e^{-2\pi i \omega \rho}\, d\omega$$

and

$$\omega_0 \approx \overline{\omega} = \frac{\displaystyle\int \omega\, S(\omega)\, e^{-2\pi i\omega\rho}\, d\omega}{\displaystyle\int S(\omega)\, e^{-2\pi i\omega\rho}\, d\omega}. \tag{6.134}$$

In Eq. (6.133), we have

$$V(\rho) \equiv \text{Fringe visibility} = \frac{|G(\rho)|}{G(0)}, \tag{6.135}$$

$$x_p = \frac{cf}{\omega_0 b}, \tag{6.136}$$

and

$$I_0(x,y) \approx G(0)(a\,h)^2\ \sin c^2(\omega_0 ax)\ \sin c^2(\omega_0 hy). \tag{6.137}$$

In other words, the "contrast" of these fringes yields the modulus $|G(\rho)|$ of the Fourier transform of the power spectrum density and the fringe shift $\frac{\sigma_2}{2\pi x_p}$ indicates the argument of the said spectrum density [1, pp. 340 and 341]. From measuring both the quantities, one can deduce $G(\rho)$ and the source spectrum $S(\omega)$ itself by means of a Fourier transformation. This is an example of *Fourier spectrometry.*

Example 6.8 Let us now look at an extended monochromatic source. This means that

$$S(x_0, y_0; \omega) = S_0(x_0, y_0)\ \delta(\omega - \omega_0) \tag{6.138}$$

and Eq. (6.127) becomes

$$I(x,y) = \iint \left| \tilde{u}\left(\left(\frac{c}{\omega_0 f}\right)[x_0 + x],\ \left(\frac{c}{\omega_0 f}\right)[y_0 + y] \right) \right|^2 S_0(x_0, y_0)\, dx_0\, dy_0\,. \tag{6.139}$$

[1, p. 341]. Again looking at a double slit experiment, let us define

$$G_0(x,y) = \iint_{-\infty}^{\infty} S_0(\nu, \mu)\, e^{\,2\pi i[\nu x + \mu y]} d\nu\, d\nu = |G_0(x,y)| e^{i\sigma_0(x,y)}. \tag{6.140}$$

Using this definition and the same procedure in the last example, Eq. (6.139) now equals

$$I(x,y) = I_0(x,y)\left[1 + \frac{\left|G_0\left(\frac{b}{\lambda}, 0\right)\right|}{G_0(0,0)} \cos\left(\frac{2\pi b}{\lambda f} x - \sigma_0\left(\frac{b}{\lambda}, 0\right) \right) \right]. \tag{6.141}$$

The function $I_0(x,y)$ again is obtained again by pulling two sinc-functions in front of the integral for the similar reasons we were able to in the last example, that is,

$$I_0(x,y) \approx G(0,0)(a\,h)^2\ \sin c^2\left(\frac{ax}{\lambda f}\right)\ \sin c^2\left(\frac{hy}{\lambda f}\right). \tag{6.142}$$

Like in the last example, the "contrast" of these fringes yields the modulus $|G(x, y)|$ of the Fourier transforms of the power spectrum density and the fringe shifts indicate the argument of the said spectrum density [1, pp. 340 and 341]. From measuring both the quantities, one can deduce $G(x, y)$ and the source spectrum $S_0(\nu, \mu)$ itself by means of a two-dimensional Fourier transformation.

As we have seen in the above examples, spatial differences between sources can affect fringe generation similar to what temporal delays can. Let us now generalize some of our previous definitions to cover case of interference from spatially and temporally separated sources.

Going back to the basics, the best descriptor for partial coherence effects is the cross-correlation given in Eq. (6.112):

$$\Gamma_{12}(\tau) = \Gamma(\boldsymbol{r}_1, \boldsymbol{r}_2; \tau) = \langle u_1(\tilde{\xi}_1, \tilde{\eta}_1; t)\, u_2^*(\tilde{\xi}_2, \tilde{\eta}_2; t + \tau)\rangle,$$

which is also related to our previously defined parameter G. Here, we have where $\boldsymbol{r}_1 = (\tilde{\xi}_1, \tilde{\eta}_1)$ and $\boldsymbol{r}_2 = (\tilde{\xi}_2, \tilde{\eta}_2)$. Specifically, we have

$$\Gamma_{12}(\tau) = \Gamma(\boldsymbol{r}_1, \boldsymbol{r}_2; \tau) = G(\boldsymbol{r}_1, \boldsymbol{r}_2; \tau). \tag{6.143}$$

Therefore, in the general case, the temporal coherence function becomes the MCF when both space and time are variables. When the optical path is much shorter than the coherence length, the light wave essentially is totally temporally coherent and the MCF can be written as

$$G(\boldsymbol{r}_1, \boldsymbol{r}_2; \tau) = G(\boldsymbol{r}_1, \boldsymbol{r}_2)\, e^{i\omega_0 \tau}, \tag{6.144}$$

where ω_0 again is the center frequency. In this case, the light is called *quasi-monochromatic* and the mutual intensity completely characterizes the spatial coherence [2, p. 415].

The complex degree of coherence also can be generalized to

$$g(\boldsymbol{r}_1, \boldsymbol{r}_2; \tau) = \frac{G(\boldsymbol{r}_1, \boldsymbol{r}_2; \tau)}{\sqrt{I(\boldsymbol{r}_1), I(\boldsymbol{r}_2)}} = \frac{\Gamma_{12}(\tau)}{\sqrt{\Gamma_{11}(0)\, \Gamma_{22}(0)}} = \gamma_{12}(\tau), \tag{6.145}$$

where again, $0 \leq g(\boldsymbol{r}_1, \boldsymbol{r}_2; \tau) = \gamma_{12}(\tau) \leq 1$. In our old notation, we have

$$G(\boldsymbol{r}_1, \boldsymbol{r}_2; 0) = \langle u_1(\boldsymbol{r}_1; t)\, u_2{}^*(\boldsymbol{r}_2; t)\rangle = G(\boldsymbol{r}_1, \boldsymbol{r}_2), \tag{6.146}$$

also defined as the *mutual intensity* and the function

$$g(\boldsymbol{r}_1, \boldsymbol{r}_2; 0) = g(\boldsymbol{r}_1, \boldsymbol{r}_2) = \frac{G(\boldsymbol{r}_1, \boldsymbol{r}_2)}{\sqrt{I(\boldsymbol{r}_1), I(\boldsymbol{r}_2)}} \tag{6.147}$$

being the *normalized mutual intensity*.

The above now lets us define a key coherence property discussed earlier, the coherence area A_C. Because we generally work with stationary processes, the special coherence of quasi-chromatic light is given by

$$|g(\boldsymbol{r}_1,\boldsymbol{r}_2)| = |g(\boldsymbol{r}_1 - \boldsymbol{r}_2)|. \tag{6.148}$$

The maximum of this function occurs when $(\boldsymbol{r}_1 = \boldsymbol{r}_2)$, decreases for nonzero spatial location differences. When the integrated area of $|g(\boldsymbol{r}_1 - \boldsymbol{r}_2)|$ decreases to some value such as 0.5 or e^{-1}, the resulting area is called the *coherence area* A_C. When the coherence areas are greater than the aperture area of an optical transmitter system, then we have $|g(\boldsymbol{r}_1,\boldsymbol{r}_2)| \approx 1$ at all points of interest and the light is considered coherent [2, p. 415]. On the other hand, if the coherence area is much smaller than the resolution of the optical system, then $|g(\boldsymbol{r}_1,\boldsymbol{r}_2)| \approx 0$ for all practical purposes and can be considered incoherent. These comments are consistent with previous discussions in Section 6.2.

From the above definitions, we would expect the *cross-spectral density*, or *cross-power spectrum*, to be given as

$$S(\boldsymbol{r}_1,\boldsymbol{r}_2;\omega) = \int_{-\infty}^{\infty} G(\boldsymbol{r}_1,\boldsymbol{r}_2;t)e^{-i\omega t}\,dt. \tag{6.149}$$

When $\boldsymbol{r}_1 = \boldsymbol{r}_2 = \boldsymbol{r}$, then the above becomes the power spectral density $S(\omega)$ defined in Eq. (6.49). The *normalized cross-spectral density* is defined as

$$s(\boldsymbol{r}_1,\boldsymbol{r}_2;\omega) = \frac{S(\boldsymbol{r}_1,\boldsymbol{r}_2;\omega)}{\sqrt{S(\boldsymbol{r}_1,\boldsymbol{r}_1;\omega)\,S(\boldsymbol{r}_2,\boldsymbol{r}_2;\omega)}}. \tag{6.150}$$

This function represents the degree of correlation of the fluctuation components at frequency ω at points $\boldsymbol{r}_1$ and $\boldsymbol{r}_2$. Light is considered *spectrally pure* when the spatial and temporal functionality of cross-spectrum density is separable, or

$$S(\boldsymbol{r}_1,\boldsymbol{r}_2;\omega) = G(\boldsymbol{r}_1,\boldsymbol{r}_2)\,s(\omega), \tag{6.151}$$

where

$$s(\omega) = \int_{-\infty}^{\infty} g(t)e^{-i\omega t}\,dt. \tag{6.152}$$

Equation (6.151) implies that $G(\boldsymbol{r}_1,\boldsymbol{r}_2;t) = G(\boldsymbol{r}_1,\boldsymbol{r}_2)\,g(t)$ for spectrally pure light [2, p. 417]. In addition, this means that

$$s(\boldsymbol{r}_1,\boldsymbol{r}_2;\omega) = g(\boldsymbol{r}_1,\boldsymbol{r}_2) \tag{6.153}$$

under these circumstances.

6.6 PARTIALLY COHERENT IMAGING THROUGH ANY OPTICAL SYSTEM

As we learned in Chapter 5, we may model the transmission of an initial source radiance distribution through any shift-invariant or isoplanatic optical system as a convolution of the form:

$$v(x,y) = u(x',y') \circledast F(x,y) = \iint_{-\infty}^{\infty} u(x',y')\, F(x-x',y-y')\, dx'dy', \qquad (6.154)$$

where $v(x, y)$ is the output from the arbitrary optical system or channel, $u(x)$ the initial source distribution, and $F(x, y)$ the coherent point spread function? Let us look at the case of partially coherent imaging.

If we use Eq. (6.154) in our definition of the mutual intensity, we find that

$$G_2(\boldsymbol{r}_1,\boldsymbol{r}_2) = \iint\iint F^*(\boldsymbol{r}_1-\boldsymbol{r}_1')\, F(\boldsymbol{r}_2-\boldsymbol{r}_2') G_1(\boldsymbol{r}_1',\boldsymbol{r}_2')\, d\boldsymbol{r}_1'\, d\boldsymbol{r}_2', \qquad (6.155)$$

where

$$G_1(\boldsymbol{r}_1,\boldsymbol{r}_2) = \langle u^*(\boldsymbol{r}_1) u'(\boldsymbol{r}_2)\rangle \qquad (6.156)$$

and

$$G_2(\boldsymbol{r}_1,\boldsymbol{r}_2) = \langle v^*(\boldsymbol{r}_1) v(\boldsymbol{r}_2)\rangle \qquad (6.157)$$

are the input and output mutual intensities.

The intensity of the output image is given by

$$I_2(r) = \iint\iint F^*(\boldsymbol{r}-\boldsymbol{r}_1')\, F(\boldsymbol{r}-\boldsymbol{r}_2')\, G_1(\boldsymbol{r}_1',\boldsymbol{r}_2')\, d\boldsymbol{r}_1'\, d\boldsymbol{r}_2'. \qquad (6.158)$$

For incoherent light, we have

$$G_1(\boldsymbol{r}_1,\boldsymbol{r}_2) = \sqrt{I_1(\boldsymbol{r}_1) I_1(\boldsymbol{r}_2)}\, \delta(\boldsymbol{r}_2-\boldsymbol{r}_1). \qquad (6.159)$$

Substituting Eq. (6.159) into Eq. (6.158) yields

$$\begin{aligned} I_2(r) &= \iint\iint F^*(\boldsymbol{r}-\boldsymbol{r}_1')\, F(\boldsymbol{r}-\boldsymbol{r}_2') \sqrt{I_1(\boldsymbol{r}_1) I_1(\boldsymbol{r}_2)}\, \delta(\boldsymbol{r}_1-\boldsymbol{r}_2)\, d\boldsymbol{r}_1'\, d\boldsymbol{r}_2' \\ &= \iint F^*(\boldsymbol{r}-\boldsymbol{r}_1')\, F(\boldsymbol{r}-\boldsymbol{r}_1')\, I_1(\boldsymbol{r}_1)\, d\boldsymbol{r}_1' = \iint P(\boldsymbol{r}-\boldsymbol{r}_1')\, I_1(\boldsymbol{r}_1)\, d\boldsymbol{r}_1' \end{aligned} \qquad (6.160)$$

with $P(\boldsymbol{r}-\boldsymbol{r}') = F^*(\boldsymbol{r}-\boldsymbol{r}')F(\boldsymbol{r}-\boldsymbol{r}') = |F(\boldsymbol{r}-\boldsymbol{r}')|^2$ being the incoherent point spread function defined in Eq. (5.35). Equation (6.160) is essentially Eq. (5.32) for incoherent imaging that we derived in Chapter 5.

Example 6.9 If one is dealing with a non-shift-invariant optical system or channel, then Eq. (6.155) is rewritten as

$$G_2(\boldsymbol{r}_1,\boldsymbol{r}_2) = \iint \iint F^*(\boldsymbol{r}_1,\boldsymbol{r}_1')\, F(\boldsymbol{r}_2,\boldsymbol{r}_2')\, G(\boldsymbol{r}_1',\boldsymbol{r}_2')\, d\boldsymbol{r}_1'\, d\boldsymbol{r}_2', \tag{6.161}$$

where $F(\boldsymbol{r}_2,\boldsymbol{r}_2')$ is the impulse response of the system or channel and the intensity of the output image is given by

$$I_2(\boldsymbol{r}) = \iint \iint F^*(\boldsymbol{r},\boldsymbol{r}_1')\, F(\boldsymbol{r},\boldsymbol{r}_2')\, G(\boldsymbol{r}_1',\boldsymbol{r}_2')\, d\boldsymbol{r}_1'\, d\boldsymbol{r}_2'. \tag{6.162}$$

Example 6.10 Let us compare the cutoff frequency performance between coherent and incoherent imaging through a simple lens system. For this example, we will assume that a plane wave is illuminating a circular lens of radius a, a focal length f to the left away from the source. The observation plane is located a focal length to the right away from the lens.

Using Eq. (2.108c) in Chapter 2, we can write the impulse response (normalized to 1) for coherent illumination of the lens as

$$F(x_p, x_p) = \frac{1}{\pi a^2} \iint \operatorname{circ}\left(\frac{r}{a}\right) e^{2\pi i k_x x' + 2\pi i k_x y'}\, dx' dy' \tag{6.163}$$

$$= \left[\frac{J_1(2\vartheta_{\text{coh}}\, \rho)}{(\vartheta_{\text{coh}}\, \rho\,)}\right], \tag{6.164}$$

where

$$\text{Circ}(r) \equiv \text{circ function}$$

$$= \begin{cases} 1\,; & |r| = \sqrt{x^2+y^2} \le 1 \\ 0; & \text{otherwise} \end{cases}, \tag{6.165}$$

$$\vartheta_{\text{coh}} = \frac{D}{2\lambda f}, \tag{6.166}$$

and

$$\rho = \sqrt{x_p^2 + y_p^2}. \tag{6.167}$$

As noted in Section 2.3 of Chapter 2, the spatial filter function $\hat{F}(\mu, v)$ is a low-pass filter for the optical system. As shown in Eq. (6.103), the Complex Pupil, or Spatial filter, Function $\hat{F}(\mu, v)$ is defined by the pupil function, which acts as a low-pass filter as noted in Chapter 2, it has the form:

$$\hat{F}(\mu, v) = \operatorname{circ}\left(\frac{\gamma}{\vartheta_{\text{coh}}}\right), \tag{6.168}$$

where

$$\gamma = \sqrt{\gamma_x^2 + \gamma_y^2}, \tag{6.169}$$

which passes two-dimensional low spatial frequency components within a circle of radius ϑ_c and zeroes the others outside of that radius.

The impulse response for incoherent illumination of a circular lens proportional to the square of Eq. (6.164), or

$$\propto \left[\frac{J_1(2\vartheta_{\text{coh}}\,\rho)}{(\vartheta_{\text{coh}}\,\rho)}\right]^2 . \tag{6.170}$$

Equations (6.164) and (6.170) have the same first zero, $1.22\,\frac{\lambda}{D}$. The same is not true for the spatial bandwidth of these filters.

Recall from Eq. (5.78) that the OTF for the incoherent illumination of a circular lens is given by

$$\hat{\Theta}\left(\frac{\vartheta}{\vartheta_c}\right) = \begin{cases} \frac{2}{\pi}\left[\cos^{-1}\left(\frac{\vartheta}{\vartheta_c}\right) - \left(\frac{\vartheta}{\vartheta_c}\right)\sqrt{1-\left(\frac{\vartheta}{\vartheta_c}\right)^2}\right]; & \text{for } \vartheta \le \vartheta_c = \frac{D}{\lambda f} \\ 0; & \text{for } \vartheta > \vartheta_c \end{cases} . \tag{6.171}$$

Comparing the cutoff frequencies between incoherent and coherent illumination of a circular lens, we see that the incoherent illumination's cutoff frequency is twice of the coherent illumination's cutoff frequency. Although incoherent has twice the spatial bandwidth of coherent, one cannot conclude the former is superior to the latter for two reasons. The first is that the OTF's contrast fall off with spatial frequencies while the impulse response is uniform until its cutoff frequency. The second is that the latter describes the imaging of source intensity while the second characterizes the imaging of the complex amplitude.

6.7 VAN CITTERT–ZERNIKE THEOREM

Let us look at the mutual intensity when the illumination is incoherent. Substituting Eq. (6.159) into Eq. (6.155) yields

$$\begin{aligned} G_2(\boldsymbol{r}_1,\boldsymbol{r}_2) &= \iint\iint F^*(\boldsymbol{r}_1-\boldsymbol{r}'_1)\,F(\boldsymbol{r}_2-\boldsymbol{r}'_2)\,\sqrt{I_1(\boldsymbol{r}'_1)I_1(\boldsymbol{r}'_2)}\,\delta(\boldsymbol{r}'_2-\boldsymbol{r}'_1)\,d\boldsymbol{r}'_1\,d\boldsymbol{r}'_2 \\ &= \iint F^*(\boldsymbol{r}_1-\boldsymbol{r})\,F(\boldsymbol{r}_2-\boldsymbol{r})\,I_1(\boldsymbol{r})\,d\boldsymbol{r} . \end{aligned} \tag{6.172}$$

Equation (6.166) implies that the transmitted light is no longer incoherent. In reality, light gains spatial coherence by the propagating large distances [2, p. 430]. Physically, one may have seen this fact indirectly by viewing the twinkling of the stars at night. Although the stars are incoherent sources, the radiance in the far field has enough coherence to be affected by atmospheric turbulences, that is, scintillation. This also happens in an optical system other than your eye. The propagation of light is a form of spatial filtering that reduces the spatial bandwidth and increases the coherence area [2, p. 430]. This coherence increase, or gain, can be seen from the following.

Recall from Eq. (2.67) that the impulse response for the quadratic approximation of the Huygens–Fresnel–Kirchhoff Integral is proportional to

$$\exp\left\{i\pi\frac{(x-x')^2+(y-y')^2}{\lambda z}\right\} = \exp\left\{i\pi\frac{x^2+y^2}{\lambda z} - 2i\pi\frac{xx'+yy'}{\lambda z} + i\pi\frac{x'^2+y'^2}{\lambda z}\right\}. \tag{6.173}$$

If we let $\boldsymbol{r} = (x, y, R)$ and $\boldsymbol{r} = (x', y', 0)$ be the input and output planes, respectively, then we have

$$F(\boldsymbol{r}_1 - \boldsymbol{r}) \propto \exp\left\{i\pi\frac{x_1^2+y_1^2}{\lambda z} - 2i\pi\frac{xx_1+yy_1}{\lambda z} + i\pi\frac{x^2+y^2}{\lambda z}\right\} \tag{6.174}$$

and

$$F(\boldsymbol{r}_2 - \boldsymbol{r}) \propto \exp\left\{i\pi\frac{x_2^2+y_2^2}{\lambda R} - 2i\pi\frac{xx_2+yy_2}{\lambda R} + i\pi\frac{x^2+y^2}{\lambda R}\right\}. \tag{6.175}$$

Substituting Eqs. (6.174) and (6.175) into Eq. (6.172), we obtain

$$\begin{aligned} G_2(\boldsymbol{r}_1,\boldsymbol{r}_2) &\propto \iint e^{-\frac{2i\pi}{\lambda R}[x(x_2-x_1)+y(y_2-y_1)]}\, I_1(x,y)\,dx\,dy \\ &= \iint e^{-2\pi i[\nu_x x+\nu_y y]}\, I_1(x,y)\,dx\,dy, \end{aligned} \tag{6.176}$$

where

$$\nu_x = \frac{(x_2-x_1)}{\lambda R} \tag{6.177}$$

and

$$\nu_y = \frac{(y_2-y_1)}{\lambda R}. \tag{6.178}$$

Using Eq. (6.176), the normalized mutual intensity can be written as

$$g_2(r_1, r_2) = \propto \frac{\iint e^{-2\pi i[\nu_x x+\nu_y y]}\, I_1(x,y)\,dx\,dy}{\iint I_1(x,y)\,dx\,dy}. \tag{6.179}$$

If we assume a uniform source of radius a, then the complex degree of coherence is equal to

$$g_2(\boldsymbol{r}_1,\boldsymbol{r}_2) \propto \left[\frac{J_1(2\vartheta_2\,\rho)}{(\vartheta_2\,\rho)}\right], \tag{6.180}$$

where

$$\vartheta_2 = \frac{D}{\lambda R}. \tag{6.181}$$

From Chapter 2, we know that the first zero occurs of Eq. (6.180) at $\rho \approx \frac{\lambda R}{D}$, which can be considered the coherence length ρ_{coh} of this normalized mutual intensity, that is, the light is coherent within this radius as we noted before. The coherence area is proportional to the square of the coherence length, namely, we have

$$A_c \approx \left(\frac{\lambda R}{D}\right)^2. \tag{6.182}$$

Equation (6.182) shows that light gains spatial coherence with the range as we highlighted above for stars.

Equations (6.176) and (6.179) are forms of the well-known Van Cittert–Zernike Theorem.[4] It was first derived by van Cittert [9] in 1934 with a simpler proof provided by Frits Zernike in 1938 [10]. The theorem describes the relationship between the spatial intensity distribution of an extended incoherent source and the first-order spatial correlation functions, which sounds very innocuous. However, the implications are significant, as suggested by Eq. (6.182). If the spatial extent of the incoherent source $I_1(x, y)$ is small and/or a long distance away, say like a star, its Fourier transform is very large. The result is the mutual intensity, normalized or not, is spread over a large area and the coherence area is large as a result.

This theorem has important implications for radio astronomy. By measuring the degree of coherence given in Eqs. (6.176) and (6.179) at different points in the imaging plane of an astronomical object, a radio, millimeter or infrared astronomer derives a characterization of the star's brightness distribution and make a two-dimensional map of its appearance. It also has applicability to adaptive optics.

6.8 PROBLEMS

Problem 6.1. If $kT \gg h\nu$, which will be true for radio and microwave frequencies, what is the mean number of photons in a single quantum state and how big is it?

Problem 6.2. Show that a source with a narrow spectral width has a coherence length given by

$$\ell_c = \frac{\lambda^2}{\Delta\lambda}.$$

Problem 6.3. Show that the coherent time τ_c defined in Eq. (6.43) is related to the spectral width $\Delta\nu$ given by Eq. (6.50) by the simple inversion formula $\tau_c = \frac{1}{2\Delta\omega}$. HINT: Use the Fourier transform relationships between $S(\omega)$ and $G(\tau)$ and Parseval's theorem.

[4] The van Cittert–Zernike theorem assumes that the medium between the source and the imaging plane is homogeneous. If the medium is not homogeneous, then light from one region of the source will be differentially refracted relative to the other regions of the source due to the difference in the travel time of light through the medium. In the case of a heterogeneous medium, one must use a generalization of the van Cittert–Zernike theorem, called *Hopkins's formula*.

Problem 6.4. Valid that the coherence time calculated by Eq. (6.43) is accurate for the following complex degrees of temporal coherence:

$$g(\tau) = \begin{cases} e^{-\frac{\tau}{\tau_c}} & (a) \\ e^{-\pi\left(\frac{\tau}{\sqrt{2}\,\tau_c}\right)^2}. & (b) \end{cases}$$

What is the drop-off rate for $|g(\tau)|$ between 0 and τ_c for each case?

Problem 6.5. Assume an MCF equal to

$$G(x_1, x_2; \tau) = J(\boldsymbol{r}_1 - \boldsymbol{r}_2)e^{2\pi i\omega_0\tau}.$$

Show that the function $J(\boldsymbol{r})$ satisfies the Helmholtz equation $\nabla^2 J + k_0^2 J = 0$, where $k_0 = \frac{\omega_0}{c}$.

Problem 6.6. Find the position of the first minimum for a single slit of width 0.04 mm on a screen of 2 m distance, when light from a He–Ne laser $\lambda = 6328$ Å is shone on the slit.

Problem 6.7. What is the irradiance at the position of the third maximum for a single slit of width 0.02 mm?

Problem 6.8. If we have a single slit of 0.2 cm wide, a screen of 1 m distance, and the second maximum occurs at a position 1 cm along the screen, what must be the wavelength of light incident on the screen?

Problem 6.9. A diffraction grating is a closely spaced array of apertures or obstacles forming a series of closely spaced slits. The simplest type in which an incoming wavefront meets alternating opaque and transparent regions (with each opaque/transparent pair being of the same size as any other pair) is called a transmission grating. Determine the angular position of the maxima of such a grating in terms of λ and a, the distance between centers of adjacent slits. If light of 5000 Å is incident of a slit containing 18,920 slits and of width 5 cm, calculate the angular position of the second maximum. HINT: Use the Fourier Transform of a shifted function to define the far-field amplitude in terms of a one period square wave (Ronchi ruling) function and a phase term, and the fact that

$$\sum_{n=-N/2}^{N/2} e^{-2\pi i\, k_x\, nd} = \frac{\sin[\pi(N+1)dk_x]}{\sin[\pi dk_x]}$$

Problem 6.10. (a) Determine the coherence area for a mercury arc lamp at 6330 Å at a distance of 1 m from the source. Assume that the output aperture is 3 mm and that the beam is diffraction-limited. (b) If a 30 μm pinhole is placed in front of the lamp, what is the effect on the results in part (a)? (c) Calculate the coherence diameter for setups (a) and (b) compare them with the spot size of a diffraction-limited beam. HINT: The coherence diameter d_{coh} is defined as the observation point separation where $|\gamma_{12}| = 0.88$, which translated into

$$d_{\mathrm{coh}} = \frac{\lambda}{(\pi\theta_{\mathrm{coh}})}$$

REFERENCES

1. Lohmann, A.W. (2006) in *Optical Information Processing* (ed. S. Sinzinger), Universitätsverlag, Ilmenau, Germany, ISBN 2-939472-00-6.
2. Saleh, B.E.A. and Teich, M.C. (2007) *Fundamentals of Photonics*, Wiley Series in Pure and Applied Optics, 2nd edn, John Wiley & Sons, New York.
3. Ross, M. (1966) *Laser Receivers: Devices, Techniques, Systems*, Wiley Series in Pure and Applied Optics, John Wiley & Sons, Inc., New York.
4. Reynolds, G.O., DeVelis, J.B., Parrent, G.B. Jr., and Thompson, B.J. (1989) *The New Physical Optics Notebook: Tutorial in Fourier Optics*, SPIE Press, Bellingham, WA.
5. Forrester, A., Gudmundsen, R.A., and Johnson, P. (1955) Photoelectric mixing of incoherent light. *Phys. Rev.*, **77** (**8**), 1691.
6. Mashaal, H., Goldstein, A., Feuermann, D., and Gordon, J.M. (2012) First direct measurement of the spatial coherence of sunlight. *Opt. Lett.*, **37** (**17**), 3516–3518 September 1.
7. Agarwal, G.S., Gbur, G., and Wolf, E. (2004) Coherence properties of sunlight. *Opt. Lett.*, **29** (**5**), 459–461 March 1.
8. Jackson, J.D. (1999) *Classical Electromagnetics*, 3er edn, John Wiley & Sons.
9. van Cittert, P.H. (1934) Die Wahrscheinliche Schwingungsverteilung in Einer von Einer Lichtquelle Direkt Oder Mittels Einer Linse Beleuchteten Ebene. *Physica*, **1**, 201–210.
10. Zernike, F. (1938) The concept of degree of coherence and its application to optical problems. *Physica*, **5**, 785–795.

7

OPTICAL CHANNEL EFFECTS

7.1 INTRODUCTION

We previously found that performance predictions of electro-optical systems can be facilitated by linear systems theory using an impulse response function or a variation of it. In Chapter 6, the mutual coherence function (MCF) was shown as one means of characterizing optical system loss for coherent or partially coherent light. Since we were able to model free space as an optical system, it is not too surprising that the degrading effects of the atmospheric, turbulent, and particulate channels also can be expressed in terms of specific channel MCFs.

We begin a discussion of radiative transfer through particulate media, then move to the development of the MCF for aerosols and molecules, and then turbulence. Finally, we provide a set of engineering equations useful in understanding and characterizing light propagation in the particulate cloud and ocean channels [1].

7.2 ESSENTIAL CONCEPTS IN RADIATIVE TRANSFER

Radiative transfer involves the energy transfer of electromagnetic radiation through a particulate medium. In general, the propagating radiation is affected by absorption, emission, and scattering processes [2, Chapter 10; 1, Chapter 6]. The equations of radiative transfer describe these interactions mathematically and cover a wide application space, which includes astrophysics, atmospheric science, and remote sensing. Analytic solutions to the radiative transfer equation (RTE) exist for simple cases,

Free Space Optical Systems Engineering: Design and Analysis, First Edition. Larry B. Stotts.

Companion website: www.wiley.com\go\stotts\free_space_optical_systems_engineering

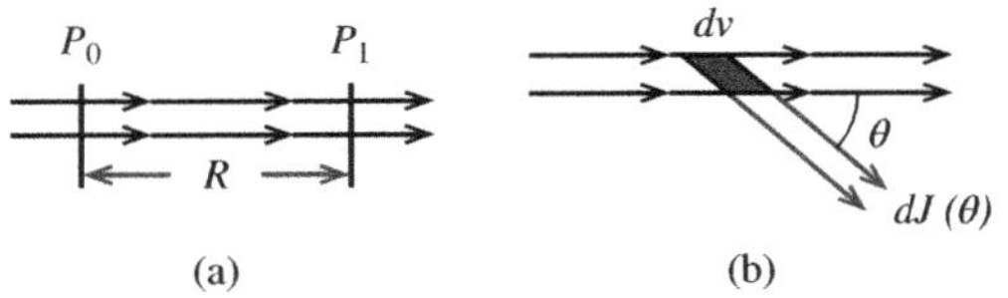

FIGURE 7.1 Key concepts in radiative transfer: (a) Beer's law and (b) volume scattering function $\beta(\theta)$.

but for more realistic environments such as the seawater, with complex multiple scattering effects, numerical methods are required [2, Chapter 10; 1 Chapter 7]. Preisendorfer postulated that there are two mutually exclusive classes of particulate medium properties that are inherent to the RTEs. They are (i) the inherent properties and (ii) the apparent properties [3, 4].

The inherent properties are ones that are independent of changes in the radiance distribution. They are the ones that characterize the power reduction at the receiver induced by the absorption and scattering by molecules and aerosols along the path and power increase created by scattered light by those particulates. For example, the transmission over a constant altitude/depth path follows Beer's law (see Figure 7.1a) and is given by

$$L_a = \frac{P_1}{P_0} = e^{-c(\lambda)R}, \tag{7.1}$$

where L_a is the atmospheric transmittance over the distance R, $c(\lambda) = a(\lambda) + b(\lambda)$ the volume extinction coefficient (m^{-1}), $a(\lambda)$ the volume absorption coefficient (m^{-1}), and $b(\lambda)$ the volume scattering coefficient (m^{-1}). All of the coefficients are functions of wavelength. The product $\tau = cR$ is called the *optical thickness or optical length* of the propagation path. When the altitude/depth is not constant, we integrate this transmission over the path length. Figure 7.2 shows the percent transmittance of various atmospheric constituents [5, p. 129].

Software packages such as LOWTRAN, FASCODE, MODTRAN, and HITRAN can be used to estimate this transmittance as a function of wavelength. Currently, LOWTRAN is no longer used because it is a low-resolution wavelength solution. MODTRAN has broadband precision at its moderate resolution and can calculate atmospheric transmission and radiance over various slant paths. Figure 7.3 illustrates a MODTRAN atmospheric transmission model example. It also has many variables, for example, visibility, time of the day, and weather, in which to do trade-offs with. FASCODE provides high-resolution atmospheric transmission estimates, while HITRAN is a database with laser line precision.

In the Earth's atmosphere, the particulate distribution of each atmospheric species is a function of altitude above ground level (AGL). The result is the volume absorption, scattering, and extinction coefficients will be a function of altitude as well. For example, transmission loss or transmissivity of a gradient atmosphere is expressed approximately by the Beer–Lambert law, that is, we have

$$L_a(\lambda) = \frac{P_1}{P_0} \approx e^{-\int_{\text{Path}} c(r;\, \lambda)dr} \tag{7.2}$$

for vertical and slant paths. The parameter $c(r; \lambda)$ now is the volume extinction coefficient along a path in the gradient atmosphere, which comprises both the volume

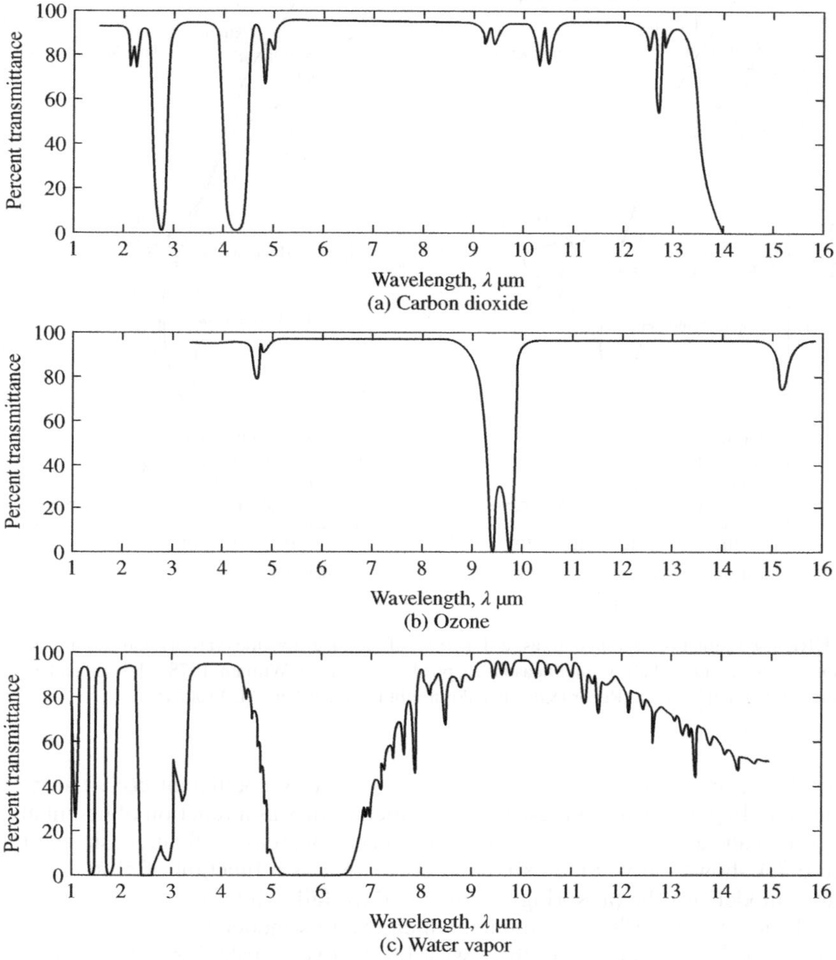

FIGURE 7.2 Percent transmittance of various atmospheric constituents [5, p. 129]. Source: Pratt, 1969 [5]. Reproduced with permission of NASA.

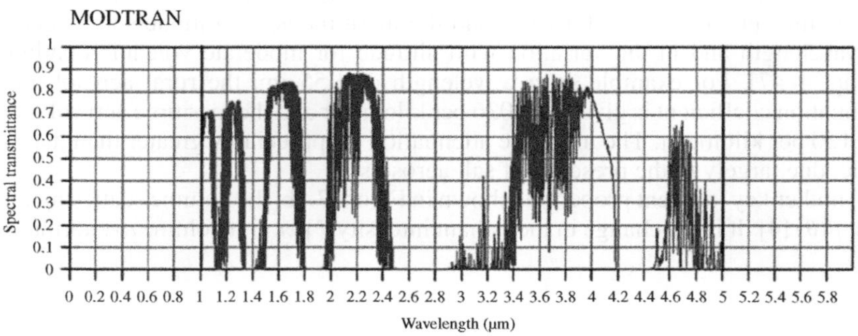

FIGURE 7.3 MODTRAN atmospheric transmission model example.

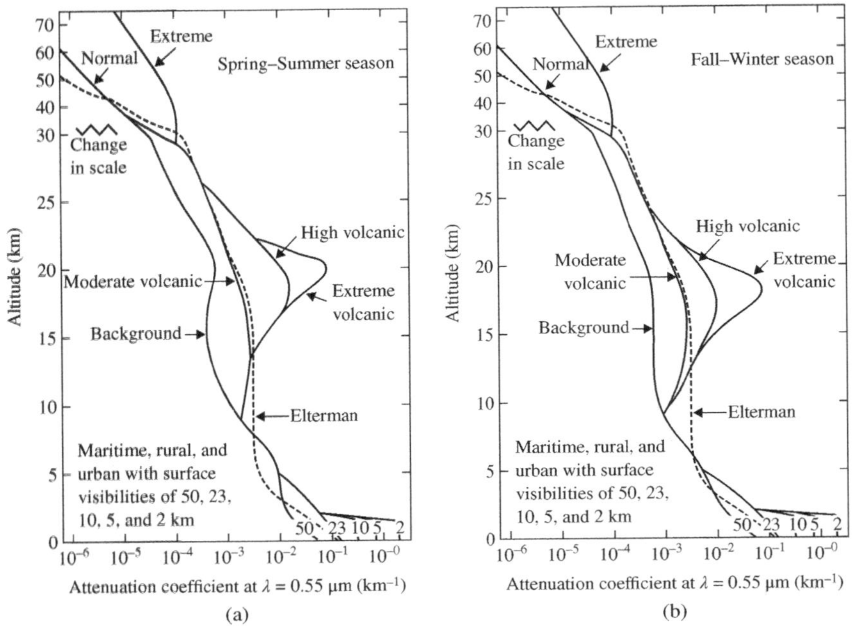

FIGURE 7.4 Extinction coefficient as a function of altitude for the various models for the (a) Spring–Summer and (b) Fall–Winter Seasons [6, pp. 4–47]. Source: William, 1978 [6]. Used under http://oai.dtic.mil/oai/oai?verb=getRecord&metadataPrefix=html&identifier=ADA100015

absorption coefficient $a(r; \lambda)$ and the volume scattering coefficient $b(r; \lambda)$ along the same path. In general, all of these three parameters vary as a function of height above the ground and are strongest close to the ground and decrease with increasing altitude. Figure 7.4 shows the volume extinction coefficient as a function of altitude for the various models for the (a) spring–summer and (b) fall–winter seasons [6, pp. 4–47]. Also depicted in the Elterman model, which exhibits moderate volcanic conditions in the 10–30 km regime [7]. In the lowest levels, five distributions are given, representing visibilities of 50, 23, 10, 5, and 2 km at ground level [6, pp. 4–46]. Figure 7.5 exhibits the transmissivity of the atmosphere [6, p. 132].

Figure 7.6 provides estimates of the volume extinction, absorption, and scattering coefficients for the rural aerosol and the maritime aerosol models as a function of wavelength and in the variation with altitude for moderate volcanic conditions [6, pp. 4–47]. For example, at a wavelength of 1.55 μm, the rural aerosol model attenuation coefficient is given as 0.036 per kilometer and the maritime aerosol model is 0.120 per kilometer. The maritime attenuation is appreciably greater than the rural value, due largely to the presence of salt aerosols.

Another key inherent property is shown in Figure 7.1b, the volume scattering function, $\beta(\theta)$ [4]. It is the change in the radiant intensity J per unit volume dv, normalized

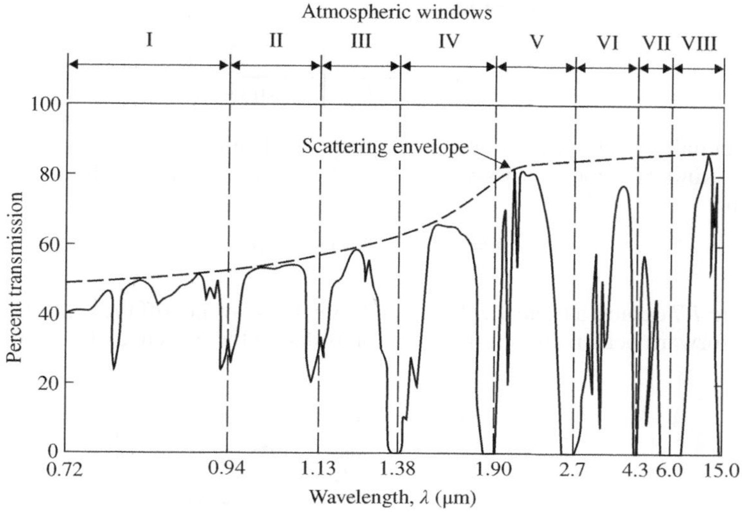

FIGURE 7.5 Transmissivity of the atmosphere [5, p. 132]. Source: Pratt, 1969 [5]. Reproduced with permission of NASA.

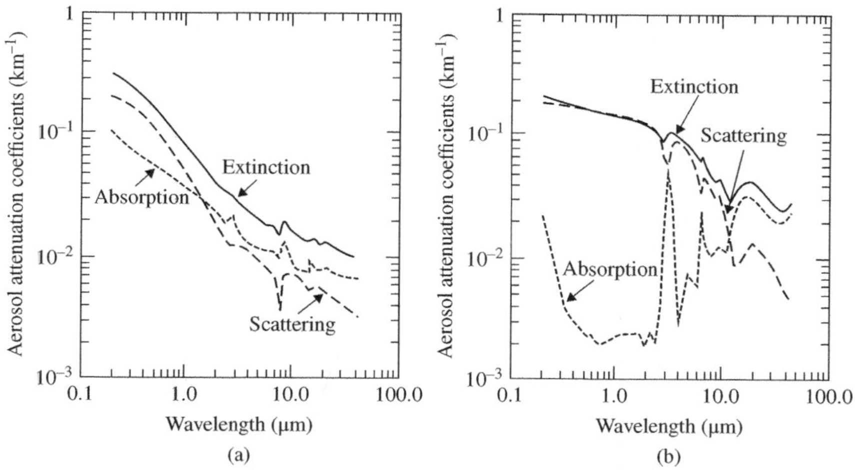

FIGURE 7.6 Volume extinction, absorption and scattering coefficients as a function of wavelength for (a) urban and (b) maritime aerosols for a moderately clear atmosphere (23 km visibility) [6, pp. 4–47]. Source: William, 1978 [6]. Used under http://oai.dtic.mil/oai/oai?verb=getRecord&metadataPrefix=html&identifier=ADA100015

to the incoming irradiance H, or mathematically,

$$\beta(\theta) = \left(\frac{1}{H}\right)\frac{dJ(\theta)}{dv}\left[\frac{1}{(\text{m-str})}\right]. \tag{7.3}$$

It essentially is the scattering impulse response from an incremental volume of particles; kind of a radar cross-sectional plot from a scattering volume. It can be shown that

$$b = 2\pi\int_0^{2\pi}\beta(\theta)\sin\theta d\theta. \tag{7.4}$$

[4]. Figure 7.7 depicts a typical volume scattering for seawater off the California coast [8]. The forward peak nature of this function is due to the presence of large diameter

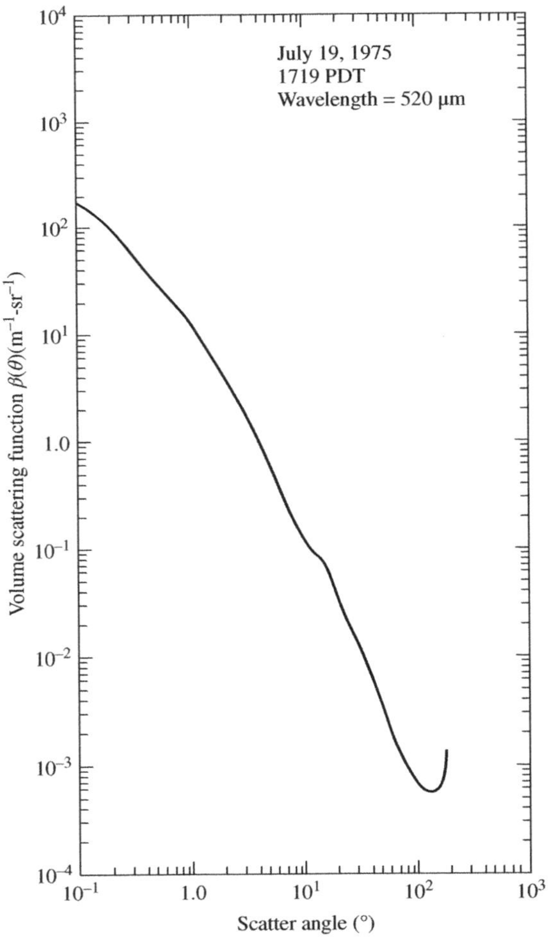

FIGURE 7.7 Typical volume scattering function of seawater off Santa Catalina Island taken on June 19, 1975 [8]. Source: Driscoll, 1976 [8]. Reproduced with permission of NAVAL ELECTRONICS LAB CENTER SAN DIEGO CA.

particulates and absorption found in seawater; similar performance can be found in maritime fogs and clouds.

The apparent properties are ones that are dependent on changes in the radiance distribution. These properties generally come from the asymptotic RTE, such as when characterizing the diffusion regime of scattering. Such properties include the radiance attenuation (loss coefficient is α) and irradiance attenuation (loss coefficient is K). For example, the latter includes the diffuse attenuation coefficient, k, for solar irradiance change in seawater. We discuss this more in a later section.

7.3 THE RADIATIVE TRANSFER EQUATION

The transfer of the field radiance in particulate medium is given by

$$\frac{dN(z,\theta,\varphi)}{dz} = -cN(z,\theta,\varphi) + N_s(z,\theta,\varphi), \tag{7.5}$$

where

$$N_s(z,\theta,\varphi) = \int_0^{2\pi}\int_0^{\pi} \beta(\theta,\varphi;\theta',\varphi')N(z,\theta',\varphi')\sin\theta' d\theta' d\varphi', \tag{7.6}$$

and the variation forms of $N(z,\theta,\varphi)$ are specific radiance distributions. Equation (7.5) is the general RTE. Its first term on the right-hand side of Eq. (7.5) represents the radiance loss by attenuation from both absorption and scattering. The second term represents the radiance gain by particulate scattering outside the scattering point, sometimes called the *path function*. Solutions to this integro-differential equation have been investigated by many researchers, most notably, Chandrasekhar, Lenoble, Duntley, and Preisendorfer among others [4]. One of the most understandable solutions is Preisendorfer's solution and it is outlined below as discussed in Jerlov's book [4]. References to seawater are made, but the solution does cover other situations where the absorption and scattering, but not emission, apply.

Let us assume a target point at depth z_t and at a distance $\boldsymbol{r}$ from an observation point at depth z. The path has the direction $(\pi-\theta,\varphi+\pi)$ where θ is the angle between nadir and the direction of the flux. Here, $z - z_t = r\cos(\theta)$. Here, we assume that the field radiance is measured by pointing a radiance meter at depth z in the direction of (θ,φ).

Given an optically homogeneous medium, the apparent radiance N_r for the target is given by the integration of Eq. (7.5), which yields

$$N_r(z,\theta,\varphi) = N_0(z_t,\theta,\varphi)e^{-cr} + \int_0^{r} N_*(z',\theta,\varphi)e^{-c(r-r')}dr'. \tag{7.7}$$

In Eq. (7.7), $N_0(z_t,\theta,\varphi)$ is the inherent radiance of the target and $z' = z_t + r'\cos\theta$. The apparent radiance $N_r(z,\theta,\varphi)$ is essentially the sum of the attenuated inherent radiance $N_0(z_t,\theta,\varphi)$ and a path radiance that consists of the scattered light into the direction $(\pi-\theta,\varphi+\pi)$ at each point of the path (z_t,θ,φ,r) that is redirected to the observation path.

Preisendorfer found that an approximate estimate of the scattered radiance $N_*(z,\theta,\varphi)$ can be obtained from the two-flow Schuster equation for irradiance

$$N_*(z,\theta,\varphi) = N_*(0,\theta,\varphi)e^{-K_* z}, \tag{7.8}$$

where K_* is independent of depth (See Figure 7.8) [4].

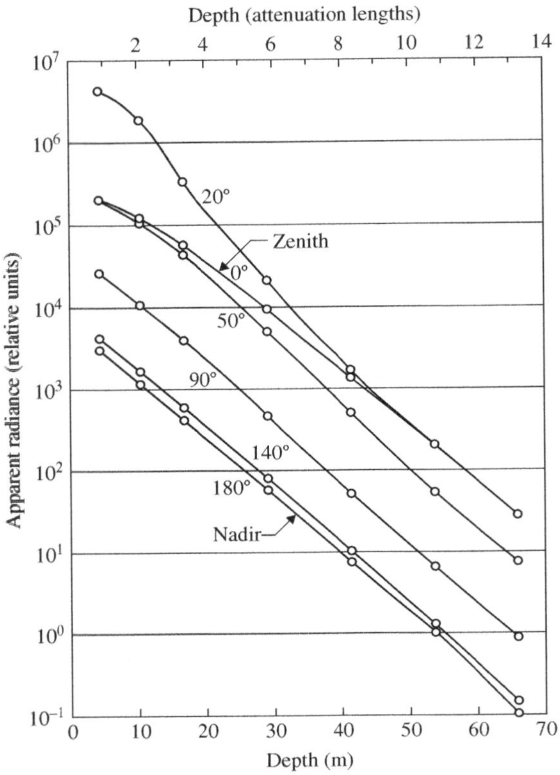

FIGURE 7.8 Depth profiles of radiance for different Zenith angles in the plane of the Sun on a clear day [9, Figure 23]. Source: Duntley, 1963 [9]. Reproduced with permission of The Optical Society of America.

Using Eq. (7.7), we have

$$N_r(z, \theta, \varphi) = N_0(z, \theta, \varphi)e^{-cr} + \frac{N_*(z, \theta, \varphi)}{c - K_* \cos\theta}(1 - e^{-(c-K_* \cos\theta)r}). \qquad (7.9)$$

This equation for observed radiance distribution was validated by Tyler on the basis of his experimental radiance measurements reported in 1961 for data obtained at Lake Pend Oreille. Figure 7.8 depicts the Tyler data (In general, the radiance and irradiance created by lake particulates is very similar to that created by seawater) [9]. What is most important about these data curves is that there is an unmistakable, systematic trend toward the formation of an asymptotic distribution of the underwater daylight radiance at depth, that is, a diffusion distribution becomes apparent at the deeper depths. We discuss this more in a later section.

The visibility of an object comes down to the perception of the radiance difference between the object and its adjacent background. There are two basic concepts that define object visibility: (i) inherent contrast (Weber) and (ii) contrast transmittance.

The inherent contrast of an object viewed against a uniform background is defined as

$$C_0 = \frac{N_0 - N_{b0}}{N_{b0}}, \tag{7.10}$$

where N_0 is object radiance and N_{b0} the uniform background radiance from zero distance away from the observer. Under this definition, the contrast varies between -1 for $N_b \gg N_0$ and ∞ for $N_b \ll N_0$, looking at the extremes of these two conditions. Examples of these extremes are very week radiant object against an ideal black background for the former and a very strong radiant object against a black background for the latter. On the other hand, the contrast transmittance comes from the relationship between inherent contrast and apparent contrast, which is as follows.

The apparent contrast of an object viewed against a background is defined as

$$C_r = \frac{N_r - N_{\mathrm{br}}}{N_{\mathrm{br}}}, \tag{7.11}$$

where N_r is object radiance at distance r away from the observer and N_{br} the background radiance from that same distance away from the observer. Assume that the object is at depth z_0, possessing a line of sight path of zenith angle θ and azimuth φ, that is,

$$z - z_0 = \cos\theta. \tag{7.12}$$

The attenuation of the radiance of sunlight, or the field radiance in a uniform medium, can be written as

$$\frac{dN(z,\theta,\varphi)}{dr} = -K(z,\theta,\varphi)N(z,\theta,\varphi)\cos\theta. \tag{7.13}$$

From the above it is assumed that $K(z, \theta, \varphi)$ is constant for certain lines of sight and becomes constant in all directions in the diffusion regime.

The transfer of radiance for an object is given by Eq. (7.5), so we can write

$$\frac{dN_t(z,\theta,\varphi)}{dr} = -cN_t(z,\theta,\varphi) + N_s(z,\theta,\varphi), \tag{7.14}$$

where N_t is the apparent target radiance. If we assume that K is constant, the integration over the entire line of sight using the last two equations and Eq. (7.9) yields

$$N_{\mathrm{tr}}(z,\theta,\varphi) = N_{t0}(z_t,\theta,\varphi)e^{-cr} + N(z_t,\theta,\varphi)e^{-Kr\cos\theta}(1 - e^{-cr+Kr\cos\theta}). \tag{7.15}$$

Repeating the same process for the background radiance, which yields a similar equation, but with b replacing t, we can subtract the two equations to give the following relationship:

$$N_{\mathrm{tr}}(z,\theta,\varphi) - N_{\mathrm{br}}(z,\theta,\varphi) = [N_{t0}(z_t,\theta,\varphi) - N_{b0}(z_t,\theta,\varphi)]e^{-cr}. \tag{7.16}$$

Equation (7.16) shows that the radiance difference between the object/target and background follows the Beer's law we defined earlier, that is,

$$L_r = e^{-cr}. \tag{7.17}$$

Using Eqs. (7.10) and (7.11) and the above, the ratio of apparent contrast to inherent contrast can be written as

$$\frac{C_r(z,\theta,\varphi)}{C_0(z,\theta,\varphi)} = L_r(z,\theta,\varphi)\frac{N_{b0}(z_t,\theta,\varphi)}{N_{br}(z_t,\theta,\varphi)}. \tag{7.18}$$

This equation represents the general solution and is true for nonuniform medium and for different levels of ambient light. The term on the right represent the contrast attenuation of the object's contrast by multiple scattering in a particulate medium. In deep water, the inherent contrast can be considered identical to the field radiance and the above becomes

$$\frac{C_r(z,\theta,\varphi)}{C_0(z,\theta,\varphi)} = e^{-cr+Kr\cos\theta}. \tag{7.19}$$

Like an optical system, we have a contrast reduction by the channel. That is, one can never do better than the inherent contrast of the target.

The application of optics to detecting objects at a distance has been around since the earliest known working telescopes appeared in 1608, credited to Hans Lippershey. S.Q Duntley and his colleagues at the Visibility Laboratory, starting at Massachusetts Institute of Technology and later at the Scripps Institute of Oceanography, were instrumental in establishing the science of visibility that the community has relied upon since the 1940s. What has become apparent is that a resolved object of interest can be seen because it has a different irradiance than the other sources of irradiance captured in that image. We now have seen that visibility contrast can be defined as the difference between the maximum irradiance and the minimum irradiance, divided by their sum (Michelson contrast for periodic patterns) or by the background irradiance (Weber contrast for general imaging).

Figure 7.9 illustrates various examples of Weber image contrasts between a uniform background and object [10, p. 432]. A contrast of 100% is the difference between pure black and pure white. The human eye can detect a minimum contrast between 0.5% and 5%, depending on scene conditions [10, p. 432].

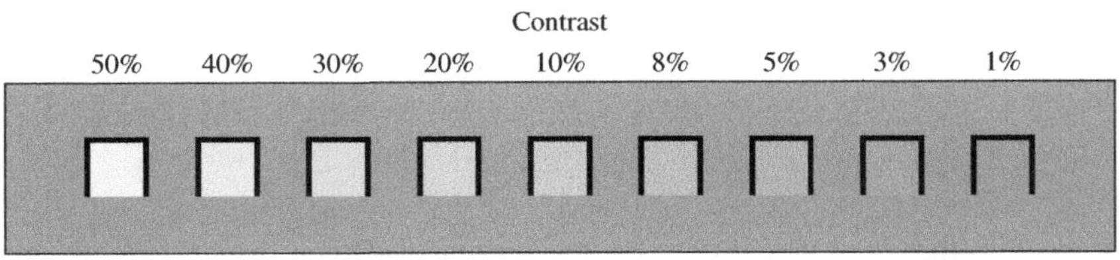

FIGURE 7.9 Example images of image contrast [10, p. 432]. Source: Smith, 1997 [10]. Reproduced with permission of Dr. Steven W. Smith.

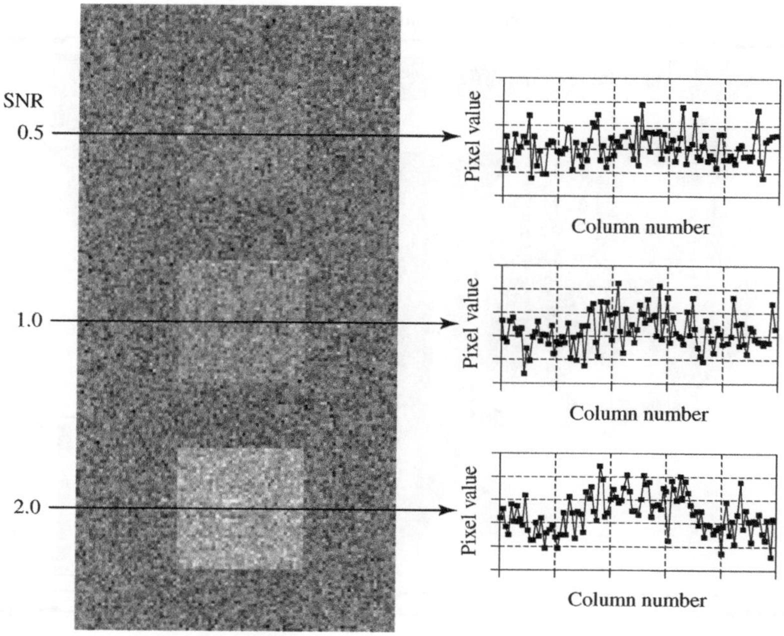

FIGURE 7.10 Example images of image contrast [10, p. 433]. Source: Smith, 1997 [10]. Reproduced with permission of Dr. Steven W. Smith.

Figure 7.10 shows noisy images with different signal-to-noise ratios (SNRs). Here, SNR is defined as the Weber contrast divided by the standard deviation of the noise [10, p. 433]. In general, the visibility of a resolved object must be $\geq 2\%$, and have an SNR ≥ 1 to be "clearly visible" to the naked eye, based on empirical testing by human subjects. The rule of thumb is that the SNR has to be 2–3 to be detectable (minimum detectable SNR).

Example 7.1 As we have seen, atmospheric scattering reduces contrast by adding a source of radiation to the line of sight that is independent of the brightness of the target. This source is integrated along the line of sight, and so is greater for a longer path.

Meteorological visibility depends on the relative difference (or contrast) between the light intensity from an object and from the intervening atmosphere in a horizontal geometry.Using Eq. (7.19) with ($\theta = \frac{\pi}{2}$), we have the equation for the horizontal visibility that follows Beer's law,

$$C(z) = e^{-cz}, \tag{7.20}$$

where $C(z)$ is the contrast and c the volume extinction coefficient of the intervening atmosphere.

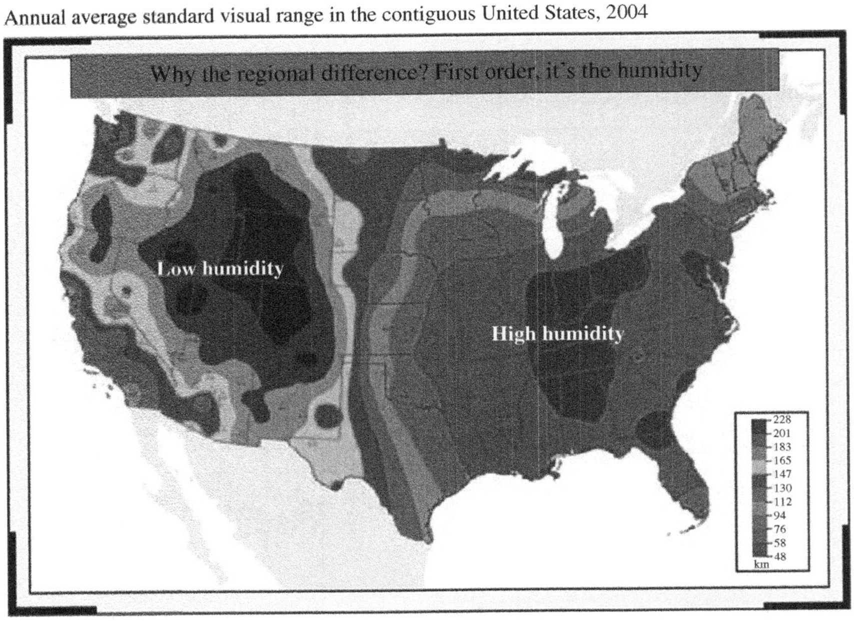

FIGURE 7.11 Visibility map of the United States [11]. Source: [11]. Used under http://nature.nps.gov/air/monitoring/vismonresults.cfm

The lowest visually perceptible brightness contrast for meteorological purposes is called the *threshold contrast*, which is typically about 2%, and its extinction coefficient is given by the Koschmieder equation:

$$c_v \approx \frac{3.912}{z_v}\left(\frac{\lambda_c}{0.55\ \mu\text{m}}\right)^{0.585 z_v^{-\frac{1}{3}}}, \tag{7.21}$$

where c_v (m^{-1}) and z_v (m) are the visibility extinction (Mie scattering) coefficient and range, respectively, and λ_c the optical wavelength of interest [5, p. 131].

Example 7.2 Figure 7.11 is a visibility map of the United States the National Park Service put out in 2007 [11]. This map illustrates the distribution of visibility conditions across the country. Large differences exist in visibility between the eastern and western United States, with western visibility generally being substantially better than eastern conditions. Climatic factors such as higher relative humidity and the greater density, quantity, and mix of emissions in the East are some of the reasons for this difference. The best visibility in the contiguous United States occurs in an area centered around Great Basin National Park, Nevada. Within the National Park System, the lowest visibilities are measured in eastern national parks such as Mammoth Cave (KY), Great Smoky Mountains (TN/NC), Shenandoah (VA), and Acadia (ME). Parks in southern California, such as Sequoia National Park, also record lower visibilities.

7.4 MUTUAL COHERENCE FUNCTION FOR AN AEROSOL ATMOSPHERE

This section summarizes the development of the aerosol MCF following Lutomirski [12]. Here, we present general expressions for the MCF for plane and spherical wave propagation, and then derive approximate formulas for use in some limiting cases. Figure 7.12 depicts the propagation geometry.

Let us assume that we have plane wave of unit amplitude and wave number k_0 entering a uniformly distributed aerosol-laden volume at $z = z_0$. Recall that the MCF is the cross-correlation function of the optical fields separated by a distance ρ in a plane normal to the direction of propagation. If we define $u(\boldsymbol{r}; z)$ as the field amplitude in the observation plane after the incident field $u_0(\boldsymbol{r}; z_0)$ traverses the aerosol medium, the aerosol MCF is given by

$$G_{\text{aero}} = \langle u^*(\boldsymbol{r}; z_1) u(\boldsymbol{r} + \boldsymbol{\rho}; z_1) \rangle, \tag{7.22}$$

where the angular brackets now denote an ensemble average over the realizations of particle locations and size distributions. Since the aerosols are assumed to be uniformly distributed within the volume, then by symmetry, the aerosol coherence function for a plane wave will be independent of absolute position of $\boldsymbol{r}$, and depend only on the magnitude of the separation $\rho = |\boldsymbol{\rho}|$ and the range z_1. This means that

$$G_{\text{aero}} = G_{\text{aero}}(\rho; z_1), \tag{7.23}$$

and its associated intensity is given by

$$I_{\text{aero}}(\boldsymbol{r}; z_1) = G_{\text{aero}}(0; z_1) = \langle u^*(\boldsymbol{r}; z_1) u(\boldsymbol{r}; z_1) \rangle. \tag{7.24}$$

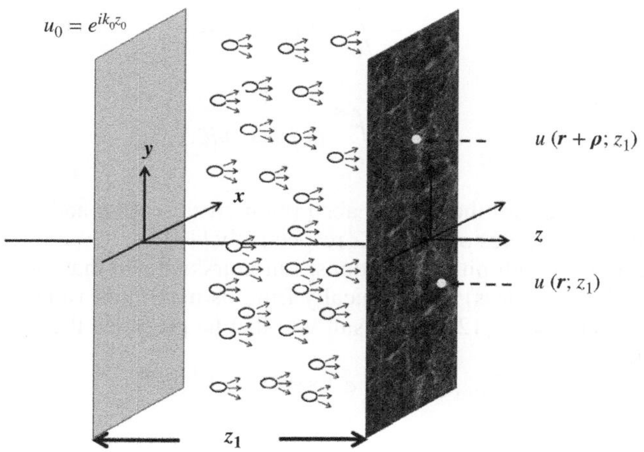

FIGURE 7.12 Geometry for propagation through an aerosol environment.

Let us see what the MCF looks like in the cases when only absorption loss occurs within the medium and when the separation distance is such that no scattered light is common to both field amplitudes. These situations are the limiting extremes for the above MCF.

If we assume that the density of aerosols is sufficient for multiple scattering to occur, but essentially at sufficiently small angles so no optical backscatter occurs, then all of the scattered light will reach the observation plane. The only reduction in the received intensity will be solely due to the absorptions by the aerosols. The result is that

$$G_{\text{aero}}(0; z_1) = \langle |u(\boldsymbol{r}; z_1)|^2 \rangle = e^{-a_{\text{aero}} z_1}, \tag{7.25}$$

where a_{aero} is the aerosol absorption coefficient [13].

If we now assume that the separation between the two points is so large that no scattered radiation from a single aerosol can reach both $\boldsymbol{r}$ and $\boldsymbol{r} + \boldsymbol{\rho}$, then the only coherence that exists is from the original plane wave, which now has been robbed of some energy because of the absorption and scattering occurring in the aerosol medium traversed. In this case, the aerosol MCF is written as

$$G_{\text{aero}}(0; z_1) = \langle |u(\boldsymbol{r}; z_1)|^2 \rangle = e^{-c_{\text{aero}} z_1}, \tag{7.26}$$

where $c_{\text{aero}}(\lambda) = a_{\text{aero}}(\lambda) + b_{\text{aero}}(\lambda)$ is the volume extinction coefficient, $a_{\text{aero}}(\lambda)$ the volume absorption coefficient, and $b_{\text{aero}}(\lambda)$ the volume scattering coefficient for aerosols.

Let $n(\zeta)$ denote the number of aerosol particles per unit volume with radii between ζ and $\zeta + d\zeta$. The total number of aerosol particles per unit volume is given by

$$n_0 = \int_0^\infty n(\zeta) d\zeta. \tag{7.27}$$

Given $n(\zeta)$, it can be shown that the aerosol volume absorption and scattering coefficients are given by

$$a_{\text{aero}} = \int_0^\infty n(\zeta)\sigma_a(\zeta) d\zeta \tag{7.28}$$

and

$$b_{\text{aero}} = \int_0^\infty n(\zeta)\sigma_s(\zeta) d\zeta, \tag{7.29}$$

where $\sigma_a(\zeta)$ and $\sigma_s(\zeta)$ are the aerosol absorption cross section and scattering cross section for aerosol particles of radius ζ, respectively [13].

Given the above two limiting equations, Lutomirski stated that the aerosol MCF must decrease (more or less) monotonically from its maximum value at $\rho = 0$ to its minimum value at $\rho = \infty$ [12]. The result was that he proposed the aerosol MCF to be of the form:

$$G_{\text{aero}}(\rho; z_1) = e^{-a_{\text{aero}} z_1 - b_{\text{aero}} z_1 [1 - f(\rho)]} \tag{7.30}$$

with $f(\rho)$ being a monotonically decreasing function of ρ. Several authors have calculated $f(\rho)$ assuming independent, multiple small-angle scattering events [12, 14, 15]. For an incident plane wave, we have

$$f(\rho) = \frac{2\pi}{n_0}\int_0^{\infty} d\zeta n(\zeta)\int_0^{\pi} d\theta \sin\theta P_{\text{aero}}(\theta;\zeta) J_0(k\rho\sin\theta) \tag{7.31}$$

and for an incident spherical wave,

$$f(\rho) = \frac{2\pi}{n_0}\int_0^{\infty} d\zeta n(\zeta)\int_0^{\pi} d\theta \sin\theta P_{\text{aero}}(\theta;\zeta)\int_0^1 du\, J_0(k\rho u\sin\theta), \tag{7.32}$$

where

$$n_0 = 2\pi\int_0^{\infty} d\zeta n(\zeta)\int_0^{\pi} d\theta \sin\theta P_{\text{aero}}(\theta;\zeta) \tag{7.33}$$

and

$$\int_0^{2\pi}\int_0^{\pi} d\Omega\ P_{\text{aero}}(\theta;\zeta) = 2\pi\int_0^{\pi} d\theta \sin\theta P_{\text{aero}}(\theta;\zeta) = 1. \tag{7.34}$$

In the above three equations, the function $P_{\text{aero}}(\theta;\zeta)$ is called the *scalar phase function* for an aerosol particle of radius ζ [1, Chapter 6; 13]. It is the optical scattering pattern created by the ensemble of particles of radius ζ within a unit volume. If we ignore polarization effects, then

$$P_{\text{aero}}(\theta;\zeta) = \frac{1}{2}[P^1_{\text{aero}} + P^2_{\text{aero}}], \tag{7.35}$$

where P^1_{aero} and P^2_{aero} are the scattering functions for the components parallel (i) and perpendicular (ii) to the scattering plane [1, Chapter 6; 11]. In real world, one rarely deals with plane waves when performing system analysis. Hence, for the remainder of this discussion, we focus solely on spherical waves.

In general, one is interested in separations close to the optical axis, which implies that the separation distance ρ is very small. Looking at Eq. (7.32), we see that

$$\int_0^1 du J_0(yu) = \int_0^1 du\left[1 - \left(\frac{yu}{2}\right)^2 + \frac{1}{4}\left(\frac{yu}{2}\right)^4 - \frac{1}{36}\left(\frac{yu}{2}\right)^6 + \cdots\right] \tag{7.36}$$

$$= \left[1 - \frac{y^2}{12} + \frac{y^4}{320} - \cdots\right]. \tag{7.37}$$

Substituting Eq. (7.37) into Eq. (7.32), we can write

$$f(\rho) = \left[1 - \overline{\sin^2\theta}\ \frac{(k\rho)^2}{12} + \overline{\sin^4\theta}\frac{(k\rho)^4}{320} - \cdots\right] \tag{7.38}$$

for $k\rho \sin\theta \ll 1$. In Eq. (7.38), $\overline{\psi(\theta)}$ represents the expected value of any function of θ, $\psi(\theta)$, averaged over the scattering function and over the particle distribution. Mathematically, we have

$$\overline{\psi(\theta)} = \frac{2\pi \int_0^\infty d\zeta n(\zeta) \int_0^\pi d\theta \sin\theta \psi(\theta) P_{\text{aero}}(\theta;\zeta)}{2\pi \int_0^\infty d\zeta n(\zeta) \int_0^\pi d\theta \sin\theta P_{\text{aero}}(\theta;\zeta)}. \tag{7.39}$$

For small forward scattering angles, Eq. (7.38) reduces to

$$f(\rho) \approx \left[1 - \overline{\theta^2}\frac{(k\rho)^2}{12}\right] = \left[1 - \theta^2_{\text{RMS}}\frac{(k\rho)^2}{12}\right] \tag{7.40}$$

for $k\rho\theta_{\text{RMS}} \ll 1$. In the above equation, $\theta^2_{\text{RMS}} = \overline{\theta^2}$ being the mean-square scattering angle for the total aerosol particle distribution. Substituting Eq. (7.40) into Eq. (7.30) yields

$$G_{\text{aero}}(\rho; z_1) \approx e^{-a_{\text{aero}}z_1 - b_{\text{aero}}z_1\left[\theta^2_{\text{RMS}}\frac{(k\rho)^2}{12}\right]} = e^{-a_{\text{aero}}z_1 - \left(\frac{\rho}{\rho_{\text{aero}}}\right)^2} \tag{7.41}$$

for $k\rho\theta_{\text{RMS}} \ll 1$ and

$$\rho_{\text{aero}} = \frac{1}{k\theta_{\text{RMS}}}\sqrt{\frac{12}{b_{\text{aero}}z_1}} = \frac{\lambda}{\pi\theta_{\text{RMS}}}\sqrt{\frac{3}{b_{\text{aero}}z_1}} \tag{7.42}$$

is the e^{-1} coherence length of the MCF for forward scattering, and $k = \frac{2\pi}{\lambda}$.

In the limit of many scattering lengths, $b_{\text{aero}}z_1 \gg 1$, Lutomirski pointed out that the argument of the second exponential in Eq. (7.30) can become large over the region $0 < k\rho\theta_{\text{RMS}} \ll 1$; that is, where $f(\rho)$ is given by Eq. (7.40). This means that the Gaussian form of Eq. (7.41) will be correct until the MCF is reduced to a very small value. Hence we obtain the important result that after many scattering lengths, the MCF becomes Gaussian independent of the form of the scattering function $P_{\text{aero}}(\theta;\zeta)$. However, he also points out that this does not imply that the important characteristics of the MCF for all applications are given by this Gaussian behavior; in particular, the minimum value of the MCF is given by Eq. (7.26), and we obtain as approximate expressions for the MCF:

$$G_{\text{aero}}(\rho; z_1) \approx \begin{cases} e^{-a_{\text{aero}}z_1 - \left(\frac{\rho}{\rho_{\text{aero}}}\right)^2} & \text{for} \quad \rho \ll \rho_{\text{aero}}\sqrt{b_{\text{aero}}z_1} \text{ and } b_{\text{aero}}z_1 \gg 1 \\ e^{-c_{\text{aero}}z_1} & \text{for} \quad \rho \gg \rho_{\text{aero}}\sqrt{b_{\text{aero}}z_1} \text{ and } b_{\text{aero}}z_1 \gg 1 \\ e^{-a_{\text{aero}}z_1} & \text{for } b_{\text{aero}}z_1 \ll 1 \end{cases}, \tag{7.43}$$

where ρ_{aero} represents a sort of aerosol cutoff distance. When the separation exceeds this cutoff, the MCF equals the atmospheric transmittance L_a. When less than the cutoff, it tends to the aerosol absorption loss as the separation goes to zero.

As a (very) approximate estimate of ρ_{aero}, we can assume that diffraction from particles of radius ζ will yield an RMS angle $\theta_{\text{RMS}} \sim \frac{\lambda}{\zeta}$. This means that after substituting Eq. (7.42), we have for ρ_{aero}:

$$\rho_{\text{aero}} \approx \zeta / \sqrt{b_{\text{aero}} z_1} \tag{7.44}$$

to within factors of order unity. Hence, after several scatterings, the transverse coherence length ρ_{aero} will be smaller than the particle sizes themselves (which might be of the order of microns).

For the intermediate case of a few scatterings, say $1 < b_{\text{aero}} z_1 < 10$, the scattering depends on the detailed shape of the scalar phase function and Eq. (7.31) must be evaluated using this function to determine the desired MCF.

As a final comment, for gradient atmospheres, the above equations following Beer's law will be rewritten as in the form of Eq. (7.3) with the loss coefficient.

7.5 MUTUAL COHERENCE FUNCTION FOR A MOLECULAR ATMOSPHERE

This section summarizes the development of the Molecular MCF following Lutomirski [12]. For molecules, the dominant scattering process is Rayleigh scattering, with a scalar phase function approximately equal to

$$P_{\text{mole}}(\theta; \zeta) = \frac{3}{16\pi}[1 + \cos^2(\theta)] \tag{7.45}$$

[16]. Lutomirski states that the scattering indicated by Eq. (7.45) is almost isotropic, and for this case no formula for the MCF clearly exists. However, he computed the molecular scattering coefficient S_{mole} by using the results from Rayleigh scattering and provided an estimated MCF for molecules.

The scattering cross section $\sigma_s(\zeta)$ for a spherical particle of size ζ small compared with the wavelength of incident light, and (complex) refractive index $n = n_r + in_i$ is given by

$$\sigma_s(\zeta) = \frac{8\pi}{3} k^4 \zeta^6 \left| \frac{n^2 - 1}{n^2 + 1} \right|^2 \tag{7.46}$$

and, for a single molecular species, we then have

$$S_{\text{mole}} = n_0 \sigma_s(\zeta) = \frac{8\pi n_0}{3} k^4 \zeta^6 \left| \frac{n^2 - 1}{n^2 + 1} \right|^2, \tag{7.47}$$

where n_0 now is the total number of molecules per unit volume. To determine the effect of several molecular species on incoming light, one can insert the particle density, radii, and refractive index for each species into Eq. (7.47), and simply add the various scattering coefficients. However, Lutomirski stated that for a given set of species, the total scattering coefficient S_m varies as k^4 and the effect of molecular scattering decreases rapidly with wavelength. He said the largest scattering coefficient listed by McClatchey [17], corresponding to a subarctic winter at sea level and a wavelength $\lambda = 0.37\,\mu m$, is $S_{mole} = 0.99\,km^{-1}$, corresponding to a scattering length of approximately 10 km. Because of the rapid decrease of S_{mole} with increasing wavelength, molecular scattering is frequently negligible compared with molecular absorption or aerosol scattering for wavelengths greater than the ultraviolet part of the spectrum and within a few kilometers from the ground [12].

He further states that the variation of the molecular absorption coefficient a_{mole} with wavelength is much more complicated. It is a highly oscillating function because of the presence of numerous molecular absorption band complexes resulting, for the most part, from minor atmospheric constituents. The responsible molecules are in the order $H_2O, CO_2, O_3, N_2O, CO,\ O_2, CH_4,$ and N_2. His conclusion was that beyond the ultraviolet portion of the spectrum ($\lambda \geq 0.3\ \mu m$), we have $b_{mole} z_1 \gg 1$, and the MCF is estimated to be

$$G_{mole}(\rho; z_1) = \langle |u(\mathbf{r}; z_1)|^2 \rangle \approx e^{-a_{mole} z_1}. \tag{7.48}$$

Similar to the aerosol discussion, for gradient atmospheres, the above equations following Beer's law will be rewritten as in the form of Eq. (7.3) with the loss coefficient.

7.6 MUTUAL COHERENCE FUNCTION FOR AN INHOMOGENEOUS TURBULENT ATMOSPHERE

In the marine and atmospheric channels, turbulence is associated with the random velocity fluctuations of the "viscous fluid" comprising that channel. Unfortunately, these fluctuations are not laminar in nature, but rather are random subflows called turbulence eddies [2, 12, 17–20], which are generally unstable. In 1941, Kolmogorov, and simultaneously Obukhov, analyzed an isotopic velocity field to derive a set of inertial-subrange predictions for the forms of the velocity vector field and velocity structure functions that bear his name [2, p. 180]. Both researchers independently found from their analyses that these eddies can be broken up into a range of discrete eddies, or turbules; in particular, a continuum of cells for energy transfer from a macroscale, known as the outer scale of turbulence L_0, down to microscale, known as the inner scale of turbulence ℓ_0. This is illustrated in Figure 7.13 showing the energy process. Near the surface (<100 m), the outer scale L_0 is on the order of the height AGL of the observation point and the inner scale ℓ_0 typically is on the order 1–10 mm. This range of turbules affects the local temperature and pressure structure, causing changes in the medium's refractive index. The twinkling of the stars at night is a manifestation of the dynamic nature of the refractive index changes induced by these cells. In this section, we calculate the various forms of the MCF for atmospheric turbulence.

The index of refraction depends mostly on the local pressure and temperature and is written as

$$n(\mathbf{r}) = 1 + 79 \times 10^{-6} \left(\frac{\mathrm{Pr}(\mathbf{r})}{\mathrm{Tmp}(\mathbf{r})} \right) = 1 + \Delta n(\mathbf{r}), \tag{7.49}$$

where $n(\mathbf{r})$ is the refractive index, $\Delta n(\mathbf{r})$ the delta-refractive index, $\mathrm{Pr}(\mathbf{r})$ the pressure, and $\mathrm{Tmp}(\mathbf{r})$ the temperature, all at 3-D location $\mathbf{r}$ [2, p. 181].

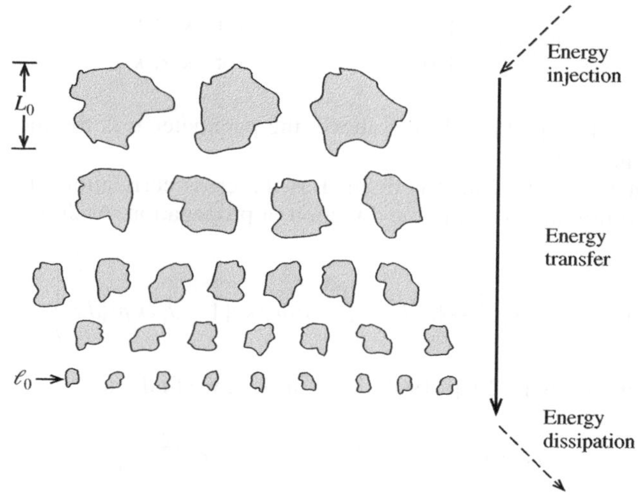

FIGURE 7.13 Geometry for propagation through a turbulent environment.

Because turbulence is a random process, it must be analyzed with statistical quantities. The turbulent fluctuations in the channel are best described by the structure function

$$D_n(\boldsymbol{r}_1,\boldsymbol{r}_2) = \langle |n(\boldsymbol{r}_1) - n(\boldsymbol{r}_2)|^2 \rangle \tag{7.50}$$

$$= D_n(|\boldsymbol{r}_1 - \boldsymbol{r}_2|) = D_n(\rho), \tag{7.51}$$

assuming homogeneous, isotropic turbulence to obtain the latter equation [2]. In the above, the parameter $\rho = |\boldsymbol{r}_1 - \boldsymbol{r}_2|$ is the modulus of the separation between vectors $\boldsymbol{r}_1$ and $\boldsymbol{r}_2$.

From the Kolmogorov–Obukhov analyses, the refractive index structure function follows the following two-thirds power law:

$$D_n(\rho) = \begin{cases} C_n^2(h)\rho^{2/3}; & \text{for } \rho \gg \ell_0 \\ C_n^2(h)\ell_0{}^{2/3}\rho^2; & \text{or } \rho \ll \ell \end{cases} \tag{7.52}$$

assuming also that $\rho < L_0$ [21, p. 110]. In the above equation, the parameter $C_n^2(h)$ is the refractive index structure parameter and h is the height above the ground. Weak turbulence typically has a structure parameter around $10^{-17}\ \text{m}^{-\frac{2}{3}}$ and strong turbulence equal to structure parameter around $10^{-13}\ \text{m}^{-\frac{2}{3}}$ or greater.

To calculate the MCF, we will need the power spectrum for the atmospheric turbulence, which is the Fourier Transform of the refractive index covariance function. Using the two-thirds power law, Kolmogorov developed the following power spectral density function for the refractive index fluctuations:

$$\Phi_n(\kappa) = \begin{cases} 0.033C_n^2\kappa^{-11/3}; & \text{for} \quad \kappa < \kappa_m \\ 0; & \text{for} \quad \kappa < \kappa_m \end{cases}, \tag{7.53}$$

where $\kappa_m = \frac{5.48}{\ell_0}$ [21, p. 110]. In the above, the parameter κ is the modulus of the spatial frequencies κ_x and κ_y.

Recall that the MCF $G(\rho; z)$ is defined as the cross-correlation of the complex fields in a direction transverse to the direction of propagation. Andrews and Phillips showed that

$$G_{\mathrm{PW}}(\rho; z_1) = \exp\left\{-4\pi^2k^2z_1 \int_0^\infty \kappa\Phi_n(\kappa)[1 - J_0(\kappa\rho)]d\kappa\right\} \tag{7.54}$$

for the plane wave MCF [20, p. 190]. For $\kappa_m\rho \ll 1$, we find that

$$G_{\mathrm{PW}}(\rho; z_1) \approx \exp\left\{-1.72k^2z_1C_n^2{\ell_0}^{-\frac{1}{3}}\rho^2\right\}. \tag{7.55}$$

Example 7.3 For $\kappa_m\rho \ll 1$, we have $[1 - J_0(\kappa\rho)] \approx \frac{(\kappa\rho)^2}{4}$, which implies that

$$4\pi^2k^2z_1 \int_0^\infty \kappa\Phi_n(\kappa)[1 - J_0(\kappa\rho)]d\kappa \approx 1.3k^2z_1C_n^2 \int_0^{\kappa_m} \frac{\kappa}{(\kappa^2 + L_0^{-2})^{\frac{11}{6}}} \frac{(\kappa\rho)^2}{4} d\kappa \tag{7.56}$$

$$\approx 0.3257k^2z_1C_n^2\rho^2 \int_0^{\kappa_m} \frac{\kappa^3}{(\kappa^2 + L_0^{-2})^{11/6}} d\kappa$$

$$\approx 0.3257k^2z_1C_n^2\rho^2 \int_0^{\kappa_m} \kappa^{-\frac{2}{3}} d\kappa \tag{7.57}$$

$$\approx 0.3257k^2z_1C_n^2\rho^2 3{\kappa_m}^{\frac{1}{3}} \tag{7.58}$$

$$\approx 1.72k^2z_1C_n^2{\ell_0}^{-\frac{1}{3}}\rho^2, \tag{7.59}$$

for $\rho \ll \ell_0$.

Although Eq. (7.53) is widely used by the turbulence community, its validity range is limited in extent and may cause its integration not to close well outside of inertial range. This fact is important when looking at the MCF over larger range of separations and ranges.

An example of a spectrum that is convenient for computational purposes is the modified von Karman spectrum, which is given by

$$\Phi_n(\kappa) = 0.033C_n^2 \frac{e^{-(\kappa\ell_0)^2}}{(\kappa^2 + L_0^{-2})^{\frac{11}{6}}}, \tag{7.60}$$

for $0 < \kappa < \infty$, which implies a flat spectrum for $\kappa \lesssim 2\pi/L_0$ [19]. Let us use Eq. (7.60) to determine the MCF for a turbulent atmosphere.

Following Lutomirski and Yura, we can write the argument of the exponential in Eq. (7.54) as

$$4\pi^2k^2z_1\int_0^\infty \kappa\Phi_n(\kappa)[1-J_0(\kappa\rho)]d\kappa = \frac{2z_1}{z_c}\left[1-\frac{2\pi^2k^2\int_0^\infty \kappa\Phi_n(\kappa)J_0(\kappa\rho)d\kappa}{2\pi^2k^2\int_0^\infty \kappa\Phi_n(\kappa)d\kappa}\right], \tag{7.61}$$

where

$$z_c^{-1} = 2\pi^2k^2\int_0^\infty \kappa\Phi_n(\kappa)d\kappa = 0.39k^2C_n^2L_0^{\frac{5}{3}} \tag{7.62}$$

for a von Karman spectrum, assuming $\kappa_m\rho \gg 1$ [22]. Here, z_1 is the propagation distance.

Example 7.4 If $\Phi_n(\kappa) = 0.033C_n^2\frac{e^{-(\kappa\ell_0)^2}}{(\kappa^2+L_0^{-2})^{\frac{11}{6}}}$ and $\kappa_m\rho \gg 1$, then the denominator of the integral ratio in Eq. (7.61) is equal to

$$\begin{aligned}2\pi^2k^2\int_0^\infty \kappa\Phi_n(\kappa)d\kappa &= 0.65k^2C_n^2\int_0^{\kappa_m}\kappa\frac{e^{-(\kappa\ell_0)^2}}{(\kappa^2+L_0^{-2})^{\frac{11}{6}}}d\kappa\\ &\approx 0.65k^2C_n^2\int_0^\infty \kappa\frac{e^{-(\kappa\ell_0)^2}}{(\kappa^2+L_0^{-2})^{\frac{11}{6}}}d\kappa\\ &\approx 0.65k^2C_n^2\int_0^\infty \frac{\kappa}{(\kappa^2+L_0^{-2})^{\frac{11}{6}}}d\kappa = 0.39k^2C_n^2L_0^{\frac{5}{3}} = z_c^{-1}.\end{aligned} \tag{7.63}$$

Substituting Eq. (7.61) into Eq. (7.54), we see that

$$G_{\mathrm{PW}}(\rho;z_1) = \exp\left\{-\frac{2z_1}{z_c}\left[1-\frac{2\pi^2k^2\int_0^\infty \kappa\Phi_n(\kappa)J_0(\kappa\rho)d\kappa}{2\pi^2k^2\int_0^\infty \kappa\Phi_n(\kappa)d\kappa}\right]\right\}. \tag{7.64}$$

It is clear from Eq. (7.64) that

$$G_{\mathrm{PW}}(0;z_1) = 1 \tag{7.65a}$$

and

$$G_{\mathrm{PW}}(\infty;z_1) = e^{-\frac{2z_1}{z_c}}. \tag{7.65b}$$

For $z_1 \ll z_c$, the latter MCF is the same as Eq. (7.65a), or

$$G_{PW}(\infty; z_1) = 1. \tag{7.66}$$

Consider the situation where $\rho \gg \ell_0$. It can be shown that the following is the turbulent MCF in the small $\left(\frac{\rho}{L_0}\right)$-limit when $z_c \ll z_1$:

$$G_{PW}(\rho; z_1) \approx \exp\left\{-1.46k^2 C_n^2 z_1 \rho^{\frac{5}{3}}\right\} \tag{7.67}$$

for $\ell_0 \ll \rho \ll L_0$.

Example 7.5 Substituting the modified von Karman spectrum into the numerator of the integral ratio in Eq. (7.61), we obtain

$$2\pi^2 k^2 \int_0^\infty \kappa \Phi_n(\kappa) J_0(\kappa\rho) d\kappa = 0.65k^2 C_n^2 \int_0^\infty \kappa \frac{e^{-(\kappa\ell_0)^2}}{(\kappa^2 + L_0^{-2})^{\frac{11}{6}}} J_0(\kappa\rho) d\kappa \tag{7.68}$$

$$\approx 0.65k^2 C_n^2 \int_0^\infty \kappa \frac{J_0(\kappa\rho)}{(\kappa^2 + L_0^{-2})^{\frac{11}{6}}} d\kappa \tag{7.69}$$

$$= 0.65k^2 C_n^2 \frac{L_0^{\frac{5}{6}} \rho^{\frac{5}{6}}}{2^{\frac{5}{6}} \Gamma\left(\frac{11}{6}\right)} K_{-\frac{5}{6}}\left(\frac{\rho}{L_0}\right) \tag{7.70}$$

using Eq. (6.565.4) in Gradshteyn and Ryzhik [23, p. 686]. In the above, $K_\alpha(x)$ is a modified Bessel Function of the second kind and is given by

$$K_\alpha(x) = \frac{\pi \csc \alpha\pi}{2} [I_{-\alpha}(x) - I_\alpha(x)] \tag{7.71}$$

$$= \frac{\pi \csc \alpha\pi}{2} \left[\sum_{k=0}^\infty \frac{1}{\Gamma(k - \alpha + 1)k!} \left(\frac{x}{2}\right)^{2k-\alpha} - \sum_{k=0}^\infty \frac{1}{\Gamma(k + \alpha + 1)k!} \left(\frac{x}{2}\right)^{2k+\alpha}\right], \tag{7.72}$$

where $I_\alpha(x)$ is the modified Bessel Function of the first kind defined in Eq. (8.402) in Gradshteyn and Ryzhik for an argument of iz [23, p. 951].

For small arguments, we have

$$K_{-\frac{5}{6}}(x) \approx \frac{-\pi \csc\left(\frac{5\pi}{6}\right)}{2} \left[\frac{1}{\Gamma\left(\frac{11}{6}\right)} \left(\frac{x}{2}\right)^{\frac{5}{6}} - \frac{1}{\Gamma\left(\frac{1}{6}\right)} \left(\frac{x}{2}\right)^{-\frac{5}{6}}\right]. \tag{7.73}$$

This implies that

$$
\begin{aligned}
2\pi^2k^2\int_0^\infty \kappa\Phi_n(\kappa)J_0(\kappa\rho)d\kappa &= -\left(\frac{0.65k^2C_n^2}{\Gamma\left(\frac{11}{6}\right)}\right)\left[\frac{\pi\csc\left(\frac{5\pi}{6}\right)}{2}\right]\left(\frac{\rho L_0}{2}\right)^{\frac{5}{6}} \\
&\quad\times\left[\frac{1}{\Gamma\left(\frac{11}{6}\right)}\left(\frac{\rho}{2L_0}\right)^{\frac{5}{6}} - \frac{1}{\Gamma\left(\frac{1}{6}\right)}\left(\frac{\rho}{2L_0}\right)^{-\frac{5}{6}}\right] \\
&\approx 0.39k^2C_n^2L_0^{\frac{5}{3}}\left[1 - \frac{\Gamma\left(\frac{1}{6}\right)}{\Gamma\left(\frac{11}{6}\right)}\left(\frac{1}{2}\right)^{\frac{5}{3}}\left(\frac{\rho}{L_0}\right)^{\frac{5}{3}}\right] \\
&\approx 0.39k^2C_n^2L_0^{\frac{5}{3}}\left[1 - 1.86\left(\frac{\rho}{L_0}\right)^{\frac{5}{3}}\right].
\end{aligned} \tag{7.74}
$$

Substituting Eqs. (7.63) and (7.74) into Eq. (7.61) yields

$$
\begin{aligned}
\frac{4z_1}{z_c}\left[1 - \frac{2\pi^2k^2\int_0^\infty \kappa\Phi_n(\kappa)J_0(\kappa\rho)d\kappa}{2\pi^2k^2\int_0^\infty \kappa\Phi_n(\kappa)d\kappa}\right] &\approx \frac{4z_1}{z_c}\left[1 - \left(1 - 1.86\left(\frac{\rho}{L_0}\right)^{\frac{5}{3}}\right)\right] \\
&= \frac{4z_1}{z_c}\left[1.86\left(\frac{\rho}{L_0}\right)^{\frac{5}{3}}\right] \\
&= 4z_1\left(0.39k^2C_n^2L_0^{\frac{5}{3}}\right)\left[1.86\left(\frac{\rho}{L_0}\right)^{\frac{5}{3}}\right] \qquad (7.75) \\
&\approx 2.914k^2C_n^2z_1\rho^{\frac{5}{3}}. \qquad (7.76)
\end{aligned}
$$

In order that, at a given range, all the transverse separations of interest lie in the inertial subrange, we must have

$$G_{\mathrm{PW}}(\ell_0; z_1) \cong 1 \tag{7.77a}$$

and

$$G_{\mathrm{PW}}(L_0; z_1) \ll 1, \tag{7.77b}$$

which is essentially the condition $z_c \ll z_1 \ll z_i$. The parameter z_i is a distance that the coherence length of the field is of the order of the inner scale, which is given by

$$z_i^{-1} = 0.39k^2 C_n^2 \ell_0^{\frac{5}{3}}. \tag{7.78}$$

For ranges greater than z_i, all the ρs of interest are small compared with the inner scale and Eq. (7.55) is valid here, that is,

$$G_{\text{PW}}(\rho; z_1) \approx \exp\left\{-1.72k^2 z_1 C_n^2 \ell_0^{-\frac{1}{3}} \rho^2\right\} \tag{7.79}$$

for $z_1 \gg z_i$.

The MCF for a spherical wave propagating through a turbulent atmosphere is

$$G_{\text{SW}}(\rho; z_1) = 1 \tag{7.80}$$

for $z_1 \ll z_c$ [10]. When $z_c \ll z_1$, we need to transform the plane wave MCF equation given in Eq. (7.54) into spherical wave MCF equation. This is done by recognizing that

$$z_1[1 - J_0(\kappa\rho)] \to \int_0^z \left[1 - J_0\left(\frac{\kappa\rho z'}{z}\right)\right] dz' = \int_0^1 [1 - J_0(\kappa\zeta\rho)]d\zeta. \tag{7.81}$$

Substituting Eq. (7.81) into Eq. (7.54) yields

$$G_{\text{SW}}(\rho; z_1) = \frac{1}{(4\pi z_1)^2} \exp\left\{\frac{ik}{z_1}\boldsymbol{p} \cdot \boldsymbol{r} - 4\pi^2 k^2 z_1 \int_0^1 \int_0^\infty \kappa\Phi_n(\kappa)[1 - J_0(\kappa\zeta\rho)]d\kappa d\zeta\right\} \tag{7.82}$$

as the general MCF expression for a spherical wave input [20, p. 191]. For $\rho \gg \ell_0$, again assuming a modified von Karman spectrum, Eq. (7.82) reduces to

$$G_{\text{SW}}(\rho; z_1) \approx \frac{1}{(4\pi z_1)^2} \exp\left\{\frac{ik}{z_1}\boldsymbol{p} \cdot \boldsymbol{r} - 0.55k^2 C_n^2 z_1 \rho^{\frac{5}{3}}\right\} \tag{7.83}$$

(see Problem 7.3). It is clear from Eq. (7.83) that the spherical wave MCF involves an unimportant phase factor that depends on the location of two points within the observation plane. This is a consequence of the phase front curvature of a spherical wave [20, p. 191].

7.7 LASER BEAM PROPAGATION IN THE TOTAL ATMOSPHERE

From the previous discussions, we expect the several entities comprising the atmosphere to reduce the total laser energy available by absorption of the various particulates and to spread the laser beam out in time, space, and angle due to various types of scattering those same particulates and also turbulence. Like the MTF, we can estimate the atmospheric degradations due to each of the constituents (molecules, turbulence, and aerosols), using the product of their associate MCFs [12]. That is, the absorption and scattering processes discussed in the previous section are essentially uncorrelated, and this means that the total atmospheric MCF for a spherical wave propagating through the atmosphere can be written as

$$G_{\text{atm}}(\rho) = G_{\text{areo}}(\rho) \cdot G_{\text{mole}}(\rho) \cdot G_{\text{turb}}(\rho), \tag{7.84}$$

where $G_{\text{mole}}(\rho)$ is the molecular MCF given by Eq. (7.48); $G_{\text{aero}}(\rho)$ the (spherical wave) aerosol MCF given generally by Eq. (7.31) and in the limit of either very few or very many scatterings, by Eq. (7.43); and $G_{\text{turb}}(\rho)$ the spherical wave turbulence MCF given by

$$G_{\text{turb}}(\rho) = \begin{cases} 1; & z \ll z_c \\ e^{-\left(\frac{\rho}{\rho_t}\right)^{\frac{5}{3}}}; & z \gg z_c \end{cases}, \tag{7.85}$$

where

$$\rho_t = (0.55k^2 C_n^2 z)^{-\frac{3}{5}}. \tag{7.86}$$

The approximate coherence lengths, defined as the transverse separations for which the respective coherence functions decay to e^{-1}, are infinity for molecules; ρ_{aero} for aerosols; and ρ_t for turbulence, subject to the conditions of Eq. (7.85). For essentially all real atmospheres and wavelengths, ρ_{aero} is of the order of particle sizes (microns) and ρ_t is on the order of millimeters to 10 s of centimeters. In the general case, using the above formulas, the total atmospheric coherence functions can be expressed approximately analytically (or numerically) in terms of the molecular and aerosol absorption coefficient, the aerosol scattering coefficient, the aerosol RMS single scattering angle (or the volume scattering function), the refractive index structure content, and the optical wavelength and path length [12]. Let us look at this in more detail.

In a vacuum, laser beam profiles are calculated by using the Huygens–Fresnel–Kirchhoff Integral as we saw in Chapters 2 and 5.

Example 7.6 If we assume a Gaussian aperture field corresponding to the TEM_{00} mode, that is,

$$F(x, y) = \sqrt{\frac{P}{\pi w_0^2}} e^{-\frac{r^2}{w_0^2}}, \tag{7.87}$$

then the ASF is given by

$$F(x, y, z) = \frac{\exp\left\{i\left[kz + \frac{\pi}{2}\right]\right\}}{\lambda z} \int\int_{-\infty}^{\infty} F(x'y') \exp\left\{i\pi \frac{(x - x')^2 + (y - y')^2}{\lambda z}\right\} dx'dy'$$

$$= \frac{\exp\left\{i\left[kz + \frac{\pi}{2}\right] + i\pi \frac{x^2+y^2}{\lambda z}\right\}}{\lambda z} \int\int_{-\infty}^{\infty} \sqrt{\frac{P}{\pi w_0^2}} e^{-\frac{r^2}{w_0^2}}$$

$$\times exp\left\{-2i\pi\frac{xx'+yy'}{\lambda z}+i\pi\frac{x'^2+y'^2}{\lambda z}\right\}dx'dy'$$

$$=\frac{\exp\left\{i\left[kz+\frac{\pi}{2}\right]+i\pi\frac{x^2+y^2}{\lambda z}\right\}}{\lambda z}\sqrt{\frac{P}{\pi w_0^2}}\int\int_{-\infty}^{\infty}e^{-\frac{r'^2}{w_0^2}\left(1-\frac{\pi i w_0^2}{\lambda z}\right)}e^{-ik\frac{\mathbf{r}\cdot\mathbf{r}'}{z}}d^2\mathbf{r}' \tag{7.88}$$

$$=\frac{\exp\left\{i\left[kz+\frac{\pi}{2}\right]+i\pi\frac{x^2+y^2}{\lambda z}\right\}}{\lambda z}\sqrt{\frac{P}{\pi w_0^2}}2\pi\int_0^{\infty}e^{-\frac{r'^2}{w_0^2}\left(1-\frac{\pi i w_0^2}{\lambda z}\right)}J_0\left(k\frac{rr'}{z}\right)r'dr' \tag{7.89}$$

$$=\frac{\exp\left\{i\left[kz+\frac{\pi}{2}\right]+i\pi\frac{x^2+y^2}{\lambda z}\right\}}{\lambda z}\sqrt{\frac{P}{\pi w_0^2}}2\pi\left(\frac{w_0^2}{2\left(1-\frac{\pi i w_0^2}{\lambda z}\right)}\right)e^{-\frac{k^2w_0^2r^2}{4z^2\left(1-\frac{\pi i w_0^2}{\lambda z}\right)}} \tag{7.90}$$

$$=\exp\left\{i\left[kz+\frac{\pi}{2}\right]+i\pi\frac{x^2+y^2}{\lambda z}\right\}\sqrt{\frac{P}{\pi w_0^2}}\left(\frac{1}{2\left(1-\frac{i\lambda z}{\pi w_0^2}\right)}\right)e^{-\frac{r^2}{w_0^2\left(1+\frac{(\lambda z)^2}{\pi^2w_0^4}\right)}}. \tag{7.91}$$

using Eq. (6.631.4) in Gradshteyn and Ryzhik [23, p. 717]. Squaring Eq. (7.91) gives the intensity of the laser beam, which specifically equals

$$I(r;z)=|F(x,y,z)|^2=\frac{P}{\pi w_0^2\left(1+\frac{(\lambda z)^2}{\pi^2w_0^4}\right)}e^{-\frac{2r^2}{w_0^2\left(1+\frac{(\lambda z)^2}{\pi^2w_0^4}\right)}} \tag{7.92a}$$

$$=\left(\frac{P}{\pi w^2(z)}\right)e^{-\frac{2r^2}{w^2(z)}}, \tag{7.92b}$$

where

$$w^2(z)=w_0^2\left[1+\frac{(\lambda z)^2}{\pi^2w_0^4}\right]=w_0^2\left[1+\frac{z^2}{z_r^2}\right] \tag{7.93}$$

and

$$z_r=\frac{\pi w_0^2}{\lambda}. \tag{7.94}$$

In Eq. (7.94), the second term in this equation is due to diffraction from the aperture. Figure 7.14 depicts an exaggerated cutaway profile of the Gaussian laser intensity given in Eqs. (7.92a) and (7.92b). Unlike conventional light beams, Gaussian

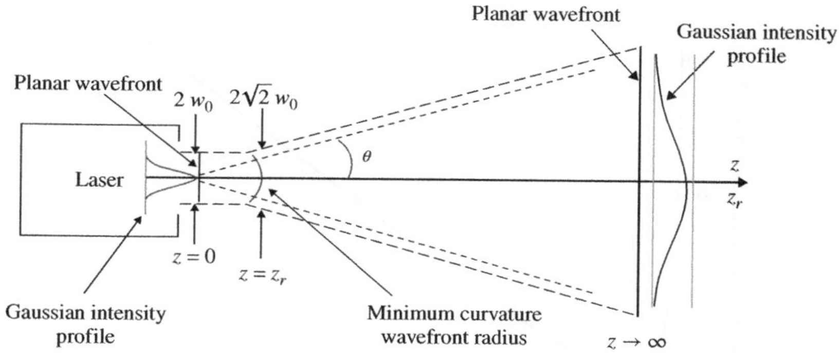

FIGURE 7.14 Growth in beam diameter as a function of distance from the laser source.

beams do not spread linearly, as can be seen in Figure 7.14. Near the laser, the divergence angle is extremely small; far from the laser, the divergence angle approaches the asymptotic limit given by

$$\theta \approx \frac{w(z)}{z} \approx \frac{\lambda}{\pi w_0} = w_0 z_r^{-1} \tag{7.95}$$

as $z \to \infty$. This value is the far-field angular radius (half-angle divergence) of the Gaussian TEM_{00} beam. The vertex of the asymptotic angular spreading lies at the center of the exiting Gaussian beam, as shown in Figure 7.14.

The parameter z_r is known as the *Rayleigh distance* or *range*. This range is the distance where the laser beam radius spreads by a factor of $\sqrt{2}$. The Raleigh range, considered to be the dividing line between near-field divergence and mid-range divergence, is the distance from the waist at which the wavefront curvature radius is a minimum. Far-field divergence (the number quoted in laser specifications) must be measured at a distance much greater than z_r (usually $> 10z_r$ will suffice) [24]. This is a very important distinction because calculations for spot size and other parameters in an optical train will be inaccurate if near- or mid-field divergence values are used. Like the far-field range for a fixed aperture illuminated by a plane wave, the far-field distance of a Gaussian beam will be measured in meters. For example, for a 6328 Å Helium Neon (HeNe) laser with a beam-waist of 0.5 mm, the Rayleigh range is 1.24 m and the above half angle approximately equals 0.402 mrad. Let us now discuss the wavefront curvature in more detail.

Even if the initial Gaussian TEM_{00} laser-beam wavefront was made perfectly flat at near $z = 0$, with all rays there moving in precisely parallel directions, it would acquire curvature and begin spreading in accordance with the equation

$$R(z) = z \left[1 + \frac{z_r^2}{z^2} \right], \tag{7.96}$$

where $R(z)$ is the wavefront radius of curvature after propagating a distance z [24]. $R(z)$ is infinite at $z = 0$, passes through a minimum at z_r and then rises again toward

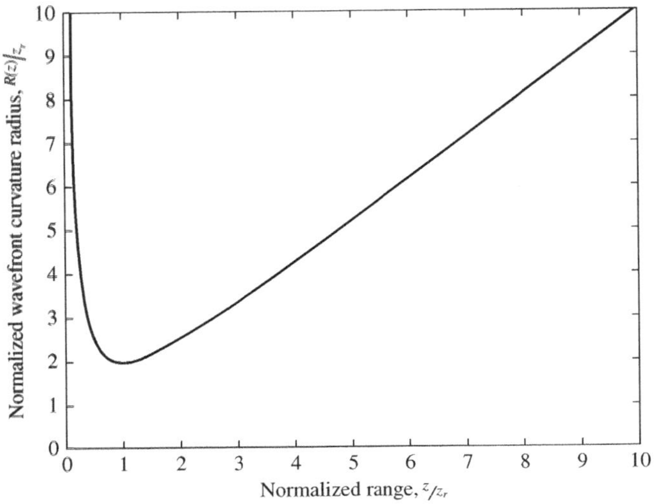

FIGURE 7.15 Plot of normalized wavefront curvature radius versus normalized range.

infinity as z is further increased, asymptotically approaching the value of z itself. This behavior is illustrated in Figure 7.15, which shows the cited minimum location of z_r. As $z \to \infty$, we find that the beam spread takes on the following asymptotic form:

$$w_{\text{asym}}(z) \approx \frac{\lambda z}{\pi w_0^2}. \tag{7.97}$$

It is clear from Eq. (7.97) the w_0^2 replaces the aperture diameter D in the angular spread equation we found for a circular aperture.

Example 7.7 If we assume a propagating Gaussian TEM_{00} electric field, the power received by an aperture of radius r_1 in the observation plane at range z is equal to

$$P(r; z) = \frac{P}{2}\left[1 - e^{-\frac{2r_1{}^2}{w^2(z)}}\right]. \tag{7.98}$$

If $r_1 = w(z)$, then the fraction of power entering the aperture equals $\frac{2P(r;z)}{P} \approx [1 - e^{-2}] = 0.8647$.

The peak intensity at an axial distance z from the beam waist can be calculated by taking the limit of the enclosed power within an aperture of radius r_1, divided by the area of said aperture. Mathematically, we have

$$I(0; z) = \lim_{r_1 \to 0} \frac{P}{2\pi r_1^2}\left[1 - e^{-\frac{2r_1{}^2}{w^2(z)}}\right] \tag{7.99}$$

$$= \lim_{r_1 \to 0} \frac{P}{2} \left[\frac{0 - (-2)(2r_1)e^{-\frac{2r_1{}^2}{w^2(z)}}}{(2\pi r_1)w_0^2} \right] = \frac{P}{\pi w_0^2} \tag{7.100}$$

using L'Hôpital's rule.

Unfortunately, the Huygens–Fresnel–Kirchhoff Integral cannot be used directly to evaluate light propagation in either turbulent or particulate media. In particular, this approach breaks down when the aperture is just a few wavelengths in dimension and/or the scattering of the light is large, for example, the RMS scattering angle of the medium is > 30° [12]. Lutomirski and Yura have shown that for those situations where the apertures are of reasonable size and the scattering by the medium is not too large, a closed form solution is possible [25]. Specifically, within those limitations, they showed that the average irradiance at a distance ρ from the optic axis of the transmitter in the observation plane located at range z is given by

$$\langle I(\rho; z)\rangle = \frac{k^2 P}{(2\pi z)^2} \int\int_{-\infty}^{\infty} G_{\text{trans}}(\rho) G_{\text{atm}}(\rho) e^{-i\frac{k}{z}\boldsymbol{\rho}\cdot\boldsymbol{r}} d^2\boldsymbol{\rho}. \tag{7.101}$$

The form of Eq. (7.101) is not too surprising that it follows the same procedure outlined in Section 11 of Chapter 5 for MTFs.

Example 7.8 Let us again assume a propagating Gaussian TEM_{00} electric field propagating in a turbulent medium. For simplicity, let us further assume that

$$G_{\text{turb}}(\rho) \approx \begin{cases} 1; & z \ll z_c \\ e^{-\left(\frac{\rho}{\rho_t}\right)^2}; & z \gg z_c \end{cases}. \tag{7.102}$$

It can be shown that

$$G_{\text{trans}}(\boldsymbol{\rho}) = \frac{1}{2} e^{-\frac{\rho^2}{2w_0^2}\left(1+\frac{\pi^2 w_0^4}{(\lambda z)^2}\right)}. \tag{7.103}$$

(See Problem 7.4). Substituting into Eq. (7.101), we have

$$\langle I(\boldsymbol{r}; z)\rangle = \frac{k^2 P}{(2\pi z)^2} \int\int_{-\infty}^{\infty} G_{\text{trans}}(\rho) G_{\text{atm}}(\rho) e^{-i\frac{k}{z}\boldsymbol{\rho}\cdot\boldsymbol{r}} d^2\boldsymbol{\rho} \tag{7.104}$$

$$= \left(\frac{k^2}{8\pi^2 z^2}\right) P \int\int_{-\infty}^{\infty} e^{-\frac{\rho^2}{2w_0^2}\left(1+\frac{\pi^2 w_0^4}{(\lambda z)^2}\right)-\frac{\rho^2}{\rho_t^2}} e^{-i\frac{k}{z}\boldsymbol{\rho}\cdot\boldsymbol{r}} d^2\boldsymbol{\rho}$$

$$= \left(\frac{k^2}{8\pi^2 z^2}\right) P 2\pi \int_0^{\infty} e^{-\frac{k^2 w_t^2 \rho^2}{8z^2}} J_0\left(k\frac{r\rho}{z}\right) \rho d\rho \tag{7.105}$$

$$= \left(\frac{k^2 P}{4\pi z^2}\right) \frac{8z^2}{2k^2 w_t^2} e^{-\frac{2r^2}{w_t^2}} = \frac{P}{\pi w_t^2} e^{-\frac{2r^2}{w_t^2}}, \tag{7.106}$$

where

$$w_t^2 = w^2(z) + \frac{2(\lambda z)^2}{(\pi^2 \rho_t^2)}. \tag{7.107}$$

Equation (7.106) suggests that when given a Gaussian beam profile, any modification from propagation in a medium where the associated MCF is also Gaussian creates another Gaussian result with a new variance.

Example 7.9 Recall that the MCF for turbulent atmosphere in the small $\left(\frac{\rho}{L_0}\right)$-limit, when $z_c \ll z_1$, is given by

$$G_{\text{PW}}(\rho; z_1) \approx \exp\left\{-1.457k^2 C_n^2 z_1 \rho^{\frac{5}{3}}\right\}. \tag{7.108}$$

For a gradient atmosphere, the above becomes

$$G_{\text{PW}}(\rho; z_1) \approx \exp\left\{-1.457k^2 \sec(\zeta) \int_{\text{Path}} C_n^2(r)dr \ \rho^{\frac{5}{3}}\right\}, \tag{7.109}$$

following what we cited for the atmospheric transmissivity [19, Chapter 12]. Using the definition in Eq. (5.177), we can rewrite the MCF in terms of the Fried parameter:

$$G_{\text{PW}}(\rho; z_1) \approx \exp\left\{-3.44\left(\frac{\rho}{r_0}\right)^{\frac{5}{3}}\right\}. \tag{7.110}$$

Example 7.10 Recall Eq. (5.1), which states that

$$v(x,y) = \int\int_{-\infty}^{\infty} u(x', y')F(x - x', y - y')dx'dy',$$

where $u(x', y')$ is the field amplitude and $F(x)$ the ASF. Fourier transforming the above equations yields

$$\hat{v}(k_x, k_y) = \int_{-\infty}^{\infty} v(x, y)e^{-2\pi i(k_x x + k_x x)}dx = \hat{u}(k_x, k_y)\hat{F}(k_x, k_y), \tag{7.111}$$

where

$$\hat{u}(k_x, k_y) = \int_{-\infty}^{\infty} u(x, y)e^{-2\pi i(k_x x + k_x x)}dxdy \tag{7.112}$$

and

$$\hat{F}(k_x, k_y) = \int_{-\infty}^{\infty} F(x, y)e^{-2\pi i(k_x x + k_x x)}dxdy, \tag{7.113}$$

where $\hat{F}(k_x, k_y)$ is the Complex Pupil Function [see Eqs. (5.6)–(5.8)].

Let us now turn to the turbulence-degraded imaging. Since one measures intensity, the quantity of interest is the square of $\hat{v}(k_x, k_y)$ and we will assume that this entity is taken under long exposure conditions, that is, in practice, this means exposures of at least a few seconds, resulting in the averaging over many different realizations of the state of the atmosphere. Under this condition, we write

$$G_v(\mathbf{r}_1', \mathbf{r}_2') = \iint \iint F^*(\mathbf{r} - \mathbf{r}_1')F(\mathbf{r} - \mathbf{r}_2')G_{\text{turb}}(\mathbf{r}_1', \mathbf{r}_2')d\mathbf{r}_1' d\mathbf{r}_2' \tag{7.114}$$

from Eq. (6.155). Fourier transforming Eq. (7.114), we obtain

$$\hat{\Theta}_v(k_x, k_y) = \hat{\Theta}_{\text{turb}}(k_x, k_y)\hat{\Theta}_F(k_x, k_y), \tag{7.115}$$

where $\hat{\Theta}_v(k_x, k_y)$, $\hat{\Theta}_{\text{turb}}(k_x, k_y)$, and $\hat{\Theta}_F(k_x, k_y)$ are the optical transfer functions of the system, of the turbulent atmosphere (neglecting the atmospheric loss terms) and of the receiver.

Quirrenbachl [26] stated that the resolving power $\mathfrak{R}$ of a telescope generally is defined by the spatial frequency integrals over the optical transfer function following Fried [27]. For a atmospheric telescope system, this means

$$\mathfrak{R}_{\text{tel}} = \iint \hat{\Theta}_v(k_x, k_y)dk_x dk_y = \iint \hat{\Theta}_{\text{turb}}(k_x, k_y)\hat{\Theta}_F(k_x, k_y)dk_x dk_y. \tag{7.116}$$

In the absence of turbulence, $\hat{\Theta}_{\text{turb}}(k_x, k_y) \equiv 1$, and we obtain the diffraction-limited resolving power of a telescope with diameter D and focal length f,

$$\mathfrak{R}_{\text{tel}} = \iint \hat{\Theta}_F(k_x, k_y)dk_x dk_y \tag{7.117}$$

$$= \int_0^{2\pi} \int_0^{\vartheta_c} \frac{2}{\pi}\left[\cos^{-1}\left(\frac{\vartheta}{\vartheta_c}\right) - \left(\frac{\vartheta}{\vartheta_c}\right)\sqrt{1 - \left(\frac{\vartheta}{\vartheta_c}\right)^2}\right]\vartheta d\vartheta$$

$$= 4\vartheta_c^2 \int_0^1 \left[x\cos^{-1}(x) - x^2\sqrt{1 - x^2}\right] dx \tag{7.118}$$

$$= 4\vartheta_c^2\left(\left[\frac{1}{2}\right]\frac{\pi}{4} + \left[\frac{x\ \sqrt[3]{1 - x^2}}{4} - \frac{x\sqrt{1 - x^2}}{8} - \frac{\sin^{-1}(x)}{8}\right]\right)_0^1$$

$$= 4\vartheta_c^2\left[\frac{\pi}{8} - \left(\frac{1}{8}\right)\frac{\pi}{2}\right] = 4\vartheta_c^2\left(\frac{1}{8}\right)\frac{\pi}{2} = \frac{\pi}{4}\vartheta_c^2 = \frac{\pi}{4}\left(\frac{D}{\lambda f}\right)^2 \tag{7.119}$$

using Eqs. (4.523.4) and (2.272.2) in Gradshteyn and Ryzhik for the first and second integrals in Eq. (7.118), respectively [23, p. 607 and p. 86]. Treating the cutoff frequency as a diameter, Eq. ((7.119) suggests that the resolving power of the telescope just is the "circular area" of the telescope's cutoff frequency. The diffraction limit of the lens is when $\hat{\Theta}_F(k_x, k_y)$ equals zero, which occurs at the cutoff frequency, that is, when

$$\vartheta_c = \left(\frac{D}{\lambda f}\right). \tag{7.120}$$

The product

$$\vartheta f = \left(\frac{D}{\lambda}\right) \tag{7.121}$$

has units of cycles per radian since the ratio $\frac{\lambda}{D}$ is the angular spread of a circular aperture diffraction defined in Eq. (2.112). If the diameter of the telescope is 10 cm(4 in.) and the wavelength of observation is 0.5 μm, the angular spread is equal to $\frac{0.5\ \mu\text{m}}{10\ \text{cm}} = 5\ \mu\text{rad}$. This implies the minimum resolving power of the telescope is

$$(\vartheta f)_{\min} = \left(\frac{1}{5\ \mu\text{rad}}\right) \approx 100 \frac{\text{line pairs}}{\text{mrad}}. \tag{7.122}$$

Let us now look at Eq. (7.116) without the telescope's effects. With perfect optics having an OTF equal to one, in this case, Eqs. (7.92a) and (7.92b) becomes

$$\Re_{\text{turb}} = \iint \hat{\Theta}_{\text{turb}}(k_x, k_y) dk_x dk_y \tag{7.123}$$

$$= 2\pi \int_0^{\infty} \vartheta \exp\left\{-3.44\left(\frac{\lambda R}{r_0}\right)^{\frac{5}{3}} \vartheta^{\frac{5}{3}}\right\} d\vartheta$$

$$= 2\pi \frac{1}{\left|\frac{5}{3}\right|} \left(3.44\left(\frac{\lambda R}{r_0}\right)^{\frac{5}{3}}\right)^{-\frac{6}{5}} \Gamma\left(\frac{6}{5}\right) = \frac{\pi}{4}\left(\frac{r_0}{\lambda R}\right)^2 \tag{7.124}$$

$$= \Re_{\max} \tag{7.125}$$

using Eq. (3.477.1) in Gradshteyn and Ryzhik [23, p. 342]. This parameter has the dimensions of cycles-squared per unit area in the observation plane. This is the limiting (best) resolution since a multiple product OTF will make the resolution worse. This is true for any optical system. It should be noted that the beam spread implied by Eq. (7.124) is given by

$$\frac{\lambda R}{r_0}, \tag{7.126}$$

which is the turbulent beam spread expression we will discuss at the beginning of the next section (see Figure 7.17). Fried first showed that Eq. (7.124) is the maximum resolving power that can be achieved in turbulence [28]. If we assume that $R = f$, then we have

$$\Re_{\max} = \frac{\pi}{4}\left(\frac{r_0}{\lambda f}\right)^2 \tag{7.127}$$

Example 7.11 If the telescope has diameter of 4 m and the atmosphere has Fried parameter equal to 10 cm, then the diffraction-limited and turbulence-induced beam spread angles at $\lambda = 0.5\ \mu m$ are:

$$\theta_{Diff} \approx \frac{\lambda}{D} = \frac{0.5\ \ \mu m}{4\ \ m} = 125\ \ \text{nrad}$$

and

$$\theta_{Turb} \approx \frac{\lambda}{r_0} = \frac{0.5\ \ \mu m}{0.1\ \ m} = 5\ \ \mu\text{rad},$$

respectively. The above implies the resolution of the telescope degrades by a factor of 40.

Let us now calculate the resolving power for the combined effects of the telescope and turbulence. Substituting the appropriate OTFs into Eq. (7.116), we have

$$\mathfrak{R} = \int_0^{2\pi}\int_0^{\vartheta_c} \frac{2}{\pi}\left[\cos^{-1}\left(\frac{\vartheta}{\vartheta_c}\right) - \left(\frac{\vartheta}{\vartheta_c}\right)\sqrt{1-\left(\frac{\vartheta}{\vartheta_c}\right)^2}\right]\exp\left\{-3.44\left(\frac{\lambda R\vartheta}{r_0}\right)^{\frac{5}{3}}\right\}\vartheta d\vartheta d\varphi, \tag{7.128}$$

where

$$\vartheta_c = \left(\frac{D}{\lambda f}\right) \tag{7.129}$$

and

$$\rho = \lambda R\vartheta. \tag{7.130}$$

19, p. 619. If we set $R = f,\ u = \left(\frac{\vartheta}{\vartheta_c}\right)$ and do the angle integration, then Eq. (7.128) reduces to

$$\mathfrak{R} = 4\vartheta_0^2\int_0^1\left[\cos^{-1}(u) - (u)\sqrt{1-(u)^2}\right]\exp\left\{-3.44\left(\frac{Du}{r_0}\right)^{\frac{5}{3}}\right\}u\,du. \tag{7.131}$$

If we divided $4\vartheta_0^2$ by the maximum resolving power for the optical system, we obtain

$$\frac{4\vartheta_0^2}{\frac{\pi}{4}\left(\frac{r_0}{\lambda f}\right)^2} = \frac{4\left(\frac{D}{\lambda f}\right)^2}{\frac{\pi}{4}\left(\frac{r_0}{\lambda f}\right)^2} = \left(\frac{16}{\pi}\right)\left(\frac{D^2}{r_0^2}\right). \tag{7.132}$$

Equation (7.132) implies that

$$\frac{\mathfrak{R}}{\mathfrak{R}_{max}} = \left(\frac{16}{\pi}\right)\left(\frac{D^2}{r_0^2}\right)\int_0^1\left[\cos^{-1}(u) - (u)\sqrt{1-(u)^2}\right]\exp\left\{-3.44\left(\frac{Du}{r_0}\right)^{\frac{5}{3}}\right\}u\,du. \tag{7.133}$$

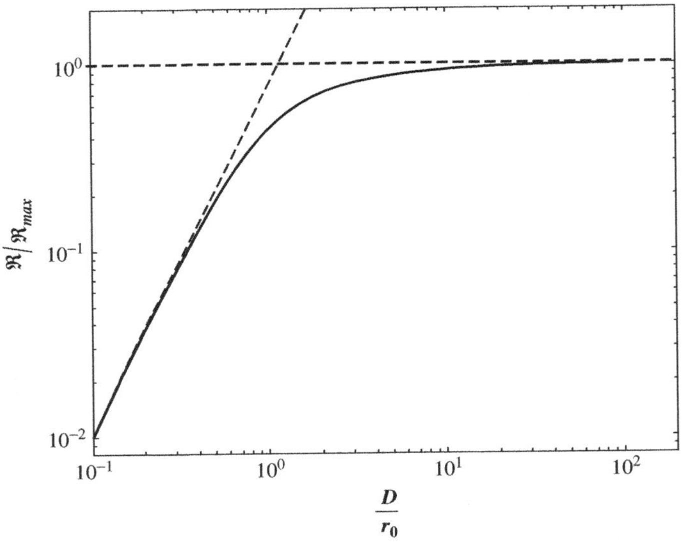

FIGURE 7.16 Normalized resolving power for long-exposure imaging as a function of $\left(\frac{D}{r_0}\right)$.

Figure 7.16 depicts the normalized resolving power for long-exposure imaging, $\frac{\Re}{\Re_{\max}}$, as function of $\frac{D}{r_0}$. The ratio $\frac{\Re}{\Re_{\max}}$ increases with increasing diameter until it is on the order of the Fried parameter, where it starts to limit out. No appreciable resolution improvement occurs for $D \gg r_0$. Like we saw in the figure depicting heterodyne efficiency, the atmospheric-induced coherence width imposes a severe limit on system performance if the receiver aperture gets to $r_0$and beyond/r_0. This curve also reflects the improvement of the normalized signal-to-noise ratio in an optical heterodyne receiver for increased receiver aperture D [27].

7.8 KEY PARAMETERS FOR ANALYZING LIGHT PROPAGATION THROUGH GRADIENT TURBULENCE

In the Earth's atmosphere, the refractive index structure parameter changes with height above the ground, that is, $C_n^2 \to C_n^2(h)$. It varies strongly with height above the ground, being largest near the ground and dramatically decreasing with increasing height. It also follows a diurnal cycle, with peaks at midday, near constant values at night and lowest near sunrise. Before we discuss two of the more popular models for $C_n^2(h)$, this section introduces a few key parameters that are utilized in system engineering analyses that will be useful. Everyone is directly, or indirectly, dependent on the vertically varying C_n^2.

When the incoming beam propagates through atmospheric turbulence, the resulting point spread function at the observation plane will be limited by both diffraction and the turbulence the beam encountered. In particular, the point spread function will be a combination of expected beam spread defined by $\left(\frac{\lambda}{D_{tx}}\right)$, D_{tx} being the transmitter

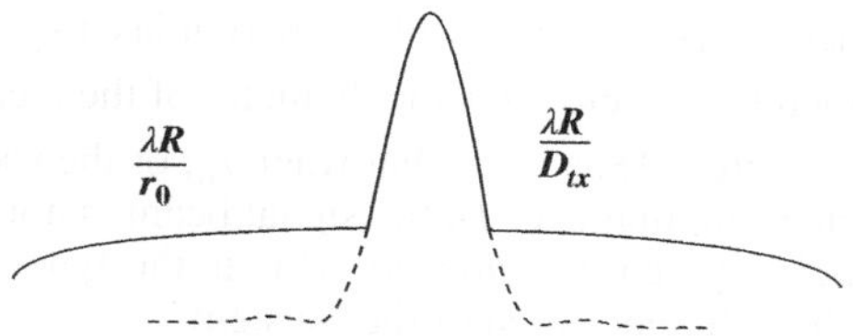

FIGURE 7.17 Spot size profile after laser beam propagation through a turbulent environment.

aperture diameter, and a "halo" about this peak with a width defined by the ratio $\left(\frac{\lambda}{r_0}\right)$. This beam shape is illustrated in Figure 7.17. The parameter r_0 is known as the transverse coherence length of the atmosphere, or more typically, *the Fried parameter*. It is given by

$$r_0 = \left[0.423k^2 \sec(\zeta) \int_{\text{Path}} C_n^2(r)dr\right]^{-\frac{3}{5}}, \tag{7.134}$$

where ζ is the zenith angle from vertical. The Fried parameter is diameter, not a radius, and is the distance, over which the phase varies no more than $\pm\pi$. At sea level, r_0 is usually 2–15 cm in daytime at visible wavelengths. More importantly, r_0 is on the order of the speckle size of the turbulence at the observation plane, which implies that $\frac{\pi r_0^2}{4}$ is on the order of the coherence area A_c. This can be seen in Figure 7.18,

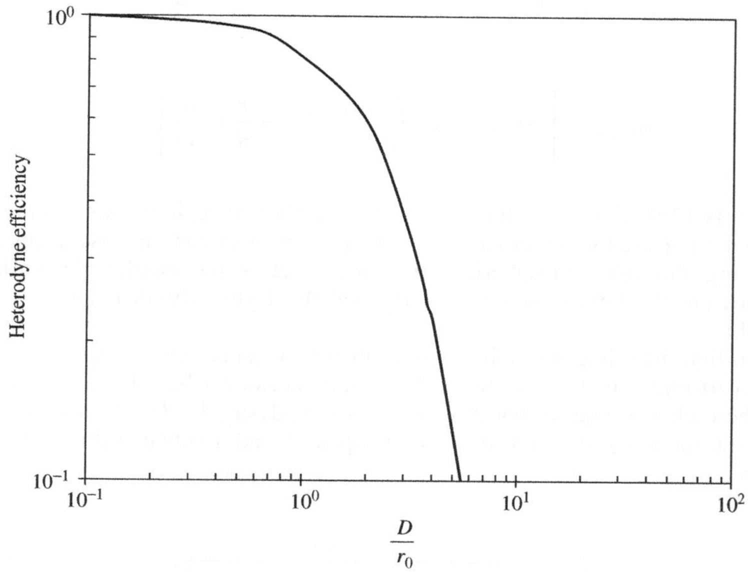

FIGURE 7.18 Heterodyne efficiency as a function of $\left(\frac{D}{r_0}\right)$.

which illustrates the heterodyne efficiency of a coherent laser system as a function of $\left(\frac{D}{r_0}\right)$ developed by Fried [27]. Here, D is the diameter of the receiver optics.

Referring again to Figure 7.18, we see that when r_0, or the speckle, is large compared to the receiver aperture diameter D, the signal beam is not distorted much and 100% heterodyne efficiency can be achieved. This is the type of performance one would expect within the coherence area of the "system."

As r_0 exceeds D (or a few speckle patches fill the aperture), the heterodyne efficiency begins to decrease rapidly. The reason is that although the amplitude errors will begin to be mitigated by aperture averaging of the speckle under this situation, the phase errors grow causing the beating between the two beams to be degraded as the turbulence moves out of the weak regime. When numerous speckles fill the receiver aperture, $\left(\frac{D}{r_0}\right) > 10$, the phase errors grow to the point that signal beam becomes essentially incoherent to its reference beam and the heterodyne efficiency goes to zero.

Since we generally use airborne and space platforms besides ground terminals in many communications and lidar/ladar applications, one can expect to have transmitters pointed from the ground to an elevated receiver and vice versa. Under these conditions, one needs a modified form of the Fried parameter to reflect the air-to-ground and ground-to-air coherence lengths for spherical waves. In many situations, one end of the systems will be in the air and the other end on the ground. In this case, the airborne and ground coherence lengths are given by

$$r_{0\,\text{air}} = \left[16.71\sec(\zeta)\int_0^L C_n^2(r)\left(\frac{r}{R}\right)^{\frac{5}{3}}\frac{dr}{\lambda^2}\right]^{-\frac{3}{5}} \tag{7.135}$$

and

$$r_{0\,\text{gnd}} = \left[16.71\sec(\zeta)\int_0^L C_n^2(r)\left(1-\frac{r}{R}\right)^{\frac{5}{3}}\frac{dr}{\lambda^2}\right]^{-\frac{3}{5}}, \tag{7.136}$$

respectively [29]. Here, L is the length of the turbulent regime. Besides the Fried parameter, there are four other parameters helpful in understanding laser propagation or imaging through atmospheric turbulence. One is the isoplanatic angle, the next ones are the Greenwood Frequency and the Rytov Number, and the last is the Strehl.

When light traveling through the atmosphere has a changing frame of reference due to platform(s) motion, there might be the occasion when the statistics of the atmosphere change due to that motion. That is undesirable. The *isoplanatic angle* defines the mean-square error of 1 radian-squared (rad^2) between the paths and is given by

$$\theta_0 = \frac{\cos^{\frac{8}{5}}(\zeta)}{\left[2.91k^2\int_{h_0}^H C_n^2(h)(h-h_0)^{\frac{5}{3}}dh\right]^{\frac{3}{5}}}, \tag{7.137}$$

where h is the height above the ground. Using Eq. (7.136) and setting $h_0 = 0$, we can rewrite Eq. (7.137) in the form suggested by Quirrenbachl; namely, we have

$$\theta_0 = \frac{0.314 r_0 \cos(\zeta)}{H} \tag{7.138}$$

with

$$H = \left(\frac{\int_0^H C_n^2(h) h^{\frac{5}{3}} dh}{\int_0^H C_n^2(h)\, dh} \right)^{\frac{3}{5}} \tag{7.139}$$

[26]. The parameter H is called the *mean effective turbulence height*. The factor in the integrals contained in Eqs. (7.137) and (7.139) imply that the isoplanatic angle is affected mostly by high-altitude turbulence [26]. At $\lambda = 0.5\ \mu m$, θ_0 is roughly 7 – 10 μrad for near vertical paths from the Earth to space.

The *Greenwood Frequency*, f_g, is the characteristic frequency of atmospheric statistical change. Big turbules move around slowly and small turbules move around much more frequently, but on the average the entirety changes at this frequency. The time interval over which turbulence remains essentially unchanged is called the *Greenwood time constant*, τ_g, and it is the inverse of the Greenwood frequency. Mathematically, we have

$$\tau_g = \frac{\cos^{\frac{3}{5}}(\zeta)}{\left[2.91 k^2 \int_{h_0}^H C_n^2(h) V^{\frac{5}{3}}(h)\ dh\right]^{\frac{3}{5}}} = \frac{1}{f_g} \tag{7.140}$$

where $V(h)$ is vertical profile model for wind speed, which will be specified shortly. This time constant is typically on the order of milliseconds and the Greenwood frequency is in the 100 s of Hertz.

One parameter not often talked about, but is very useful in systems analysis, is the *Rytov Number*. It represents the log amplitude variance and is an indicator of the strength of turbulence along the integrated path. A companion to the Rytov Number is the variance of the log-intensity known as the *Rytov Variance*. It comes from one of the most widely used approximations for solving the wave equation, the Rytov approximation.[1]

For plane wave propagation, the Rytov number is given by

$$\sigma_\chi^2 = 0.56 k^{\frac{7}{6}} \int_0^L C_n^2(r)(L-r)^{\frac{5}{6}} dr \tag{7.141}$$

[1]The solution of the wave equation in this case presents the log amplitude χ and wave front phase φ as weighted linear sums of contributions from all random index fluctuations in the semi-infinite volume between the source and receiver. Two other popular approximations, the near field and Born approximations, are special cases of the Rytov approximation.

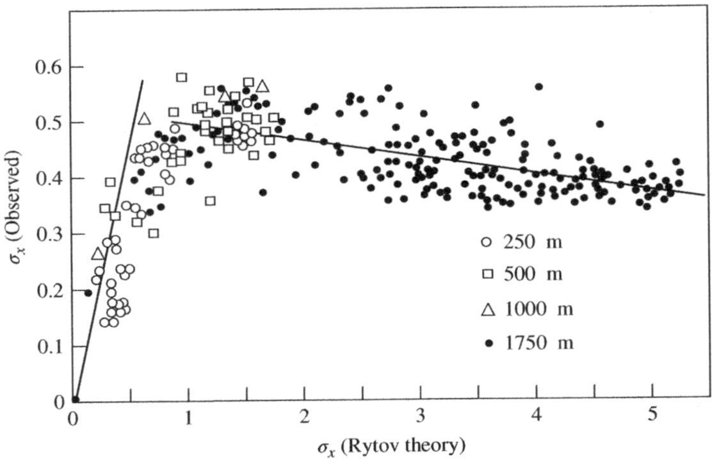

FIGURE 7.19 Observed scintillation versus predicted scintillation.

and for spherical wave propagation, we have

$$\sigma_\chi^2 = 0.56k^{\frac{7}{6}} \int_0^L C_n^2\left(\frac{r}{L}\right) r^{\frac{5}{6}}(L-r)^{\frac{5}{6}} dr. \tag{7.142}$$

In both these equations, L is the length of the propagation path where $\ell_0^2 \ll \lambda L \ll L_0^2$. These equations predict that the Rytov Number increases without limit as $C_n^2(r)$, or as the path length increases. In reality, this is not true and there is a limit. Figure 7.19 depicts observed scintillation versus predicted scintillation from Rytov theory [29, 30].

It is clear from this figure that the scintillation peaks when the Rytov Number reaches a value of 0.3. When the square root of the Rytov number is around 0.3 or less, little scintillation is to be expected. Above 0.3, scintillation (fading) is likely. When it exceeds 1.0, we are in the saturation regime where wave optics simulations often fail to converge and the high scintillation actually goes down slightly for increasing Rytov Number. Physically, we can see what is happening from Figure 7.20.

Referring to Figure 7.20a, when turbulence is mild [σ_χ(Rytov) < 0.3], the signal will experience wavefront tilt/tip from larger turbules (large r_0), which are microradian deflections causing minor scintillation. Under higher turbulence condition [$0.3 < \sigma_\chi$(Rytov) < 1.0], the signal beam begins to propagate through numerous small turbules (small r_0) which has the possibility of one, or more, of them deflecting their portion of the beam into the same array element. This is shown in Figure 7.20b. The scintillation increases, but still is in the linear regime.

When we have strong to severe turbulence, the situation in Figure 7.20b is exacerbated and many small beams are mixed over the receiver plane, causing the asymptotically decreasing scintillation shown for σ_χ(Rytov) > 1.

As noted in an earlier chapter, the Strehl ratio is a good measure of the quality of the incoming irradiance created by an optical system/channel. The concept was

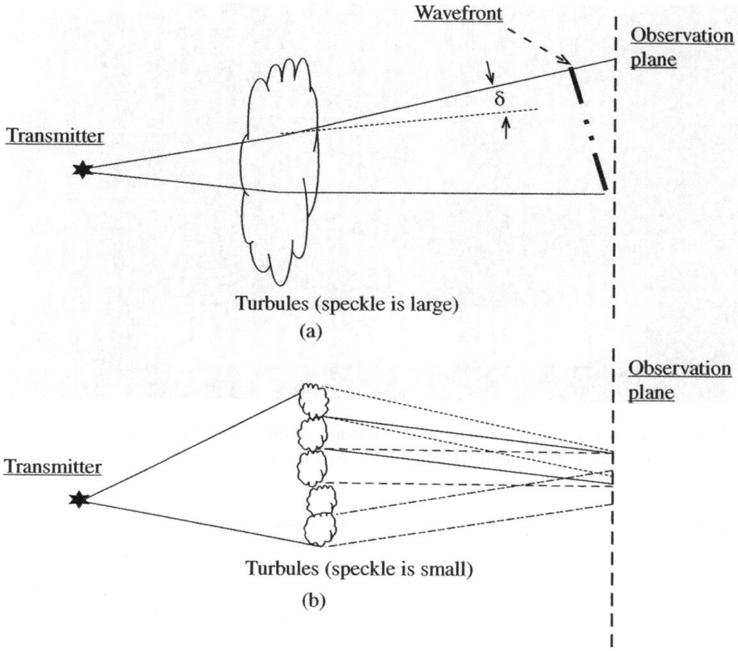

FIGURE 7.20 Idealistic example of turbulent scatter.

introduced by Strehl in 1902 [31] and amplified by Mahajan in the 1980s [32, 33]. Figure 7.21 shows an adaptive optics-corrected PSF after daytime turbulence degradation as compared to a perfect PSF [34]. This figure shows both image blurring and beam wander. The citation "TT" stands for Tip-Tilt only correction.

Dividing the normalized resolving power given in Eq. (7.133) by $\left(\frac{D^2}{r_0^2}\right)$, the resulting equation is the Strehl ratio expression developed by Andrews and Phillips:

$$\mathrm{SR} = \left(\frac{16}{\pi}\right)\int_0^1 \left[\cos^{-1}(u) - (u)\sqrt{1-(u)^2}\right] \exp\left\{-3.44\left(\frac{Du}{r_0}\right)^{\frac{5}{3}}\right\} u\, du \quad (7.143)$$

[19, p. 408]. To simplify future calculations, those authors showed that Eq. (7.143) is closely approximated by the following equation:

$$\mathrm{SR} \approx \left[1 + \left(\frac{D}{r_0}\right)^{\frac{5}{3}}\right]^{-\frac{6}{5}} \quad (7.144)$$

[19, 20]. Figure 7.22 depicts these two equations as a function of $\frac{D}{r_0}$ and clearly demonstrates their close agreement.

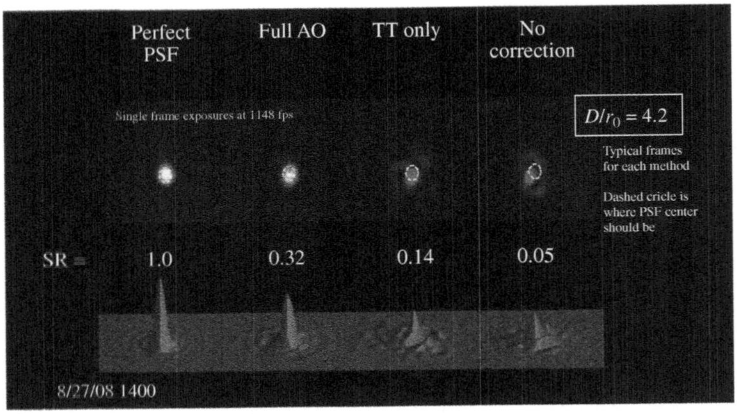

FIGURE 7.21 The AO-corrected PSF after turbulence degradation as compared to a perfect PSF. Source: Karp and Stotts, 2013 [34]. Reproduced with permission of IEEE.

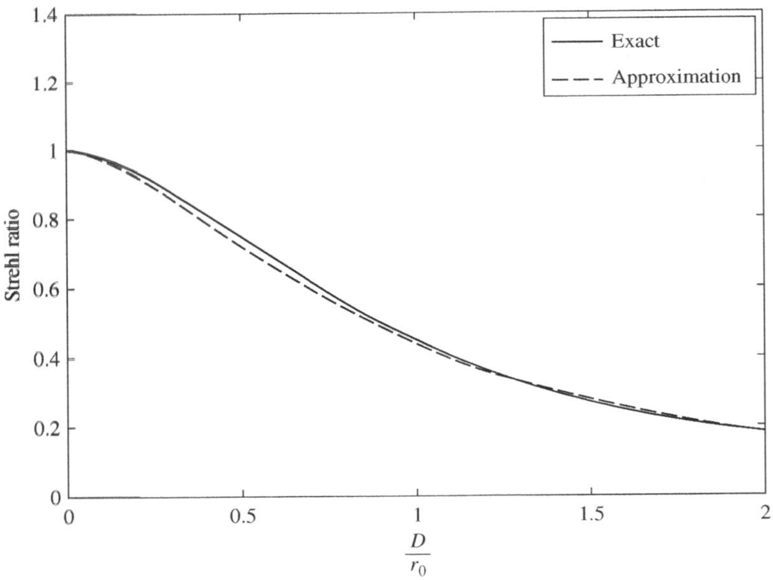

FIGURE 7.22 Comparison of exact Strehl ratio calculation for turbulence with Andrews and Phillips' approximate equation.

7.9 TWO REFRACTIVE INDEX STRUCTURE PARAMETER MODELS FOR THE EARTH'S ATMOSPHERE

One of the most accepted refractive index structure parameter models is the Hufnagel–Valley, or HV, $\frac{5}{7}$ model because of its traceability to statistical link availability [2, Chapter 10]. Here, the term $\frac{5}{7}$ means that for a wavelength of 0.5 μm, the value of 5 represents a Fried parameter of 5 cm and the value of 7 represents an isoplanatic angle for a receiver on the ground looking up of 7 μrad.

Mathematically, the HV $\frac{5}{7}$ profile is given by

$$C_n^2(h) = 0.00594\left(\frac{w}{27\ \text{m/s}}\right)^2 \left(10^{-5}\ \text{h}\right)^{10} e^{-\left(\frac{h}{1\,\text{km}}\right)} + 2.7\times 10^{-16} e^{-\left(\frac{h}{1.5\,\text{km}}\right)} + Ae^{-\left(\frac{h}{100\,\text{m}}\right)}. \quad (7.145)$$

In the above, $A = C_n^2(0)$ is the ground level value of the structure parameter and w the RMS-wind speed given by the equation

$$w = \sqrt{\frac{1}{15\times 10^3}\int_{500}^{20,000} V_{\text{HV}}^2(h)dh} \quad (7.146)$$

$$= \sqrt{\frac{1}{15\times 10^3}\int_{500}^{20,000} \left[w_s h + w_g + 30e^{\left\{-\left[\frac{h-9400\,\text{m}}{4800\,\text{m}}\right]^2\right\}}\right]^2 dh}, \quad (7.147)$$

where $V_{\text{HV}}^2(h)$ is the vertical profile for the wind speed, w_g the ground wind speed, and w_s the beam slew rate associated with a satellite moving with respect to an observer on the ground.

Figure 7.23 depicts the HV $C_n^2(h)$ profiles for two values of $C_n^2(0)$ $(= 1.7\times 10^{-14}\,\text{m}^{-\frac{2}{3}}$ for HV $\frac{5}{7}$) and three upper atmosphere wind pseudo-wind speeds

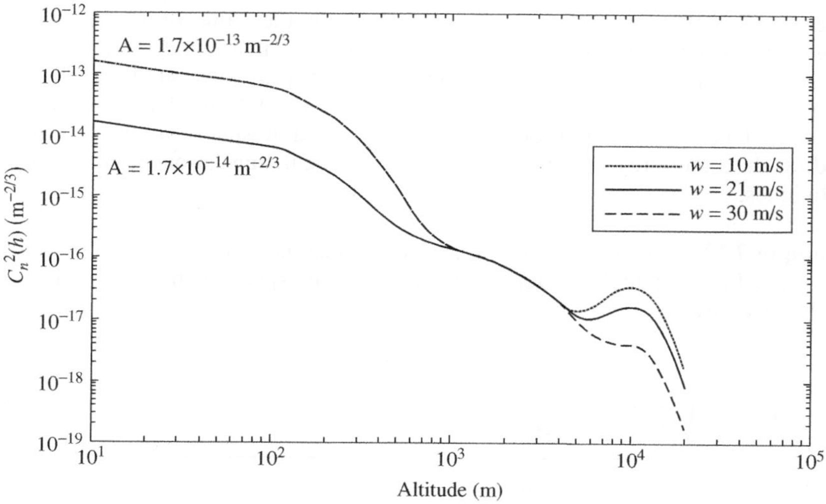

FIGURE 7.23 Vertical distribution of atmospheric turbulence C_n^2 for the HV model.

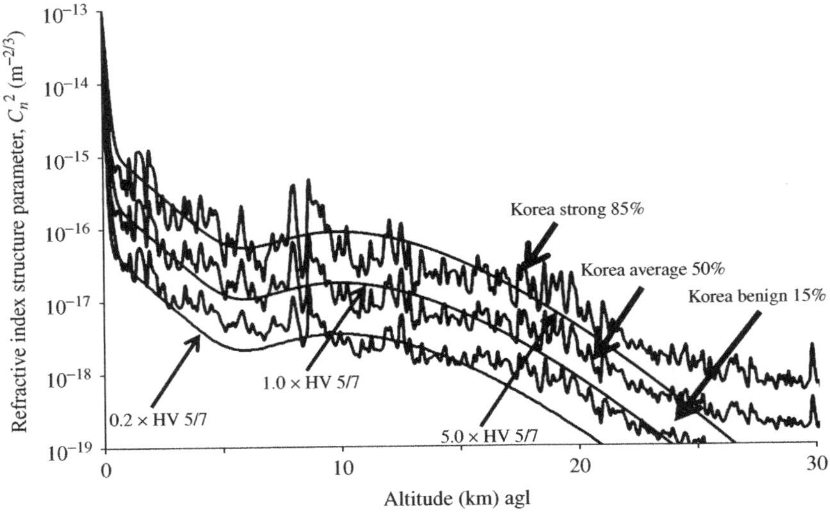

FIGURE 7.24 Multiples of HV 5/7 model compared to Korean turbulence statistics.

($w = 21$ m/s for HV $\frac{5}{7}$). It is clear in Figure 7.23 that the value of $C_n^2(0)$ affects the structure parameter below 1 km altitude and has no effect on it above that height. It is also clear that the value of the pseudo-wind speed has little effect on the structure parameter below 5 km altitude, but affects it above that height. Figure 7.24 compares multiples of HV $\frac{5}{7}$ model compared with annual Korean turbulence statistics. In this figure, we have 0.2, 1, and 5 × HV $\frac{5}{7}$ model values plotted against turbulence occurrence statistics of 15%, 50%, and 85%, respectively (the percentages in the legend reflect the amount of time during the year the value of $C_n^2(h)$ occurred). This figure shows an emerging convention, which is that some researchers characterize the refractive index structure parameter in terms of a multiple of the HV $\frac{5}{7}$ model rather than in varying $C_n^2(0)$ and w in the HV model as noted above [35, 36]. One needs to pay attention to how the atmosphere is characterized in reading the literature as there is not a mathematical translation between the two techniques.

Example 7.12 It should be noted that some researchers feel that above estimated wind speed w is too high and replace it with the wind speed estimated by the Bufton model given by

$$w_{\text{Bufton}} = \sqrt{\frac{1}{15 \times 10^3} \int_{500}^{20,000} V_{\text{Bufton}}^2(h)dh} \tag{7.148}$$

$$= \sqrt{\frac{1}{15 \times 10^3} \int_{500}^{20,000} \left[w_s h + w_g + 20e^{\left\{ -\left[\frac{h - 12,000\,\text{m}}{6000\,\text{m}} \right]^2 \right\}} \right]^2 dh}, \tag{7.149}$$

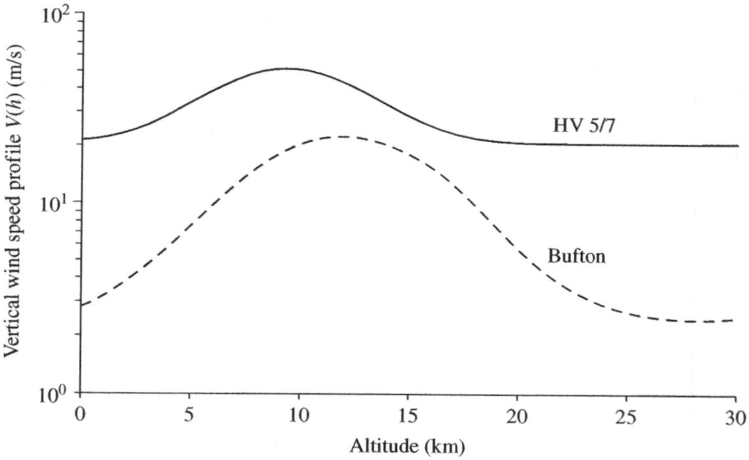

FIGURE 7.25 Comparison of HV and Bufton wind speed profiles as a function of altitude.

where $V^2_{\text{Bufton}}(h)$ is the Bufton vertical profile for the wind speed and $w_g = 2.5\ \text{m/s}$. Figure 7.25 compares the HV $\frac{5}{7}$ wind model with the Bufton Model, which clearly shows the latter's smaller wind profile. Although the former is generally used, the above is provided for completeness.

It has long been recognized that the advantage of the HV model over other atmospheric models is its inclusion of the wind speed, w, and the ground level refractive index structure parameter, $C_n^2(0)$ Their inclusion permits variations in high-altitude wind speed and local near-ground turbulence conditions to better model real-world profiles over a large range of geographic locations. It also provides a model consistent with measurements of the transverse coherence length or Fried parameter, r_0, and the isoplanatic angle, θ_0. However, Andrews and Phillips noted that the last exponential term in HV model describes the near-ground turbulence conditions that slowly decrease in $C_n^2(h)$ with altitude up to approximately 1 km [37]. This is in conflict with the $C_n^2(0)$ behavior of $h^{-\frac{4}{3}}$ noted by Walters and Kunkel, and supported by a number of other early measurements [38]. The result is that Andrews and Phillips modified the HV to reflect the power law dependence at low altitude. This new model is called the Hufnagel–Andrews–Phillips (HAP) model, and is given by

$$C_n^2(h) = \text{M}\left[0.00594\left(\frac{w}{27\,\text{m/s}}\right)^2 \left(10^{-5}[h+h_s]\right)^{10} e^{-\left(\frac{[h+h_s]}{1000\,\text{m}}\right)} + 2.7\times 10^{-16} e^{-\left(\frac{[h+h_s]}{1500\,\text{m}}\right)} + C_n^2(0)\left[\frac{h_0}{h}\right]^{-\frac{4}{3}}\right] \quad (7.150)$$

for $h > h_0$ [37]. Comparing Eqs. (7.145) and (7.150), we see that (i) the last exponential in the former equation has been replaced in the latter by a term that reflects

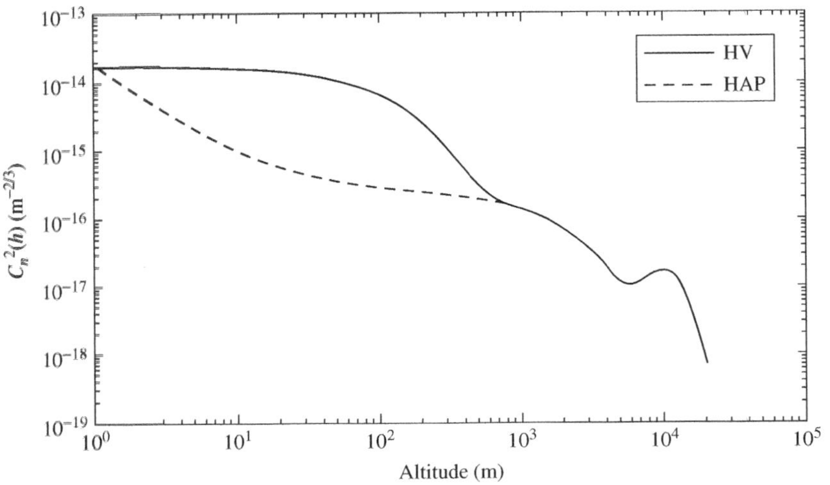

FIGURE 7.26 Comparison of Hufnagel–Valley (HV) and Hufnagel–Andrews–Phillips (HAP) models.

the observed behavior by Walters and Kunkel, (ii) a reference height of the ground above sea level, h_s, has been added, and (iii) a scaling factor, M, also added to represent the strength of the average high altitude background turbulence. Figure 7.26 compares the HV and HAP models for $h_s = 0$ m, $h_0 = 1$ m, $M = 1$, $w = 21$ m/s, and $C_n^2(0) = 1.7 \times 10^{-14}$ $\text{m}^{-\frac{2}{3}}$. For the first few hundred meters, there is considerable difference between the models. At approximately 1 km and higher altitudes, the models are essentially the same.

As a final comment on the HAP model, because the power law behavior as a function of altitude changes from $h^{-\frac{4}{3}}$ during the day to $h^{-\frac{2}{3}}$ at night, it is clear that there must be a transition period between day and night Andrews and his colleagues recently have developed a transition model that varies like h^{-p}, where p is dependent upon the temporal hour of the day. A temporal hour is defined as $\frac{1}{12}$ the number of hours between sunrise and sunset. This more general model therefore makes use of the actual sunrise and sunset times, and particular time of day under which experiments are performed, to determine the value of p.

7.10 ENGINEERING EQUATIONS FOR LIGHT PROPAGATION IN THE OCEAN AND CLOUDS

In Section 7.3, we introduced the radiative transfer process for light traveling through absorptive particulates. This discussion introduced the inherent and apparent properties of the optical channel. In this section, we expand this discussion, focusing on the detailed effects on light propagation through the atmospheric and maritime scatter channels, and their engineering equation modeling.

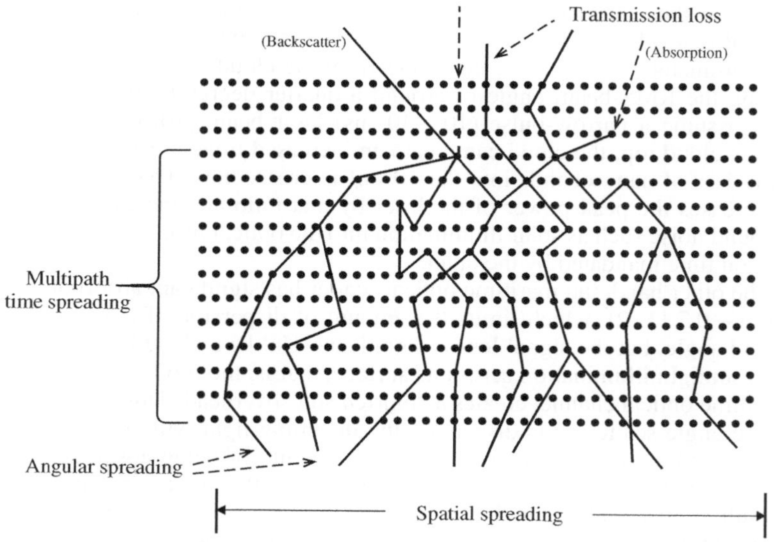

FIGURE 7.27 Various light beam effects from particulate scattering process.

The four major effects that a particulate medium has on optical beams are as follows:

- Angular spreading
- Spatial spreading
- Temporal spreading
- Transmission loss (including absorption and backscatter).

We illustrated these effects graphically in Figure 7.27. For most system applications requiring high link availability, one must consider clouds with large optical thickness, $\tau = cz$, on the order of 100 or more, and at depths where both regimes are in the diffusion regimes of scattering. Bucher [39] and Lee *et al.* [G.M. Lee, C. M. Ciany and C. Tranchita, McDonnell-Douglas Astronautics, private communications] provided a very useful set of engineering models for characterizing the above effects for clouds and water using Monte Carlo simulations. These equations are very useful in creating link budgets for the optical communications. The angular distribution of light emerging from the bottom, or top, of a cloud, or from the ocean surface, follows essentially a Lambertian distribution. Let us now turn to some general observation for light propagation in clouds and the ocean. Each has its own unique characteristic.

The atmospheric optical scatter channel has negligible absorption (single scatter albedo $\omega_0 = \frac{b}{c} > 0.99$ [G.M. Lee, C. M. Ciany and C. Tranchita, McDonnell-Douglas Astronautics, private communications], dispersing the signal in time space, time, and angle. Assume that we are dealing with solar illumination of the Earth for the moment. On the stormiest day, the maximum solar irradiance loss is essentially only 10 dB or so. That is the good news. Scattered radiation is available for possible use. Unfortunately, the radiance emerging from the cloud is Lambertian in nature, increasing the RMS-beam angle of around a quarter degree to 10 s of degrees. If we now assume a narrow-pulse (10 − 40 ns) laser beam projecting a large spot on a thick cloud top, the cloud can create an increased time pulse spreading, going from the 10 s of nanoseconds into the 10 s of microseconds. That is the bad news. This decreases the peak power of the laser by that ratio and increases the external background noise such as solar illumination into the optical receiver by the increased integration time mandated by the latter.

On the other hand, the maritime optical scatter has significant absorption, that is, $\omega_0 \sim 0.6 - 0.7$ [1, 2], which limits the amount of dispersion of the signal in time space and angle, but the signal loss is significant. Unlike a cloud layer that exhibits uniform sunlight illumination across its surface, that is, the profile of the sun is lost, the maritime optical channel creates a "blurred sun" overhead. This is because while that large angle scattering is dispersing the incoming light within the channel, the same scattered light also has an increased probability of being absorbed in current and future interactions with particles before it hits the optical receiver aperture. The result is that only light that is scattered a few times, mostly down, makes it to the optical receiver aperture with the aforementioned radiance distribution (Fermat's Principle) [1, 2]. We will look at light propagation in ocean and lake water first.

As we noted earlier, beam attenuation in a particulate medium involves the absorption and scattering of light from the interaction with the particulates with the source beam. Duntley investigated the angular distribution of the source radiance at different distances from a uniform, spherical underwater lamp in 1963 [9]. Figure 7.28 shows the results of his experimentation at the Diamond Island Field Station. This figure shows that the resulting distribution of light as a combination of unscattered and scattered radiance. The unscattered light is given by the expected equation

$$N_{\text{unscat}}(r) = \frac{e^{-cr}}{r^2}. \tag{7.151}$$

On the other hand, the latter shows a pronounced forward scattering angular distribution profile that is common to natural seawater, namely, a "glow" around the unscattered light distribution.

The apparent radiance is exponentially attenuated by a particulate medium:

$$N(r) = N_0 e^{-\alpha r}. \tag{7.152}$$

$$H(r) = J\left(\frac{e^{-Kr}}{r^2}\right) \approx J\left(\frac{e^{-kr}}{r^2}\right). \tag{7.153}$$

The unscattered irradiance, $H(r)$, produced on a surface at normal incidence by a radiant intensity, $J(r)$, at distance r is attenuated exponentially. Diffusion theory (based on isotropic scattering) suggests that the scattered irradiance, $H_s(r)$, produced on a surface from the unscattered irradiance at a distance r is given by

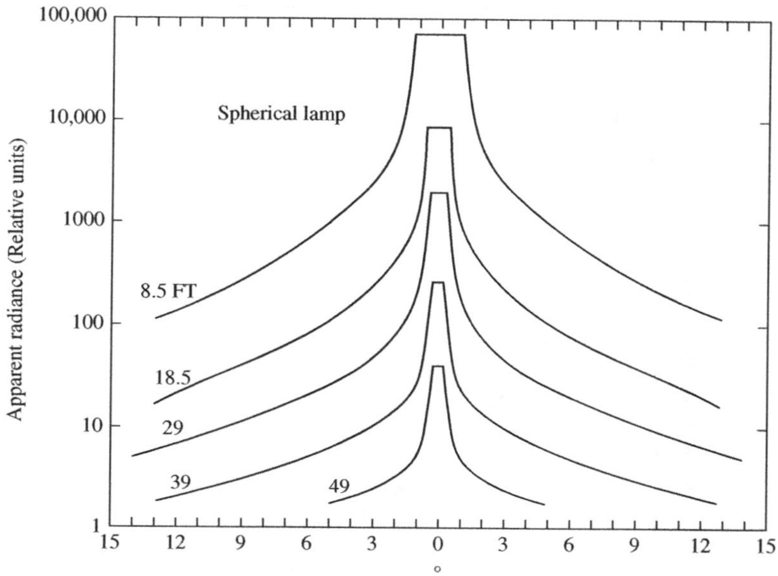

FIGURE 7.28 Angular distribution of the apparent radiance produced by a uniform, spherical underwater lamp at distance of 7.5, 17.5, 29, and 39 ft [9, Figure 15]. Source: Duntley, 1963 [9]. Reproduced with permission of The Optical Society of America.

$$H_s(r) = J\,K\left(\frac{e^{-Kr}}{4\pi r^2}\right) \approx Jk\left(\frac{e^{-kr}}{4\pi r^2}\right). \tag{7.154}$$

Let us look at some more data.

Figure 7.29 shows some additional measurements taken by Duntley at the Diamond Island Field station [9]. In this figure, the sum of Eqs. (7.126) and (7.128) fits data from 10 – to 70 ft lamp distances. For lamp distances beyond 70 ft, Eq. (7.128) must be modified to read:

$$H_s(r) = 2.5(1 + 7e^{-Kr})JK\left(\frac{e^{-Kr}}{4\pi r^2}\right) \tag{7.155}$$

in order for the theory and data to agree within experimental error.

We saw in Section 7.3 that Tyler's data indicates the measured radiance distribution takes on a diffusive nature at sufficient depth and greater. In other words, the radiance's angular distribution is the same no matter what the depth, but the total power falls off exponentially.

Figure 7.30 exhibits in-water lake radiance distributions in the plane of the Sun obtained experimental data, showing this effect. This is known as the "manhole"

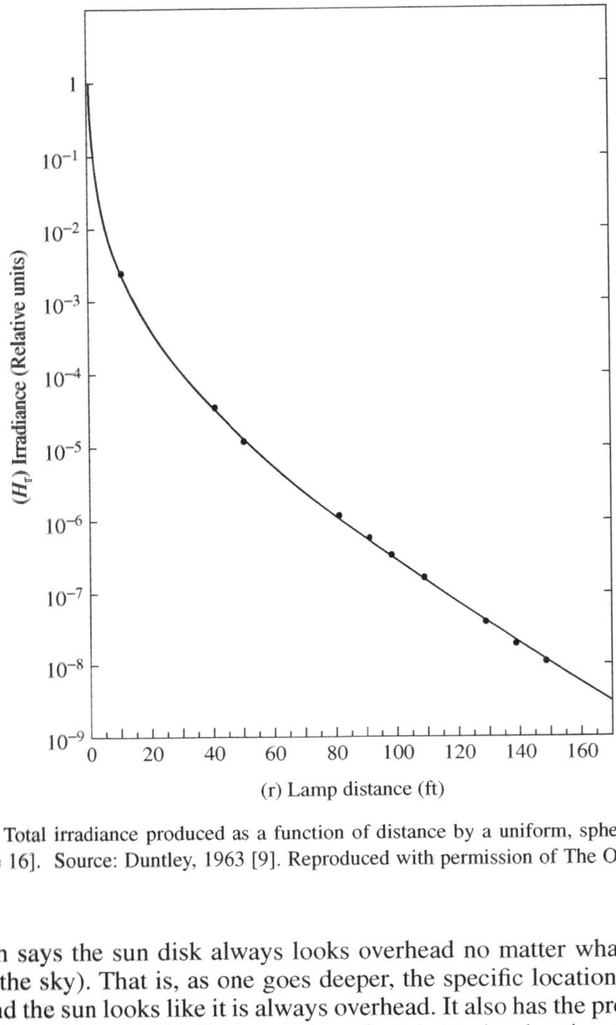

FIGURE 7.29 Total irradiance produced as a function of distance by a uniform, spherical underwater lamp [9, Figure 16]. Source: Duntley, 1963 [9]. Reproduced with permission of The Optical Society of America.

effect, which says the sun disk always looks overhead no matter what time of day (position in the sky). That is, as one goes deeper, the specific location of the sun in sky is lost and the sun looks like it is always overhead. It also has the previously cited asymptotic shape. Figure 7.31 illustrates these facts better by showing these radiance profiles normalizing their absolute power to the same value. Let us look at all this mathematically.

In an unbounded medium with uniform scattering and absorption properties, the asymptotic radiance distribution for plane wave illumination can be shown to be given by

$$N_{\infty}(u) \approx S_1 N(u) e^{-K_d z} + S_2 N(u) e^{+K_u z}, \tag{7.156}$$

where $u = \cos\theta$, $S_1 \equiv$ "diffusion stream" in positive z-direction (down), $S_2 \equiv$ "diffusion stream" in negative z-direction (up), $K_d \equiv$ downwelling irradiance attenuation coefficient, $K_u \equiv$ upwelling irradiance attenuation coefficient and $N_{\infty}(u) \equiv$ asymptotic diffusion pattern (derived from the medium's scalar phase function, $\frac{P(\theta)}{4\pi}$, and its single-scatter albedo, $\omega_0 = b/c$.

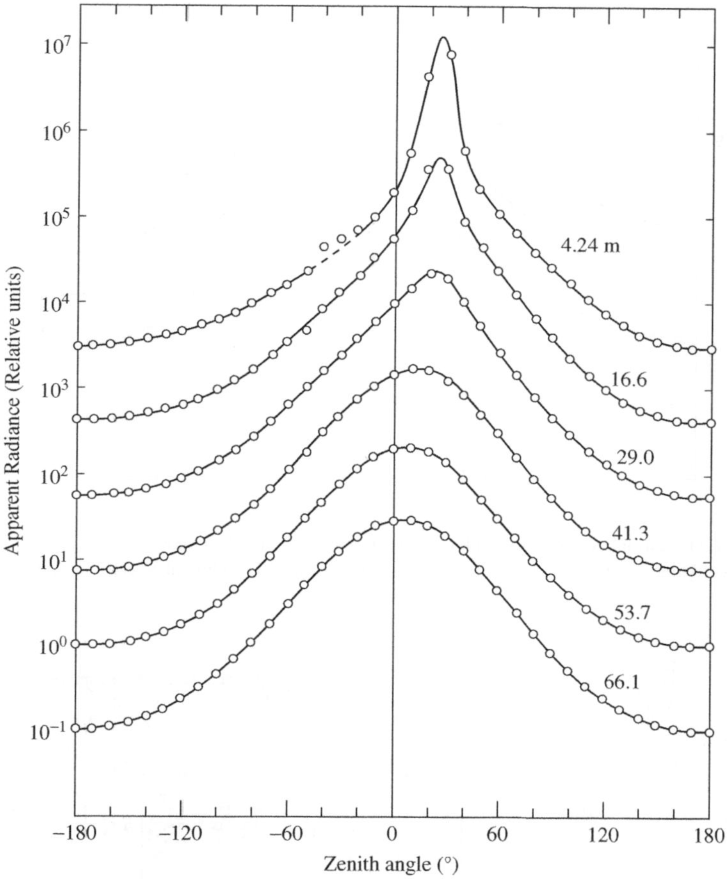

FIGURE 7.30 In-water radiance distributions in the plane of the Sun obtained experimental data [9, Figure 26]. Source: Duntley, 1963 [9]. Reproduced with permission of The Optical Society of America.

This solution can be derived from the equation

$$(1 - ku)N(u) = -\int \int_{-1}^{+1} h(u, v)N(v)dv, \tag{7.157}$$

where $h(u, v)$ is redistribution function defined as the average over the azimuth of the medium's scalar phase function. Here, k is the smallest eigenvalue, and $N(u)$, its corresponding eigenfunction, both derivable from Eq. (7.157). From the above discussion, it is clear that

$$K_d = k = K_u$$

in the asymptotic limit. In optical oceanography, k is called the diffuse attenuation coefficient. Note that this loss rate relates to spherical beams that approximate a plane

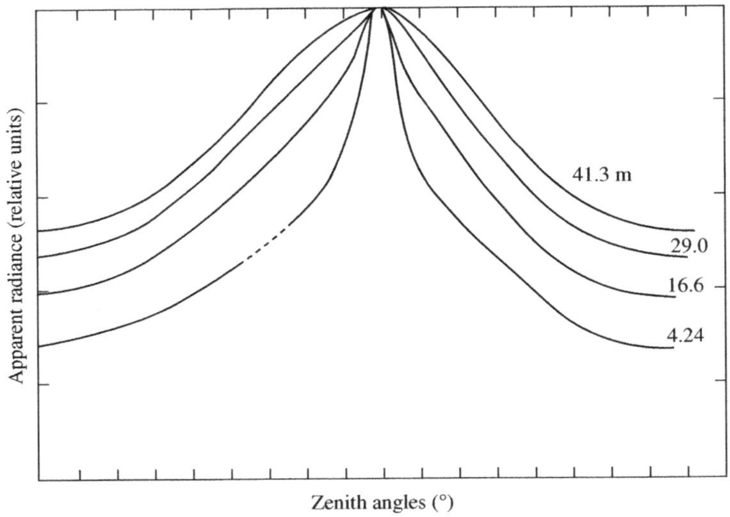

FIGURE 7.31 Normalized in-water radiance distributions for the data shown in Figure 7.27 [9, Figure 27]. Source: Duntley, 1963 [9]. Reproduced with permission of The Optical Society of America.

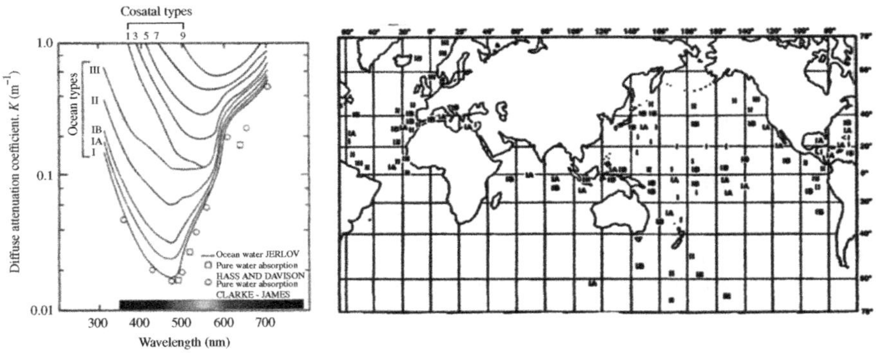

FIGURE 7.32 Diffuse attenuation coefficient for the waters of the world (Jerlov water types) [1]. Source: Karp *et al.*, 1987 [1]. Reproduced with permission of Springer Science + Business Media.

wave like the sunlight does hitting the surface of the Earth, that is, no spatial spreading is negligible. It cannot be used to characterize transmission loss for narrow light beams.

The most popular model for describing diffuse attenuation coefficient is the Jerlov water types shown in Figure 7.32, which shows both a graph for the function of wavelength and a world map where these water types are located. The oceanic Jerlov water types are as follows: Type 1A is the cleanest (Sargasso Sea); Type 1B is the next cleanest (Hawaii and Caribbean); Type II become less clean as it has more chlorophyll-based critters than IB (San Diego); Type III has more

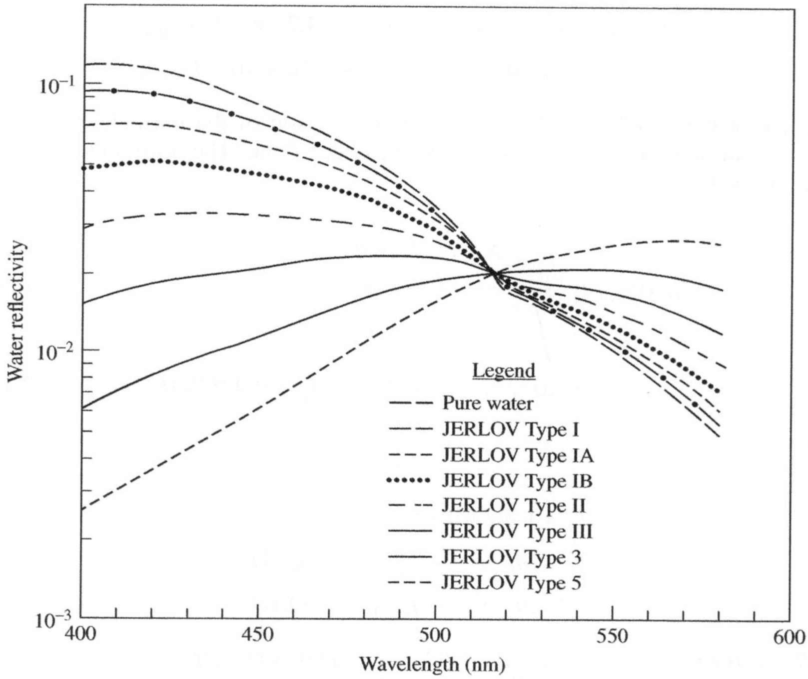

FIGURE 7.33 Water reflectivity as a function of wavelength for the Jerlov water types [1]. Source: Karp *et al.*, 1987 [1]. Reproduced with permission of Springer Science + Business Media.

critters (Puget Sound); and the coastal water are the dirtiest with a lot of things like earth runoff, and so on. Figure 7.33 shows the water reflectivity ρ as a function of wavelength for the Jerlov ocean and coastal water types, and pure water, as devised by Petzold [40]. It is apparent from this figure that the clearest waters have the highest reflectivity for the ocean's best transmission wavelengths and the dirtiest waters, the lowest (Hawaii vice, Seattle). To first order, the water reflectivity ρ is inversely proportional to diffuse attenuation coefficient k.

Lee *et al.* investigated optical propagation in the diffusive regime of seawater using Monte Carlo techniques to generalize the solution to real seawater environments and solar sun angles (Lee *et al.*, McDonnell-Douglas Astronautics, private communications). Using typical Jerlov IB/II water characteristics and forward-biased volume scattering functions, they found that angular spreading reaches an asymptotic state because of the increased probabilities of light being absorbed from the additional path incurred from the particulate scattering process. This asymptotic radiance model for these environments is given by

$$N_{\infty}(\theta) \approx N_0 F_w(\theta_{\text{solar}}) \frac{e^{-\left(\frac{\theta^2}{\sigma^2}\right)}}{\pi\sigma^2}, \tag{7.158}$$

where $\sigma = 0.6537$ rad ($37.8°$) and

$$F_w(\theta_{\text{solar}}) = 1 - 9.72 \times 10^{-4}\theta_{\text{solar}} - 4.28 \times 10^{-4}\theta_{\text{solar}}{}^2$$
$$+6.04 \times 10^{-6}\theta_{\text{solar}}{}^3 - 4.28 \times 10^{-8}\theta_{\text{solar}}{}^4. \tag{7.159}$$

To check the validity of Eq. (7.158), a comparison of the normalized received power for an underwater receiver at depth has been done. The normalized received power is given by

$$f_{sg}(\theta_{\text{fov}}) = \frac{\int_0^{\theta_{\text{fov}}} N_\infty(\theta')d\cos\theta'}{\int_0^{\frac{\pi}{2}} N_\infty(\theta')d\cos\theta'} \tag{7.160}$$

$$= 1.077311I_0 - 0.178888I_1 + 0.008198I_2, \tag{7.161}$$

where

$$I_0 = 1 - e^{\frac{\theta_{fov}^2}{(0.43525)^2}}, \tag{7.162a}$$

$$I_1 = (\theta_{\text{fov}})^2(1 - I_0) + 0.435204I_0, \tag{7.162b}$$

$$I_2 = (\theta_{fov})^4(1 - I_0) + 0.435204I_1, \tag{7.162c}$$

and θ_{fov} is the receiver field-of-view (FOV) half angle in water.

Figure 7.34 compares normalized received power from an oceanic field experiment with Eq. (7.133) as function of receiver FOV half angle [1, 2]. Figure 7.35 compares experimental lake data from Baker and Smith [42] as a function of depth and solar

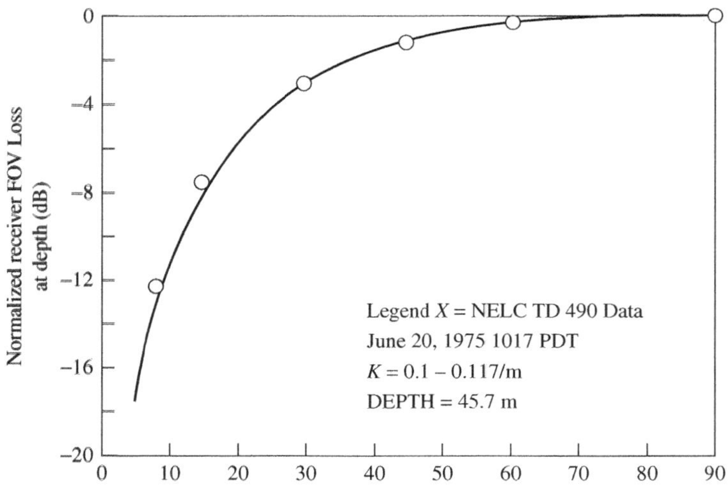

FIGURE 7.34 Comparison of normalized receiver field-of-view loss with experimental data [1]. Source: Karp *et al.*, 1987 [1]. Reproduced with permission of Springer Science + Business Media.

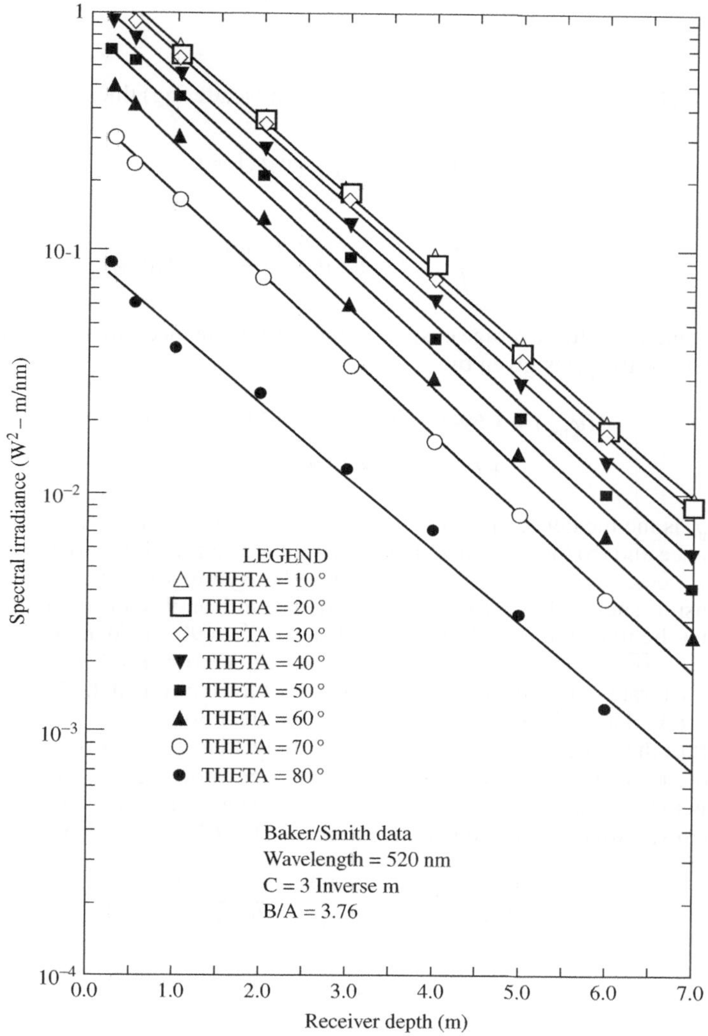

FIGURE 7.35 Comparison of Monte Carlo-based radiance model with experimental data [41]. Source: Figure published with permission from the SPIE.

incidence angle against predictions using Eq. (7.159). We see that both figures show very good fit between data and theory.

Expressions for cloud transmittance and spatial spreading were developed by Danielson *et al.* [43] and van de Hulst [44], validated by Bucher [39]. The specific equations are given by

$$L_T = \frac{A(\theta_i)}{1.42 + \tau_d} \tag{7.163}$$

and

$$r = 0.92z\tau_d^{-0.08}, \tag{7.164}$$

respectively, for $\tau_d > 3$. In the above, the diffusion (scattering) thickness τ_d equals

$$\tau_d = bz\left(1 - \overline{\cos\theta}\right) = \omega_0\tau(1 - g), \tag{7.165}$$

where

$$g = \overline{\cos\theta} = \int_0^{2\pi}\int_0^{\pi} \cos\theta \frac{P(\theta,\ \varphi)}{4\pi} \sin\theta d\theta d\varphi \tag{7.166}$$

is the asymmetry factor (average cosine) of the particulate medium and the incidence angle function $A(\varphi_{\text{inc}})$ is given by

$$\begin{aligned} A(\varphi_{\text{inc}}) &= 1.69–0.5513\varphi_i + 2.7173\varphi_i^2–6.9866\varphi_i^3 \\ &+7.1446\varphi_i^4–3.4249\varphi_i^5 + 0.6155\varphi_i^6, \end{aligned} \tag{7.167}$$

where φ_{inc} is the incidence angle on cloud layer. Figure 7.36 illustrates the (a) cloud transmittance and (b) the normalized average radius as a function of optical and scattering thicknesses, respectively. In these figures, the asymmetry factor is 0.875 and the cloud transmittance is shown for various incident zenith angles. From Figure 7.36(a), we see that the transmittance decreases 6 dB from 0° to 90° incident angles, as well as follows $\sqrt{\cos\varphi_{\text{inc}}}$ for $\varphi_{\text{inc}} < 85°$. The form of Eq. (7.164) implies that the spatial spreading saturates and even decreases as a cloud of fixed physical thickness becomes more optically dense. This is clearly shown in Figure 7.36b.

As part of the aforementioned Monte Carlo study, Lee *et al.* investigated the effect of absorption on the plane-wave multiple-scattering processes in both clouds and ocean water (Lee *et al.*, McDonnell-Douglas Astronautics, private communications). Their first result was that the cloud transmittance in Eq. (7.163) needed to be modified as follows:

$$L_T = \frac{A(\theta_i)}{1.42 + \tau_d}\left[\frac{2k'(\tau + g_1)\exp\{-k'(\tau + g_1)\}}{1 - \exp\{-k'(\tau + g_1)\}}\right], \tag{7.168}$$

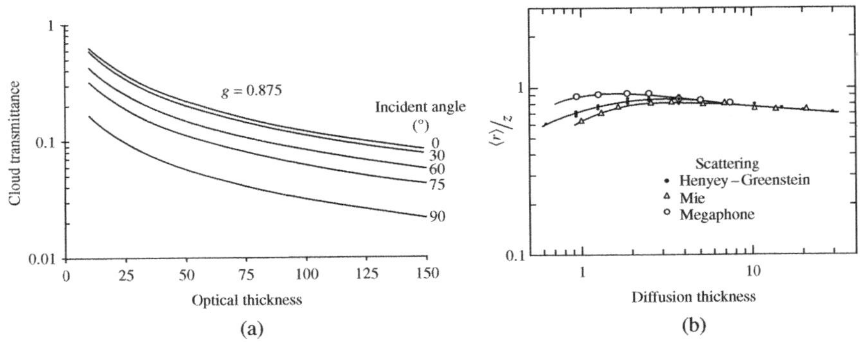

FIGURE 7.36 Graphs of (a) channel transmittance and (b) normalized spatial spread as a function of optical and diffusion thickness, respectively.

where

$$k' = \sqrt{3(1 - g_1)(1 - \omega_0)} \tag{7.169}$$

and

$$g_1 = \frac{1.42}{(1 - g)}. \tag{7.170}$$

In addition, they investigated the effect of particulate scattering on laser pulse propagation and developed the following equation for multipath time spreading:

$$\Delta t = \frac{1}{c_0} \frac{0.58D\left(\frac{\tau_d^{1.5}}{(1-g)}\right)}{0.49 + (1 - \omega_0)\left(\frac{\tau_d^{1.1}}{(1-g)}\right)}, \tag{7.171}$$

where c_0 is the speed of light (=299,792.458 km/s or $\sim 3.0 \times 10^8$ m/s) and the parameter $D\left(=\frac{1}{b}\right)$ the scattering length of the medium (Lee *et al.*, McDonnell-Douglas Astronautics, private communications).

Figure 7.37 depicts cloud transmittance and temporal spreading as function of optical thickness for selected single scatter albedos. It is clear that a little absorption can significantly affect transmission loss for optical thicknesses greater than an optical thickness of 10. It gets much lower for smaller values of ω_0 as the optical thickness increases. Similarly, the temporal spreading is affected significantly for optical thickness greater than 20 as the absorption in the cloud increases.

Arnush [45] showed that spatial spreading in water for beam increases as the depth-cubed, which Stotts [46, 47] showed was valid when in diffusion regime, creating a large loss mechanisms relative to free space and clouds. Mathematically, this

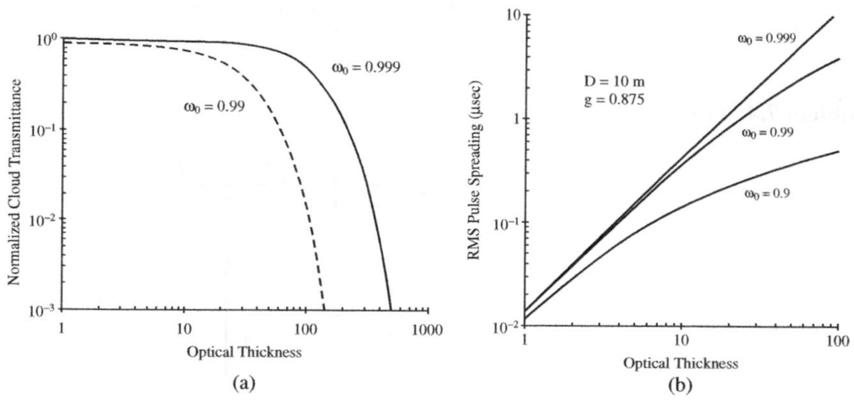

FIGURE 7.37 Graphs of (a) the normalized transmittance of absorptive clouds as function of optical thickness for two single scatter albedos and (b) the RMS pulse spreading in absorptive clouds as function of optical thickness for various single scatter albedos[1]. Source: Karp *et al.*, 1987 [1]. Reproduced with permission of Springer Science + Business Media.

spreading is given by

$$\bar{r} \approx \sqrt{\frac{2}{3}\omega_0 \tau z^2 \gamma_0^2}, \tag{7.172}$$

where γ_0 is the RMS scatter angle of the particulate medium.

7.11 PROBLEMS

Problem 7.1. Given the Koschmieder equation, calculate the atmospheric extinction coefficient for visibilities of 5, 10, 15, 23, 50, and 100 km. Plot the atmospheric loss for the range of 0 – 200 km. Assume the wavelength of interest is 1.55 μm.

Problem 7.2. Assume a horizontal viewing geometry. If the visibility is 35 km, what is the Koschmieder extinction coefficient if we assume the wavelength of interest is 0.55 μm? Given that extinction coefficient, what is the contrast of an object located 12 km away?

Problem 7.3. Given

$$G_{SW}(\rho; z_1) = \frac{1}{(4\pi z_1)^2} \exp \times \left\{ \frac{ik}{z_1} \boldsymbol{p} \cdot \boldsymbol{r} - 4\pi^2 k^2 z_1 \int_0^1 \int_0^\infty \kappa \Phi_n(\kappa)[1 - J_0(\kappa \zeta \rho)] d\kappa d\zeta \right\},$$

assuming a von Karman spectrum, show that for small arguments, we have

$$G_{SW}(\rho; z_1) \approx \frac{1}{(4\pi z_1)^2} \exp \left\{ \frac{ik}{z_1} \boldsymbol{p} \cdot \boldsymbol{r} - 0.55 k^2 C_n^2 z_1 \rho^{\frac{5}{3}} \right\}.$$

Problem 7.4. Given

$$F(x, y; z) = \exp \left\{ i \left[kz + \frac{\pi}{2} \right] + i\pi \frac{x^2 + y^2}{\lambda z} \right\} \times \sqrt{\frac{P}{\pi w_0^2}} \left(\frac{1}{2\left(1 - \frac{i\lambda z}{\pi w_0^2}\right)} \right) e^{-\frac{k^2 w_0^2 r^2}{4z^2\left(1 - \frac{\pi i w_0^2}{\lambda z}\right)}},$$

show that

$$G_{\text{trans}}(\rho) = \frac{1}{2} e^{-\frac{\rho^2}{2w_0^2}\left(1 + \frac{\pi^2 w_0^4}{(\lambda z)^2}\right)}.$$

Problem 7.5. Calculate

$$\langle I(\mathbf{r};z)\rangle = \frac{k^2P}{(2\pi z)^2}\int\int_{-\infty}^{\infty} G_{\text{trans}}(\rho)G_{\text{turb}}(\rho)e^{-i\frac{k}{z}\boldsymbol{\rho}\cdot\mathbf{r}}d^2\boldsymbol{\rho},$$

assuming that

$$G_{\text{turb}}(\rho) \approx \begin{cases} 1; & z \ll z_c \\ e^{-\left(\frac{\rho}{\rho_t}\right)^2}; & z \gg z_c \end{cases}.$$

Problem 7.6. Assume that we have a horizontal 10 km laser link at height h_0 above the ground where the Fried parameters for the air and ground terminals are given by

$$r_{0\text{ air}} = \left[16.71\sec(\zeta)\int_0^L C_n^2(r)\left(\frac{r}{R}\right)^{\frac{5}{3}}\frac{dr}{\lambda^2}\right]^{-\frac{3}{5}}$$

and

$$r_{0\text{ gnd}} = \left[16.71\sec(\zeta)\int_0^L C_n^2(r)\left(1-\frac{r}{R}\right)^{\frac{5}{3}}\frac{dr}{\lambda^2}\right]^{-\frac{3}{5}},$$

respectively. Assume that the refractive index structure parameter at altitude is

$$C_n^2(h_0) = 3.0\times 10^{-14}\ \ \text{m}^{-\frac{2}{3}}.$$

Let the wavelength equal 1.55 μm and the air and ground receiver aperture diameters be 10 cm. What is the value of the two Fried parameters? What is the approximate value of the two Strehl ratios?

Problem 7.7. Graphically show that $\sqrt{\cos\varphi_{\text{inc}}}$ approximates Eq. (7.163) for φ_{inc}.

Problem 7.8. Using Eq. (7.167), plot the multitime spread for asymmetry factors of $g = 0.8, 0.9$, and 0.95 for the optical thickness range from 10 to 100. Assume that $c = 0.2286$, $b = 0.16$, and $a = 0.0686$. Explain the results.

Problem 7.9. Using Eq. (7.167), plot the multitime spread for single scatter albedos of $\omega_0 = 0.5, 0.65$, and 0.0.8 for the optical thickness range from 10 to 100. Assume that $g = 0.875$. Explain the results.

REFERENCES

1. Karp, S., Gagliardi, R.M., Moran, S.E., and Stotts, L.B. (1988) *Optical Channels: Fiber, Atmosphere, Water and Clouds*, Springer Science + Business Media (was Plenum Publishing Corporation), New York.
2. Karp, S. and Stotts, L.B. (2013) *Fundamentals of Electro-Optic Systems Design: Communications, Lidar, and Imaging*, Cambridge Press, New York.

3. Preisendorfer, R.W. (1976) *Hydrologic optics*vol. 5, in *Properties*, U.S. Government Printing Office, Boulder, Colorado.
4. Jerlov, N.G. (1976) *Marine Optics*, Elsevier, Amsterdam.
5. Pratt, W.K. (1969) *Laser Communications Systems*, John Wiley & Sons, Inc., New York.
6. Wolfe, W.L. and Zissis, G.J. (1978) *The Infrared Handbook*, The Infrared Information and Analysis Center of the Environmental Research Institute of Michigan for the Office of Naval Research, Washington, DC.
7. Elterman, L. (1968) UV, Visible and IR Attenuation for Altitudes up to 50 km. *Air Force Cambridge Research Laboratories Technical Report, Report Number AFRCL-67-0153.*
8. Driscoll, R.N., Martin, J.N., and Karp, S. (June 1, 1976) OPTSATCOM Field Measurements. Technical Document 490, Naval Electronic Laboratory Center.
9. Duntley, S.Q. (1963) Light in the sea. *J. Opt. Soc. Am.*, **53**, 214–233.
10. Smith, S.W. (1997) *The Scientist's and Engineer's Guide to Digital Signal Processing*, Special Imaging Techniques, California Technical Publishing, San Diego, CA.
11. The Map and Information is Taken from http://nature.nps.gov/air/monitoring/vismon results.cfm.
12. Lutomirski, R.F. (1978) Atmospheric degradation of electro-optical system performance. *SPIE Proc. Opt. Prop. Atmos.*, **142**, 120–129.
13. Diermendjian, D. (1969) *Electromagnetic Scattering on Spherical Polydispersions*, Elsevier, New York.
14. Wells, W.H. (1969) Loss of resolution in water as a result of multiple small-angle scattering. *J. Opt. Soc. Am.*, **59** (**6**), 686.
15. Lutomirski, R.F. (1975) Tech. Note PSR -N73, Pacific -Sierra Research Corp., May.
16. van de Hulst, H.C. (1957) *Light Scattering by Small Particles*, John Wiley & Sons, Inc., New York, p. 470.
17. McClatchey, R.A. (January, 1974) AFCRL Atmospheric Absorption Line Parameters Compilation. *Air Force Cambridge Research Laboratories Technical Report, AFCRL –TR-74-003.*
18. Majumdar, A. and Ricklin, J. (2007) *Free Space Laser Communications: Principles and Advances*, Optical and Fiber Communications Series, Springer Science + Business Media, New York.
19. Andrews, L.C. and Phillips, R.L. (2005) *Laser Beam Propagation Through Random Media*, 2nd edn, SPIE Press, Washington.
20. Andrews, L.C. (2004) *Field Guide to Atmospheric Optics*, SPIE Press.
21. Tatarski, V.I. (1961) *Wave Propagation in a Turbulent Atmosphere* translated by R.A. Silverman, McGraw-Hill, New York.
22. Lutomirski, R.F. and Yura, H.T. (1971) Wave structure function and mutual coherence function of an optical wave in a turbulent atmosphere. *J. Opt. Soc. Am.*, **61** (**4**), 482–487.
23. Gradshteyn, I.S. and Ryzhik, I.M. (1965) *Table of Integrals, Series and Products*, Fourth Edition prepared by Y.V. Geronimus and M.Y. Tseytlin, Translated by A. Jeffery edn, Academic Press, New York.
24. IDEX Optics and Photonics/CVI Laser Technology *Gaussian Beam Optics*, https://marketplace.idexop.com/store/SupportDocuments/All_About_Gaussian_Beam_Optics WEB.pdf.

25. Lutomirski, R.F. and Yura, H.T. (July 1971) Propagation of a finite optical beam in an inhomogeneous medium. *Appl. Opt.*, **10** (**7**), 1654.

26. Cassen, P., Guillot, T., and Quirrenbachl, A. (2006) The effects of atmospheric turbulence on astronomical observations, in *Extrasolar Planets*, vol. **31**, Swiss Society for Astrophysics and Astronomy, Saas-Fee Advanced Course, Chapter 7, Springer, New York, pp. 137–152, ISBN: 978-3-540-29216-6 (Print) 978-3-540-31470-7 (Online).

27. Fried, D.L. (January 1967) Optical heterodyne detection of an atmospherically distorted signal wavefront. *Proc. IEEE*, **55** (**1**), 57–67.

28. Fried, D.L. (1966) Optical resolution through a randomly inhomogeneous medium for very long and very short exposures. *J. Opt. Soc. Am.*, **56** (**10**), 1372–1379.

29. Beland, R.R. (1996) Propagation through atmospheric optical turbulencevol. 2, Chapter 2, in *Infrared and Electro-Optical Systems Handbook*, Environmental Research Institute of Michigan.

30. Gracheva, M.E. (1967) Research into the statistical properties of the strong fluctuation of light when propagated in the lower layers of the atmosphere. *Izv. Vuz. Radio fiz.* Moscow, USSR, **10**, 775–787.

31. Strehl, K. (1902) *Z. Instrumentenkd.*, **22**, 213.

32. Mahajan, V.N. (1982) Strehl ratio for primary aberrations: some analytical results for circular and annular pupils. *J. Opt. Soc. Am.*, **72**, 1257–1266.

33. Mahajan, V.N. (1983) Strehl ratio for primary aberrations in term of their aberration variance. *J. Opt. Soc. Am.*, **73**, 860–861.

34. Karp, S. and Stotts, L.B. (2013) Visible/infrared imaging on the rise. *IEEE Digital Signal Process. Mag.*, **30** (**6**), 177–182.

35. Stotts, L.B., Kolodzy, P., Pike, A. *et al.* (2010) Free space optical communications link budget estimation. *Appl. Opt.*, **49** (**28**), 5333–5343.

36. Andrews, L.C., Phillips, R.L., Crabbs, R., Wayne, D., Leclerc, T., and Sauer,P. (February 2010) *Atmospheric Channel Characterization for ORCA Testing at NTTR*. Atmospheric and Oceanic Propagation of Electromagnetic Waves IV, Proceedings of SPIE Volume: 7588, Editor: Olga Korotkova, 15.

37. Walters, D.L. and Kunkel, K.E. (1981) Atmospheric modulation transfer function for desert and mountain locations: the atmospheric effects on r_0. *J. Opt. Soc. Am.*, **71**, 397–405.

38. Andrews, L.C., Phillips, R.L., Wayne, D. *et al.* (2009) Near-ground vertical profile of refractive-index fluctuations. *Proc. SPIE*, **7324**, Atmospheric Propagation VI, p 732402.

39. Bucher, E.A. (1973) Computer simulation of light pulse propagation for communications through thick clouds. *Appl. Opt.*, **12**, 2391–2400.

40. Petzold, T.J. (September 18, 1981) Prediction of Ocean Water Reflection Factor from "Water Types" of Diffuse Attenuation Coefficient. *Science Application, Inc., Interim Report No. 33*.

41. Stotts, L.B., Lee, G.M., and Stadler, B. (2008) *Free Space Optical Communications: Coming of Age*. Proceedings of the SPIE Conference on Atmospheric Propagation V, G. Charmaine Gilbreath; Linda M. Wasiczko, Editors, 69510W, 18 April.

42. Baker, K.S. and Smith, R.C. (1980) Quasi-inherent characteristics of the diffuse attenuation coefficient for irradiance. *Proc. Photo-Opt. Instrum. Eng.: Ocean Optics VI*, **208**, 60–63.

43. Danielson, R.E., Moore, D.R., and Van de Hulst, H.C. (1969) The transfer of visible radiation through clouds. *J. Atmos. Sci.*, **26**, 1077–1087.

44. van de Hulst, H.C. (1977) Multiple Light Scattering in Atmospheres, Oceans, Clouds and Snow. Institute for Atmospheric Optics and Remote Sensing, Short Course No. 420, Williamsburg, Virginia, December 4–8.

45. Arnush, D. (1969) Underwater light-beam propagation in the small-angle scattering approximation. *J. Opt. Soc. Am.*, **59**, p686.

46. Stotts, L.B. (1979) Limitations of approximate Fourier techniques in solving radiative transfer problems. *J. Opt. Soc. Am.*, **69** (**12**), 1719.

47. Stotts, L.B., Stadler, B., Hughes, D., Kolodzy, P., Pike, A., Young, D.W., Sluz, J., Juarez, J., Graves, B., Dougherty, D., Douglass, J., and Martin, T. (August 2009) *Optical Communications in Atmospheric Turbulence*. Proceedings of the SPIE Conference on Free-Space Laser Communications IX, Majumdar, A.K. and Davis, C. (eds) vol. 7091, 2-6.

8

OPTICAL RECEIVERS

8.1 INTRODUCTION

Free-space optical communications and remote sensing systems are designed to receive signals from a distant source and yield useful information to the user [1–5]. Unfortunately, the quality of the derived information is subject to the efficiency of converting light into photoelectrons, the corrupting system and external noise degrading it, the direct-beam transmission loss, and any negative effects incurred from the medium and electronics. Previous chapters discussed the channel effects. We now focus on what happens at light detection and beyond. Because all of these effects typically are statistical in nature, the theory of signal detection and estimation was created to provide an analytical means to quantify the quality of this information transfer. Chapters 9 and 10 address the detection process. This chapter provides an overview of the various optical detector mechanisms and devices following Pratt [4], Kopeika [5], and Saleh and Teich [6]. We then look at the possible noise sources in an optical receiver that influence the quality of signal reception [4, 5]. When these detector and noise mechanisms are combined with the received signal, we obtain the arguably key parameter in detection theory, the electrical signal-to-noise ratio (SNR). The construction of this particular metric connects it with radio frequency (RF) engineering, so the synergism between the two areas can be easily exploited. Unfortunately, this metric has not been universally adopted by the optical community. One finds that other definitions of the SNR, plus other performance metrics, are prevalent within certain group in the community. We discuss them and include some detection sensor/receiver examples to illustrate its various forms.

Free Space Optical Systems Engineering: Design and Analysis, First Edition. Larry B. Stotts.

Companion website: www.wiley.com\go\stotts\free_space_optical_systems_engineering

As a final comment, Chapter 9 outlines the basics of signal detection and estimation theory metric. Most of our introductory material assumes signal and noise being additively combined. This assumption is not always true. Specifically, in optical imagery, one can be faced with "target or background (clutter) noise present" (replacement model) hypothesis testing rather than "signal-plus-noise or noise-only" (additive model) hypothesis testing cited above. Chapter 9 introduces the new "replacement" detection model necessary for resolved target detection in background clutter so the reader can employ the theory appropriate for the problem(s) they are tackling.

8.2 OPTICAL DETECTORS

Optical detectors can take many forms [4, Chapter 5]. One can have a single detector element like an avalanche photodiode (APD), or a focal plane array like charge-coupled device (CCD) detector array. Photodetectors also do utilize various ways for converting received photons to photoelectrons. They typically fall into two classes: thermal and photon detectors. In addition, the latter can be further categorized in terms of four specific photon conversion mechanisms. They are as follows:

1. *Photoemissive* – Detectors in which an electron receives sufficient energy from an incident photon to physically escape a material (*photoelectric effect*); photoelectron collection occurs in an external circuit, for example, the electron is attracted to and collected by a positively charge anode.
2. *Photoconductive* – Detectors in which photonic excitation causes electrons to move from the valence to the conduction band; the result is that material conductivity changes as a function of the incoming irradiance level.
3. *Photovoltaic* – Detectors in which incoming photon absorbed in semiconductor p–n junction causes a diffusion of electron–hole pairs, resulting in an output voltage change.
4. *Photoelectromagnetic* – Detectors in which its photoconductive material is immersed in a magnetic field orthogonal to the propagation direction of the incident light; they emitted photoelectrons via the photoelectric effect when the incoming radiation is incident on the detector.

Looking at the above definitions, the photoemissive effect relates to the emission of electrons from a vacuum tube cathode/microchannel plate in response to incoming light. The other three effects involve semiconductor detectors where the photon conversion changes the charge carrier concentration in the solid-state material. This section focuses on the first three and imaging arrays.

8.2.1 Performance Criteria

No matter what optical detector one chooses, they all have a set of common performance criteria, some of which have been introduced in previous chapters. They are (1) its quantum efficiency η, (2) its spectral responsivity $\mathcal{R}_\lambda$, (3) its frequency response, (4) its impulse response, and (5) its dark current.

The *quantum efficiency* η of an optical detector is the ratio of the average number of photoelectrons created to the average number of photons incident on the detector element.

The *responsivity* $\mathcal{R}_\lambda$ of a photodetector is usually expressed in units of either amperes or volts per watt of incident radiant power. Many common photodetectors respond linearly as a function of the incident power, but some exhibit nonlinear behavior. In general, the responsivity is a function of the wavelength of the incident radiation and of the detector properties, such as the bandgap of the material of which the photodetector is made. One general expression for the responsivity $\mathcal{R}_\lambda$ is given by

$$\mathcal{R}_\lambda = \left(\frac{\eta q}{h\nu}\right) = \left(\frac{\eta\lambda}{123{,}985\,\mu\text{m}}\right), \tag{8.1}$$

where η is the above defined quantum efficiency, q the electronic charge, h Planck's constant, ν the frequency of the incoming light, and λ the wavelength of the incoming light in microns; Eq. (8.1) has units of amperes per watt (A/W).

Frequency response is the variation in output current as the sinusoidally modulated light signal varies in temporal frequency. Similar to what we saw for spatially modulated inputs to an optical system, the amplitude peak for the output decreases as the modulation frequency increases; in this case, the output current rather than the output image. In general, the frequency response is defined as that modulation frequency where the peak output current is one-half its maximum value.

A detector's *impulse response* corresponds to the resulting current temporal profile induced by a very narrow optical pulse incident on an optical detector. Mathematically, the output current is given by

$$i(t) = \mathcal{R}_\lambda P_{\text{rec}}(t) \circledast h(t), \tag{8.2}$$

where $P_{\text{rec}}(t)$ is the received optical pulse waveform and $h(t)$ the detector's impulse response. In some circles, one specifies the full width-half maximum (FWHM) bandwidth of the impulse response $h(t)$ as specification of the impulse response.

Alternative to the impulse response is its Fourier transform $H(f)$, a complex entity called the *transfer function*. This entity too can be specified by a single parameter; namely, the frequency where the $H(f)$ decreases 3 dB from its dc value. This is the called the *optical bandwidth*. Another definition is to specify the frequency where the square of the transfer function, $H^2(f)$ decreases 3 dB from its low frequency average. This is called the *electrical bandwidth*.

Dark current is the relatively small electric current that flows through photosensitive devices such as a photomultiplier tube (PMT), photodiode, or charge-coupled device even when no photons are entering the device. It is referred to as reverse bias leakage current in nonoptical devices and is present in all diodes. Physically, dark current is due to the random generation of electrons and holes within the depletion region of the device. It is one of the main sources for noise in image sensors such as charge-coupled devices. In this case, the pattern of different dark currents can result in a fixed-pattern noise. Although dark frame subtraction can remove an estimate of the mean fixed pattern, there still remains a temporal noise component because the dark current itself has a shot noise.

8.2.2 Thermal Detectors

Thermal detectors absorb radiation, which produces a temperature change that in turn changes a physical or electrical property of the detector. They include thermocouples, thermistors, bolometers, and thermopiles and pyroelectric detectors. Since a change in temperature is required for photoelectron generation, thermal detectors are generally slow in response and have relatively low sensitivity compared to other detectors. They also lack the sensitivity of semiconductor and other detectors, but they can provide better wavelength response in certain applications as well as a variety of options in size, weight, and power (SWaP).

Primary noise sources for a thermal detector are white noise (the noise associated with a blackbody) and Johnson noise (noise due to random thermal fluctuations).

8.2.3 Photoemissive Detectors

Photoemissive detectors were the workhorse detector technologies prior to the semiconductor revolution and still find much utility today, especially in photon counting applications. This section provides an overview of this detector type.

The basis for photoemissive detectors is called the *external photoelectric effect*. If the energy of the photon illuminating a material in vacuum is sufficiently large, a photoelectron can overcome the potential barrier of the surface of the photoemissive material and escape into vacuum as a free electron [4, 6]. This latter aspect is the reason the photoelectric effect is deemed "external." (If the photoelectron is excited to the conduction band, but remains within the material to contribute to the output current, the situation is called the *internal photoelectric effect*.) If the energy of the photon is $h\nu$, then the maximum energy for any emitted electron from a metal is given by

$$\mathcal{E}_{\max} = h\nu - \mathcal{W}, \tag{8.3}$$

where $\mathcal{W}$ is the *photoelectric work function* expressed in electron volts (eV). Equation (8.3) is known as the *Einstein Photoemission Equation*. Figure 8.1a illustrates the photoelectric emission process for a metal. This figure shows that $\mathcal{W}$ is the energy difference between the Vacuum and Fermi levels. Only if the electron is in the Fermi level can the maximum energy in Eq. (8.3) can be achieved; otherwise, more energy is needed to move any electron at a lower level to the Fermi level and this reduces the kinetic energy of the emitted photoelectron [6, p 750]. The lowest work function for metal is found in Cesium (Cs) and is approximately 2 eV.

The photoemission process for an intrinsic semiconductor is depicted schematically in Figure 8.1b. Unlike a metal, the photoelectrons are emitted from the valence band. For this process, we have

$$\mathcal{E}_{\max} = h\nu - \mathcal{W} = h\nu - (E_g + \chi), \tag{8.4}$$

where E_g is the bandgap energy and χ the electron affinity of the material, defined as the energy difference between vacuum and the bottom of the conduction band. The smallest value for $(E_g + \chi)$ is around 1.4 eV in certain materials. Examples of photocathodes most commonly used in photomultipliers are silver–oxygen–cesium (Ag–O–Cs), cesium-antimony (Cs_3Sb), multialkali or trialkali ($N_2KSb : Cs$), and

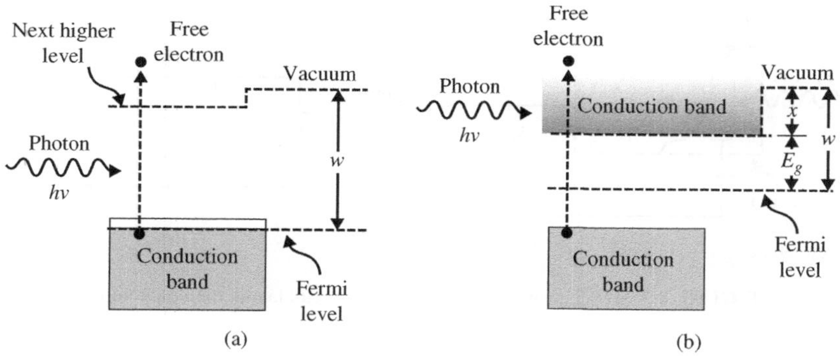

FIGURE 8.1 Photoelectric effect for (a) a metal and (b) an intrinsic semiconductor.

bialkali (K_2CsSb). Typical spectral response curves for these materials are shown in Figure 8.2 [7].

It should be noted that negative-electron-affinity (NEA) semiconductors have been developed [6]. This type of material has the conduction-band edge lying above the conduction vacuum level in the bulk material. The result is that the

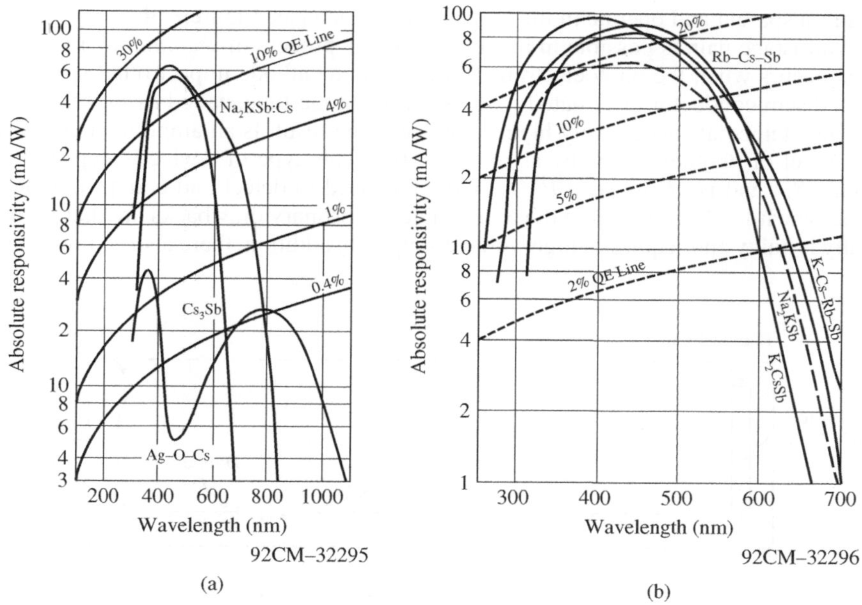

FIGURE 8.2 Plots of (a) typical spectral response curves, with 0080 lime-glass window for silver–oxygen–cesium (Ag–O–Cs), cesium-antimony (Cs_3Sb), and multialkali or trialkali (Na_2KSb : Cs) and (b) for various photocathodes useful in scintillation counting applications. (The variation in (b) of the cutoff at the low end is due to the use of different envelope materials). Source: [7], Reproduced with permission of Photonis USA Pennsylvania, Inc.

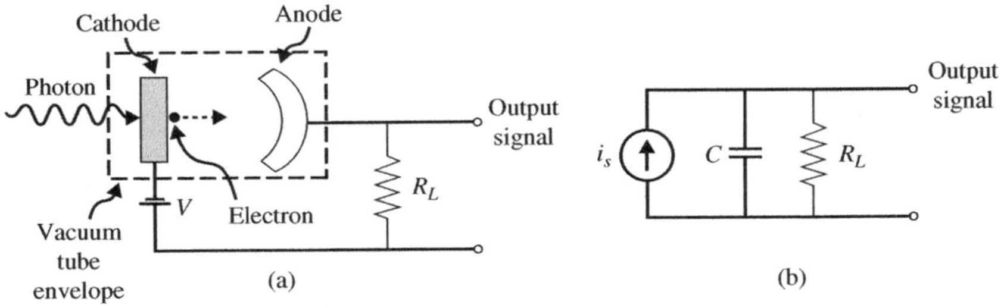

FIGURE 8.3 (a) Schematic and (b) equivalent circuit layout for a phototube.

incoming photon energy only needs to exceed E_g for photoelectrons to be released. For example, Cesium-coated GaAs chips can be responsive to slightly longer near-infrared wavelengths at higher quantum efficiency and with reduced dark current.

Figure 8.3 depicts (a) the principle elements of a vacuum tube photoemissive detector, commonly called a *phototube* and (b) its equivalent circuit layout. Photoemissive material is coated onto the cathode and releases photoelectrons when illuminated by the light of the requisite wavelength. These electrons are attracted to and collected by the anode, thereby producing a current through a load resistor R_L. The cathode can be opaque and create light in the reflection mode [not pictured], or semitransparent and operate in the transmission mode [see Figure 8.3b].

Multistage gain is possible in phototubes by the process of *secondary emissions*. This occurs when the photoelectrons from the photocathode impact other specially placed semiconductor or cesiated-oxide surfaces in the tube, called *dynodes*, which are maintained at successively higher voltages. The result is an amplification of the initial set of photoelectrons by as much as 10^8. This type of device is depicted in Figure 8.4 and is known as a *PMT*. A PMT is used to detect and count individual photons while offering a large dynamic range. Its primary drawbacks are that it normally is bulky and requires a high-voltage supply. In addition, there also is time delay

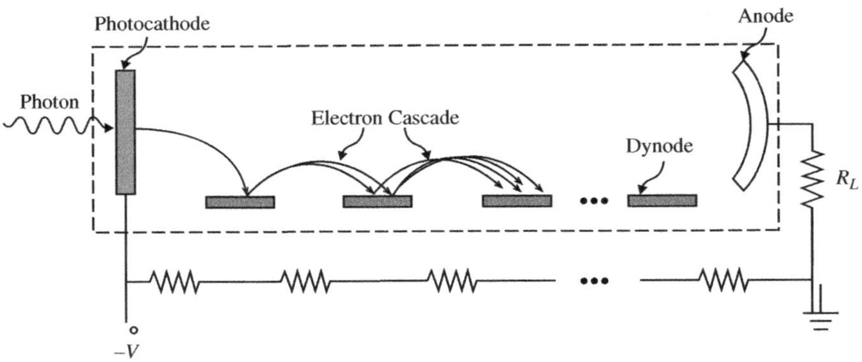

FIGURE 8.4 Electron multiplication in a photomultiplier tube (PMT) with a semitransparent photocathode operated in the transmission mode.

on the order of 50–100 ns from the initial light impingement to the observation of anode photocurrent [4, p. 94]. This may or may not be viewed as a drawback, but it exists. Fortunately, this delay does not affect the frequency response of the device. The frequency response of a PMT is due to the variation in transit times of the emitted electrons and the tube's capacitance [6]. Several techniques have been proposed and used to phototubes and PMTs allowing the devices to be used at GHz rates, for example, traveling wave tubes and dynamic cross-field electron multipliers.

In photoemissive detectors, dark current is created by thermal emissions, field emissions, and current leakage in the detector. The latter two usually are negligible by proper construction practices and operating conditions [4, p. 91]. The average dark current from thermal emissions in the cathode obeys Richardson's law [8]; mathematically, his law is written as

$$i_{\text{dark current}} = A_d A_g\, T^2 e^{-\frac{\mathcal{W}}{kT}}, \tag{8.5}$$

where $i_{\text{dark current}}$ is the dark current, A_d the photocathode area, T the photocathode temperature (K), and $\mathcal{W}$ the work function of the photocathode, and

$$A_g = \varsigma_R A_0. \tag{8.6}$$

In Eq. (8.6), ς_R is the Richardson's constant, a material-specific correction factor, and A_0 a universal constant given by

$$A_0 = \frac{4\pi m k^2 q}{h^3} = 1.2 \times 10^6\,\text{A/m}^2\,\text{K}^2 \tag{8.7}$$

with m and q being the mass and charge of an electron, and h Planck's constant. Table 8.1 shows work functions and Richardson's constants for a set of specific materials [8, 9]. This table shows that ς_R varies from about 32 to 160 A/cm^2 K^2 for pure (polycrystalline) metals and over a much greater range for oxide and composite surfaces [8]. However, for molybdenum, tantalum, tungsten, and barium, we see that $\varsigma_R \approx 0.5$ and for nickel, $\varsigma_R \approx 0.25$.

Example 8.1 Subsequent work has suggested that one must take into account the portion r_{av} of the outgoing electrons that would be reflected as they reached the emitter surface. The result is that A_g now is written as

$$A_g = \varsigma_B (1 - r_{\text{av}}) A_0, \tag{8.8}$$

introducing a new correction factor $\varsigma_B(1 - r_{\text{av}})$ replacing ς_R. However, in many books and papers, this additional correction is ignored.

8.2.4 Semiconductor Photodetectors

The most sensitive types of photon detectors normally are fabricated from semiconductor materials. Like photoemissive devices, photoconductive, photovoltaic (photodiode) and photoelectromagnetic detectors all rely on the photoelectric effect, but the electrons (and positively charged holes for that matter) remain internal to the material. This section provides an overview of semiconductor photodetectors following Liu [9], and Saleh and Teich [6].

TABLE 8.1 List of Work Functions and Richardson's Constants for a Set of Specific Materials

Cathode Material	W (eV)	$\varsigma_R A_0$ (A/(cm^2 K^2))
Molybdenum	4.15	55
Nickel	4.61	30
Tantalum	4.12	60
Tungsten	4.54	60
Barium	2.11	60
Cesium	1.81	160
Iridium	5.4	170
Platinum	5.32	32
Rhenium	4.85	100
Thorium	3.38	70
Ba on W	1.56	1.5
Cs on W	1.36	3.2
Th on W	2.63	3
Thoria	2.54	3
BaO + SrO	0.95	$\sim 10^{-2}$
Cs-oxide	0.75	$\sim 10^{-2}$
TaC	3.14	0.3
LaB_6	2.7	29

8.2.4.1 Photodiode Device Overview Figure 8.1b illustrates the photoelectric effect for an intrinsic semiconductor. Absorbed photons generate free electrons. In addition, a positive charged hole is created in the material when the electron moves up to the conduction band. The result is an *electron–hole photogeneration* with the photoexcited electrons and holes remaining within the material, increasing the electrical conductivity of the material. This type of effect is known as the *internal photoelectric effect*. The transport of the free electrons and holes upon an electric field results in a current.

For good detector performance, the bandgap of the semiconductor material should be slightly less than the photon energy corresponding to the longest operating wavelength in the system's optical band-pass filter. This gives a sufficiently high absorption coefficient to ensure a good response and limits the number of thermally generated carriers, that is, low "dark current." Figure 8.5 shows the optical absorption coefficient α as a function of wavelength for selected semiconductor materials.

There are two mechanisms that exist for photoexcitation of electron–hole pairs in intrinsic semiconductors. One is called *direct optical transition*, where the absorbed phonon directly leads to an electron–hole pair. This is the process we discussed above. The other process is called *indirect optical transition*. This occurs when the absorbed light is assisted by a phonon to create the carriers. Specifically, the absorption process is sequential, with the excited electron–hole pair thermalizing within their respective energy bands by releasing energy/momentum via phonons. The consequence of this effect is that the indirect absorption is less efficient than direct absorption. Table 8.2 lists photodetector semiconductors materials, and their direct and indirect bandgaps at 300 K. This table includes the associated wavelengths for the bandgaps

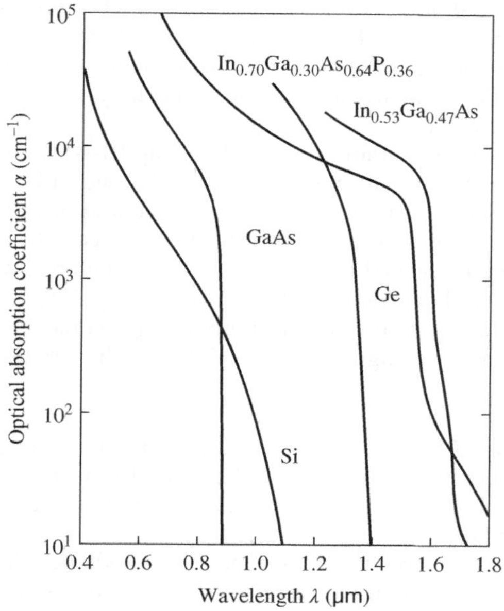

FIGURE 8.5 Optical absorption coefficient for selected semiconductors versus wavelength.

TABLE 8.2 List of Photodetector Semiconductors Materials, and Their Bandgaps and Wavelengths at 300 K

Semiconductor Material	Bandgap at 300 K Direct/Indirect (eV)	Wavelength (μm)
Si	4.10/1.14	0.3/1.09
Ge	0.81/0.67	1.53/1.86
GaAs	1.43	0.87
InP	1.35	0.92
$In_{0.53}Ga_{0.47}As^a$	0.75	1.66
InAs	0.35	3.56
InSb	0.18	6.93
$Hg_{1-x}Cd_xTe$	$0 < E_g < 1.44$	$0.86 < \lambda < \infty$

[a]Lattice matched to InP.

listed. Referring to Figure 8.5, Si is only weakly absorbing over the wavelength band 0.8–0.9 μm relative to the other materials. This results in the indirect absorption mechanism being the dominate process. The threshold for indirect absorption (long wavelength cutoff) occurs at 1.09 μm.

On the other hand, Ge is another semiconductor material for which the lowest energy absorption takes place by indirect optical transitions. Indirect absorption will occur up to a threshold of 1.85 μm. However, the threshold for direct absorption occurs at 1.53 μm. For shorter wavelengths, germanium becomes strongly absorbing. Unfortunately, Ge photodiodes have relatively large dark currents because of their narrow bandgaps in comparison to other semiconductor materials. This negates the use of Ge photodiodes in many applications, especially when shorter wavelengths below 1.1 μm are involved.

For longer wavelength operations, direct-bandgap III–V compound semiconductors will be the better material choice. Specifically, one tailors their bandgaps to specific wavelengths by changing the relative concentrations of their constituents, which results in lower dark current in the resulting device. These mixed compound materials also can be fabricated into heterojunction structures, which enhance their high-speed operations. For example, the lattice of $In_{0.53}Ga_{0.47}As$ can be matched to InP substrates, creating photoexcitation up to approximately 1.7 μm. This range encompasses two key wavelengths, 1.3 and 1.55 μm, that are used in many popular optical systems.

8.2.4.2 Photodiode Physics A semiconductor photodiode is a crystalline p–n-junction structure that augments its basic performance through the internal photoelectric effect. Let us begin with the discussion of normal p–n-junction diodes following reference [10, 11].

Figure 8.6a depicts the structure of the junction photodiode. The p–n junction is the continuous boundary region, or interface, between two types of doped semiconductor material, one being *p-type* and the other *n-type*, inside a single crystal substrate of base semiconductor material. These two type regions are created by doping the substrate by either ion implantation, diffusion of dopants, or by epitaxy (growing a layer of crystal doped with one type of dopant on top of a layer of crystal doped with another type of dopant). Donor impurities such as Antimony (n-type doping) will create regions where the mobile charges are primarily electrons. On the other hand, acceptor impurities such as Boron (p-type doping) will create regions where the mobile charges are mainly holes. Let us first look at this in more detail.

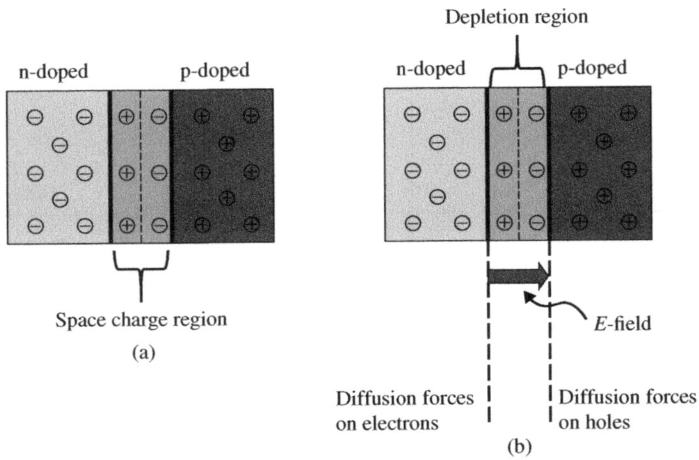

FIGURE 8.6 Pictures of (a) p–n-junction diode structure and (b) its diffusion forces and electric field.

In Figure 8.6a, we have assumed a step junction in which the doping concentration is uniform in each doped region and there is an abrupt change in doping at the junction, initially in regions. As a result, majority carrier electrons in the n-doped region then will migrate across the junction to the p-doped side and vice versa. If there are no external connections to the semiconductor, then these migration processes will begin to diminish to zero after a while.

As electrons move from the n-doped region, positively charged donor atoms are left behind.

Similarly, as holes transient from the p-doped region, they uncover negatively charged acceptor atoms. The net positive and negative charges in the n- and p-doped regions induce an electric field in the region near the p–n junction. This field runs from the positive to the negative charge, or from the n to the p region. The net positively and negatively charged regions within the junction are shown in Figure 8.6. The two combined regions are referred to as the *space charge region*. Essentially, all mobile electrons and holes are swept out of the space charge region by the electric field. Since the space charge region is depleted of any mobile charge, this region also is referred to as the *depletion region*. Density gradients still exist in the majority carrier concentrations at each edge of the space charge region. We can think of a density gradient as producing a "diffusion force" that acts on the majority carriers. These diffusion forces, acting on the electrons and holes at the edges of the space charge region, are shown in Figure 8.6b.

The electric field in the space charge region produces another force on the electrons and holes that is in the opposite direction to the diffusion force for each type of particle. This junction is in thermal equilibrium, which means that the diffusion force and the *E*-field force exactly balance each other.

When the p–n-junction diode is shorted, it is known to be *Zero Biased*, that is, no external energy source is applied. Figure 8.7a and b shows the circuit diagrams for the device structure shown in Figure 8.6 and with its electronic circuit symbol, respectively. In both cases, the cathode and anode terminals are shown connected to ground. Since there is no voltage across the p–n junction, the junction again is in thermal equilibrium and the Fermi energy level is constant throughout the entire system. The conduction and valance band energies, E_c and E_v, respectively, must decrease as we move from the p-side to the n-side since the relative position of the conduction

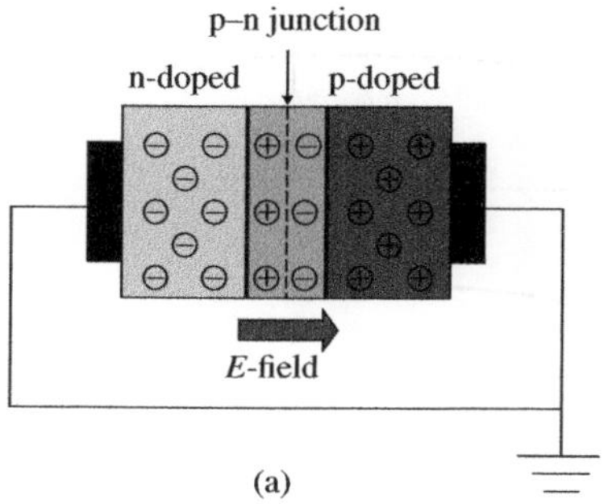

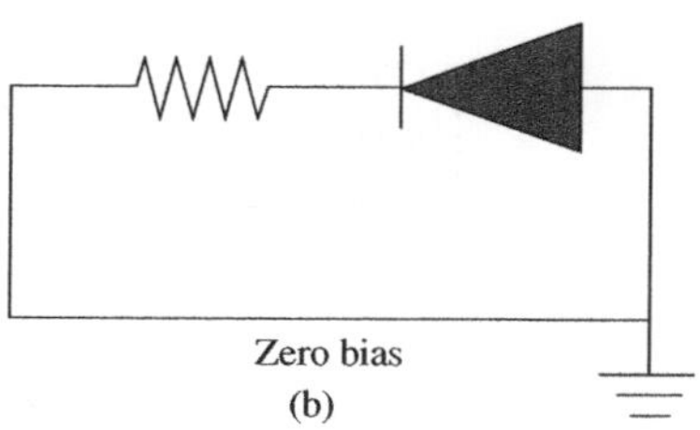

FIGURE 8.7 Schematics of zero-biased circuits using (a) a p–n-junction structure and (b) the electronic component symbol.

and valence bands with respect to the Fermi energy changes between p- and n-doped regions. Electrons in the conduction band of the n region see a potential barrier in trying to move into the conduction band of the p-doped region. This potential barrier is referred to as the *built-in potential barrier* and is denoted by V_{bi}. It is approximately 0.5–0.7 V for silicon diodes and approximately 0.3 V for germanium diodes. This barrier maintains an equilibrium between majority carrier electrons in the n region and minority carrier electrons in the p region, and also between majority carrier holes in the p region and minority carrier holes in the n region. The potential V_{bi} maintains equilibrium, so no current is produced by this voltage.

The intrinsic Fermi level is equidistant from the conduction band edge through the junction; thus, the built-in potential barrier can be determined as the difference between the intrinsic Fermi level in the p and n regions. The barrier potential is given by

$$V_{\text{bi}} = |\varphi_{Fp}| + |\varphi_{Fn}| \tag{8.9}$$

with the potentials φ_{Fp} and φ_{Fn} defined in Figure 8.8.

From previous work, the electron population in the n-doped region is given by

$$n_0 = N_c\, e^{-(E_c - E_F)/kT} \tag{8.10a}$$

$$= n_i\, e^{(E_F - E_{Fi})/kT}, \tag{8.10b}$$

where the potentials n_i and E_{Fi} are the intrinsic carrier population and the intrinsic Fermi energy, respectively. This means that the potential φ_{Fn} equals

$$q\varphi_{Fn} = E_{Fi} - E_F, \tag{8.11}$$

which implies that

$$n_0 = n_i e^{-\frac{\varphi_{Fi}}{kT}}. \tag{8.12}$$

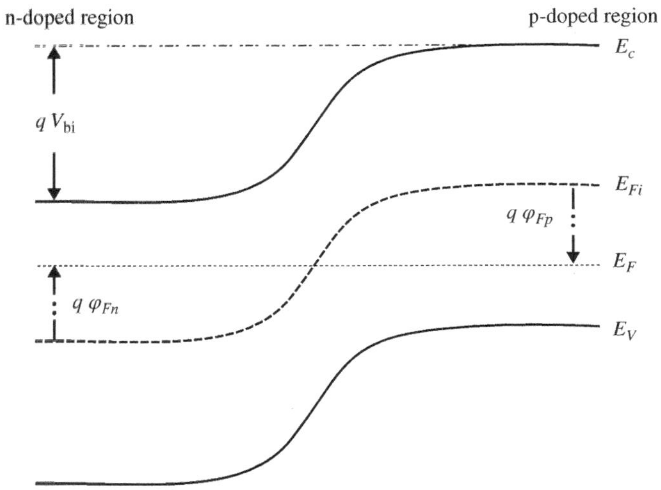

FIGURE 8.8 The energy-band diagram for the p–n junction in thermal equilibrium.

If we define $n_0 = N_d \equiv$ Donor population, then it is easy to show that

$$\varphi_{Fn} = -\frac{kT}{q} \ln\left(\frac{N_d}{n_i}\right). \tag{8.13}$$

Similarly, we can show that

$$\varphi_{Fp} = +\frac{kT}{q} \ln\left(\frac{N_a}{n_i}\right), \tag{8.14}$$

where N_a is the Acceptor population since

$$q\varphi_{Fp} = E_{Fi} - E_F \tag{8.15}$$

and

$$p_0 = n_i e^{\frac{(E_{Fi}-E_F)}{kT}}. \tag{8.16}$$

Substituting Eqs. (8.13) and (8.14) into Eq. (8.9) shows that barrier potential is equal to

$$V_{bi} = \frac{kT}{q} \ln\left(\frac{N_d N_a}{n_i^2}\right) = V_T \ln\left(\frac{N_d N_a}{n_i^2}\right), \tag{8.17}$$

where

$$V_T = \frac{kT}{q}. \tag{8.18}$$

The term V_T is known as the *thermal voltage.*

The electric field associated with the built-in potential obviously comes from the separation between positive and negative space charge densities in the p- and n-doped regions, respectively. Assume that the x-axis parallels the p–n device such that the p–n junction is at $x = 0$, the n-doped region is to the right of that junction, and the p-doped region is to the right of that junction. The electric field is determined from Poisson's equation, which for a one-dimensional analysis is given by

$$\frac{d^2\varphi(x)}{d^2x} = -\frac{\rho(x)}{\varepsilon_s} = \frac{dE(x)}{dx}, \tag{8.19}$$

where $\varphi(x)$ is the electric potential, $E(x)$ the electric field, $\rho(x)$ the volume charge density, and ε_s the permittivity of the semiconductor material. For this discussion, we further assume that the space charge region abruptly stops on the donor side at $x = +x_n$ and abruptly stops on the acceptor side at $x = -x_p$. Given these facts, we can write the charge densities as

$$\rho(x) = qN_d \quad 0 < x < x_n \tag{8.20}$$

and

$$\rho(x) = -qN_a \quad x_p < x < 0. \tag{8.21}$$

By integrating Eq. (8.19), we obtain for the electric field in the acceptor region,

$$E(x) = \int \frac{\rho(x)}{\varepsilon_s} dx = -\int \frac{qN_a}{\varepsilon_s}\, dx = -\frac{qN_a}{\varepsilon_s}\, x + C_a \tag{8.22}$$

where C_a is a constant of integration. Since acceptor region is in thermal equilibrium, there is no current is flowing, and the electric field can be assumed to be 0 for $x < -x_p$. In addition, the fact that there are no surface charge densities within the p–n-junction structure implies that the electric field can be assumed a continuous function within this part of the p–n-junction structure. Finally, the constant of integration is determined by setting $E(x) = 0$ at $x = x_p$. Hence, Eq. (8.22) becomes

$$E(x) = -\frac{qN_a}{\varepsilon_s}(x + x_p) \tag{8.23}$$

for $-x_p < x < 0$.

Similarly, we can show that the electric field in the donor region equals

$$\begin{aligned} E(x) &= \int \frac{\rho(x)}{\varepsilon_s} dx = \int \frac{qN_d}{\varepsilon_s} dx = \frac{qN_d}{\varepsilon_s} x + C_d \\ &= -\frac{qN_d}{\varepsilon_s}(x_n - x). \end{aligned} \tag{8.24}$$

for $0 < x < x_n$. In Eq. (8.24), C_d is a constant of integration. As with the case above, the electric field in this part of the structure also is a continuous function. Since the field is continuous, the fields on both sides of the p–n junction equal each other at $x = 0$. Equating Eqs. (8.23) and (8.24) at $x = 0$ yields

$$E(0) = -\frac{qN_a}{\varepsilon_s}\, x_p = -\frac{qN_d}{\varepsilon_s}\, x_n \tag{8.25}$$

or

$$N_a\, x_p = N_d\, x_n. \tag{8.26}$$

Equation (8.26) states that the number of negative charges per unit area in the p-doped region is equal to the number of positive charges per unit area in the n-doped region. The electric field direction is from the n to the p region, that is, in the negative x, as seen in Figure 8.7a. For the uniformly doped p–n junction, the E-field is a linear function of distance through the junction, and the maximum (magnitude) electric field occurs at the metallurgical junction, $x = 0$. In other words, an electric field exists in the depletion region even when no voltage is applied between the p- and n-doped regions.

The potential in the p–n junction is found by integrating the electric field. For the acceptor region, we have

$$\begin{aligned} \varphi(x) &= -\int E(x)dx = \int \frac{qN_a}{\varepsilon_s}(x + x_p)dx \\ &= \frac{qN_a}{\varepsilon_s}\left(\frac{x^2}{2} + x_p x\right) + C'_a, \end{aligned} \tag{8.27}$$

where C'_a is a constant of integration. The potential difference through the p–n junction is the important parameter, rather than the absolute potential, so we may arbitrarily set the potential equal to zero at $x = x_p$. Using this fact, we find that

$$C'_a = \frac{qN_a}{2\varepsilon_s} x_p^2. \tag{8.28}$$

Substituting Eq. (8.28) into Eq. (8.27) allows us to write the potential in the acceptor region as

$$\varphi(x) = -\frac{qN_a}{2\varepsilon_s}(x + x_p)^2 \tag{8.29}$$

for $-x_p < x < 0$.

In the donor region, we can write

$$\varphi(x) = \int \frac{qN_d}{\varepsilon_s}(x_n - x)\,dx = \frac{qN_d}{\varepsilon_s}\left(x_n x - \frac{x^2}{2}\right) + C'_d. \tag{8.30}$$

The potential is a continuous function, so setting Eq. (8.30) equal to Eq. (8.29) when $x = 0$ yields

$$C'_d = \frac{qN_a}{2\varepsilon_s} x_p^2 \tag{8.31}$$

and the potential in the donor region now given by

$$\varphi(x) = \frac{qN_d}{\varepsilon_s}\left(x_n x - \frac{x^2}{2}\right) + \frac{qN_a}{2\varepsilon_s} x_p^2. \tag{8.32}$$

for $0 < x < x_n$. Figure 8.9 shows a plot of the potential through the p–n junction, clearly exhibiting the quadratic dependence on the distance. The magnitude of the potential at $x = x_n$ is equal to the built-in potential barrier. From Eq. (8.32), we have

$$V_{\text{bi}} = |\varphi(x = x_n)| = \frac{qN_d}{2\varepsilon_s}(N_d x_n^2 + N_a x_p^2). \tag{8.33}$$

The potential energy of an electron is given by $E(x) = -q\varphi(x)$, which means that the electron potential energy also varies as a quadratic function of distance through the space charge region.

It can be shown that

$$x_n = \sqrt{\left(\frac{2\varepsilon_s V_{\text{bi}}}{q}\right)\left(\frac{N_a}{N_d}\right)\left(\frac{1}{N_a + N_d}\right)} \tag{8.34}$$

and

$$x_p = \sqrt{\left(\frac{2\varepsilon_s V_{\text{bi}}}{q}\right)\left(\frac{N_d}{N_a}\right)\left(\frac{1}{N_a + N_d}\right)}. \tag{8.35}$$

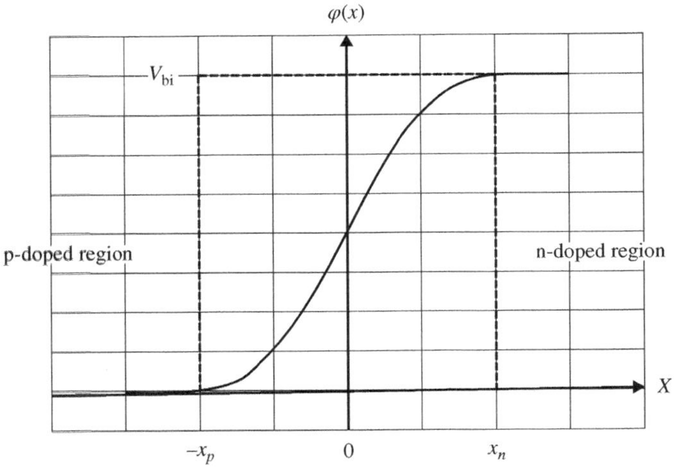

FIGURE 8.9 Electric potential through the space charge region of a uniformly doped p–n junction.

Equations (8.34) and (8.35) give the space charge width, or the width of the depletion region, x_n, extending into the n-doped region and x_n, extending into the p-doped region, respectively, for the case of zero applied voltage. The total depletion or space charge width W_{sc} is the sum of these two components, or more specifically,

$$W = \sqrt{\left(\frac{2\varepsilon_s V_{bi}}{q}\right)\left[\frac{N_a + N_d}{N_d N_a}\right]} \tag{8.36}$$

When a junction diode is *reverse biased*, we have the situation shown in Figure 8.10. Figure 8.10a and b depicts schematics of reverse-biased circuits using a p–n-junction structure and the electronic component symbol, respectively. In this case, the width of the depletion regions increases to W_{rb} and the diode acts like an open circuit blocking any current flow, (only a very small leakage current) [10]. This situation is illustrated in Figure 8.10c as an open-circuit condition.

Figure 8.10a indicates that there is an electric field in the space charge region. This field is created by the electric field E_R induced by the applied voltage V_R [10]. Figure 8.11 shows the revised energy-band diagram for the p–n junction under reversed bias conditions, which shows that V_R is replaced by $V_{total} = V_{bi} + V_R$. The Fermi energies shift accordingly. This applied voltage does not induce any electric fields in the adjacent semiconductor regions just like in the thermal equilibrium case, that is, any electric fields in those regions are essentially zero, or at least very small. This means that the magnitude of the electric field in the space charge region must increase above the thermal equilibrium value again due to the applied voltage. From Figure 8.10a, we see that the electric field originates on positive charge and terminates on negative charge. The number of positive and negative charges must increase if the electric field increases. For given impurity doping concentrations, this will only occur if the space charge width W grow larger to accommodate this increase and this growth is dependent on the reverse-bias voltage V_R. The result is

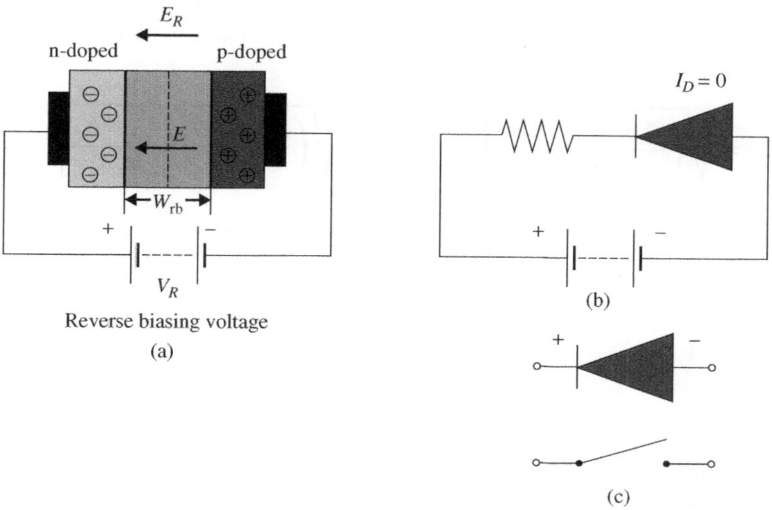

FIGURE 8.10 Schematics of reverse-biased circuits using (a) a p–n-junction structure and (b) the electronic component symbol, and (c) its open-circuit condition.

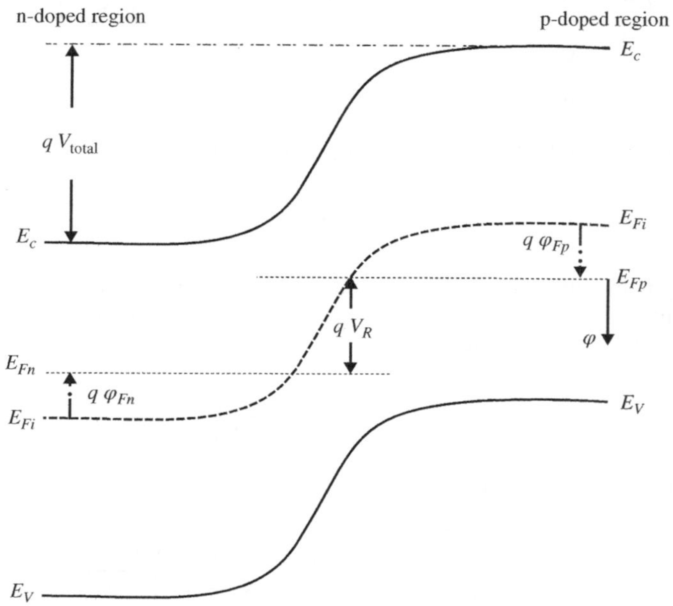

FIGURE 8.11 The revised energy-band diagram for the p–n junction under reversed bias conditions.

that in all of the previous equations, the built-in potential barrier V_{bi} can be replaced by the total potential barrier V_{total}. The total space charge width under reversed voltage bias is

$$W_{\mathrm{rb}} = \sqrt{\left(\frac{2\mathcal{E}_s(V_{\mathrm{bi}} + V_R)}{q}\right)\left[\frac{N_a + N_d}{N_d N_a}\right]} = \sqrt{\left(\frac{2\mathcal{E}_s V_{\mathrm{total}}}{q}\right)\left[\frac{N_a + N_d}{N_d N_a}\right]} \tag{8.37}$$

with

$$x_n = \sqrt{\left(\frac{2\mathcal{E}_s V_{\mathrm{total}}}{q}\right)\left(\frac{N_a}{N_d}\right)\left(\frac{1}{N_a + N_d}\right)} \tag{8.38}$$

and

$$x_p = \sqrt{\left(\frac{2\mathcal{E}_s V_{\mathrm{total}}}{q}\right)\left(\frac{N_d}{N_a}\right)\left(\frac{1}{N_a + N_d}\right)}. \tag{8.39}$$

The electric field equations given in Eqs. (8.24) and (8.23) still are valid here, except the values of x_n and x_p in those two equations now are given by Eqs. (8.38) and (8.39), respectively. The maximum electric field is still at the p–n junction at $x = 0$ and we find in this situation that

$$E(0) = -\frac{qN_a}{\mathcal{E}_s}x_p = -\frac{qN_d}{\mathcal{E}_s}x_n = -\sqrt{\left(\frac{2\mathcal{E}_s V_{\mathrm{total}}}{q}\right)\left[\frac{N_a + N_d}{N_d N_a}\right]} \tag{8.40}$$

$$= -\frac{2V_{\mathrm{total}}}{W_{\mathrm{rb}}}. \tag{8.41}$$

The *differential junction capacitance* C_{rb} is defined as

$$dC_{\mathrm{rb}} = \frac{dQ_{\mathrm{rb}}}{dV_R}, \tag{8.42}$$

where dQ_{rb} is the differential charge in units of coulomb per square centimeter and the differential capacitance dC_{rb} is in units of farads per square centimeter. Since $dQ_{\mathrm{rb}} = qN_d\,dx_n = qN_a\,dx_p$, we have

$$dC_{\mathrm{rb}} = \frac{dQ_{\mathrm{rb}}}{dV_R} = qN_d\frac{dx_n}{dV_R}, \tag{8.43}$$

which using Eq. (3.38) yields

$$dC_{\mathrm{rb}} = qN_d\frac{dx_n}{dV_R} = qN_d\sqrt{\left(\frac{2\,\mathcal{E}_s}{q}\right)\left(\frac{N_a}{N_d}\right)\left(\frac{1}{N_a + N_d}\right)}\frac{d\sqrt{V_{\mathrm{total}}}}{dV_R}$$
$$= \frac{qN_d}{2}\sqrt{\left(\frac{2\mathcal{E}_s}{qV_{\mathrm{total}}}\right)\left(\frac{N_a}{N_d}\right)\left(\frac{1}{N_a + N_d}\right)} = \sqrt{\left(\frac{q\mathcal{E}_s}{2\,V_{\mathrm{total}}}\right)\left(\frac{N_a N_d}{N_a + N_d}\right)} \tag{8.44}$$

The junction capacitance also is referred to as the *depletion layer capacitance*. Using Eq. (8.44), the above junction capacitance can be rewritten as

$$dC_{\mathrm{rb}} = \frac{\varepsilon_s}{W_{\mathrm{rb}}} \tag{8.45}$$

Equation (8.45) is the same as the equation for the capacitance per unit area of a parallel plate capacitor [10]. Multiplying Eqs. (8.44) and (8.45) by the junction area A_{pn} gives the capacitance C_j of the p–n-junction device. Since the space charge width is a function of the reverse-bias voltage, the junction capacitance also is a function of that circuit voltage.

When a junction diode is *forward biased*, we have the situation shown in Figure 8.12. Figure 8.12a and b depicts schematics of forward-biased circuits using a p–n-junction structure and the electronic component symbol, respectively. In this case, the width of the depletion regions reduces to W_{fb} and the diode acts like a short circuit allowing full current to flow at its maximum value through the resistor. This is seen in Figure 8.12b and c. Figure 8.13 shows the revised energy-band diagram for the p–n junction under forward bias conditions, which shows that V_R is replaced by $V_{\mathrm{total}} = V_{\mathrm{bi}} - V_F$. The Fermi energies shift accordingly. The smaller potential barrier reduces the electric field in the depletion region, and the electrons and holes are no longer held back in the n- and p-doped regions. In other words, there will be a diffusion of holes from the acceptor region across the space-charge region where they now will flow into the donor region (*reverse current*). Similarly, there will be a diffusion of electrons from the donor region across the space-charge region where they will flow into the acceptor region (forward current).

With this background, the mathematical derivation or the p–n-junction current–voltage relationship is considered in the following section.

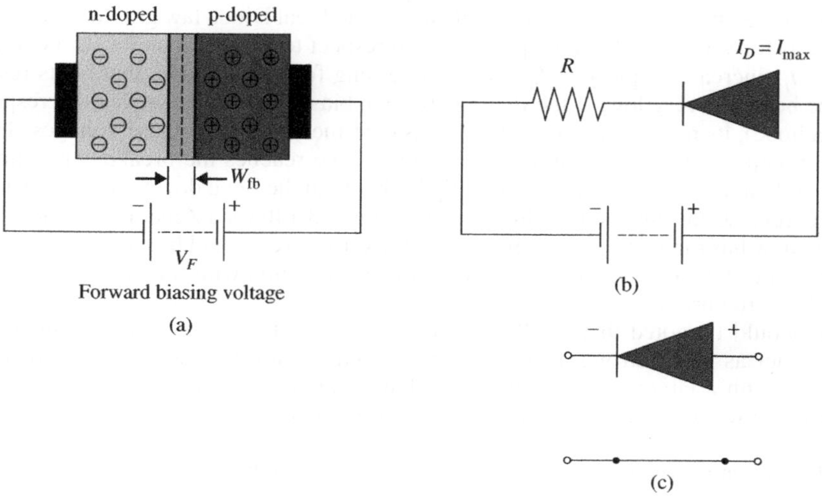

FIGURE 8.12 Schematics of forward-biased circuits using (a) a p–n-junction structure and (b) the electronic component symbol, and (c) its short-circuit condition.

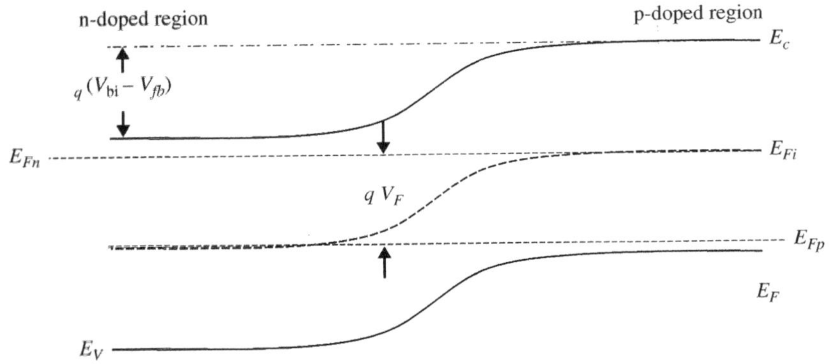

FIGURE 8.13 The revised energy-band diagram for the p–n junction under forward bias conditions.

8.2.4.3 The Diode Laws The diode equation gives an expression for the current through a diode as a function of voltage. There are two definitions in use. The first is the *Ideal Diode law*, which is given by

$$I_{ph} = I_0 \left(e^{\frac{qV}{kT}} - 1\right) = I_0 \left(e^{\frac{V}{V_T}} - 1\right), \tag{8.46}$$

where

$I \equiv$ the net current flowing through the diode

$I_0 \equiv$ dark saturation current (the diode leakage current in the absence of light)

$V \equiv$ applied voltage across the terminals of the diode.

The thermal voltage V_T is approximately 25.85 mV at 300 K, a temperature close to "room temperature." Figure 8.14a illustrates the Ideal Diode law graphically. If the diode is forward biased (anode positive with respect to cathode), its forward current ($I_D = i_f$) increases rapidly to I_{max} with increasing forward voltage, that is, its resistance becomes very low. If the diode is reverse biased (anode negative with respect to cathode), its reverse current ($-I_D = i_r$) is extremely low for reverse voltages. This last situation is only valid until the reverse voltage reaches the breakdown voltage V_{BR}. When the reverse voltage is slightly higher than the breakdown voltage, a sharp rise in reverse current results. This is seen in Figure 8.14b for a Zener p–n diode. This diode now has three regions of operations: forward, reverse, and breakdown. The latter two are shown in Figure 8.14b; "forward" is the region with forward bias voltage and forward current.

It should be noted that the "dark saturation current" is an extremely important parameter as it is a strong measure of the recombination in a device. In particular, a diode with a larger recombination will have a larger I_0. In general, the dark current increases as the temperature T increases and decreases as the material quality improves.

The second is the *nonideal diode law*, which is given by

$$I_{ph} = I_0 \left(e^{\frac{qV}{\varepsilon kT}} - 1\right) = I = I_0 \left(e^{\frac{V}{\varepsilon V_T}} - 1\right), \tag{8.47}$$

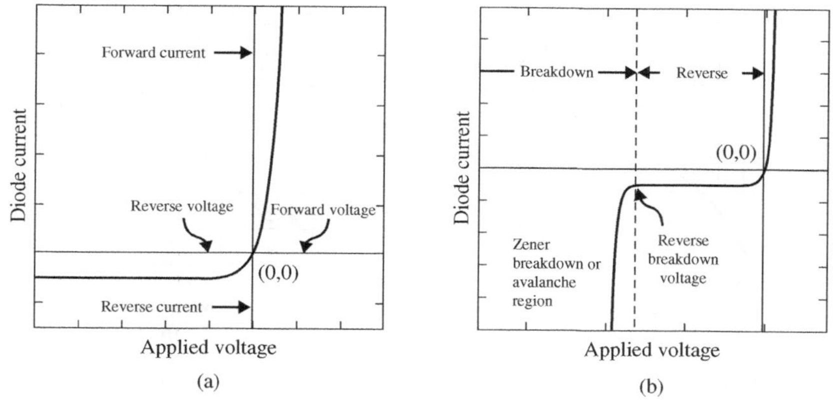

FIGURE 8.14 Current versus voltage (I–V) characteristics of (a) ideal and (b) Zener p–n-junction diodes.

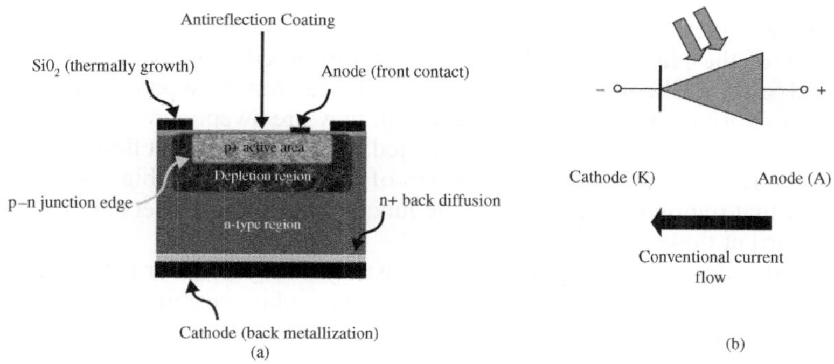

FIGURE 8.15 Pictures of (a) p–n-junction diode structure and (b) its electronic symbol.

where ε is the ideality factor, a number between 1 and 2, which typically increases as the current decreases. The graphs and comments are similar to the ideal case but adjusted to the ideality factor. This is the law usually associated with real diodes.

8.2.4.4 Junction Photodiodes A *semiconductor photodiode detector* is a p–n-junction structure similar to the above but also responds to the photoelectric effect [11]. Figure 8.15 depicts (a) its basic structure and (b) one of its symbol representation. Referring to Figure 8.15a, we see that the device consists of a shallow diffused p–n junction, normally a p-on-n configuration.[1] In this p-on-n planar diffused configuration, the junction edge is located on the boundary between the p-region and depletion, where it is passivated by a thermally grown silicon-dioxide layer. The p–n junction and the depletion act to first order like the photodiode operation described above.

When photons of energy greater than E_g (e.g., the bandgap of semiconductor material) fall on the device, they are absorbed and electron–hole pairs are created.

[1] "P-type" devices (n-on-p) are available for enhanced responsivity in the 1 μm region.

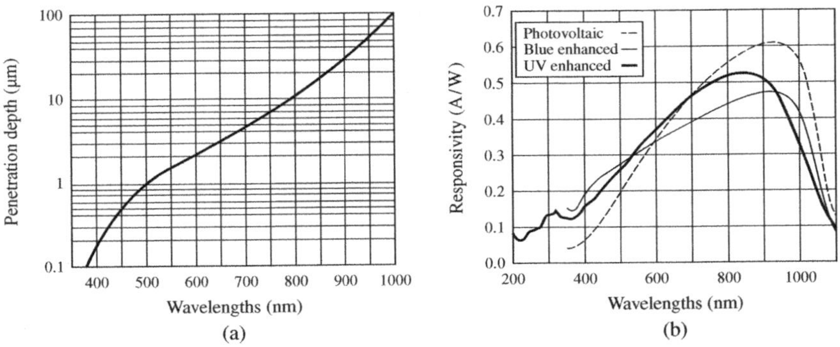

FIGURE 8.16 Example plots of (a) penetration depth (e^{-1}) of light into silicon substrate for various wavelengths and (b) spectral responsivity of several different types of planar diffused photodiodes. Source: OSI Optoelectronics [12]. Reproduced with permission of OSI Optoelectronics.

The depth at which the photons are absorbed depends upon their energy; the lower the energy of the photons, the deeper they are absorbed. This is clearly shown in Figure 8.16a for a silicon-based device [12]. The electron–hole pairs drift apart, and when the minority carriers reach the junction, they are swept across by the electric field. If the two sides are electrically connected, an external current flows through the connection. If the created minority carriers of that region recombine with the bulk carriers of that region before reaching the junction field, the carriers are lost and no external current flows.

As noted earlier in this chapter, the responsivity of any detector is the measure of the sensitivity to light. It is defined as the ratio of the photocurrent I_{ph} to the incident light power P_{in} at a given wavelength:

$$R_\lambda = \frac{I_{ph}}{P_{in}} \tag{8.48}$$

In other words, it is a measure of the effectiveness of the conversion of the optical power into electrical current. It varies with the wavelength of the incident light (see Figure 8.16b for the spectral responsivity of a silicon photodiode) as well as applied reverse bias and temperature [12].

The equivalent circuit of a junction photodiode is shown in Figure 8.17. In this figure, we have

$$C_j \equiv \text{junction capacitance (F)}$$

$$I_D \equiv \text{dark current (A)}$$

$$I_{ph}(= R_\lambda P_{in}) \equiv \text{light-induced current (A)}$$

$$I_{noise} \equiv \text{noise current (A)}$$

$$R_{sh} \equiv \text{Shunt resistance } (\Omega)$$

$$R_s \equiv \text{Series resistance } (\Omega)$$

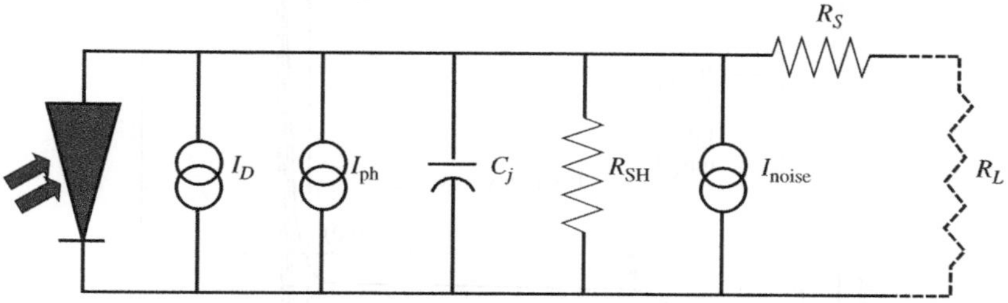

FIGURE 8.17 Equivalent circuit for the junction photodiode.

and

$$R_L \equiv \text{load resistance} \quad (\Omega)$$

The photodiode behaves as a current source when illuminated. When operated without bias, this current is distributed between the internal shunt resistance and external load resistor. In this mode, a voltage develops that creates a forward bias, thus reducing its ability to remain a constant current source. When operated with a reverse voltage bias, the photodiode becomes an ideal current source. Let us be more specific.

If a junction of cross-sectional area A_{pn} is uniformly illuminated by photons with $h\nu > E_g$, a photogeneration rate G_{pg} (Electron–Hole Pairs per cm^{-3} s or EHP/cm^{-3} s) gives rise to a photocurrent. The number of holes created per second within a diffusion length L_h of the depletion region on the n side is $A_{pn}\,L_h\,G_{pg}$; the number of electrons created per second within a diffusion length L_e of the depletion region on the p side is $A_{pn}\,L_e\,G_{pg}$. The forward and reverse currents then are given by

$$I_{fb} = I_0\left(e^{\frac{V_{fb}}{V_T}} - 1\right) - qA_{pn}\,(L_h + L_e)G_{pg} = I_0\left(e^{\frac{V_{fb}}{V_T}} - 1\right) - I_{ph} \tag{8.49}$$

and

$$I_{rb} \approx I_0 + qA_{pn}(L_h + L_e)G_{pg} = I_0 - I_{ph}, \tag{8.50}$$

respectively [11]. Figure 8.18 depicts the *I–V* curve of a p–n-junction photodiode with an added photocurrent $-\,I_{ph}$ proportional to the incoming optical power level. It is clear from this figure that the *I–V* curve we saw in Figure 8.14b lowered by an amount proportional to the incoming light power level, P_{in}. For zero illumination, we obtain dark current level depicted in Figure 8.14b as expected. The *open-circuit photovoltage* V_{oc} for zero current and nonzero illumination levels is given by

$$V_{oc}(I_{ph}) = V_T \ln\left[\left(\frac{I_{ph}}{I_0}\right) + 1\right]. \tag{8.51}$$

from this figure. The open-circuit voltage increases only logarithmically $\left(V_{oc}(I_{ph}) \propto \left[\frac{I_{ph}}{I_0}\right]\right)$ and is limited by the *equilibrium contact potential.* Similarly, we see from that the *short-circuit photocurrent* for zero voltage and nonzero

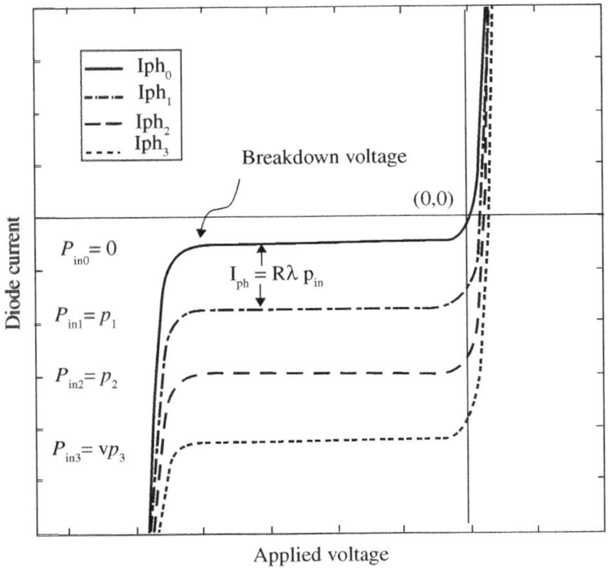

FIGURE 8.18 Characteristic *I–V* curves of a junction photodiode for photoconductive and photovoltaic modes of operation. $P_{in0} - P_{inj}$, $j = 1, 2, 3$, etc., indicate different incoming light power levels.

illumination equals the values of the photocurrent derived from the incoming illumination level, for example, $I_{sc1} = R_{\lambda}P_{in3}$. In this case, as the light intensity increases, the short-circuit current increases linearly ($I_{sc1} \propto P_{in3}$).

Given the above, there essentially are two modes of operation for a junction photodiode. One is called the *photoconductive* mode, which is where the photodiode is reverse biased. In other words, this mode operates in the third (left lower) quadrant of its current–voltage characteristics (external reverse bias, reverse current), including the short-circuit condition on the vertical axis for $V = 0$ (acting as a current source). In this case, the series load resistor sets the load line. The output voltage must be less than the bias voltage to keep the photodiode reversed biased, which means that it must be sufficiently large to be useful. Under these conditions and before it saturates, a photodiode provides the following linear response, $V_{out} = (I_{ph} + I_0)R_L$.

The other mode is called the *photovoltaic* mode and operates in the fourth (right lower) quadrant (internal forward bias, reverse current), including the open-circuit condition on the horizontal axis for $I = 0$ (acting as a voltage source with output voltage limited by the equilibrium contact potential). This mode does not require a bias voltage but does require a large load resistance.

A *p–i–n photodiode*, also called *PIN photodiode*, is a photodiode with an intrinsic (i.e., undoped) region in between the n- and p-doped regions. (In practice, the intrinsic region does not have to be truly intrinsic but only has to be highly resistive, i.e., lightly doped p or n region.)

In general, the depletion layer is almost completely defined by the intrinsic region. Most of the photons are absorbed in the intrinsic region, and carriers generated therein can efficiently contribute to the photocurrent. Figure 8.19 illustrates a schematic drawing of PIN structure. In this figure, we have the cathode as a flat

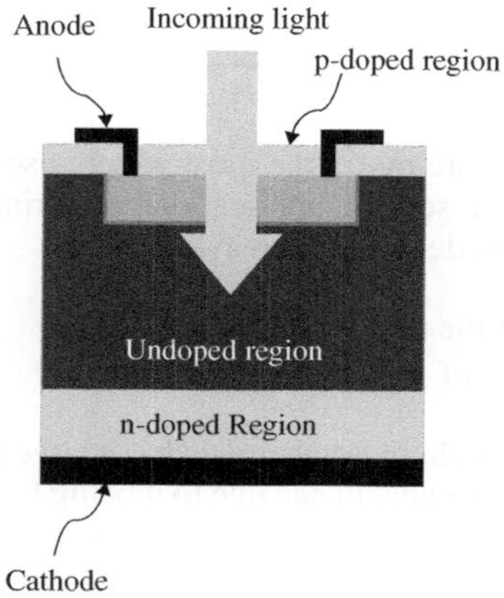

FIGURE 8.19 Schematic drawing of a PIN photodiode.

electrode at the bottom of the device and the anode as the form of a ring (we see a cross section of the rings in the figure). The positive pole of the (reverse) bias voltage is connected to the cathode. On top of the p region, there is an antireflection coating.

The depletion-layer width W_{PIN} in a PIN diode does not vary significantly with bias voltage but is essentially fixed by the thickness, d_{int}, of the potential region, that is, $W_{\text{PIN}} \approx d_{\text{int}}$.

The internal capacitance of a PIN diode then equals

$$C_i = C_j = \frac{\varepsilon_s A_{pn}}{W_{\text{PIN}}} \approx \frac{\varepsilon_s A_{pn}}{d_{\text{int}}}. \tag{8.52}$$

This capacitance is essentially independent of the bias voltage, remaining constant in operation.

The advantages of a PIN photodiodes are as follows:

- Increasing the width of the depletion layer (where the generated carriers can be transported by drift) increases the area available for capturing light.
- Increasing the width of the depletion layer reduces the junction capacitance and thereby the RC time constant. Yet, the transit time increases with the width of the depletion layer.
- Reducing the ratio between the diffusion length and the drift length of the device results in a greater proportion of the generated current being carried by the faster drift process.

8.2.4.5 Photodiode Response Time The bandwidth, $f_{3\text{dB}}$, and the 10–90% rise-time response, τ_r, both are determined by the values of the diode capacitance C_j and the load resistance R_L. As a general rule of thumb, we can write

$$f_{3\text{dB}} = \frac{1}{2C_j R_L}. \tag{8.53}$$

and

$$\tau_r = \frac{0.35}{f_{3\text{dB}}}. \tag{8.54}$$

High-speed photodiodes are by far the most widely used photodetectors in many communications and remote sensing applications requiring high speed. In general, the speed of a photodiode is determined by two factors:

- The response time of the photocurrent
- The RC time constant of its equivalent circuit.

Photodiodes operating in photovoltaic mode has a large RC time constant because of the large internal diffusion capacitance due to it being forward biased in this mode. This essentially eliminates it from high-speed application. In general, photovoltaic photodiodes are employed in solar cell/energy harvesting applications. Thus, only photodiodes operating in a photoconductive mode are suitable for high-speed or broadband applications.

As an alternate to the above equations, we can assume the RC frequency dependence, where $\tau_{\text{rc}} = C_j R_L$. This means we can write the RC circuit frequency response as

$$R_c^2(f) = R_c^2(0)\,[1 + 4\pi^2 f^2 \tau_{\text{rc}}^2] \tag{8.55}$$

where the 3-dB bandwidth now equals

$$f_{\text{rc,3dB}} = \frac{1}{2\pi\tau_{\text{rc}}} = \frac{1}{2\pi C_j R_L}. \tag{8.56}$$

Combining the photocurrent response and the circuit response, the total output power spectrum of an optimized photodiode operating in photoconductive mode

$$R^2(f) = R_c^2(0)\,[1 + 4\pi^2 f^2 \tau_{\text{rc}}^2]\left[\frac{\sin^2(\pi f \tau_{\text{rc}})}{(\pi f \tau_{\text{rc}})^2}\right]. \tag{8.57}$$

Example 8.2 For a voltage step input of amplitude V_{st}, the output voltage waveform $V_{\text{out}}(t)$ as a function of time t is

$$V_{\text{out}}(t) = V_{st}\left[1 - e^{-\left(\frac{t}{RC}\right)}\right]. \tag{8.58}$$

The 10–90% rise-time response for the circuit is equal to

$$\tau_r = 2.2\ \text{RC} = \frac{2.2}{2\pi f_{\text{rc,3dB}}} = \frac{0.35}{f_{\text{rc,3dB}}}, \tag{8.59}$$

which yields the following transfer function for this circuit

$$|H(\omega)|^2 = \frac{1}{1 + (\omega \text{RC})^2}. \tag{8.60}$$

8.2.5 Photodiode Array and Charge-Coupled Devices

Multichannel detectors such as the linear photodiode array (PDA) and the charge-coupled device (CCD) enabled new imaging systems to be developed for ultraviolet (UV)–visible (Vis) spectrophotometers and various other optical remote sensing systems [13]. A PDA is linear or array of discrete PN or PIN photodiodes on an integrated circuit chip, whose number of elements can run from 128 to 1024, with some devices now up to 4096. On the other hand, a CCD is an integrated circuit chip that contains an array of photosensitive capacitors that can store charge when incoming light creates EHP with the said element. The charge accumulates independently in each element and then all elements are read out serially in a fixed time interval.

In his paper on the advantages of PDAs, [13], Choi compares RDAs and CCDs to each other and his summary is as follows:

Under optimum conditions, these devices can detect as many wavelengths simultaneously as their number of individual diodes, resolution elements or pixels. Stray light and background illumination are common to both, but CCDs experience very low dark currents as compared to RDAs. They also require less of an electrical charge than their competitor, with higher charge-to-voltage conversion efficiency. This makes CCDs ideal for low-light-level detection such as Raman and Luminescence.

In general, PDAs are more suited for applications where the light level is relatively high. Because the photon saturation charge of CCD is small, the light intensity can be saturated in this device under many lighting conditions. These PDAs are preferred for strong light levels since their photon saturation charge is greater than CCD. In other words, the detection range of PDAs is larger than that of CCDs. Furthermore, PDAs deliver lower noise than CCD. That is why PDAs appear to be better suited for those applications where higher output accuracy is needed.

8.3 NOISE MECHANISMS IN OPTICAL RECEIVERS

Figure 8.20 illustrates an example layout for an optical receiver system. Some portion of the light from an optical source is captured by a lens systems and focused onto a detector or detector array. There may be optical spectral filters or predetection light amplifier in this optical train to minimize interfering background radiation from the sun, terrain, or from whatever, and boost incoming signal irradiance. The resulting received signal is sent to a postdetection amplifier and then onto the signal processing electronics. Ignoring any inherent statistical fluctuation of the incoming light, this figure shows an example set of receiver noise sources that inhibit optical receipt of the incoming signal. These include the following:

- *Photon noise* – The most fundamental source of noise is associated with the random arrivals of the photons (usually described by Poisson statistics).
- *Photoelectron noise* – a single photon generates an electron–hole pair with probability. The photocarrier-generation process is random.
- *Gain noise* – The amplification process that provides internal gain in certain photodetectors is stochastic.

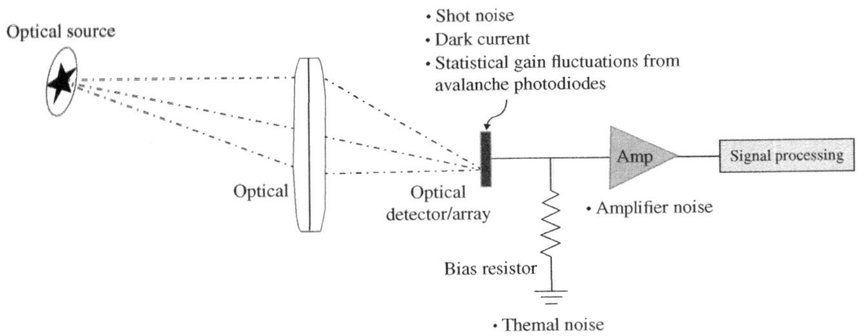

FIGURE 8.20 An example of the basic layout for an optical receiver system.

- *Receiver circuit noise* – Various components in the electrical circuitry of an optical receiver, such as thermal noise in resistors.

This section summarizes the possible noise sources in an optical receiver that influence the quality of signal reception.

8.3.1 Shot Noise

Lohmann once said that light was like Voltaire: "Voltaire was born a Catholic, lived a Protestant and died a Catholic; Light is born a particle, lives as a wave and dies a particle" [14]. Although light obeys the Maxwell equations while it propagates, its generation and detection are both "particulate generation" processes. The light received by an optical detector or focal plane array turns into electrical current (i.e., a stream of photoelectrons) proportional to the received optical power. This occurs when the incoming light to the detector/focal plane array originates from a laser, thermal emissions from background terrain, reflected sunlight, or any other external optical source.[2]

A direct effect from this detection process is when a photon is absorbed, the time of the associated photoelectron emission is not immediate but delayed due to some later random time. This phenomenon comes from the quantum mechanical nature of light and is predicted by the Heisenberg Uncertainty Principle. The result is a statistical self-noise component known as *quantum or signal shot noise*. The resulting photoelectron distribution obeys Poisson statistics. It generally is identified with those applications involving photon counting where signal reception is associated with the particulate nature of light. However, this source also emerges in background-limited system analysis where the magnitude of the external noise source, say reflected sunlight, is large [3, Chapter 11].

[2]Several researchers are trying to create "optical antenna" to directly detect the field amplitude of the incoming light. This is still a research topic at the moment, for example, Bharadwaj, P., Deutsch, B., and Novotny, L. (2009) Optical antennas, *Advances in Optics and Photonics*, **1** (3), 438–483 and Novotny, L. and Van Hulst, N. (2011) Antennas for light. *Nature Photonic*, **5** (2), 83–90. Since this book is on the engineering aspects of the field, we do not address this area this time and leave it to a later edition when the technology matures.

8.3.1.1 Quantum Shot Noise If one ran the Young's slit experiment and then turn down the light intensity, one could see the resulting diffraction pattern even when counting single photoelectrons coming out of the detector. The time occurrence of these photoelectrons would not track with the absorption of incoming photons by the detector. Rather, the emission of the photoelectrons would happen as random times subject to Poisson statistics. The result is the creation of a unique noise source dependent on the incoming light. This type of noise is called *quantum or signal shot noise*. The importance of this noise is that even if all other noise sources are suppressed, the detection process would not be noise-free. On the basis of our earlier discussion of radiation, we begin by illustrating optical shot noise statistics at either one of the two extremes. They are (1) incoherent and (2) fully coherent noise conditions, respectively. This discussion will closely follow Pratt [4, Chapter 1].

Mandel was credited with showing that the probability of obtaining k photoelectron counts in the finite time interval t to $t + \Delta t$ due to a general optical source is a Poisson distribution given by

$$\Pr(U_{R,\Delta t} = k; t) = \frac{1}{k!} \left[\int_t^{t+\Delta t} g\, i(t')dt' \right]^k e^{-\int_t^{t+\Delta t} g\, i(t')dt'}, \tag{8.61}$$

where $U_{R,\tau}$ is the random variable that represents the number of photoelectron counts, $i(t)$ the instantaneous intensity, measured in watts per second, of a quasi-monochromatic (narrowband compared to the center frequency) light centered at frequency ν_c that is incident on a detector/detecting element, and

$$g = \left(\frac{\eta}{E_{\text{photon}}} \right) = \left(\frac{\eta}{h\nu} \right) \tag{8.62}$$

the proportionality constant dependent on the photodetection mechanism for most photodetectors [5]. Here, η is the detector quantum efficiency. As noted in an earlier section, the time interval, Δt, is related to the electrical bandwidth of the filter following the detector.

The average number of photoelectron counts in the detection interval Δt of the ensemble average of $\Pr(U_{R,\Delta t} = k; t)$ is

$$\langle k_{R,\Delta t} \rangle = \left\langle \sum_{n=0}^{\infty} k \Pr(U_{R,\tau} = k; t), \right\rangle \tag{8.63}$$

which alternately may be written as the mean of a Poisson distribution given by

$$\langle k_{R,\Delta t} \rangle = \left\langle \int_t^{t+\Delta t} g i(t')\, dt' \right\rangle = g \langle i(t) \rangle, \tag{8.64}$$

where $i(t)$ is the time-averaged optical intensity over the time interval τ. Recognizing that the probability distribution $\Pr(U_{R,\Delta t} = k; t)$ is a random function of time since

$i(t)$ is a random function, we can calculate the time-averaged $\Pr(U_{R,\Delta t} = k; t)$ from the equation

$$\Pr(U_{R,\tau} = k) = \frac{1}{k!} \int_0^{\infty} \left[\int_t^{t+\tau} g\, i(t')dt' \right]^k e^{\left\{ -\int_t^{t+\tau} g\, i(t')dt' \right\}} \Pr(i)di, \tag{8.65}$$

where $\Pr(k)$ is the probability distribution of the intensity of the optical wave incident on the detector/detecting element. In general, $\Pr(U_{R,\Delta t} = k)$ is not a Poisson distribution.

There are two limiting cases of interest, defined by the coherence time, τ_c, of the optical radiation. The first case is where the coherence time is much smaller than the integration period ($\tau \gg \tau_c$) and the second case, the opposite ($\tau_c \gg \Delta t$). The former represents the situation for incoherent radiation while the latter is the situation for laser radiation. Let us look at these situations in more detail.

In the first case, the incoherent radiation is made quasi-monochromatic by an optical filter before hitting a detector. As noted in the last section, this type of radiation can be considered to be composed of a multitude of waves added together in random phase. By the Central Limit Theorem, the wave amplitude of the radiation has been found to be Gaussian random variable. By a transformation of variables, the probability distribution of the instantaneous intensity is the exponential distribution

$$\Pr(i_b) = \frac{1}{\langle i_b \rangle} e^{-i_b/\langle i_b \rangle}, \tag{8.66}$$

where $\langle i_b \rangle$, as in eq. 8.66 is the average incoherent radiation intensity over the time interval Δt. Using this distribution, the variance of the photoelectron counting distribution is then

$$\sigma^2(U_{R,\Delta t}) = g \langle i_b \rangle \tau + (g\, i_b \tau)^2 \left(\frac{\tau_c}{\Delta t} \right) \left(\frac{A_c}{A_d} \right) \tag{8.67}$$

In this situation, the coherence time for background radiation emitted by thermal sources is approximately 10^{-12} s and the ratio of $\left(\frac{A_c}{A_d} \right)$ is usually less than 10^{-3}. Thus, for all practical purposes, we have

$$\sigma^2(U_{R,\Delta t}) = g \langle i_b \rangle \Delta t \tag{8.68}$$

for incoherent light and the resulting distribution $\Pr(U_{R,\Delta t} = k)$ is Poisson.

For the second case, this condition is met when the laser emits at a single frequency – a single-mode laser – and operates far above its threshold of oscillation. The probability distribution for laser intensity then is assumed to be the ideal form

$$\Pr(i_c) = \delta(i_c - \langle i_c \rangle), \tag{8.69}$$

where i_c is the average laser intensity over the time interval Δt. The resulting photoelectron counting distribution is then a stationary Poisson distribution given by

$$\Pr(U_{R,\Delta t} = k) = \frac{1}{k!} [m_c]^k e^{-m_c} \tag{8.70}$$

with the mean and variance both are equal to $m_c = g\langle i_c\rangle \Delta t$. The variance term in this case is often called *signals shot noise* and has a flat spectrum.

It should be noted that background radiation from the sun, moon, stars, reflected solar radiance, and other natural and man-made optical sources can enter the receiver. Even if this radiation is constant, it may inhibit system performance. In the case of constant background optical sources, this radiation will create random fluctuations like above. These fluctuations are known as *background short noise*. It too has a flat spectrum.

The optical energy incident upon a photodetector is given by

$$E(\Delta t, t) = \int_t^{t+\Delta t} i(t')\, dt' \tag{8.71}$$

and the counting distribution $\Pr(U_{R,\Delta t} = k)$ may be investigated through its cumulants. By this technique, the average light intensity is small and $\Pr(U_{R,\Delta t} = k)$ is Poisson. This is important because some electro-optical systems have to operate at low received power levels because of channel degradation on incoming laser beams.

As a final note to this discussion, when the average light intensity is large, the distribution $\Pr(U_{R,\Delta t} = k)$ approaches the distribution of $gqE(\Delta t, t)$, which is the integrated photodetector current. The parameter $q(= 1.60217646 \times 10^{19}\,\mathrm{C})$ represents the electrical charge of an electron. Therefore, what does that mean? It indicates that the average output current of a photodetector, illuminated by a general optical radiation, is given by

$$i_p = g\langle qi\rangle = \left(\frac{\eta q}{h\nu}\right) P_{\mathrm{rx}} \tag{8.72}$$

where P_{rx} is the average received power. This equation represents the optical power-to-current conversion model of photodetection. The effect is that the electrical power created by this current is proportional to optical power-squared!

Example 8.3 Taking the logarithm of Eq. (8.70), we obtain

$$\begin{aligned}\ln\left[\Pr(U_{R,t} = k)\right] &= k\ln[m_c] - m_c - \ln[k!] \approx k\ln[m_c] - m_c - \ln[k^k e^{-k}\sqrt{2\pi k}]\\ &= k\ln\left[\frac{m_c}{k}\right] + (k - m_c) - \frac{1}{2}\ln[2\pi k]\end{aligned} \tag{8.73}$$

for $k \gg 1$, using Stirling's formula.

Letting $k = m_c + y$, Eq. (8.73) becomes

$$\begin{aligned}\ln[\Pr(U_{R,t} = k)] &= -(m_c + y)\ln\left[1 + \frac{y}{m_c}\right] + y - \frac{1}{2}\ln[2\pi(m_c + y)]\\ &\approx -\frac{y^2}{2m_c} + \frac{y^3}{6(m_c)^2} - \frac{1}{2}\ln[2\pi(m_c + y)]\\ &\approx -\frac{y^2}{2m_c} - \frac{1}{2}\ln[2\pi m_c]\end{aligned} \tag{8.74}$$

as $m_c \gg y$ and the cubic term is much smaller than the quadratic term. Exponentiation of Eq. (8.74) yields

$$\Pr(U_{R,t} = k) = \frac{1}{\sqrt{2\pi m_c}} e^{-\frac{(k-m_c)^2}{2m_c}}, \tag{8.75}$$

which is a Gaussian distribution with mean and variance equal to m_c

8.3.1.2 Dark Current Shot Noise The concept of electrical shot noise was first introduced in 1918 by Walter Schottky who studied fluctuations of current in vacuum tubes [15]. This type of noise may be dominant when the finite number of "particles" that carry energy (such as photo-electrons in an electronic circuit) is small so that uncertainties in the received signal level is significant. It is something we need to still consider today in optical receivers.

In any photoemissive or photovoltaic detector, a dark current is created in the absence of any external optical source and this in turn creates shot noise within the receiver [5, pp. 148–150]. This also is time-dependent Poisson process. In particular, the probability that the number of dark current electrons emitted in a time period Δt is exactly the integer k is given by

$$P(U_{\mathrm{dc},\Delta t} = k) = \frac{(\mu_{\mathrm{dc},\Delta t})^k e^{-\mu_{\mathrm{dc},\Delta t}}}{k!}, \tag{8.76}$$

where $\mu_{\mathrm{dc},\Delta t} = \frac{i_{\mathrm{dc}}\Delta t}{q}$ is the average number of dark current electrons released by the detector in a time period Δt. Here q is the charge of an electron and i_{dc} is the dark current. It can be shown that the noise power spectrum density for this noise component is given by

$$G_{i_{\mathrm{dc}}}(\nu) = G^2 q i_{\mathrm{dc}} + G^2 i_{\mathrm{dc}}^2\, \delta(\nu) \tag{8.77}$$

with G being the postdetection gain current [5]. This equation implies that the noise power spectrum is composed of a flat spectrum defined by the first part of the equation and direct current component by the second part of the equation. The total *dark current shot noise power* at a load resistor, R_L, created by the simple resistor–capacitor filter of bandwidth B can then be written as

$$P_{i_{\mathrm{dc}}} = 2G^2 q i_{\mathrm{dc}} B R_L. \tag{8.78}$$

8.3.2 Erbium-Doped Fiber Amplifier (EDFA) Noise

Many optical communications systems use an erbium-doped fiber amplifier (EDFA) to amplify the received signal. Its inherent noise sources are a little more complicated as they involve spontaneous and stimulated emissions in the fiber amplifier but are well understood [16].

The signal–spontaneous beat noise is equal to

$$\sigma_{s\text{-sp}}^2 = 2G i_s i_{\mathrm{sp}} \left(\frac{B_e}{B_0}\right) \tag{8.79}$$

and the spontaneous–spontaneous beat noise is equal to

$$\sigma^2_{\text{sp–sp}} = \frac{1}{2}\, i^2_{\text{sp}} \left(\frac{B_e(2B_0 - B_e)}{B_0^2} \right) \tag{8.80}$$

where

B_e ≡ Bandwidth of the electrical filter used in the electrical receiver circuit

B_0 ≡ Optical Bandwidth

G ≡ Amplifier gain

i_s ≡ Photocurrent generated by received bit "1" signal in the absence of a fiber amplifier

$$= \left(\frac{q\eta}{h\nu} \right) P_s \tag{8.81}$$

P_s ≡ Receiver bit "1" optical power before the amplifier

i_{sp} ≡ Photocurrent generated by the spontaneous emissions (including both polarizations) at the output of the detector

$$= 2qn_{\text{sp}}(G - 1)B_0 \tag{8.82}$$

n_{sp} ≡ Population inversion factor (or inversion parameter)

$$= \frac{N_2}{N_2 - N_1} \tag{8.83}$$

N_1 ≡ Electron population in state 1

and

N_2 ≡ Electron (elevated) population in state 2.

In order to obtain source gain, the stimulated emission from state 2 to state 1 of source must be greater than the absorption from state 1 to state 2. This condition implies that the population in state 2 must be maintained at a greater level than that of state 1, that is, population inversion. The degree of population inversion is expressed by the population inversion factor given in Eq. (8.83).

8.3.3 Relative Intensity Noise

Relative intensity noise (RIN) describes the instability in the power level of an optical source like a laser or EDFA. For example, RIN can be generated from cavity vibration, fluctuations in the laser gain medium, or simply from transferred intensity noise from a pump source. It is known as *pink noise*.

RIN is obtained from the autocorrelation integral of the optical power fluctuations, divided by the total power squared [17]. As we saw before, these temporal fluctuations can be described by their frequency spectrum, which is derivable from the Fourier transform of said autocorrelation integral. That is, an optical source of output $P(t)$ and fluctuations $\delta P(t)$ has a total RIN,

$$\text{RIN}_T = \frac{\langle \delta P(t)^2 \rangle}{\langle P(t) \rangle^2}, \tag{8.84}$$

where the time-averaged parameter $\delta P(t)^2$ comes from the autocorrelation function $\delta P(t)\delta P(t+\tau)$ evaluated at $\tau = 0$. The total RIN can be expressed in terms of frequency via the equation

$$\mathrm{RIN}_T = \int_0^\infty \hat{R}_{\mathrm{RIN}}(\nu)\, d\nu, \tag{8.85}$$

where $\hat{R}_{\mathrm{RIN}}(\nu)$ denotes the RIN spectral density and ν is the optical frequency in hertz. The RIN spectral density can be defined from the Wiener–Khintchine theorem as the autocorrelation function $\kappa(\tau)$. Specifically, equating Eqs. (8.85) and (8.86) yields

$$R_{\mathrm{RIN}}(\nu) = \int_{-\infty}^\infty \kappa(\tau) e^{2\pi i \nu \tau}\, d\tau \tag{8.86}$$

The above implies that the RIN essentially is $\hat{R}_{\mathrm{RIN}}(\nu)$ and the total RIN its integral. Obarski and Splett noted that for classical states of light, the spectral density of the amplitude fluctuations has a minimum of the standard Poisson limit, to which is added some excess noise [17]. In other words, the RIN spectral density equates to the sum of a Poisson RIN and excess RIN. The former depends on the system losses and the latter is unchanged throughout the receiver train. Thus, the total RIN is the spectral integral of these two RINs.

Example 8.4 For light that obeys a Poisson distribution, its variance is proportional to the mean number of photons. This implies that the square of the optical power fluctuations is proportional to the optical power. When it is presented by a single-sided noise spectrum, the Poisson RIN at a single frequency is given by $\frac{2h\nu}{P_0}$, where P_0 is the optical power at that frequency [17]. In electrical units, we can write $\frac{2q}{i_0}$, where q again is the electron charge and $i_0 = \frac{\eta q P_0}{h\nu}$ is the photocurrent and η is the quantum efficiency of the photodetector. This implies that the Poisson RIN in a real detector circuit increases as $\frac{1}{\eta}$ [4].

Example 8.5 Obarski and Splett discussed a specific RIN measurement experiment and it will be provided here [17]. Figure 8.21 depicts their RIN measurement system layout. A laser beam propagates through an attenuating medium and part of the resulting optical beam is captured by a photodetector and processed.

Referring to this figure, we see that the total RIN is obtained in reference plane A. The measurement will define $\hat{R}_{\mathrm{RIN}}(\nu)$ and the total RIN, its integral. At reference plane B, the Poisson RIN increases because of the beam losses. Similarly, at reference plane C, the Poisson RIN further increases because of the photodetector's quantum efficiency not being in unity and its receiving optics having losses. This coined the loss parameter at reference plane C as η_T, which is defined as

$$\eta_T = \eta(1 - \gamma_{\mathrm{opt}}), \tag{8.87}$$

where γ_{opt} is the fractional loss before the detecting element. Turning to the excess RIN referenced above, it propagates unperturbed through the attenuating medium and detector optics.

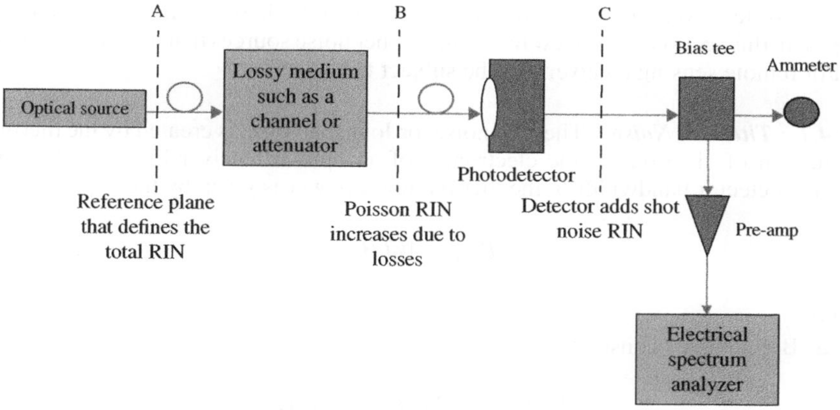

FIGURE 8.21 Basic RIN measurement setup.

Let $R_C(\nu)$ be the measured RIN determined in at reference plane C. Using the results from Example 8.4, the RIN at reference plane A is given by

$$\hat{R}_C(\nu) = \hat{R}_C(\nu) - \left(\frac{2q}{i_0}\right)(1 - \eta_T). \tag{8.88}$$

The excess RIN can be found by subtracting shot noise RIN, $\frac{2h\nu}{P_0}$, from Eq. (8.28), which means that

$$\hat{R}_{\text{excess}}(\nu) = \hat{R}_C(\nu) - \left(\frac{2q}{i_0}\right), \tag{8.89}$$

where P_0 is the laser power at reference plane A and ν the laser frequency.

Obarski and Splett highlighted that measuring RIN in the electrical domain involves using a bias tee to send the dc photocurrent to an ammeter while the ac noise is amplified and then sent to an RF spectrum analyzer [17]. They denoted the electrical frequency as f, referenced to the baseband. The RIN then is the noise power per unit bandwidth, $\delta P_e(f)$, weighted by the electrical frequency-dependent calibration function $\kappa_c(f)$ of the detection system and divided by the electrical dc power P_e. This implies that

$$R(f) = \frac{\kappa_c(f)\delta P_e(f)}{P_e}, \tag{8.90}$$

where $\delta P_e(f)$ is the noise after subtracting the thermal noise. Note that $\kappa_c(f)$ is proportional to the frequency response of the system and can be obtained by inputting a broadband, flat source of known RIN into the system.

8.3.4 More Conventional Noise Sources

Besides the above, there are a number of other noise sources that will plague the optical systems engineer, many generated inside the receiver electronics, under the various system configurations possible. These include thermal, flicker, and current noises. In this section, we review the various other noise sources that a communication and/or remote sensing receiver may be subject to.

8.3.4.1 Thermal Noise Thermal noise, or Johnson noise, is created by the thermal fluctuation of electrons in the electronics of an optical receiver [5, pp. 145–148]. Given a detector bandwidth B, the thermal noise power is given by the expression

$$P_{\mathrm{TH}} = 4kTB_e \tag{8.91}$$

where
k ≡ Boltzmann's constant =

$$1.3806504(24) \times 10^{-23}\,\mathrm{J/K} \tag{8.92}$$

T ≡ Receiver temperature in degrees kelvin (K)
B_e ≡ Detector bandwidth in hertz (Hz) or s^{-1}.

Equation (8.91) is the same form for thermal noise one has with RF and microwave systems, which is not too surprising. It follows Gaussian statistics.

8.3.4.2 Flicker Noise The fluctuations in the emissions from spots on a vacuum tube cathode create *flicker noise* [5, p. 152]. The spectrum of this noise is inversely proportional to frequency for frequencies down to 1 Hz and to the square of the photocathode current, i_K. Its power spectrum density is given by

$$G_{i_{\mathrm{dc}}}(\nu) = \frac{\alpha_F G^2 i_K^2}{\nu}, \tag{8.93}$$

where α_F is a proportionality constant. This expression is not valid down to zero frequency since the noise power must be bounded. Flicker noise is associated mainly with thermionic emission, so it is not a significant noise factor in photoemissive tubes.

8.3.4.3 Current Noise Semiconductor detectors carrying a steady current exhibit a noise effect, called *current of 1/f noise (current noise)*, which is a one-sided power spectrum inversely proportional to frequency below 1 Hz [5, pp. 152–153]. The noise spectrum dependents upon the square of the postdetector amplified current and the operational environment – most notably upon the humidity but not upon the temperature of the device. The physical mechanism producing current noise is believed to be the trapping of charge carriers near the surface of a material. The power spectrum density of current noise is written as

$$G_{i_{\mathrm{dc}}}(\nu) = \frac{\alpha_C G^2 i_P^2}{\nu} \tag{8.94}$$

where α_C is a proportionality constant and i_P the steady current in the semiconductor detector.

8.3.4.4 ***Phase Noise*** In a heterodyne, or homodyne, optical receiver in which optical mixing occurs, the spectral (line) shape of the transmitter and local oscillator lines becomes significant because the laser lines are essentially shifted intact to a lower radio frequency called an intermediate frequency (IF) [5, pp. 153–155]. *Phase noise* arises from the random frequency shifting of the carrier and local oscillator lasers as result of spontaneous emission quantum noise and environmental vibrations of the laser cavity mirrors. The quantum noise is a fundamental irreducible cause of frequency fluctuations. In most applications, the quantum noise is hidden by the much stronger environmental noise that comes from turbulent atmospheric signal fading. Carrier frequency fluctuations due to atmospheric turbulence are another major cause of phase noise.

8.4 PERFORMANCE MEASURES

Prior to World War II (WWII), it was believed that one just needed more power, i.e., larger microwave tubes, to detect targets or to communicate. Most of the work in that era was strictly on making better tubes [18]. WWII accelerated radar research significantly and it was found that power did not necessarily guarantee excellent target detection at range. Sometimes they could see that targets are larger ranges than they expected and other times they had trouble getting good results at short ranges. The trouble was noise. Kilbon's history on RCA's radar research noted that

> Noise is a perennial and annoying problem in electronic systems. In this sense, the word refers not simply to audible phenomena, but to all forms of interference that result from the operating characteristics of a tube or circuit. This type of noise is evidence of the effort being made by the tube or circuit to do its job. It is objectionable when it becomes great enough to have an appreciable effect on the output signal, whether the signal happens to be a television picture, a radar pulse, or the sound of a radio. This accounts for the constant preoccupation of the electronic specialist with "signal-to-noise ratio," and his continued effort to find ways of increasing the signal relative to the interfering noise.

They found that a new performance measure, not just transmitter power, was needed to quantify any system performance involving electromagnetic wave reception; specifically, it was the SNR. North determined via classified work during WWII that the maximum possible SNR is obtained when the receiving system used a matched filter for the transmitted waveform [19]. This is still true today. However, there are additional performance measures that help us understand how well the system will work. The following are the most popular ones:

1. *Signal-to-Noise Ratio (*SNR *or* S/N*)* – ratio of square mean of the signal to its noise variance
2. *Minimum Detectable Power* – the mean signal power level that yields SNR = 1
3. *Probability of False Alarm/Bit Error* – related to the probabilities of saying signal is present when it is not, or saying it is not when it is
4. *Receiver Sensitivity* – the signal that corresponds to a prescribed value of the SNR.

This section discusses parameters (1), (2), and (4), beginning with SNR and its various definitions in optics, and then move to the other two. Item (3) is discussed in Chapter 9.

8.4.1 Signal-to-Noise Ratio

Signal-to-Noise Ratio (denoted by SNR or S/N) is a measure used in science and engineering that compares the level of a desired signal to the level of undesired signal, which is known as *noise*. By normal convention, it is defined as the ratio of signal power to the noise power

$$\mathrm{SNR} = \frac{\text{Signal Electrical Power}}{\text{Noise Electrical Power}} = \frac{\text{Signal Mean-Squared}}{\text{Noise Variance}} = \frac{i^2_{\text{signal}}}{i^2_{\text{noise}}}$$

$$= \frac{\left(\frac{\eta q}{h\nu}\right)^2 P^2_{\text{signal}}}{\sigma^2_{\text{noise}}} \tag{8.95}$$

where i is the average current, η the detector quantum efficiency, q the electronic charge, h the Planck's constant, ν the frequency of the incoming light, P the average optical power, and σ^2_{noise} the noise variance from internal and external sources. A ratio higher than 1:1 indicates more signal than noise. The required SNR value needed usually comes from the specification of the probabilities of detection and false alarm for remote sensing applications, and the probabilities of detection and bit errors for communications applications. This is shown in the next chapter.

Example 8.6 The electrical SNR for a direct detection is given by the relation

$$\mathrm{SNR} \approx \frac{\left(\frac{Gq\eta}{h\nu}\right)^2 R_L {P_{\text{rec}}}^2}{2qB_eG^2\left[\left(\frac{q\eta}{h\nu}\right)(P_{\text{rec}} + P_b) + i_D\right] R_L + 4kTB_e}, \tag{8.96}$$

where $P_{\text{rec}} \equiv$ received optical signal power before the detector (Watts) (includes optical losses) and $P_b \equiv$ received optical background noise power before the detector (Watts).

When dark current and thermal noise are negligible, then Eq. (8.96) can be written as

$$\mathrm{SNR} \approx \frac{\left(\frac{Gq\eta}{h\nu}\right)^2 {P_{\text{rec}}}^2}{2qB_eG^2\left(\frac{q\eta}{h\nu}\right)(P_{\text{rec}} + P_b)} = \left(\frac{\eta}{2h\nu B_e}\right)\frac{{P_{\text{rec}}}^2}{(P_{\text{rec}} + P_b)}$$

$$= \left(\frac{\eta}{2h\nu B_e}\right)\frac{P_{\text{rec}}}{\left(1 + \frac{1}{\mathrm{CNR}}\right)}, \tag{8.97}$$

where

$$\mathrm{CNR} \equiv \text{Carrier-to-Noise Ratio} = \frac{P_{\text{rec}}}{P_b} \tag{8.98}$$

CNR is the optical SNR at the input of the first detector of the optical receiver. If CNR is very large, then Eq. (8.97) reduces to

$$\mathrm{SNR} \approx \left(\frac{\eta P_{\mathrm{rec}}}{2h\nu B_e} \right), \tag{8.99}$$

which represents *signal shot-noise-limited, or quantum-limited, SNR*. This is the limiting case for an optical system when all external and internal noise sources are eliminated. It represents the uncertainty in the number of photoelectrons emitted by the detector. As it turns out, Eq. (8.99) is valid for both very high (Gaussian statistics) and very low (Poisson statistics) received power levels.

On the other hand, when the background noise source is strong compared to all other noise sources, Eq. (8.96) reduces to

$$\mathrm{SNR} \approx \frac{\left(\frac{Gq\eta}{h\nu}\right)^2 {P_{\mathrm{rec}}}^2}{2qB_e G^2 \left(\frac{q\eta}{h\nu}\right) P_b} = \left(\frac{\eta}{2h\nu B_e} \right) \frac{{P_{\mathrm{rec}}}^2}{P_b}. \tag{8.100}$$

Equation (8.100) is known as the *background-limited SNR* and is where the background noise dictates the performance of the receiver. In this situation, narrowing the optical band-pass and the receiver electrical bandwidth of the receiver to the laser signal characteristics will enhance the probability of signal detection as the system operates under a peak power detection mode.

Example 8.7 Figure 8.22 depicts the most common optical receiver for high-speed optical communications, which utilizes a PIN-photodiode. An incoming signal $P_s(t)$ impinges an optical detector and is changed into current that gets electrically amplified in an RF amplifier. At this stage, we denote its gain by G, its load resistor as R_L and its *Noise Figure* by F_{rf}. This current is then subjected to postdetection filtering with impulse response $h_{\mathrm{rf}}(t)$ and effective electrical bandwidth B_e. This filtered results then passes through sample-and-hold circuitry before yielding the final result, Data. The electrical SNR in this detection scheme is given by [1, p 182]

$$\mathrm{SNR}_{\text{DD-PIN}} \approx \frac{\left(\frac{q\eta}{h\nu}\right)^2 R_L {P_{\mathrm{rec}}}^2}{2qB_e \left[\left(\frac{q\eta}{h\nu}\right) P_{\mathrm{rec}} + I_{\mathrm{D}}\right] R_L + 4kTF_{\mathrm{rf}}B_e}. \tag{8.101}$$

The detector responsivity $R_\lambda = \left(\frac{q\eta}{h\nu}\right) = 1\ \mathrm{mA/mW}$ when $\lambda_c = 1.55\,\mu\mathrm{m}$ and $\eta \sim 0.8$. If the signal shot noise is much greater than the thermal noise, Eq. (8.101) reduces to the quantum-limited SNR given in Eq. (8.99).

Example 8.8 APD detectors can increase communications performance over PIN detectors by providing internal gain within the detection process. That is, the photocurrent is multiplied by an average gain factor of M. That is the good news. The bad news is that amplification process creates additional noise like it always does. The resulting SNR for this type of system is equal to

$$\mathrm{SNR}_{\text{DD-APD}} \approx \frac{\left(\frac{q\eta}{h\nu}\right)^2 R_L {P_{\mathrm{rec}}}^2}{2qB_e F_{\mathrm{ADP}} \left[\left(\frac{q\eta}{h\nu}\right) P_{\mathrm{rec}} + i_D\right] R_L + \frac{4kTF_{\mathrm{rf}}B_e}{M^2}} \tag{8.102}$$

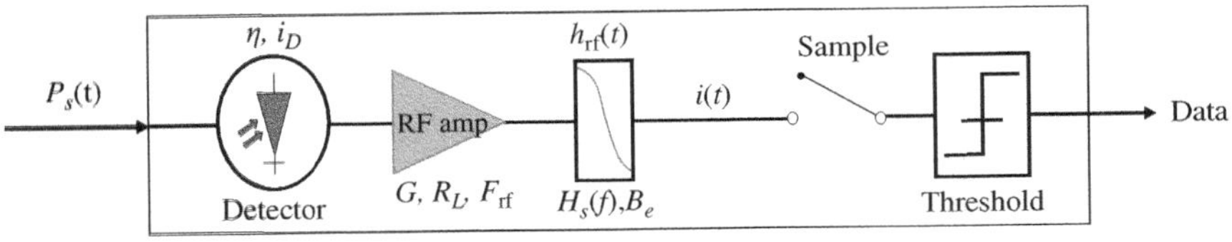

FIGURE 8.22 Schematic of a PIN-based optical receiver.

where F_{ADP} is the excess noise factor associated with the variations in M. It is clear in this equation that the thermal noise term is reduced by a factor of M-squared and the shot noise term is increased by F_{ADP}.

The excess noise factor is a function of both the gain and the APD's effective ionization coefficient ratio (k_{ion}) [20]. It is calculated using a formula derived by Mclntyre, which equals

$$F_{\text{ADP}}(M, k_{\text{ion}}) = M\left[1 - (1 - k_{\text{ion}})\left(\frac{M-1}{M}\right)^2\right] \tag{8.103}$$

[21]. Equation (8.103) assumes an avalanche medium with uniform characteristics and an impact-ionization process that is independent of carrier history [21]. Thus, for $M \approx 10$ and $k_{\text{ion}} \approx 0.2$, we have $F_{\text{ADP}} \approx 3.52$. By application of the Burgess variance theorem (used to find the variance of a multiplied quantity) and an extension of Milatz's theorem (used to relate a rate of occurrence to a spectral density), we have the commonly used formula for an APD's noise spectral intensity:

$$S_{\text{ADP}} = 2qM^2 F_{\text{ADP}}(M, k_{\text{ion}})\left[\left(\frac{q\eta}{h\nu}\right)P_{\text{rec}} + I_D\right] \tag{8.104}$$

in amperes-squared per hertz, neglecting minor 1/*f*-noise and thermal-noise components [20]. Equation (8.104) is most appropriate for those noise analysis cases when the signal is encoded in the instantaneous power level of an optical signal and one integrates S_{ADP} over the bandwidth of the receiver to find the APD's contribution to the variance of the current fluctuations.

8.4.2 The Optical Signal-to-Noise Ratio

Outside of classical electrical engineering, there are many alternate definitions for SNR that are accepted and used extensively. This is especially true in the field of optics. For example, Eq. (8.95) also can be written as

$$\text{SNR} = \frac{\left(\frac{\eta q}{h\nu}\right)^2 P^2_{\text{signal}}}{\sigma^2_{\text{noise}}} = (\text{OSNR})^2 \tag{8.105}$$

with

$$\text{OSNR} = \frac{\left(\frac{\eta q}{h\nu}\right) P_s}{\sqrt{\sigma^2_{\text{noise}}}} = \frac{\left(\frac{\eta q}{h\nu}\right) P_s}{\sigma_{\text{noise}}} \tag{8.106}$$

Equation (8.105) starts out with power in electrical watts and ends with a SNR in terms of optical power. Both equations are functions of power (watts), but their entities are very different. This book has adopted the electrical engineering convention, which is given in Eq. (8.95), so the reader can relate things between electrical engineering and optical engineering easily. However, the OSNR is used by a lot of people and deserves a little discussion.

Optical signals have a carrier frequency that is much higher than the modulation frequency (about 200 THz and more). This way, the noise covers a bandwidth that is much wider than the signal itself. The resulting signal influence relies mainly on the filtering of the noise. To describe the signal quality without taking the receiver into account, OSNR is used by certain researchers to characterize system performance. As defined above, the OSNR is the ratio between the signal power and the noise power in a given bandwidth. Most often, a reference bandwidth of 0.1 nm is used. This bandwidth is independent of the modulation format, the frequency and the receiver. Therefore, we see that this definition of SNR has its own bandwidth that directly relates to the optical power in a spectral, and not electrical, sense. This bandwidth generally is chosen to minimize the effect of direct or reflected sunlight into the detector. The conclusion is that there are two separate bandwidths used in any link analysis. One directly relates to the received OSNR and the other directly to SNR.

Let us look at a specific example where SNR and OSNR are proportional to each other to establish this dependence on both.

Example 8.9 Many modern communications systems employ high-power amplifiers such as an EDFA in both the transmitter and receiver subsystems because of their high electrical-to-optical conversion efficiency. It probably is the most popular high-power amplifier in use today. However, when an EDFA is involved, the SNR is not proportional to optical power-squared.

In Section 8.3.2, we learned about the additional noise mechanisms in ERDAs. The electrical SNR for an EDFA-based system is given by

$$\mathrm{SNR} = \left(\frac{G^2 {i_s}^2}{{\sigma_{\text{total}}}^2} \right) \tag{8.107}$$

where the total noise variance is given by

$${\sigma_{\text{total}}}^2 \approx {\sigma_{\text{shot}}}^2 + {\sigma_{\text{thermal}}}^2 + {\sigma_{s\text{–sp}}}^2 + {\sigma_{\text{sp–sp}}}^2. \tag{8.108}$$

Figure 8.23 illustrates a typical plot of various amplifier noise powers, their total, and the received SNR as a function of amplifier gain. In this figure, $\lambda = 1.55\,\mu\text{m}$, B_e is 8.5 GHz, B_0 is 75 GHz, $n_{\text{sp}} = 1.4$, T is 290 K, $\eta = 0.7$, and the received power is −30 dBmW. It is clear from this graph that thermal noise typically is the major system noise source when the gain is low. As the gain increases, the in-band beat noise (postdetection) from the ss–sp and from the sp–sp emissions emerges as the dominant noise sources. This is one of the times that signal-induced noise does not dominate. These sources affect the form of the electrical SNR under high amplifier gain operation.

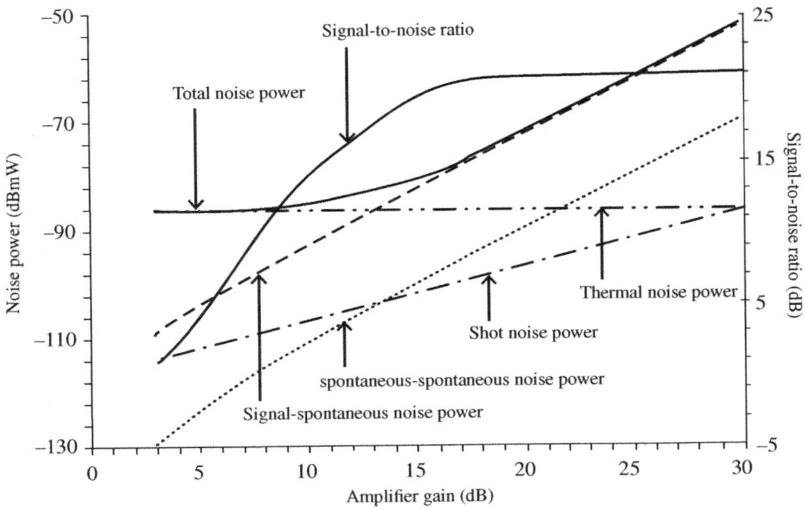

FIGURE 8.23 Noise powers and signal-to-noise ratio for an erbium-doped fiber preamplifier system as a function of amplifier gain.

Let us now look at SNR when the in-band beat noise dominates. Under this situation, we have $\sigma^2_{\text{shot}} + \sigma^2_{\text{thermal}} \ll \sigma^2_{s\text{–sp}} + \sigma^2_{\text{sp–sp}}$ and Eq. (8.107) becomes

$$\text{SNR} = \left(\frac{G^2\, i_s{}^2}{\sigma_{s\text{–sp}}{}^2 + \sigma_{\text{sp–sp}}{}^2} \right). \tag{8.109}$$

Substituting in Eqs. (8.79) and (8.80), we obtain

$$\text{SNR} = \left(\frac{G^2 i_s{}^2}{2 G i_s i_{\text{sp}} \left(\frac{B_e}{B_0}\right) + \frac{1}{2} i_{\text{sp}}{}^2 \left(\frac{B_e (2B_0 - B_e)}{B_0{}^2}\right)} \right) \tag{8.110}$$

If $2B_0 \gg B_e$, which is normally true, then

$$\text{SNR} = \left(\frac{\text{OSNR}_0^2 \left[\frac{B_0}{2B_e}\right]}{\text{OSNR}_0 + \frac{1}{2}} \right), \tag{8.111}$$

where

$$\text{OSNR} = \left(\frac{G i_s}{i_{\text{sp}}} \right) \tag{8.112}$$

If OSNR $\gg \frac{1}{2}$, which also usually is true, the above equation further reduces to

$$\text{SNR} = \text{OSNR}\left[\frac{B_0}{2B_e}\right] \tag{8.113}$$

Equation (8.113) shows that electrical SNR is proportional to the optical SNR when the nonsignal-related noise from the high-power amplifier dominates the receiver system noise. Writing out the optical electrical signal-to-noise ratio, we obtain

$$\text{OSNR} = \left(\frac{G\left(\frac{\eta q}{h\nu}\right)P_{\text{rec}}}{2qn_{\text{sp}}(G-1)B_0}\right) = \left(\frac{G\left(\frac{\eta}{h\nu}\right)P_{\text{rec}}}{2n_{\text{sp}}(G-1)B_0}\right) \tag{8.114}$$

Substituting equation (8.114) into the electrical SNR equation yields

$$\text{SNR} = \left(\frac{G\left(\frac{\eta}{h\nu}\right)P_{\text{rec}}}{2n_{\text{sp}}(G-1)B_0}\right)\left[\frac{B_0}{2B_e}\right] \tag{8.115}$$

$$= \left(\frac{G\left(\frac{\eta}{h\nu}\right)P_{\text{rec}}}{4n_{\text{sp}}(G-1)B_e}\right) \tag{8.116}$$

$$= \frac{\eta P_{\text{rec}}}{4n_{\text{sp}}h\nu B_e} = \frac{\text{SNR}_{\text{Signal Shot Noise}}}{4n_{\text{sp}}} \tag{8.117}$$

for $G \gg 1$. Equation (8.117) shows that the electrical SNR for amplifier noise-limited communications is proportional to the "signal shot-noise (quantum limited)" SNR, scaled by the inverse of four times the inversion parameter under high amplifier gain conditions. It says that the SNR is constant under high gain conditions, which is consistent with Figure 8.23 where the SNR reaches an asymptote for $G > 16$ dB.

Example 8.10 Let us compare the SNR for PIN and EDFA receivers. Let us assume that $I_s = \left(\frac{q\eta}{h\nu}\right)P_{\text{rec}}$, $\left(\frac{q\eta}{h\nu}\right) = 0.5$, $P_{\text{rec}} = 3.2\,\mu\text{W}$, $\lambda = 1.55\,\mu\text{m}$, $G = 1556$, $n_p = 2.25$, $B_e = 6.2\,\text{GHz}$, $B_0 = 12.4\,\text{GHz}$, $T = 300\,\text{K}$, and $R_L = 50\,\Omega$. Using these numbers, we find that the signal shot noise is equal to

$$\sigma^2_{\text{shot}} = 2qB_eI_s = 2\,(1.6\times10^{-19})[6.2\times10^{9}](0.5\times3.2\times10^{-6})\,\text{A}^2 = 3.174\times10^{-15}\,\text{A}^2$$

and the Johnson noise variance is given by

$$\sigma^2_{\text{thermal}} = \frac{4kTB_e}{R_L} = \frac{4\,(1.38\times10^{-23})[300](6.2\times10^{9})}{50} = 2.053\times10^{-12}\,\text{A}^2.$$

In this situation, Johnson noise is much larger than the signal shot noise. This implies that the signal-to-noise ratio is given by

$$\text{SNR} \approx \frac{(\eta P_{\text{rec}})^2}{\frac{4kTB_e}{R_L}} = \frac{(0.5\times3.2\times10^{-6}\,\text{A}^2)^2}{2.053\times10^{-12}\,\text{A}^2} = \frac{2.56\times10^{-12}\,\text{A}^2}{2.053\times10^{-12}\,\text{A}^2} = 1.247\ \ (0.96\text{dB})$$

Let us now turn to the SNR for the preamplifier systems. For the above parameter set, the spontaneous current i_{sp} equals

$$i_{sp} = 2qn_{sp}(G - 1)B_0 = 2(1.6 \times 10^{-19})[2.25](1555)[12.4 \times 10^9] = 1.39 \times 10^{-5}\,\text{A}.$$

This implies that the signal–spontaneous beat noise equals

$$\sigma^2_{s\text{–sp}} = 2Gi_s i_{sp}\left(\frac{B_e}{B_0}\right) = 2(1556)(0.5 \times 3.2 \times 10^{-6})[1.39 \times 10^{-5}](0.5)\,\text{A}^2$$
$$= 3.46 \times 10^{-8}\,\text{A}^2$$

and the spontaneous–spontaneous beat noise equals

$$\sigma^2_{sp\text{–sp}} = \frac{1}{2}\,i^2_{sp}\left(\frac{B_e(2B_0 - B_e)}{B_0^2}\right) \approx i^2_{sp}\left(\frac{B_e}{B_0}\right)$$
$$= 0.5(1.39 \times 10^{-5}\,\text{A})^2 = 9.66 \times 10^{-11}\,\text{A}^2$$

Clearly, signal–spontaneous beat noise is the larger of the two. This means that

$$\text{SNR} \approx \frac{(Gi_s)^2}{\sigma^2_{s\text{–sp}}} = \frac{(1556 \times 0.5 \times 3.3 \times 10^{-6}\,\text{A})}{(3.46 \times 10^{-8}\,\text{A}^2)} = \frac{6.198 \times 10^{-6}\,\text{A}^2}{3.46 \times 10^{-8}\,\text{A}^2} = 179.1\ (22.5\,\text{dB})$$

Alternately, we could have calculated the SNR using Eq. (8.113), which yields the same result:

$$\text{SNR} \approx \left(\frac{Gi_s}{i_{sp}}\right)\left[\frac{B_0}{2B_e}\right] = \frac{2.49 \times 10^{-3}\,\text{A}^2}{1.39 \times 10^{-5}\,\text{A}^2} = 179.1\ \ (22.5\,\text{dB}).$$

This comparison shows why a number of communications systems use an optical preamplifier.

Example 8.11 High-speed free-space optical communication systems have recently used EDFAs as predetection amplifiers [3, Chapter 10]. The received laser beam in such a system must be coupled into a single-mode fiber at the input of the receiver terminal. However, propagation through atmospheric turbulence degrades the spatial coherence of a laser beam and limits the fiber-coupling efficiency. Dikmelik and Davidson derived the following equation for the fiber-coupling efficiency under these conditions:

$$\eta_{fc} = 8a^2 \int_0^1 \int_0^1 e^{-\left(a^2 + \frac{A_R}{A_c}\right)(x_1^2 + x_2^2)} I_0\left(\frac{A_R}{A_c} x_1 x_2\right) x_1 x_2\, dx_1\, dx_2, \tag{8.118}$$

where η_{fc} is the fiber-coupling efficiency, a the ratio of the receiver lens radius to the radius of the back propagated fiber mode, $A_R = \frac{\pi D_R^2}{4}$ the area of the receiver aperture,

D_R the diameter of the receiver aperture, $A_c = \frac{\pi \rho_{pw}^2}{4}$ the spatial coherence area of the incident plane wave, and ρ_{ps} the plane wave spatial coherence radius, which given by

$$\rho_{pw} = \frac{r_0}{2.1} = \left[1.46k^2 \int_0^L C_n^2(z')\,dz'\right]^{-\frac{3}{5}} \tag{8.119}$$

(unlike the Fried Parameter, this parameter is a radius and not a diameter) [22]. The coherence area A_c also is called its speckle size, so the ratio $\frac{A_R}{A_c}$ represents the number of speckles over the receiver aperture area. Figure 8.24 depicts fiber-coupling efficiency (solid curve) as a function of the number of speckles, $\frac{A_R}{A_c}$, over the receiver aperture, with $a = 1.12$ [22].

Equation (8.118) shows that the coupling efficiency depends on the coupling geometry through a single parameter a, which is given by

$$a = \left(\frac{D_R}{2}\right)\left(\frac{\pi W_m}{\lambda f}\right) \tag{8.120}$$

where W_m is the fiber-mode field radius at the fiber end face [23], λ the wavelength of light, and f the lens focal length. The fiber mode profile was approximated by a Gaussian function in the derivation of Eq. (8.118). This approximation is commonly

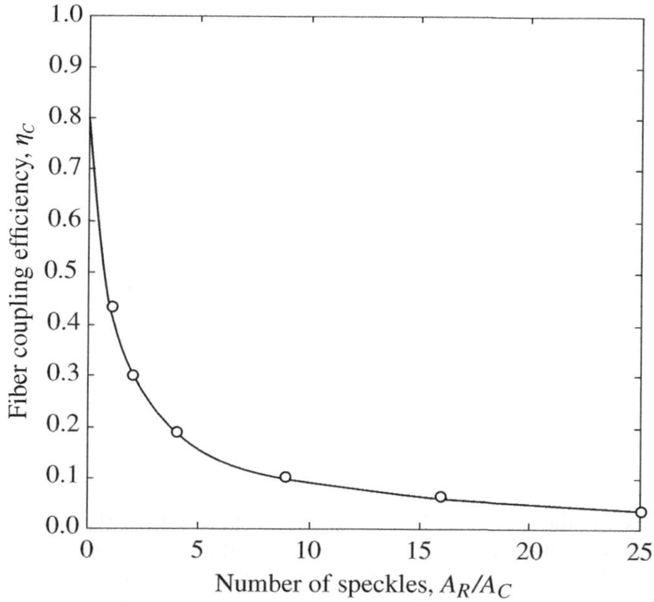

FIGURE 8.24 Fiber-coupling efficiency as a function of the number of speckles, $\frac{A_R}{A_c}$, over the receiver aperture. The coupling geometry parameter a equals 1.12 in this curve. The circles in this graph represent the coupling efficiency for optimized values of a derived for six specific number of speckles $\frac{A_R}{A_c}$.

used in calculations of fiber coupling efficiency and does not lead to an appreciable loss of accuracy [22]. In the absence of turbulence, the incident plane wave is fully coherent and the optimum value of a is 1.12 [24]. With turbulence, the optimum value of the coupling-geometry parameter a that maximizes the coupling efficiency depends on the value of the number of speckles $\frac{A_R}{A_c}$. Dikmelik and Davidson compared the coupling efficiencies for six optimum values with the efficiency calculated with the nonturbulence value of 1.12. Figure 8.24 has the resulting data points plotted with the $a = 1.12$ curve [22]. These results suggest that one can employ $a = 1.12$ for all fiber-coupling efficiency calculations with turbulence.

Recall from Chapter 6 that the spherical beam wave approach to link characterization was deemed more appropriate in real FSOC applications than the plane beam wave approach. This means that we should replace Eq. (8.118) with

$$\rho_{\text{sw}} = \left[0.55k^2 \int_0^L C_n^2(z')dz'\right]^{-\frac{3}{5}} \tag{8.121}$$

in calculating the fiber-coupling efficiency. This coherence radius is ~1.8 times bigger than the one given in Eq. (8.118). Andrews did a detailed investigation of Gaussian-beam wave propagation and came up with the following equation for its spatial coherence radius

$$\rho_c = \frac{1.80\rho_{\text{pw}}}{\left[a_0 + 0.62\Lambda^{\frac{11}{6}}\right]^{\frac{3}{5}}}, \tag{8.122}$$

where

$$a_0 = \begin{cases} \dfrac{\left(1-\Theta^{\frac{8}{3}}\right)}{(1-\Theta)} & \text{for } \Theta \geq 0 \\ \dfrac{\left(1-|\Theta|^{\frac{8}{3}}\right)}{(1-\Theta)} & \text{for } \Theta < 0 \end{cases}, \tag{8.123}$$

$$\Theta = 1 - \frac{L}{f}, \tag{8.124}$$

$$\Lambda = \frac{2L}{kW^2}, \tag{8.125}$$

$$\Lambda_0 = \frac{2L}{kW_0^2}, \tag{8.126}$$

and W_0 represents the Gaussian beam radius [25, pp. 196–199]. Figure 8.25 shows the normalized spatial coherence radius ρ_c of a collimated Gaussian-beam wave as a function of the Fresnel ratio Λ_0. When $\Lambda_0 \ll 1$, that is, wide Gaussian beams, the ratio is around one, indicating that the plane wave approach is valid. This corresponds to the near-field diffraction regime. On the other hand, when $\Lambda_0 \gg 1$, that is, collimated Gaussian beam, the ratio is approximately 1.8, indicating that the spherical wave approach is valid. This region is in the far-field diffraction regime. When

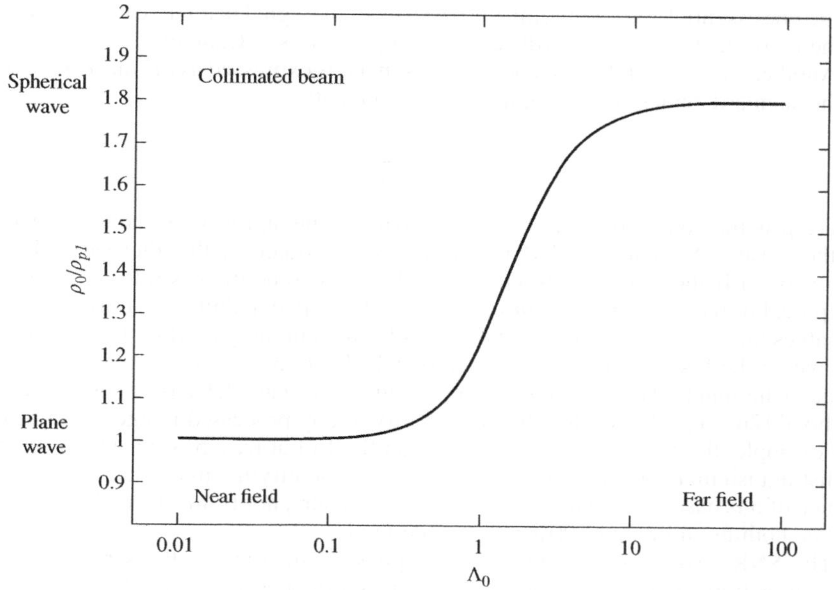

FIGURE 8.25 Spatial coherence radius ρ_c of a collimated Gaussian-beam wave, scaled by the plane wave coherence radius ρ_{pw} and plotted as a function of the Fresnel ratio Λ_0.

$0.6 < \Lambda_0 < 8$, the ratio has a 1.46 average value. This corresponds to the transition regime between the near- and far-field diffraction regimes. Since most real FSOC are in the far field, this suggests that Eq. (8.121) should be used when evaluating the fiber-coupling efficiency for real applications.

8.4.3 The Many Faces of the Signal-to-Noise Ratio

The optics community uses other definitions for the SNR because the optical problem depends on different performance criteria. For example, some researchers may be more interested in subjected image recognition rather than deterministic signal detectability by the array. For completeness, we introduce some other definition that the readers may run across during their careers.

In imaging applications, the SNR often written as

$$\text{SNR} = \frac{C^2}{\frac{\sigma^2_{\text{noise}}}{\mu_b^2}}, \tag{8.127}$$

where C is the Weber contrast, σ^2_{noise} the noise variance, and μ_B the mean background level. The derivation of this equation is shown in Chapter 9.

If the signal is zero-mean, then SNR is rewritten as

$$\text{SNR} = \frac{\sigma^2_{\text{signal}}}{\sigma^2_{\text{noise}}}, \tag{8.128}$$

where one essentially is comparing the variance of the signal in relative to the variance of the noise fluctuations. We will see an example of this in Chapter 9.

Another variation of the SNR definition is that it is expressed as the ratio of median of the standard deviation of a signal or measurement:

$$\mathrm{SNR} = \frac{\mu}{\sigma}, \tag{8.129}$$

where μ is the signal median or 50% occurring value and σ the standard deviation of the noise.[3] (Sometimes SNR is defined as the square of the alternative definition above.) If the statistics are Gaussian, then this sometimes is referred to as the reciprocal of the *coefficient of variation*. This alternative definition is only useful for variables that are always nonnegative. It has been commonly used in image processing where the SNR of an image is usually calculated as the ratio of the mean pixel value to the standard deviation of the pixel values, both calculated over a given neighborhood [26–31]. The result is that the quality of any processed image is subjected. For example, the *Rose criterion* states that an SNR of at least 5 is needed to be able to distinguish image features at 100% certainty in identifying image details [32]. The choice of 5 is based on the judgment of the researchers, not from a formal derivation, that is, nothing in life has 100% absolute certainty.

The SNR also is used in imaging as a physical measure of the sensitivity of a (digital or film) imaging system. Industry standards measure and define sensitivity in terms of the ISO film speed equivalent; SNR:32.04 dB = excellent image quality and SNR:20 dB = acceptable image quality. This is too much of *niche* topic, so we do not go into detail here. We leave it to the interested reader learn more about this by independent library searching.

8.4.4 Noise Equivalent Power and Minimum Detectable Power

Photodetector sensitivity is a convenient, even necessary, metric by which the performance of a particular photodetector can be quantified and compared with other detectors. The noise equivalent power (NEP) is the most common metric that quantifies a photodetector's sensitivity or the power generated by a noise source. It is the required incident power on the detector necessary for unity RMS signal-to-noise ratio at the output for a specific center frequency and electrical bandwidth. For blackbodies, it depends on the source temperature over an electronic filter's center frequency and its electrical bandwidth. Mathematically, we can write

$$\mathrm{SNR} \approx \frac{\left(\frac{q\eta}{h\nu}\right)^2 {P_{\mathrm{signal}}}^2}{2qB_e\ \left[\left(\frac{q\eta}{h\nu}\right)(P_{\mathrm{signal}}) + i_D\right] + \frac{4kTB_e}{R_L}} = 1,$$

which leads to

$$R_\lambda P_{\mathrm{signal}} = \sqrt{B_e} \times \sqrt{2q[\ R_\lambda P_{\mathrm{signal}} + i_D] + \frac{4kT}{R_L}}.$$

[3]Median and average are sometimes confused. For non-Gaussian statistics, the former is the correct term to use to define the 50% occurring value. When one has Gaussian statistics, then they are interchangeable as they are equal to one another.

Rewriting the above, the definition of NEP emerges:

$$\frac{R_\lambda P_{\text{signal}}}{\sqrt{B_e}} = \frac{\sqrt{2q[R_\lambda P_{\text{signal}} + i_D] + \frac{4kT}{R_L}}}{R_\lambda} \equiv \text{NEP} \tag{8.130}$$

Its general formula is

$$\text{NEP} = \text{NEP}_{\min}\left[\frac{R_{\max}}{R_\lambda}\right], \tag{8.131}$$

where $\text{NEP}_{\min}$ is the NEP as given in the specifications, $R_{\max}$ the maximum responsivity of the detector, and R_λ the responsivity of the detector at wavelength λ. All of these parameters can be found in the manufacturer's detector specification sheet or operating manual. Let us assume a wavelength of interest of 532 nm (2×Nd:YAG). From a specific detector data sheet, we have $\text{NEP}_{\min} = 0.01\,\text{pW}/\sqrt{\text{Hz}}$, $R_{\max} = 0.47\,\text{A/W}$ at $\lambda = 830\,\text{nm}$, and $R_\lambda = 0.27\,\text{A/W}$ at $\lambda = 532\,\text{nm}$. The NEP then is $0.017\,\text{pW}/\sqrt{\text{Hz}}$. The minimum detectable power is given by

$$P_{\min} = \text{NEP}\sqrt{BW} \tag{8.132}$$

If the measurement bandwidth is 800 MHz, then $P_{\min} = 480\,\text{pW}$.

A term one sometimes finds in the literature is the *noise equivalent input (NEI)*. The NEI is the irradiance at the detector that produces a unity SNR, which parallels, and is related to, the definition of NEP. Mathematically, we find that

$$\text{NEI} = \frac{\text{NEP}}{A_{\text{d}}} \tag{8.133}$$

The specific (normalized) detectivity (D^*) is derived from the NEP and the active detector area A_{det}. mathematically, we have

$$D^* = \frac{\sqrt{A_{\text{d}}BW}}{\text{NEP}} \tag{8.134}$$

D^* is used in cases when the noise scales as the square root of the area, such as shot noise. The advantage of using specific detectivity D^* is that it allows for the comparison of detectors that have different active areas.

8.4.5 Receiver Sensitivity

The sensitivity of an electro-optical device, such as a communications or remote sensing receiver, is the minimum value of input optical power required to produce a specified output signal with a specified SNR, or other specified criteria. In communications, this is the minimum optical power for the required SNR set by a specific bit-error-rate (BER), say BER-10^{-9}.

Because optical receiver sensitivity indicates how weak an optical beam can be to be successfully detected by the receiver under the above criteria, the lower the

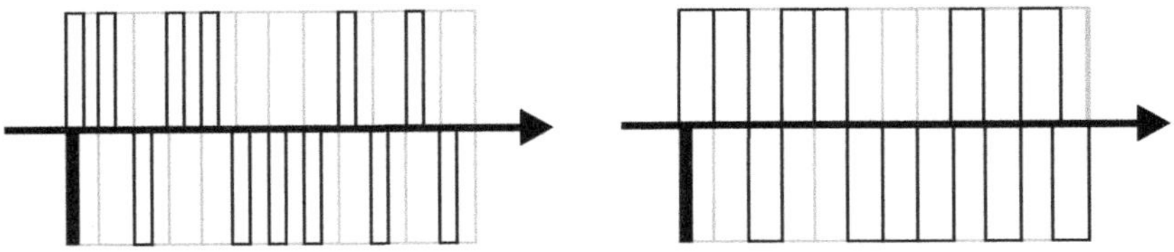

FIGURE 8.26 The binary signal is encoded using rectangular pulse amplitude modulation with (a) polar return-to-zero code and (b) polar non-return-to-zero code.

power level, the better. Lower power for a given SNR_{REQ} means better sensitivity since the receiver's contribution is smaller. When the optical power is expressed in decibel milliwatts, the larger the absolute value of the received signal (signals at range always register negative decibel milliwatts), the better the optical receiver sensitivity. For example, a receiver sensitivity of −60 dBmW is better than a receive sensitivity of −55 dBmW by 5 dB, or about a factor of 3. In other words, at a specified data rate, a receiver with a −60 dBmW sensitivity can detect a "1" signal that are one-third the power of a "1" signal recorded by a receiver with a −55 dBmW receiver sensitivity.

For example, Juarez *et al.* compared the FSOC receiver sensitivities for two modulation schemes: (1) a 10 Gbps non-return-to-zero (NRZ) on–off key (OOK) and (2) a 10 Gbps return-to-zero (RZ) different phase shift key (DPSK) [33]. Figure 8.26 illustrates the modulation formats of RZ and NRZ. An EDFA was used in both cases as a part of a predetection optical automatic gain control (OAGC) device. In addition, Reed–Solomon forward error correction (FEC) coding was employed in both the approaches. Total noise is a combination of different noise contribution mechanisms such as thermal noise, signal and dark current shot noise, RIN noise (depends of RIN value of optical sources such as lasers), and signal–spontaneous and spontaneous–spontaneous beat noise from the EDFA.

Figure 8.27 depicts the performance of both approaches. The performance improvement shown in this figure is the result of changing the signal coding scheme as well as adjusting certain receiver parameters [33]. Clearly, the latter system performed better with the former by ~ 8 dB. This is not a surprise. This figure also shows that adding the FEC improved both the NRZ–OOK and RZ–DPSK communications performance by another ~ 8 dB.

Example 8.12 Figure 8.28 illustrates the Eye Diagram for the ORCA 10 Gbps NRZ–OOK system with, and without, an OAGC (see Appendix B for instructions on how to interpret Eye Diagrams) [34]. The left-hand figure is an Eye Diagram without an OAGC and was set to operate at a BER ~ 10^{-6}. The right-hand figure is an Eye Diagram with the OAGC and the BER ~ 10^{-12}. This figure shows that OAGC creates better signal detectability against the noise, as exhibited by the sharper pattern.

Figure 8.29 shows the optical 10 Gbps stream of "1"s from an Adaptive Optical Telescope, just before the OAGC (PIF) and the resulting received stream after the OAGC and FEC (POF) for the NRZ–OOK ORCA system cited above [33, 34]. This data is from 185 km link run during a 5× HV 5/7 turbulence condition [34]. The

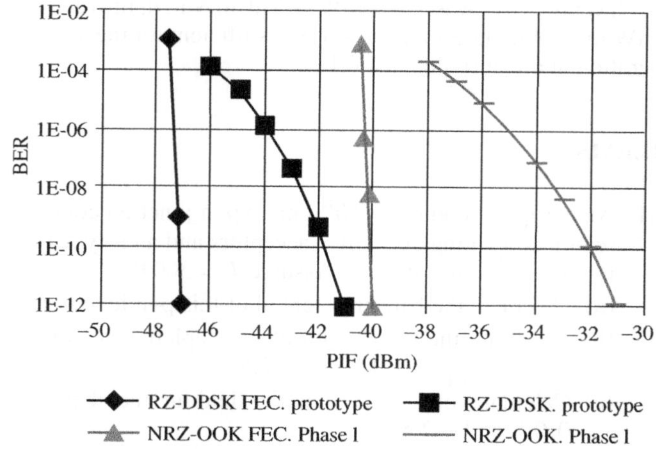

FIGURE 8.27 Measured DPSK transceiver and OAGC BER baselines as a function of received optical power into the OAGC with a 100 GHz drop filter. Note: These measurements include the input losses into the OAGC.

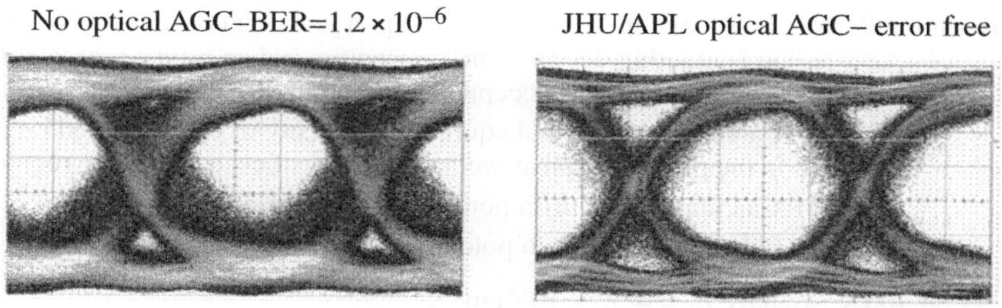

FIGURE 8.28 Eye Diagrams for ORCA NRZ–OOK system with, and without, the OAGC.

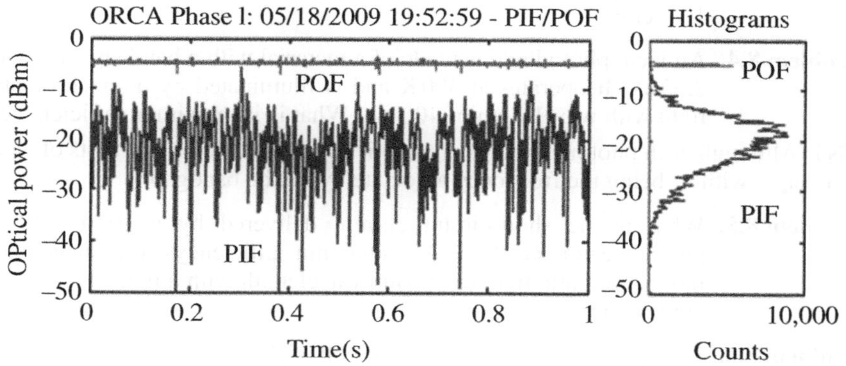

FIGURE 8.29 Example of received power dynamics and operation of first-generation OAGC during ORCA flight tests.

required signal level for a "1" was −5 dBmW (−5 dBmW in figure) at a BER $\sim 10^{-12}$. The −41 dBmW minimum detectable power was sufficient for the system to generate error-free operation at this turbulence level.

8.5 PROBLEMS

Problem 8.1. An abrupt silicon ($n_i = 10^{10}\,\text{cm}^{-3}$) p–n junction consists of a p-type region containing $10^{16}\,\text{cm}^{-3}$ acceptors and an n-type region containing $5 \times 10^{10}\,\text{cm}^{-3}$ donors. Assume $T = 300\,\text{K}$.

(a) Calculate the barrier potential of this p–n junction.

(b) Calculate the total width of the depletion region if the applied voltage $V_{\text{app}} = 0, 0.5, \text{and} - 2.5$.

(c) Calculate maximum electric field in the depletion region at $0, 0.5, \text{and} - 2.5$.

(d) Calculate the potential across the depletion region in the n-type semiconductor at $0, 0.5, \text{and} - 2.5$.

HINT: The permittivity of abrupt silicon, ε_s, is equal to 1.054× pico-farads per centimeter (pF/cm) [35].

Problem 8.2. An abrupt silicon ($n_i = 10^{10}\,\text{cm}^{-3}$) p–n junction consists of a p-type region containing $2 \times 10^{16}\,\text{cm}^{-3}$ acceptors and an n-type region containing also $10^{16}\,\text{cm}^{-3}$ acceptors in addition to $10^{17}\,\text{cm}^{-3}$ donors.

(a) Calculate the thermal equilibrium density of electrons and holes in the p-type region as well as both densities in the n-type region.

(b) Calculate the built-in potential of the p–n junction.

(c) Calculate the built-in potential of the p–n junction at 400 K.

HINT: at 400 K, we have $n_i = 4.52 \times 10^{12}\,\text{cm}^{-3}$.

Problem 8.3. Consider an abrupt p–n diode with $N_a = 10^{18}\,\text{cm}^{-3}$ and $N_a = 10^{16}$. Calculate the junction capacitance at zero bias. The diode area equals $10^{-4}\,\text{cm}^2$.

Problem 8.4. An ideal photodiode is made of a material with a bandgap energy of 2.35 eV. It operates at 300 K and is illuminated by monochromatic light with wavelength of 400 nm. What is its maximum efficiency?

HINT: Although each photon has an energy $h\nu$ it will produce only $h\nu_g$ units of electric energy, with ν_g being the frequency associated with the bandgap.

Problem 8.5. What is the short-circuit current delivered by a 10 cm × 10 cm photodetector (with 100% quantum efficiency) illuminated by monochromatic light of 400 nm wavelength with a power density of $1000\,\text{W/m}^2$.

Problem 8.6. Under different illumination, the cell of Problem 8.4 delivers 5 A into a short circuit. The reverse saturation current is 100 pA. Disregard any internal resistance of the photodiode. What is the open-circuit voltage at 300 K?

Problem 8.7. A GaAs p–n junction has a 100 μm × 100 m cross section and a width of the depletion layer $W = 440$ nm. Consider the junction in thermal equilibrium without bias at 300 K. Find the junction capacitance.

HINT: $\varepsilon_s = 13.18\varepsilon_0$ for GaAs and $\varepsilon_0 = 8.854 \times 10^{-12}$ F/m.

Problem 8.8. Assume the photodiode of this problem as an ideal structure with 100% quantum efficiency and area 1 cm^2(= 1 cm × 1 cm). In addition, assume it is illuminated by monochromatic light with a wavelength of 780 nm and with a power density of 1000 W/m^2. At 300 K, the open-circuit voltage is 0.683 V.

(a) What is its reverse saturation current, I_0?

(b) What is the load resistance that allows maximum power transfer?

(c) What is the efficiency of this photodiode with the load above?

Problem 8.9. A photodiode is exposed to radiation of uniform spectral power density

$$S = \frac{dP}{d\nu} \equiv \text{constant}$$

(covering the range from 300–500 THz. Outside this range there is no radiation. The total power density is 2000 W/m^2. Assume that the photodiode has 100% quantum efficiency.

(a) What is the short-circuit photocurrent of a diode having an active area of 1 cm^2(= 1 cm × 1 cm)?

(b) When exposed to the radiation in Part a of this problem, the open-circuit voltage delivered by the diode is 0.498 V. A 1.0 V voltage is applied to the diode (which is now in darkness) in the reverse conduction direction (i.e., in the direction in which it acts almost as an open circuit). What current circulates through the device? The temperature of the diode is 300 K.

Problem 8.10. The power density of monochromatic laser light (586 nm) is to be monitored by a silicon photodiode with area equal to 1 mm^2 (= 1 mm × 1 mm). The quantity observed is the short-circuit current generated by the silicon. Assume that the diode is an ideal device.

(a) What current would you obtain if the light level is 230 W/m^2?

(b) Explain how the temperature of the semiconductor affects this current. (Note: the temperature has to be lower than that which will destroy the device. For example, some 150 °C for silicon).

(c) Instead of being shorted, the diode now is connected to a load compatible with maximum electric output. Estimate the load voltage.

Problem 8.11. What is the theoretical efficiency of a photodetector with a 2.5 eV bandgap when exposed to 100 W/m^2 solar radiation through with transmittance,

$$\gamma_r = \begin{cases} 1; & \text{if } 600\,\text{nm} \leq \lambda \leq 1000\,\text{nm} \\ 0; & \text{if } \lambda < 600\,\text{nm or } \lambda > 1000\,\text{nm} \end{cases}$$

Problem 8.12. Consider a small silicon photodiode with a $100\,\text{cm}^2 = 10\,\text{cm} \times 10\,\text{cm}$ area. When 2 V of reversed bias is applied, the reverse saturation current is 30 nA. When the photodiode is short-circuited and exposed to blackbody radiation with a power density of $1000\,\text{W/m}^2$, a short-circuit current circulates. Assume 100% quantum efficiency where each photon creates one electron–hole pair and all pairs are separated by the p–n junction of the diode.

(a) What is the value of this current?

(b) What is the open-circuit voltage of the photodiode at 300 K under the above illumination?

Hint: The sun can be modeled as essentially a 6000 K black body. When the power density of such radiation is $1000\,\text{W/m}^2$, the total photon flux is 4.46×10^{21} photons/m^2-s. It can be shown that essentially half of these photons have energy equal or larger than 1.1 eV, the bandgap energy, W_g, of silicon.

Problem 8.13. Referring to Problem 8.11, consider an ideal photodiode with no internal resistance.

(a) Under an illumination of $1000\,\text{W/m}^2$, at 300 K, what is the maximum power the photodiode can deliver to a load. What is the efficiency? Do this by trial and error and be satisfied with three significant figures in your answer.

(b) What is the load resistance that leads to maximum efficiency?

Hint: Observe that the I–V characteristics of a photodiode are very steep at the high current end. In other words, the best operating current is only slightly less that the short-circuit current.

Problem 8.14. Assume that we have an EDFA FSOC communications receiver where

$$\sigma_{\text{shot}}{}^2 + \sigma_{\text{thermal}}{}^2 \ll \sigma_{s\text{–sp}}{}^2 + \sigma_{\text{sp–sp}}{}^2$$

under high amplifier gain levels. For $\lambda = 1.55\,\mu\text{m}$, $B_0 = 75\,\text{GHz}$, $B_e = 7.5\,\text{GHz}$, $n_p = 1.4$, $\eta = 0.7$, and $P_s = -30\,\text{dBmW}$, calculate SNR and OSNR?

Problem 8.15. Assume that we have a horizontal 10 km FSOC link at height h_0 above the ground. If the refractive index structure function at altitude is $C_n{}^2(h_0) \approx 3 \times 10^{-14}\,\text{m}^{-\frac{2}{3}}$, the wavelength of the FSOC laser is 1.55 μm, what is the spatial coherence radius ρ_{sw}? What is the value of the Fried Parameter r_0?

Problem 8.16. A photomultiplier (PMT) has a dark count noise level of 100 counts per seconds. Calculate the minimum optical power ($SNR = 1$) at 500 nm that can be detected within a 10 s integration time. Assume a quantum efficiency of 0.2 at 500 nm.

Problem 8.17. A p–i–n photodiode has a responsivity of 0.8 A/W at 1.55 μm. The effective resistance of the detector circuit is 50 Ω. The photodiode dark current density is $1\,\text{pA}/\sqrt{\text{Hz}}$. Calculate the NEP of the detector at 300 K. Include both shot noise and Johnson noise in your calculation. If the area of the detector is 1 mm^2, what is the NEI?

Problem 8.18. Referring to problem 8.16, what would the NEP be if the detector is now feeding an amplifier with a noise temperature of 600 K?

Hint: The Noise Figure F_{rf} is related to its noise temperature T_{amp}:

$$F_{rf} = 1 + \frac{T_{amp}}{T_0}$$

REFERENCES

1. Majumdar, A.K. (2014) *Advanced Free Space Optics (FSO): A Systems Approach*, Springer Series in Optical Sciences, Volume 186, Editor-in-Chief William T. Rhodes, Springer, New York.
2. Khatib, M. (ed.) (2014) , InTech (Open Access Publishing), Contemporary Issues in Wireless Communications. doi: 10.5772/58482. ISBN: 978-953-51-1732-2
3. Karp, S. and Stotts, L.B. (2013) *Fundamentals of Electro-Optic Systems Design: Communications, Lidar, and Imaging*, Cambridge Press, New York.
4. Pratt, W.K. (1969) *Laser Communications Systems*, John Wiley & Sons, Inc., New York.
5. Kopeika, N.S. (1998) *A System Engineering Approach to Imaging*, SPIE Optical Engineering Press, Bellingham, WA.
6. Saleh, B.E. and Teich, M.C. (2007) *Fundamentals of Photonics*, Wiley-Interscience, John Wiley and Sons, New York.
7. Burle Industries Inc. *Photomultiplier Tube Handbook*, http://psec.uchicago.edu/links/Photomultiplier_Handbook.pdf (accessed 22 November 2016).
8. Richardson, O.W. (1901) On the negative radiation from hot platinum. *Philosophical of the Cambridge Philosophical Society*, **11**, 286–295.
9. Liu, J.-M. (2005) *Photonic Devices*, Chapter 14, Cambridge Press, New York.
10. *PN Junction Diode and Diode Characteristics Tutorial*, http://www.electronics-tutorials.ws/diode/diode_3.html (accessed 26 September 2016).
11. Dervos, C.T., Skafidas, P.D., Mergos, J.A., and Vassiliou, P. (2004) p–n junction photocurrent modelling evaluation under optical and electrical excitation. *Sensors*, **5**, 58–70.
12. OSI Optoelectronics "*Photodiode Characteristics and Applications*" Fact Sheet.
13. Choi, H., *Advantages of Photodiode Arrays*, Source: http://www.hwe.oita-u.ac.jp/kiki/ronnbunn/paper_choi.pdf (accessed 26 September 2016)
14. Lohmann, A.W. (2006) in *Optical Information Processing* (ed. S. Sinzinger), Universitätsverlag, Ilmenau, Germany. ISBN: 4-939474-00-6
15. Schottky, W. (1918) Über spontane Stromschwankungen in verschiedenen Elektrizitätsleitern. *Annalen der Physik*(in German), **57**, 541–567.
16. Becker, P., Olsson, N., and Simpson, J. (1999) *Erbium-Doped Fiber Amplifiers Fundamentals and Technology*, Academic Press, New York.
17. Obarski, G.E. and Splett, J.D. (2001) Transfer standard for the spectral density of relative intensity noise of fiber sources near 1550 nm. *Journal of the Optical Society of America*, **18 (6)**, 750–761.

18. Kilbon, K., *A Short History of the Origins and Growth of RCA Laboratories, Radio Corporation of America, 1919 to 1964*, August 1964, Sarnoff Library. Source: davidsarnoff .org/kil.html (accessed 26 September 2016).
19. North, D.O. (1963) An analysis of the factors which determine signal/noise discrimination in pulsed-carrier systemsRCA Tech. Rept., PTR-6C, June 25, 1943 (ATI 14009). Reprinted in . *Proceedings of the IEEE*, **51**, 1016–1027.
20. Williams, G.M., "*Avalanche Photodiode Receiver Performance Metrics Introduction*," VOXTEL Technical Note V803. Source: http://voxtel-inc.com/files/2012/07/TECH-NOTE-V803-Avalanche-Photodiode-and-Receiver-Performance-Metrics-By-GM-Williams-VOXTEL_B.pdf (accessed 26 September 2016).
21. Mclntyre, R.J. (1966) Multiplication noise in uniform avalanche diodes. *IEEE Transactions on Electron Devices*, **ED13**, 161–168.
22. Dikmelik, Y. and Davidson, F.M. (2005) Fiber-coupling efficiency for free-space optical communication through atmospheric turbulence. *Applied Optics*, **44** (**23**), 4946–4951.
23. Buck, J.A. (1995) *Fundamentals of Optical Fibers*, Wiley.
24. Winzer, P.J. and Leeb, W.R. (1998) Fiber coupling efficiency for random light and its applications to lidar. *Optics Letters*, **23**, 986–988.
25. Andrews, L.C. and Phillips, R.L. (2005) *Laser Beam Propagation through Random Media*, 2nd edn, SPIE Press.
26. Schroeder, D.J. (1999) *Astronomical Optics*, 2nd edn, Academic Press, New York, p. 433. ISBN: 978-0-12-629810-9
27. Bushberg, J.T., Seibert, J.A., Leidholdt, E.M. Jr., and Boone, J.M. (2006) *The Essential Physics of Medical Imaging*, Lippincott Williams & Wilkins, 2nd edn, p. 280.
28. González, R.C. and Woods, R.E. (2008) *Digital Image Processing*, Prentice Hall, New York, p. 354. ISBN: 0-13-168728-X
29. Stathaki, T. (2008) *Image Fusion: Algorithms and Applications*, Academic Press, New York, p. 471. ISBN: 0-12-372529-1
30. Raol, J.R. (2009) *Multi-Sensor Data Fusion: Theory and Practice*, CRC Press. ISBN: 1-4398-0003-0
31. Russ, J.C. (2007) *The Image Processing Handbook*, CRC Press. ISBN: 0-8493-7254-2. 9780306307324..
32. Rose, A. (1973) *Vision – Human and Electronic*, Plenum Press, p. 10. ISBN: 9780306307324
33. Juarez, J.C., Young, D.W., Sluz, J.E., and Stotts, L.B. (2011) High-sensitivity DPSK receiver for high-bandwidth free-space optical communication links. *Optics Express*, **19** (**11**), 10789–10796.
34. Bagley, Z.C., Hughes, D.H., Juarez, J.C. *et al.* (2012) Hybrid optical radio frequency airborne communications. *Optical Engineering*, **51**, 055006-1–055006-25. doi: 10.1117/1.OE.51.5
35. Leroy, C. and Rancoita, P.-G. (2012) *Silicon Solid State Devices and Radiation Detection*, World Scientific, p. 184. ISBN: 9814397385

9

SIGNAL DETECTION AND ESTIMATION THEORY

9.1 INTRODUCTION

Statistical detection and estimation theory revolutionized the electrical engineering fields in the last century, beginning with Stephen Rice's seminal paper entitled, "The Mathematical Analysis of Random Noise" [1]. His work had immense scientific and engineering influence not only on radio frequency (RF) communications but also on other fields of engineering where random processes are important such as radar and sonar. However, it was not recognized until the 1960s by educator/researchers such as Mandel [2], Helstrom [3], Lohmann [4], and Papoulis [5] that optics would benefit from this theory as well. They felt that the merger of physical optics and communications theory would enable a new understanding of old optical schemes and hopefully facilitate significant advances in communications and data processing technologies.

In this chapter, we review the classical statistical detection theory and then show its applicability to optical communications and remote sensing. However, this chapter also provides the theory for unresolved target detection. In this case, the signal-plus-additive-noise hypothesis testing approach is no longer valid. A theory for tackling the replacement model problem where hypotheses involve whether either a target or background clutter is present, not both, is presented.

Free Space Optical Systems Engineering: Design and Analysis, First Edition. Larry B. Stotts.

Companion website: www.wiley.com\go\stotts\free_space_optical_systems_engineering

9.2 CLASSICAL STATISTICAL DETECTION THEORY

This section follows the classic textbooks by Helstrom [6, 7] and McDonough and Whalen [8], drawing on their rich developments. To introduce RF statistical hypothesis testing, we focus on the case of a signal measurement x that will be used to decide between two hypotheses, H_0 (null hypothesis) and H_1 (signal-present hypothesis). Taking into account that any measurement can create errors, we treat the outcome of our testing strategy as a random variable that is governed by the probability densities functions (PDFs) $p_0(x)$ and $p_1(x)$ for Hypotheses H_0 and H_1, respectively. In general, the PDFs for Hypotheses H_0 and H_1 can be written as

$$p_m(x) = \frac{\exp\{-(x-\mu_m)^2\}}{\sqrt{2\pi\sigma_m^2}} \quad m = 0,1, \tag{9.1}$$

respectively. In the above equation, the parameters $\{\mu_m; m = 0, 1\}$ and $\{\sigma_m^2; m = 0,1\}$ are the means and variances for Gaussian PDFs of the two possible hypotheses. For our purposes, we will assume $\mu_0 < 0 < \mu_1$ and $\sigma_0^2 = \sigma_1^2 = \sigma^2$. That we have a common noise distribution and the difference between the two hypothesis's PDF is the offset of the Gaussian distribution. Figure 9.1a illustrates a comparison of the two PDFs.

Hypothesis H_0's PDF implies that the measured value for x comes from the noise distribution and has an averaged value of $|\mu_0|$. Equation (9.1) implies that x will vary too close to that averaged value most of the time (values near the peak), but also will be extremely different occasionally (values at the tails). One will see similar measurement deviations occurring with Hypothesis H_1's PDF, but around an average value of $|\mu_1|$, rather than $|\mu_0|$. The question one might ask is how often each PDF occurs? The fraction of time that H_0 happens is ζ and is called the *prior probability* of this hypothesis. Alternately, the fraction of time that H_1 happens is consequently $(1-\zeta)$ and is called the *prior probability* of that hypothesis. It is clear that there is

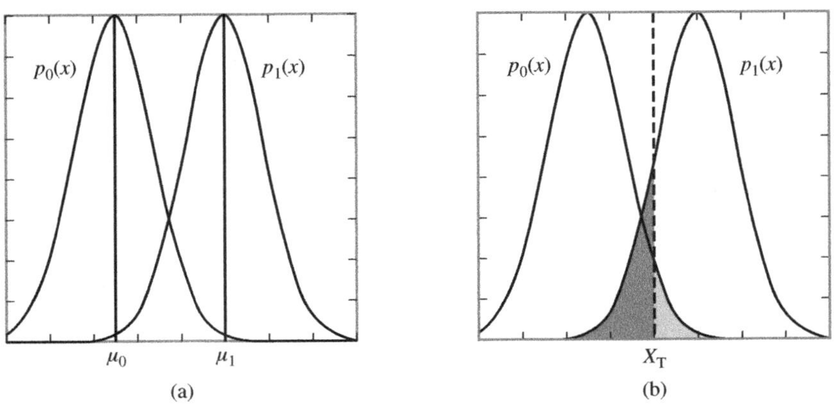

FIGURE 9.1 (a) Probability density functions (PDFs) for Hypotheses H_0 and H_1, pinpointing mean levels, μ_0 and μ_1, respectively, and (b) same PDFs with threshold x_T.

100% probability that either H_0 and H_1 occurs; that is, no other hypothesis outcome is possible.

Given all of the above, what is the strategy for making a decision between Hypotheses H_0 and H_1 given a measurement x? The obvious choice is that a decision maker picks a value of x, say x_T, and then chooses Hypothesis H_0 when $x \leq x_T$ and Hypothesis H_1 for $x > x_T$ (see Figure 9.1b). That makes senses until you wonder, "How did we pick x_T?" As noted earlier, the presence of noise causes mistakes to be made in the decision process. This suggests that the decision maker would like the value of x_T to minimize any negative effects caused by making mistakes. How do we determine the desired value for x_T? Let us begin by looking at the possible mistakes we can make.

The first possible mistake is "choosing H_1 when H_0 is true." The probability of that happening is defined as Q_0 and is mathematically given by the expression:

$$Q_0 = \int_{x_T}^{\infty} p_0(x)\,dx. \tag{9.2}$$

This is the so-called error of the first kind. Referring to Figure 9.1b, it is the area under the curve of $p_0(x)$ to the right of x_T ["lighter" colored portion of $p_0(x)$]. The second possible mistake is "choosing H_0 when H_1 is true." The probability of that happening is defined as Q_1 and is mathematically given by the expression:

$$Q_1 = \int_{-\infty}^{x_T} p_1(x)\,dx. \tag{9.3}$$

Q_1 is called an "error of the second kind." Referring again to Figure 9.1b, it is the area under the curve of $p_1(x)$ to the left of x_T ["darker" colored portion of $p_0(x)$].

With those definitions under our belt, let us now try to quantify the negative effect that making an error creates. Let us define the effect to be the *cost* to the decision maker that these mistakes create. In particular, let C_0 be the cost (e.g., money, time) to the decision maker that an error of the first kind creates and C_1 be the cost that an error of the second kind affects. The product C_0Q_0 is called the *risk* associated with Hypothesis H_0; the product C_1Q_1 is called the *risk* associated with Hypothesis H_1. We now need to combine these two risk entities into a meaningful equation that total cost for each decision. Mathematically, we define the average "risk" per decision as:

$$\overline{C}(\zeta; x_T) = \zeta C_0 Q_0 + (1-\zeta) C_1 Q_1 \tag{9.4a}$$

$$= \zeta C_0 \int_{x_T}^{\infty} p_0(x)\,dx + (1-\zeta) C_1 \int_{-\infty}^{x_T} p_1(x)\,dx. \tag{9.4b}$$

Therefore, given Eq. (9.4b), how do we choose the value of x_T that minimizes the average risk per decision, the desired state for any decision maker?

The solution is to differentiate Eq. (9.4b) and then set the derivative to 0. The result from this mathematical operation is

$$\frac{p_1(x)}{p_0(x)} = \frac{\zeta C_0}{(1-\zeta) C_1} = \Lambda_T. \tag{9.5}$$

Taking the natural logarithm of Eq. (9.5), we find that

$$x_T = \frac{(\mu_1 + \mu_0)}{2} + \frac{\sigma^2}{(\mu_1 - \mu_0)} \ln \left[\frac{\zeta C_0}{(1-\zeta) C_1} \right]. \tag{9.6}$$

For $\zeta = (1 - \zeta) = \frac{1}{2}$ and $C_0 = C_1$, the decision point becomes $x_T = \frac{(\mu_1 + \mu_0)}{2}$, the average of the two distribution means.

As the reader would probably guess, this is a very simple decision process and more elaborate ways to decision have been developed. In the following sections, we review the classical decision processes.

9.2.1 The Bayes Criterion

The rule that the decision strategy is chosen to minimize the average risk is known as the "Bayes Criterion." It is typically employed when a decision maker must make a large number of decisions under similar circumstances. In the above section, we used a cost structure that only emphasized the penalty for errors. This is a limiting case for the Bayes criterion.

In general, one should take into account the cost of each type of decision, which usually forms the elements of a cost matrix $\boldsymbol{C}$:

$$\boldsymbol{C} = \begin{bmatrix} C_{00} & C_{01} \\ C_{10} & C_{11} \end{bmatrix}, \tag{9.7}$$

where C_{mn} is the cost of "choosing H_m when H_n is true," for $(m, n = 0.1)$. The relative cost of an error of the first kind is written as $(C_{10} - C_{00})$ and that of an error of the second kind is given by $(C_{01} - C_{11})$. Both of the quantities in parentheses are positive. In the previous section, the cost matrix we used was

$$\boldsymbol{C} = \begin{bmatrix} 0 & C_1 \\ C_0 & 0 \end{bmatrix}. \tag{9.8}$$

In that case, the costs depend on the action taken after each decision and on the consequences of the said action.

Using Eq. (9.7), we find that the average risk per decision is given by

$$\begin{aligned} \overline{C}(\zeta; x_T) = \zeta & \left[C_{00} \int_{-\infty}^{x_T} p_0(x)\,dx + C_{10} \int_{x_T}^{\infty} p_0(x)\,dx \right] \\ & + (1-\zeta) \left[C_{01} \int_{-\infty}^{x_T} p_1(x)\,dx + C_{11} \int_{x_T}^{\infty} p_1(x)\,dx \right]. \end{aligned} \tag{9.9}$$

Let us now define the minimum risk. Following a similar procedure as found in the last section, we compute the quantity

$$\Lambda(x) = \frac{p_1(x)}{p_0(x)} \tag{9.10}$$

for each measurement x taken. Equation (9.10) is called the *likelihood ratio*. Again taking the derivative of Eq. (9.9) and setting it to 0, we find that the decision point can be written as

$$x_T = \frac{\zeta(C_{10} - C_{00})}{(1-\zeta)(C_{01} - C_{11})}. \tag{9.11}$$

This strategy is known as the *Bayes solution* of the decision problem and the minimum risk $\overline{C}_{min}(\zeta; x_T)$ is called the *Bayes Risk*.

One also can develop the Bayes solution using conditional probabilities for the two hypotheses. The conditional probability for Hypothesis H_0 is given by

$$p(H_0|x) = \frac{\zeta p_0(x)}{p(x)} \tag{9.12}$$

and the conditional probability for Hypothesis H_1 is

$$p(H_1|x) = \frac{(1-\zeta)p_1(x)}{p(x)}, \tag{9.13}$$

where

$$p(x) = \zeta p_0(x) + (1-\zeta)p_1(x) \tag{9.14}$$

represents the total probability density function of the outcome x in all trials. The "conditional risk" for choosing H_0 is written as

$$C(H_0|x) = C_{00}\, p(H_0|x) + C_{01}\, p(H_1|x) \tag{9.15}$$

and for choosing H_1,

$$C(H_1|x) = C_{10}\, p(H_0|x) + C_{11}\, p(H_1|x). \tag{9.16}$$

The decision maker selects the hypothesis that has the smaller conditional risk, given the outcome x from the measurement. Equations (9.15) and (9.16) produce a similar strategy as above. The conditional risk $C(H_0|x)$ is the average cost per decision if Hypothesis H_0 is picked in all cases where the measurement lies in a small region in the neighborhood of the value "x"; $C(H_1|x)$ is defined similarly, but H_1 replaces H_0.

If the relative costs of the errors of the first kind and second kind are equal, that is,

$$(C_{01} - C_{11}) = (C_{10} - C_{00}),$$

then the conditional probability that is greater dictates the hypothesis chosen. The above conditional probabilities $p(H_0|x)$ and $p(H_1|x)$ are often called the "inverse" probabilities or the *posterior* probabilities of the two hypotheses. When the hypothesis with the greater posterior probability is always selected, the total probability of error

$$P_E = \zeta Q_0 + (1-\zeta)Q_1$$

is minimized.

9.2.2 The Minimax Criterion

It is clear from above that knowing the prior probabilities is key to creating the Bayes solution. If they are not available, then an alternate strategy for decision-making must be utilized. The approach generally used is to replace the minimum average risk criterion by what is known as the "minimax criterion." This approach uses the Bayes strategy appropriate for that value of ζ for which the Bayes risk is maximum.

Example 9.1 Let us take a first-order look at the minimax criterion method via an example. Here we assume the same decision situation in the introduction where the cost matrix is given in Eq. (9.8) and the following parameters: $\mu_0 = -1$, $\mu_1 = +1$, $\sigma = 2$, $C_0 = 2$, and $C_1 = 1$.

To begin with, we will plot $\overline{C}_{\min}(\zeta)$ as function of ζ using the above parameters. This is the solid curve shown in Figure 9.2. From Eq. (9.6), we know that the decision maker would choose Hypothesis H_0 when

$$x < x_T = 2\ln\left[\frac{2\zeta}{(1-\zeta)}\right].$$

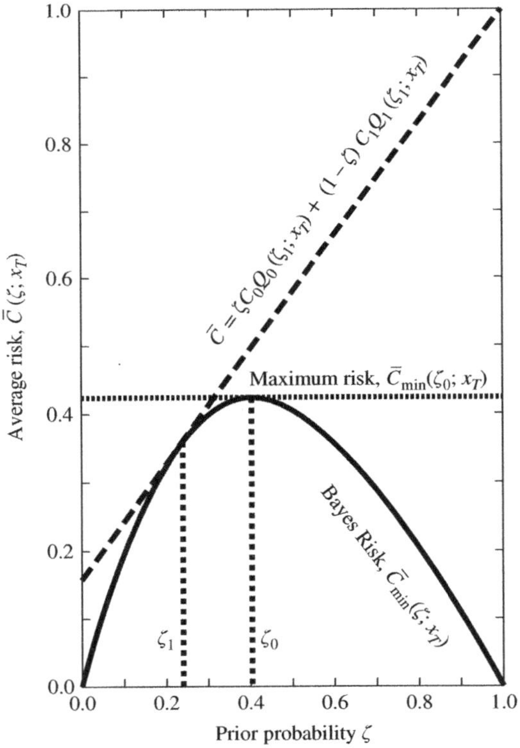

FIGURE 9.2 Plot of the Bayes and minimax risks.

Therefore, what is the value of ζ that we should use in the above equation? At the point chose $\zeta = \zeta_1$, the average loss will be given by the straight line tangent to the point $\overline{C}_{min}(\zeta_1)$. See Figure 9.2. Looking at all the tangents that can occur on the curve, it is clear that the tangent is horizontal at the point $\zeta = \zeta_0$ where Bayes risk is maximum. In this curve, we find that $\zeta_0 = 0.40$ and that leads to $x_T = 0.60$. The equation for the tangent in this case is given by

$$\overline{C}(\zeta_0; x_T) = \zeta[C_{00}(1 - Q_0(\zeta_0; x_T)) + C_{10}Q_0(\zeta_0; x_T)] + (1 - \zeta)[C_{01}Q_1(\zeta_0; x_T) + C_{11}(1 - Q_1(\zeta_0; x_T))] \tag{9.17}$$

$$= \zeta C_0 Q_0(\zeta_0; x_T) + (1 - \zeta)C_1 Q_1(\zeta_0; x_T). \tag{9.18}$$

The value of the maximum risk, $\overline{C}_{min}(\zeta_0; x_T)$, then is 0.42 using Eq. (9.18). The Bayes strategy at $\zeta = \zeta_0$ is called the "minimax strategy" and $\overline{C}_{min}(\zeta_0; x_T)$ is referred to as the "minimax risk." In reality, no one would make decisions at the maximum risk level, which is the result of our example. However, the outlined minimax approach offers a way to determine a reasonable decision without knowing the actual prior probability for the decision problem.

As final comment, we recognize that the straight line specified by Eq. (9.17) intercepts the vertical lines $\zeta = 0$ and $\zeta = 1$ in general with risks $C_{00}(1 - Q_0(\zeta_0; x_T)) + C_{10}Q_0(\zeta_0; x_T)$ and $C_{01}Q_1(\zeta_0; x_T) + C_{11}(1 - Q_1(\zeta_0; x_T))$, respectively. At the minimax solution point where the slope is horizontal, these risks are equal. The resulting risk where

$$C_{00}(1 - Q_0(\zeta_0; x_T)) + C_{10}Q_0(\zeta_0; x_T) = C_{01}Q_1(\zeta_0; x_T) + C_{11}(1 - Q_1(\zeta_0; x_T))$$

is the minimax risk.

9.2.3 The Neyman–Pearson Criterion

The Neyman–Pearson criterion is used for the case when both the prior probabilities and cost matrix are not known. This typically is true for sonar, radar, optical, and electro-optical systems doing real world surveillance where Hypothesis H_1 happens rarely. Our problem now becomes one of "anomaly detection." In other words, the problem reduces to the dealing strictly with mistakes in presence of very little detection. Specifically, the decision maker will most likely be faced with a penalty for reacting to a "false alarm" and must minimize overreaching to those possible events. Strategy-wise, the decision maker must set a value of Q_0 that he/she can live with and create an effective decision strategy that can achieve that quantity consistently. Constraining this strategy is the fact that the probability Q_1 also must be minimized at the same time. This strategy is said to meet the "Neyman–Pearson Criterion." In the theory of hypothesis testing, Q_1 is called the "power" of the hypothesis test and usually is denoted by the symbol β. The probability Q_0 is referred to as the "size" of the hypothesis test and denoted by the symbol α. For the engineer and scientist, a more intuitive strategy is required to specify a surveillance problem. Namely, the more realistic strategy essentially is to maximize the detection of a target for a given false alarm probability. In this case, the "detection probability" is defined as $(1 - Q_1)$

and it is maximized rather than the probability Q_1 being minimized. The "false alarm probability" is defined as Q_0.

To implement this criterion based on a single measurement x, we calculate the likelihood ratio defined in Eq. (9.10) using that outcome and compare the result to some threshold value Λ_T. If $\Lambda(x) \leq \Lambda_T$, the decision maker picks Hypothesis H_0; if $\Lambda(x) > \Lambda_T$, the decision maker selects Hypothesis H_1.[1] The computed likelihood ratio is a random variable and has a probability density function $P_0(\Lambda)$ under Hypothesis H_0. It is related to the known PDF as follows:

$$P_0(\Lambda)\,d\Lambda = p_0(x)\,dx \quad \text{for } \Lambda(x) = \frac{p_1(x)}{p_0(x)}.$$

The probability Q_0 therefore can be written as

$$Q_0 = \int_{\Lambda_T}^{\infty} P_0(\Lambda)\,d\Lambda. \tag{9.19}$$

The value of Λ_T is established by preassigning a value to Q_0 and inverse computing it. The probability of detection is then determined by calculating

$$(1 - Q_1) = 1 - \int_{0}^{\Lambda_T} P_1(\Lambda)\,d\Lambda = \int_{\Lambda_T}^{\infty} P_1(\Lambda)\,d\Lambda, \tag{9.20}$$

where $P_1(\Lambda)$ is the PDF under Hypothesis H_1. The results of these computations generally are portrayed in a figure, plotting the detection probability ("power of the test") as a function of the false alarm probability ("size of the test). This curve often called the "Receiver Operating Characteristic" ("operating characteristic"), or "ROC" curve [8, p. 157]. Figure 9.3 shows an example of such a plot (solid curve).

Example 9.2 Let us now look how the Bayes solution found previously for $\zeta = 0.23$ relates to the Neyman–Pearson criterion. We again use the complete parameter set described in the previous section. The slope for any point on the ROC curve depicted in Figure 9.3 is given by

$$\left.\frac{dQ_d}{dQ_0}\right|_{\Lambda=\Lambda_T} = \Lambda_T = \frac{p_1(x_T)}{p_0(x_T)}. \tag{9.21}$$

This equation is obtained by differentiating Eqs. (9.19) and (9.20) and using the fact that

$$\Lambda = \frac{P_1(\Lambda)}{P_0(\Lambda)} = \frac{p_1(x)}{p_0(x)}. \tag{9.22}$$

[1]Our selection criterion has the equal sign associated with choosing Hypothesis H_0. Some researchers have it the other way, associating the equal with Hypothesis H_1. Both are acceptable. The author is biased with the former convention. However, when discussing the work of others, the reader may find the latter following those researchers' usage. From an engineering point of view, it is your choice.

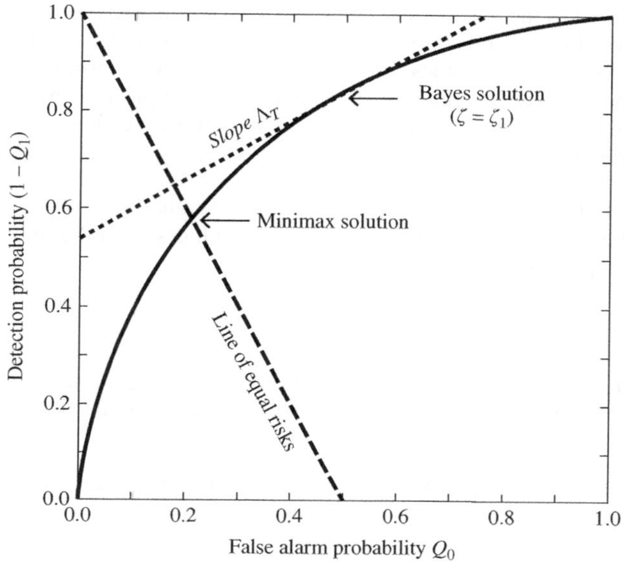

FIGURE 9.3 Example of a receiver operator characteristic (ROC) curve.

Calculating the threshold x_T, we find that

$$x_T = 2\ln\left[\frac{2\zeta}{(1-\zeta)}\right] = 2\ln\left[\frac{2(0.23)}{(1-0.23)}\right] = 2\ln\left[\frac{0.46}{0.77}\right] = -1.03 \approx -1.$$

Substituting into Eq. (9.21), we have

$$\Lambda_T = \frac{p_1(x_T)}{p_0(x_T)} = e^{-\frac{1}{2}} = 0.6065.$$

As expected, the tangent curve at the point Λ_T will be a linear line with slope Λ_T. Because the value of Q_d is 0.845 for Q_0 equal to 0.5060, we can easily derive the linear slope line to be

$$\text{Linear Slope} = a + bQ_0 = 0.5831 + 0.6065Q_0.$$

This slope line is the dotted line shown in Figure 9.3 centered at the point (0.5060, 0.845). The point on the ROC curve where the tangent line is located represents the Bayes solution we found previously. This solution provides a reasonable detection probability ("knee of the curve"), but the false alarm probability is rather high.

Example 9.3 Let us now look how the minimax solution relates to the Neyman–Pearson criterion. To do this, we use the fact that the risks are equal, that is,

$$C_{00}(1 - Q_0(\zeta_0; x_T)) + C_{10}Q_0(\zeta_0; x_T) = C_{01}Q_1(\zeta_0; x_T) + C_{11}(1 - Q_1(\zeta_0; x_T)).$$

This line is shown as the dashed line labeled "Line of Equal Risks" in Figure 9.3 and its intersection with the ROC curve is the minimax solution. That is, it intersects the ROC curve at the minimax values for the detection and false alarm probabilities. The slope at that point gives the value of Λ_T, from which the prior probability ζ_0 can be determined using previously used equations. This solution reduces the detection and false alarm probabilities by a factor of 33% and 60%, respectively.

Example 9.4 Let us assume a known (deterministic) signal, say $s(t)$, to be detected in Additive White Gaussian Noise (AWGN). Helstrom has shown that the test statistics in this case is equal to

$$G = \int_0^{T_1} s(t)v(t)\,dt, \tag{9.23}$$

where G is the decision test statistics, T_1 the receiver integration time, and $v(t)$ the input voltage [7, pp. 102–108]. Using this statistics, we find that

$$\mathcal{E}(G|H_1) = \int_0^{T_1} s(t)v(t)\,dt|_{v(t)=s(t)} = \int_0^{T_1} |s(t)|^2\,dt = E_s \tag{9.24}$$

and

$$\mathcal{E}(G|H_0) = \int_0^{T_1} s(t)\;v(t)\,dt|_{v(t)=n(t)} = 0, \tag{9.25}$$

where E_s is the energy of the received signal and $n(t)$ is the receiver noise in the absence of a coming signal. The variance of G is given by

$$\begin{aligned}\operatorname{var} G &= \int_0^{T_1}\int_0^{T_1} s(t_1)s(t_2)\,\overline{n(t_1)\,n(t_2)}\;dt_1\,dt_2 \\ &= \frac{N_0}{2}\int_0^{T_1}\int_0^{T_1} s(t_1)s(t_2)\delta(t_2-t_1)\,dt_1\,dt_2 \\ &= \frac{N_0}{2}\int_0^{T_1} |s(t)|^2\,dt_1 = \frac{NE_s}{2}.\end{aligned} \tag{9.26}$$

Using the above equations, we can write the PDFs for Hypotheses H_1 and H_0 as

$$p_1(G) = \frac{1}{\sqrt{\pi N_0 E_s}}\,e^{-\frac{(G-E_s)^2}{N_0E_s}} \tag{9.27}$$

and

$$p_0(G) = \frac{1}{\sqrt{\pi N_0 E_s}}\,e^{-\frac{G^2}{N_0E_s}}, \tag{9.28}$$

respectively.

The false alarm and detection probabilities then are given by

$$Q_0 = \operatorname{erfc}(x_T) \tag{9.29}$$

and

$$Q_1 = \mathrm{erfc}(x_T - \sqrt{\mathrm{SNR}}), \tag{9.30}$$

respectively, with the threshold being

$$x_T = G_0\sqrt{\frac{2}{N_0 E_s}}. \tag{9.31}$$

In Eqs. (9.29) and (9.30), erfc(y) is the complementary error function defined as

$$\mathrm{erfc}(y) = \frac{1}{\sqrt{2\pi}}\int_y^{\infty} e^{-\frac{q^2}{2}}\,dq$$

(see Eq. (2.1.4) in Helstrom [7, p. 86]). In Eq. (9.30), we have

$$\mathrm{SNR} = \frac{[\mathcal{E}(G|H_1)]^2}{\mathrm{var}\,G} = \frac{2E_s}{N_0} \tag{9.32}$$

equal to electrical signal-to-noise ratio (SNR).

9.3 TESTING OF SIMPLE HYPOTHESES USING MULTIPLE MEASUREMENTS

As the reader might expect, the hypothesis testing theory described in the previous section can easily be extended to the case of decision-making using multiple measurements [6–8]. Instead of one outcome x, we now have a set of measurements, or "measurement vector," given by the relation

$$\boldsymbol{x} = \{x_1, x_2, x_3, \ldots, x_n\}, \tag{9.33}$$

where $\boldsymbol{x}$ represents a measurement vector and x_i is the ith measurement. These measurements can be either from unprocessed or processed image frames in time or color, or from successive received laser pulses. They can be described statistically by their "joint probability density functions" $p_0(\boldsymbol{x}) = p_0(x_1, x_2, x_3, \ldots, x_n)$ and $p_1(\boldsymbol{x}) = p_1(x_1, x_2, x_3, \ldots, x_n)$ for Hypotheses H_0 and H_1, respectively. If these joint PDFs do not contain any unknown parameters, then the corresponding hypotheses are considered "simple."

Helstrom stated it was convenient to consider the measurement vector as a point in an n-dimensional Cartesian space and we follow his development here [7]. The decision strategy then can be separated into two regions $\boldsymbol{R}_0$ and $\boldsymbol{R}_1$ such that the decision maker chooses Hypotheses H_0 when the measurement vector $\boldsymbol{x}$ lies in region $\boldsymbol{R}_0$ and H_1 when $\boldsymbol{x}$ is found in region $\boldsymbol{R}_1$. These two regions are separated by a surface D referred to as the "decision surface." Alternatively, one could describe the decision strategy by a set of inequalities on the n variables $\{x_i; 1 \le i \le n\}$, but the geometrical interpretation is easy to understand and more intuitive. Figure 9.4 depicts the decision

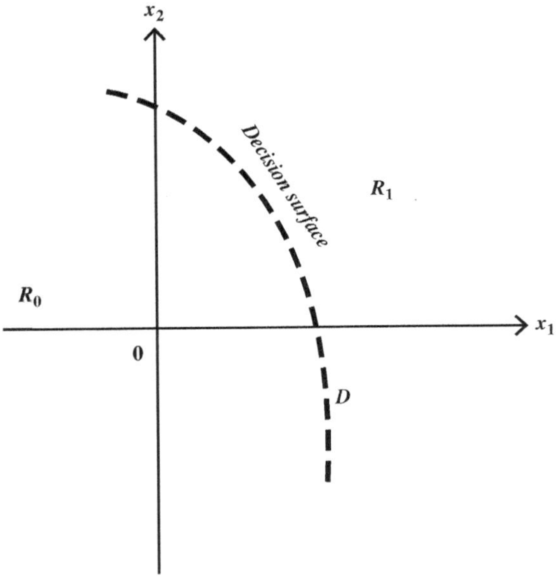

FIGURE 9.4 Decision regions for a two measurement data set.

regions and decision surface for a two measurement hypothesis test. The decision surface is determined by the particular criterion used.

For example, if the decision maker can establish a cost matrix $\boldsymbol{C}$ that specifies the costs for each possible decision under the two hypotheses, as well as knows the prior probability ζ, then he/she can apply the Bayes criterion for the minimum average risk we found previously, but defined over a multidimensional space [7]. That is, we have for the average risk per decision

$$\overline{\boldsymbol{C}} = \zeta \left[C_{00} \int_{\boldsymbol{R}_0} p_0(\boldsymbol{x})\, d^n\boldsymbol{x} + C_{10} \int_{\boldsymbol{R}_1} p_0(\boldsymbol{x})\, d^n\boldsymbol{x} \right] + (1-\zeta) \left[C_{01} \int_{\boldsymbol{R}_0} p_1(\boldsymbol{x})\, d^n\boldsymbol{x} + C_{11} \int_{\boldsymbol{R}_1} p_1(\boldsymbol{x})\, d^n\boldsymbol{x} \right], \tag{9.34}$$

where $d^n\boldsymbol{x} = dx_1, dx_2, dx_3, \ldots, dx_n$ is the volume element in the n-dimensional space of possible outcomes $\boldsymbol{x}$. The first integral in first bracket of Eq. (9.34) is the probability when Hypothesis H_0 is true when measurement vector $\boldsymbol{x}$ can be found as a point in region $\boldsymbol{R}_0$. The decision maker therefore makes the correct decision of H_0. The integral is multiplied by the cost C_{00}. Conversely, the second integral involves making the mistake of choosing H_0, with its accompanying penalty C_{10}. These two entities comprise the risk associated with the Hypothesis H_0. This risk then is multiplied by the prior probability ζ that the hypothesis is true. A similar discussion holds for the two integrals in the second bracket of Eq. (9.34) concerning the selection of Hypothesis H_1.

Just like in the last section, the Bayes criterion requires the decision surface D to be selected so that average risk given above is a minimum. This implies that we use a multidimensional likelihood ratio similar to what we defined before in Eq. (9.10), but utilizing joint PDFs rather than the single dimension type. Specifically, we define our new likelihood ratio for multiple-outcome hypothesis testing as

$$\Lambda(\boldsymbol{x}) = \frac{p_1(\boldsymbol{x})}{p_0(\boldsymbol{x})} = \frac{p_1(x_1, x_2, x_3, \ldots, x_n)}{p_0(x_1, x_2, x_3, \ldots, x_n)}. \tag{9.35}$$

We then compare the result to a decision threshold Λ_T and decide which of the two hypotheses, H_0 and H_1, to choose; $\Lambda(\boldsymbol{x}) \leq \Lambda_T$ for H_0 and $\Lambda(\boldsymbol{x}) > \Lambda_T$ for H_1, respectively. It is clear from this discussion that equation for the decision surface is $\Lambda(\boldsymbol{x}) = \Lambda_T$, for region $\boldsymbol{R}_0$ is $\Lambda(\boldsymbol{x}) \leq \Lambda_T$ and for region $\boldsymbol{R}_1$ is $\Lambda(\boldsymbol{x}) > \Lambda_T$. The minimum average risk $\overline{\boldsymbol{C}}_{\min}$ defined in Eq. (9.34) when the Bayes decision regions defined above are used.

As before, we find for hypothesis testing is achieved when the cost matrix is known, but prior probability ζ is unknown, we turn to the minimax criterion for decision-making. For each value of the prior probability ζ, the minimum average risk $\overline{\boldsymbol{C}}_{\min}(\zeta)$ can be computed using the Bayes solution strategy outlined above. The minimax strategy then is the one where we find that the prior probability $\zeta = \zeta_0$ creates the largest Bayes risk; that is, where $\overline{\boldsymbol{C}}_{\min}(\zeta_0)$ is maximum.

In this situation, we view the quantity $\Lambda(\boldsymbol{x})$ as random variable with probability density functions $P_0(\Lambda)$ and $P_1(\Lambda)$ under Hypotheses, H_0 and H_1, respectively. How do we calculate the PDFs? We begin by defining a region in an n-dimensional space $\boldsymbol{R}(\lambda)$ for which $\Lambda(\boldsymbol{x}) = \frac{p_1(\boldsymbol{x})}{p_0(\boldsymbol{x})}$ is less than λ [7]. Under Hypothesis H_0, the probability that the measurement vector lies in $\boldsymbol{R}(\lambda)$ is given by

$$\Pr\{\boldsymbol{x} \in \boldsymbol{R}(\lambda) | H_0\} = \int_{\boldsymbol{R}(\lambda)} p_0(\boldsymbol{x})\, d^n\boldsymbol{x} = \int_0^{\lambda} P_0(\Lambda)\, d\Lambda.$$

The above function of λ is the cumulative distribution of the likelihood ratio. By differentiating the above by λ, the PDF $P_0(\lambda)$ can be obtained. We can formulate a similar expression for $\Pr\{\boldsymbol{x} \in \boldsymbol{R}(\lambda)|H_1\}$ and this will allow us to derive the PDF $P_1(\lambda)$. The errors of the first and second kind therefore are

$$Q_0 = \int_{\boldsymbol{R}_1} p_0(\boldsymbol{x})\, d^n x = \int_{\Lambda_T}^{\infty} P_0(\Lambda)\, d\Lambda \tag{9.36}$$

and

$$Q_1 = \int_{\boldsymbol{R}_0} p_0(\boldsymbol{x})\, d^n\boldsymbol{x} = \int_0^{\Lambda_T} P_1(\Lambda)\, d\Lambda, \tag{9.37}$$

respectively. The risks c_0 and c_1 associated with our two hypotheses can be written as

$$c_0 = C_{00}(1 - Q_0) + C_{10}Q_0 \tag{9.38}$$

and

$$c_1 = C_{01}Q_1 + C_{11}(1 - Q_1), \tag{9.39}$$

respectively, in terms of the false alarm and detection probabilities. Just like in the last section, the minimax strategy is calculated under the presumption that these two risks are equal. Using the above four equations, an equation for Λ_T can be derived that allows the likelihood to be compared to. The common value of the risks c_0 and c_1 facilitates the computation of $\zeta = \zeta_0$ that creates a minimum average risk $\overline{C}_{\min}(\zeta_0)$ that is maximum.

The Neyman–Pearson criterion requires the decision maker to choose the location on the decision surface D that minimizes the probability Q_1 for a fixed value of Q_0. Again, this criterion is employed when the prior probability and cost matrix are unknown and/or undefinable. To find the decision surface under this criterion, Helstrom [7] proposed the use of the Lagrange multipliers, forming the following linear combination:

$$\Gamma(D) = Q_1 + \lambda Q_0 = \int_{\boldsymbol{R}_0} p_1(\boldsymbol{x})\, d^n\boldsymbol{x} + \lambda \int_{\boldsymbol{R}_1} p_0(\boldsymbol{x})\, d^n\boldsymbol{x}. \tag{9.40}$$

The decision surface D is obtained by moving it around until Eq. (9.40) is minimal; the resulting surface will be function of the parameter λ. Comparing Eqs. (9.34) and (9.40), we find that they are identical when

$$\boldsymbol{C} = \begin{bmatrix} 0 & 2 \\ 2\lambda & 0 \end{bmatrix}$$

and the prior probability is 0.5. This implies that the equation for the decision surface that minimizes $\Gamma(D)$ is given by

$$\frac{p_1(\boldsymbol{x})}{p_0(\boldsymbol{x})} = \Lambda(\boldsymbol{x}) = \lambda. \tag{9.41}$$

The optimum decision surface therefore is one of a family of surfaces generated by varying λ in Eq. (9.41). That value of λ is derived from the false alarm probability

$$Q_0 = \int_{\lambda}^{\infty} P_0(\Lambda)\, d\Lambda$$

that we have preassigned a value to. In other words, λ equals Λ_T. As noted earlier, our strategy is to compare likelihood Λ to a decision threshold Λ_T and decide which of the two hypotheses, H_0 or H_1, to choose; again, if $\Lambda(\boldsymbol{x}) \leq \Lambda_T$, we select H_0 and if $\Lambda(\boldsymbol{x}) > \Lambda_T$, we select H_1. The Bayes and minimax solutions can be obtained employing the same procedures we discussed in Examples 9.2 and 9.3, respectively.

An important aspect of hypothesis testing is that one also can make the hypothesis selection based on any monotone function $G = G(\Lambda)$ of the likelihood ratio Λ, not just on the likelihood ratio itself [7]. Without loss of generality, we can assume that G is an increasing function. The decision criterion in this case is that if $G(\Lambda) \leq G_T$, we select H_0 and if $G(\Lambda) > G_T$, we select H_1. The inequality decision procedure remains the same. Under the Neyman–Pearson criterion, the decision threshold G_T is picked so that the false alarm probability takes on a preassigned value. In our new operating space, we have

$$Q_0 = \int_{G_T}^{\infty} p_0(G)\, dG, \tag{9.42}$$

where $p_0(G)$ is the PDF of G under Hypothesis H_0. Similarly, we have

$$Q_d = \int_{G_T}^{\infty} p_1(G)dG, \tag{9.43}$$

where $p_1(G)$ is the PDF of G under Hypothesis H_1.

The entity G is a function of the measurement vector $\boldsymbol{x} = \{x_1, x_2, x_3, \ldots, x_n\}$ and contains the entire set of information necessary to make a reasonable decision using any, or all, of the criterion we have discussed to this point. Such a function is called a "statistic." As you would expect, the likelihood ratio is a statistic. The implication from the above is that if one can measure the statistic G directly rather than having to measure individually all the elements in the measurement vector $\boldsymbol{x}$, then one can make decisions with the same efficacy. This is because the likelihood ratio depends solely on them and G depends on the likelihood ratio. This implies that we now have likelihood ratio of the form

$$\Lambda(G) = \frac{p_1(G)}{p_0(G)} \tag{9.44}$$

and under the Bayes solution, we have

$$\Lambda(G_T) = \frac{p_1(G_T)}{p_0(G_T)} = \Lambda_T = \frac{\zeta(C_{10} - C_{00})}{(1 - \zeta)(C_{01} - C_{11})}. \tag{9.45}$$

The points on the ROC now are indexed by the values of G_T or Λ_T, and we have

$$Q_d = Q_d(G_T) \quad \text{and} \quad Q_0 = Q_0(G_T).$$

using Eqs. (9.42) and (9.43). Clearly, we can apply the same procedures for all the criterions discussed in the last section in this analytic framework.

At this point, the reader may be wondering what is an example of statistic G? One of the most popular statistics used in surveillance applications is the natural-logarithmic likelihood ratio, or

$$g = \ln \Lambda(\boldsymbol{x}) = \ln \left[\frac{p_1(\boldsymbol{x})}{p_0(\boldsymbol{x})} \right]. \tag{9.46}$$

If one is dealing with a Gaussian PDF for each element of the measurement vector $\boldsymbol{x}$, then the above statistic takes on a particular simple form. In particular, when the data is statistically independent Gaussian distributed random variables with a common variance, the natural log of this likelihood ratio is proportional to the sum of the elements, normalized by the common variance.

Example 9.5 This example illustrates this last noted property of the natural-log-likelihood ratio. Suppose the measurement vector again is comprised of statistically independent random variables that are Gaussian distributed with common variance σ^2. Define a_0 and a_1 to be the expected values of the Gaussian PDFs under hypothesis H_0 and H_1, respectively. Then the joint PDFs can be written as

$$p_k(\boldsymbol{x}) = \sqrt[-n]{2\pi\sigma^2} \exp\left\{ \frac{-\sum_{i=1}^{n}(x_i - a_k)^2}{\sigma^2} \right\}, \quad k = 0,1. \tag{9.47}$$

The log-likelihood ratio easily can be shown to equal

$$g = \ln\left[\frac{p_1(\boldsymbol{x})}{p_0(\boldsymbol{x})}\right] = \frac{(a_1 - a_0)}{\sigma^2}\sum_{i=1}^{n} x_i - \frac{n(a_1^2 - a_0^2)}{2\sigma^2}. \tag{9.48}$$

The decision maker will pick H_0 if $\Lambda(\boldsymbol{x}) \leq \Lambda_T$, where Λ_T depends on the criterion employed. Given the second term in Eq. (9.48) is a constant, as well as the multiplier to the summation, the decision can be based strictly on the following equation:

$$X = \frac{1}{n}\sum_{i=1}^{n} x_i, \tag{9.49}$$

which is the "sample mean" of the measurements. The threshold for this entity is given by

$$X_T = \frac{\sigma^2}{(a_1 - a_0)}\left(g + \frac{n(a_1^2 - a_0^2)}{2\sigma^2}\right), \tag{9.50}$$

assuming $a_1 > a_0$. One now chooses H_0 if $X \leq X_T$, H_1 if $X > X_T$. If the Bayes criterion is used, then X_T depends on the costs and prior probability through the threshold Λ_T previously found. In this example, the decision surface D is defined by the equation

$$Q_d = \int_{X_T}^{\infty} \frac{1}{\sqrt{\frac{2\pi\sigma^2}{n}}} \exp\left\{-\frac{(x - a_1)^2}{\left(\frac{2\sigma^2}{n}\right)}\right\} dx, \tag{9.51a}$$

where

$$\sum_{i=1}^{n} x_i = nX_T.$$

The probabilities of detection and false alarm are equal to

$$= 0.5\,\mathrm{erfc}\left[\frac{(X_T - a_1)}{\left(\frac{\sigma}{\sqrt{n}}\right)}\right] \tag{9.51b}$$

and

$$Q_0 = \int_{X_T}^{\infty} \frac{1}{\sqrt{\frac{2\pi\sigma^2}{n}}} \exp\left\{-\frac{(x - a_0)^2}{\left(\frac{2\sigma^2}{n}\right)}\right\} dx \tag{9.52a}$$

$$= 0.5\,\mathrm{erfc}\left[\frac{(X_T - a_0)}{\left(\frac{\sigma}{\sqrt{n}}\right)}\right], \tag{9.52b}$$

respectively.

It is interesting to note in both of the above equations that the original standard deviation of the measurement vector elements is reduced by the number of the said elements. This is a nice benefit from the use of this statistic. If one wants to utilize the Neyman–Pearson criterion picks a threshold X_T that creates the preselected fixed false alarm rate.

Example 9.6 Let us assume we have an optical image containing a multipixel target additively embedded in background clutter and system noise that has been vectorized (see Appendix C). Let N denote the total number of pixels comprising the target. The two hypotheses in this case are defined as follows:

Hypothesis 1: Target plus background and system noise

and

Hypothesis 0: Background and system noise.

The PDF for Hypothesis H_0 is given by

$$p_0 = \left(\frac{1}{2\pi\sigma_T^2}\right)^{\frac{N}{2}} \exp\left[-\sum_{n=1}^{N}\left[\frac{(i_n - \mu_b - \mu_t)^2}{2\sigma_T^2}\right]\right], \tag{9.53}$$

where

i_n ≡ nth pixel photo-counts in image vector
σ_T^2 ≡ total photo-counts noise variance
σ_t^2 ≡ system noise photo-counts variance

$$\sigma_T^2 = \sigma_b^2 + \sigma_t^2 \tag{9.54}$$

μ_t ≡ system noise mean photo-counts
and
μ_b ≡ background noise mean photo-counts.

The PDF for Hypothesis H_1 is given by

$$p_1 = \left(\frac{1}{2\pi\sigma_T^2}\right)^{\frac{N}{2}} \exp\left[-\sum_{n=1}^{N}\left[\frac{(i_n - s_n - \mu_b - \mu_t)^2}{2\sigma_T^2}\right]\right]. \tag{9.55}$$

Redefining the current as $m_n = i_n - \mu_b - \mu_t$, the above two PDFs can be rewritten as

$$p_0(\boldsymbol{m}) = \left(\frac{1}{2\pi\sigma_T^2}\right)^{\frac{N}{2}} \exp\left[-\sum_{n=1}^{N}\left[\frac{m_n^2}{2\sigma_T^2}\right]\right] \tag{9.56}$$

and

$$p_1(\boldsymbol{m}) = \left(\frac{1}{2\pi\sigma_T^2}\right)^{\frac{N}{2}} \exp\left[-\sum_{n=1}^{N}\left[\frac{(m_n - s_n)^2}{2\sigma_T^2}\right]\right]. \tag{9.57}$$

Using the above, the likelihood ratio test is given by

$$\Lambda(\boldsymbol{m}) = \left(\frac{p_1(\boldsymbol{m})}{p_0(\boldsymbol{m})}\right) = \exp\left[-\sum_{n=1}^{N}\left[\frac{(m_n - s_n)^2}{2\sigma_T^2} - \frac{m_n^2}{2\sigma_T^2}\right]\right]. \tag{9.58}$$

Taking the logarithm of Eq. (9.58) yields

$$\ln\Lambda(\boldsymbol{m}) = \ln\left(\frac{p_1(\boldsymbol{m})}{p_0(\boldsymbol{m})}\right) = -\sum_{n=1}^{N}\left[\frac{(m_n - s_n)^2}{2\sigma_T^2} - \frac{m_n^2}{2\sigma_T^2}\right] \tag{9.59}$$

$$= \sum_{n=1}^{N}\left[\frac{(2m_n s_n - s_n^2)}{2\sigma_T^2}\right]. \tag{9.60}$$

Equation (9.60) implies that we can write the log-likelihood ratio test as

$$q = \frac{1}{N}\sum_{n=1}^{N} s_n m_n > \frac{\sigma_T^2}{N}\ln\Lambda\sum_{n=1} + \frac{s_n^2}{2N} = q_0 \tag{9.61}$$

to decide that H_1 is present. The opposite inequality

$$q = \frac{1}{N}\sum_{n=1}^{N} s_n m_n \leq q_0 \tag{9.62}$$

decides that H_0 is present. The above equations simply state that the sampled mean of the data, q, is to be compared against a threshold and a decision is made based on whether the mean is greater than, or equal to, q_0, or less than q_0.

Since q is the sum of Gaussian variables, it is Gaussian as well. Calculating the means for the two hypotheses, we see that

$$\mathcal{E}(q|H_0) = \frac{1}{N}\sum_{n=1}^{N}\mathcal{E}(s_n m_n|H_0) = 0 \tag{9.63}$$

and

$$\mathcal{E}(q|H_1) = \frac{1}{N}\sum_{n=1}^{N}\mathcal{E}(s_n m_n|H_1) = \frac{1}{N}\sum_{n=1}^{N} s_n 2 = E. \tag{9.64}$$

The variance for the two hypotheses is given by

$$\sigma_q^2 = \mathcal{E}\left((q - \mathcal{E}(q))^2|H_i\right) = E\frac{\sigma_T^2}{N}. \tag{9.65}$$

The probability of false alarm then is

$$Q_{fa} = \int_{q_0}^{\infty} p_0(q)\,dq \tag{9.66}$$

$$= \frac{1}{\sqrt{2\pi\left(\frac{\sqrt{E}\sigma_T^2}{N}\right)}} \int_{q_0}^{\infty} e^{-\frac{Nq^2}{2\sqrt{E}\sigma_T^2}}\,dq \tag{9.67}$$

$$= \frac{1}{\sqrt{2\pi}} \int_{\left(\frac{\sqrt{N}q_0}{\sqrt{E}\sigma_T}\right)}^{\infty} e^{-\frac{w^2}{2}}\,dw = 1 - P\left(\frac{\sqrt{N}q_0}{\sqrt{E}\sigma_T}\right), \tag{9.68}$$

where

$$P(x) = \frac{1}{\sqrt{2\pi}} \int_{-\infty}^{x} e^{-\frac{q^2}{2}}\,dq = 0.5\frac{2}{\sqrt{\pi}} \int_{-\infty}^{x} e^{-t^2}\,dt = 0.5\,\mathrm{erfc}(x) \tag{9.69}$$

is the integral of the Normal, or Gaussian, Probability Function, Eq. (26.2.2) in Abramowitz and Stegun [9, p. 931]. The last part of Eq. (9.69) for the complementary error function comes from Eq. (7.1.2) in Abramowitz and Stegun [9, p. 297].[2] From this equation, the threshold can be determined for any specified probability of false alarm, for example, 10^{-3}, either numerically or from a table of Gaussian error probabilities, for example, Abramowitz and Stegun [9, pp. 925–996].

This implies that Eq. (9.68) alternately can be rewritten as

$$Q_{\mathrm{fa}} = 0.5\,\mathrm{erfc}\left(\sqrt{\frac{N}{E}}\,\frac{q_0}{\sigma_T}\right). \tag{9.70}$$

The probability of detection can be written as

$$Q_d = \int_{(q_0-E)}^{\infty} p_1(q)\,dq \tag{9.71}$$

$$= \frac{1}{\sqrt{2\,\pi}} \int_{\sqrt{\frac{N}{E\sigma_T^2}}(q_0-E)}^{\infty} e^{-\frac{q^2}{2}}\,dq \tag{9.72}$$

$$= 0.5\,\mathrm{erfc}\left(\sqrt{\frac{N}{E}}\,\frac{(q_0-E)}{\sigma_T}\right). \tag{9.73}$$

The reader probably has recognized that the above detection and false alarm probabilities are identical to those found in Example 9.5 if $a_0 = 0$, $a_1 = S$. Equations (9.70)

[2] As you can see from above, there are a couple of ways to write the complementary error function. This is true for tables as well. The reader needs to make sure what version is being used in any paper or text.

and (9.73) are the matched filter results showing that this filter increases the received signal level (or reduces the noise standard deviation) by the square root of the number of total target pixels.

If we let

$$x_T = q_0\sqrt{\frac{N}{E\sigma_T^2}} \tag{9.74}$$

and

$$\text{SNR} = \frac{NE}{\sigma_T^2}, \tag{9.75}$$

then Eqs. (9.70) and (9.73) can be rewritten as

$$Q_{\text{fa}} = 0.5\,\text{erfc}\,[x_T] = Q_0 \tag{9.76}$$

and

$$Q_d = 0.5\,\text{erfc}(x_T - \sqrt{\text{SNR}}) = Q_1, \tag{9.77}$$

respectively. It is clear that Eqs. (9.76) and (9.77) are the same as Eqs. (9.29) and (9.30). Here we see that Eq. (9.75), the SNR has an increased energy level, or reduced noise variance, by a factor of N.

9.4 CONSTANT FALSE ALARM RATE (CFAR) DETECTION

If the background against which targets are to be detected is constant with time and space, then a fixed threshold level can be chosen that provides a specified probability of false alarm under the Neyman–Pearson criterion, governed by the probability density function of the noise, which is usually assumed to be Gaussian. The probability of detection then is a function of the target's SNR. However, in most fielded systems, unwanted clutter and interference sources mean that the noise level changes both spatially and temporally. In this case, a changing threshold can be used, where the threshold level is raised and lowered to maintain a constant probability of false alarm. This variable threshold technique is known as "constant false alarm rate (CFAR) detection".

Example 9.7 In most simple CFAR detection schemes, the threshold level is calculated by estimating the statistics of background clutter in a specific configuration of pixels. Let us look at the latter situation as an example [10]. An illustrative situation is shown in Figure 9.5a. In this case, we assume that we have a signal pixel target (recall that we are dealing with signal-plus-noise versus noise-only hypothesis testing), one takes a block of pixels around our possible target pixel and calculates the average intensity level. To avoid corrupting this estimate with intensity from the pixel of interest, pixels immediately adjacent to that pixel are normally ignored (and referred to as "guard cells"). As shown in Figure 9.5b, we are computing the mean and variance of the intensities contain in a 1- or 2-pixel square-annulus separated 1–3 pixel from the pixel of in both the x and y directions. This sometimes is referred to as the "gate." We subtract the mean from the pixel of interest's signal level, square the result, and then normalize it to the calculated variance. A target is declared present if the square root of that value is greater than the threshold for the chosen Q_{fa}.

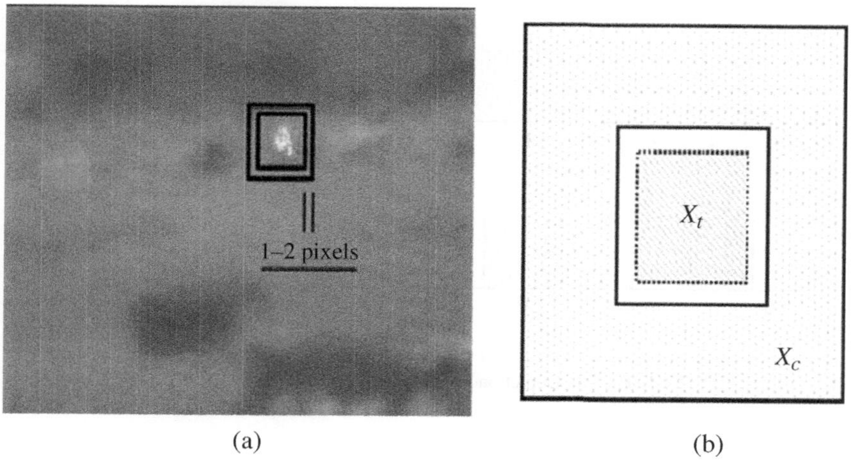

FIGURE 9.5 (a) Example of a gate for calculating local image statistics and (b) an illustration of the two windows used the RX algorithm.

9.5 OPTICAL COMMUNICATIONS

Optical communications have been an active research area since lasers were invented in 1964. There are many good books on the subject [10–18] and dozens of papers are published in the various society journals. Fortunately, the above theory is directly applicable to quantizing optical communications system performance, but with slightly different nomenclature. This section provides the basics to its application. This section shows how Sections 9.1 and 9.2 translate into forms used to quantify communications system performance.

9.5.1 Receiver Sensitivity for System Noise-Limited Communications

In most practical optical receivers, either thermal or preamplifier noises dominates.[3] The general figure of merit of how well the receiver can do this task is the received electrical SNR. As we indicated in the last chapter, the bit error probability and bit error rate (BER) also are appropriate performance metric for optical communications. This section outlines the basic theory for characterizing general optical communication systems.

Digital optical communications systems rely on the finite probability of error that characterizes the detection and estimation mechanism of incoming bits; in their simplest form, determining whether bit "1" or bit "0" was sent [16]. There are many

[3]We need to note that the developments to come assume Gaussian noise statistics, which does not reflect the statistics of a preamplified signal detected by a square-law photodetector. The actual statistics are Rayleigh and Rician statistics for bits "1" and "0" transmission, respectively, after the square-law processing of the Gaussian noise statistics. However, for most practical situations, a Gaussian PDF can be used to approximate these two statistical functions with the appropriate arguments without much loss of accuracy (≤ 2 dB).

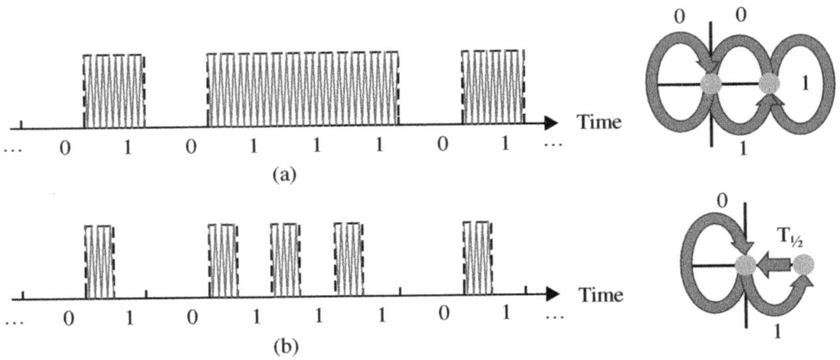

FIGURE 9.6 Depiction of (a) NRZ-OOK and (b) RZ-OOK.

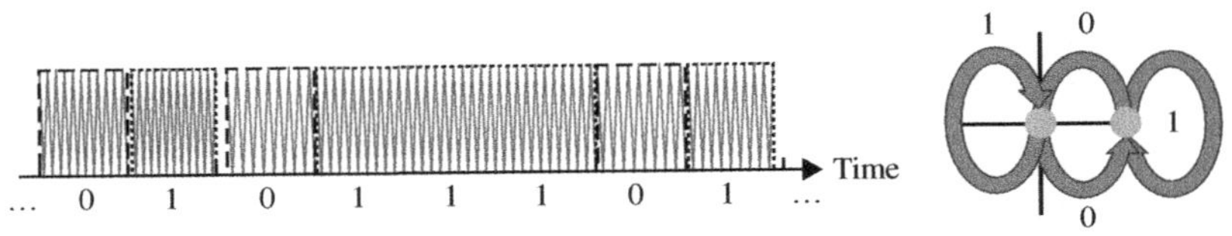

FIGURE 9.7 Depiction of DPSK.

ways to encode these bits on an optical laser beam. For example, Amplitude Shift Keying (ASK), also known as On–Off Keying (OOK), signal modulation involves turning the laser on and off to convey a "1" or a "0", respectively. Figure 9.6 illustrates OOK (one bit per symbol) signal format for (a) Non-Return-to-Zero (NRZ) OOK and (b) Return-to-Zero (RZ) OOK. This NRZ-OOK format has been the modulation of choice for 10 Gb/s and below. However, it is not self-clocking and needs the data stream to be scrambled to increase number of transitions. Alternately, the RZ-OOK format not only improves clocking recovery but also needs data scrambling to avoid long streams of "0"s. Simple detection can be used.

On the other hand, a Differential Phase Shift Keying (DPSK) signal modulation differentially encodes a "1" or a "0" onto one of two phase-states of the carrier. Figure 9.7 illustrates DPSK. Binary information is encoded onto two carrier streams, with "0" have no phase change and "1" have a π-phase difference with the "0" carrier. (DPSK compares the phase change possibility between adjacent pulses.)

Figure 9.8 depicts an OOK data input under a threshold decision process. In general, an incoming signal is determined via some mechanism during a synchronized time interval T_1 to establish whether either a "1" or a "0" is transmitted; this specific mechanism is defined by the signal modulation format used. For example, an OOK incoherent receiver measures the received optical energy using an integrate-and-hold circuit with integration time T_1 and a bandwidth that is given by the reciprocal data rate. On the other hand, a coherent detection system beats the weak incoming signal against with the strong signal from the local oscillator (LO) to yield a more robust signal. Due to the nonlinear nature of the photodetector, the resulting

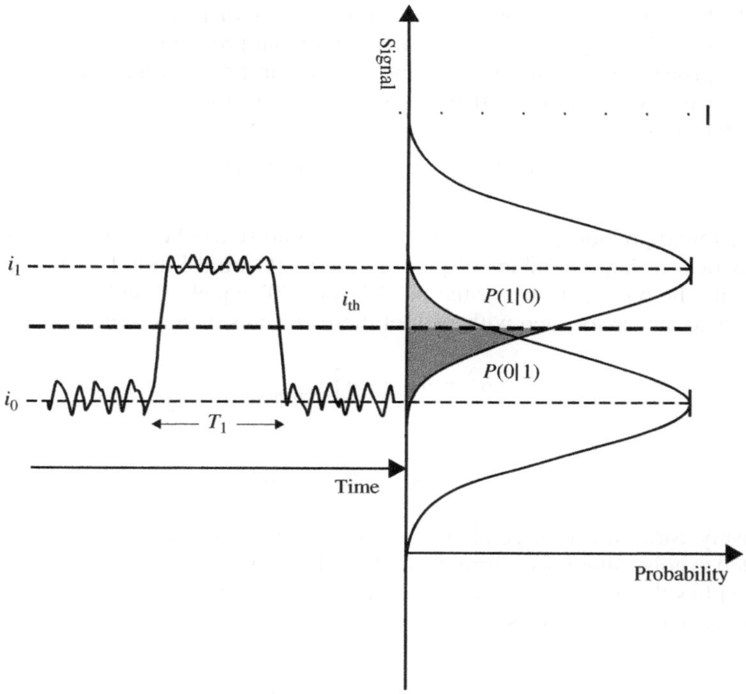

FIGURE 9.8 Conception of a OOK determination of bit transmitted.

signal sometimes will be at a different "carrier" frequency, commonly known as an *intermediate frequency* (*IF*). In general, a coherent system with a zero IF is called a homodyne system while a system with a finite IF is called a heterodyne system. No matter what the system approach is, the optical channels corrupt the incoming signal, dropping its energy and coherence attributes significantly. The implication is that sophisticated means must be employed to mitigate the effects and improve the detection of the various bits [10–18]; this is discussed in a later chapter. Let us continue our discussion of the basics, starting with two key definitions.

The *BER* is the number of bit errors per unit time. That is, the BER is the number of bit errors divided by the total number of transferred bits during a studied time interval. It is a unitless performance measure, often expressed as a percentage.

On the other hand, the *bit error probability* P_e is the expected value for the *BER*. The *BER* can be considered as an approximate estimate of the bit error probability. This estimate is accurate for a long time interval and a high number of bit errors. We do not distinguish between the two in the discussions to come.

On the basis of the material presented in the first two sections, we most likely would use for communications, a *maximum a posteriori* (*MAP*) optical receiver that has a *BER* given by

$$\text{BER} = \Pr\{1\}\ \Pr\{1|0\} + \Pr\{0\}\ \Pr\{0|1\}, \tag{9.78}$$

where Pr{1} and Pr{0} are the *a priori* probabilities of receiving a "1" or a "0", respectively; and Pr{1|0} and Pr{0|1} are conditional probabilities representing the respective probability of falsely declaring the first argument true, when the second argument actually is true [19]. If we assume equal *a priori* probabilities, then Eq. (9.78) reduces to

$$\mathrm{BER} = \frac{1}{2}[\Pr\{1|0\} + \Pr\{0|1\}]. \tag{9.79}$$

As we found previously, both shot and thermal noise can be modeled with Gaussian statistics for large numbers of photoelectrons (Central Limit Theorem). Let us assume that the photocurrent for the bits "1" and "0" equals i_1 and i_0, respectively. The total variance associated with each of these currents then is written as

$$\sigma_1^2 = \sigma_{1\text{-ss}}^2 + \sigma_{1\text{-thermal}}^2 \tag{9.80}$$

and

$$\sigma_0^2 = \sigma_{0\text{-ss}}^2 + \sigma_{0\text{-thermal}}^2, \tag{9.81}$$

respectively. Since the average photocurrent is different for the bits "1" and "0", the associated average shot noise current and its variance will be different, that is, $i_1 > i_0$, which implies that $\sigma_1^2 \neq \sigma_0^2$. Let i_{th} represent the decision threshold. The conditional probabilities can be written as

$$\Pr\{1|0\} = \frac{1}{\sqrt{2\pi\sigma_0^2}} \int_{i_{\text{th}}}^{\infty} e^{-\frac{(i-i_0)^2}{2\sigma_0^2}} \, di = \frac{1}{2}\,\mathrm{erfc}\left(\frac{(i_{\text{th}} - i_0)}{\sqrt{2}\sigma_0}\right) \tag{9.82}$$

and

$$\Pr\{0|1\} = \frac{1}{\sqrt{2\pi\sigma_1^2}} \int_{-\infty}^{i_{\text{th}}} e^{-\frac{(i-i_1)^2}{2\sigma_1^2}} \, di = \frac{1}{2}\,\mathrm{erfc}\left(\frac{(i_1 - i_{\text{th}})}{\sqrt{2}\sigma_1}\right) \tag{9.83}$$

under the Gaussian statistics assumption. Substituting Eqs. (9.82) and (9.83) into Eq. (9.79) yields

$$\mathrm{BER} = \frac{1}{4}\left[\mathrm{erfc}\left(\frac{(i_{\text{th}} - i_0)}{\sqrt{2}\sigma_0}\right) + \mathrm{erfc}\left(\frac{(i_1 - i_{\text{th}})}{\sqrt{2}\sigma_1}\right)\right]. \tag{9.84}$$

Example 9.8 Let us determine the threshold that minimizes the BER given in Eq. (9.84). Differentiating this equation with respect to i and setting the result to zero yields

$$\left.\frac{d\mathrm{BER}}{di}\right|_{i=i_{\text{th}}} = 0 = \left[\frac{1}{\sqrt{2\pi\sigma_0^2}} e^{-\frac{(i-i_0)^2}{2\sigma_0^2}} - \frac{1}{\sqrt{2\pi\sigma_1^2}} e^{-\frac{(i_1-i)^2}{2\sigma_1^2}}\right]_{i=i_{\text{th}}}.$$

This means that

$$\frac{1}{\sqrt{2\pi\sigma_0^2}}e^{-\frac{(i_{\text{th}}-i_0)^2}{2\sigma_0^2}} = \frac{1}{\sqrt{2\pi\sigma_1^2}}e^{-\frac{(i_1-i_{\text{th}})^2}{2\sigma_1^2}}.$$

Taking the natural log of the above yields

$$\frac{(i_{\text{th}}-i_0)^2}{2\sigma_0^2} = \frac{(i_1-i_{\text{th}})^2}{2\sigma_1^2} - \frac{1}{2}\ln\left(\frac{\sigma_1^2}{\sigma_0^2}\right).$$

Since thermal or preamplifier noises are the largest components in both σ_1^2 and σ_0^2, the second term on the right can be ignored and the above reduces to

$$\frac{(i_{\text{th}}-i_0)}{\sigma_0} \approx \frac{(i_1-i_{\text{th}})}{\sigma_1}.$$

Using the results of Example 9.8, the minimum BER occurs when

$$\frac{(i_{\text{th}}-i_0)}{\sigma_0} \approx \frac{(i_1-i_{\text{th}})}{\sigma_1} \equiv Q \tag{9.85}$$

or

$$i_{\text{th}} \approx \left(\frac{\sigma_0 i_1 + \sigma_1 i_0}{\sigma_0 + \sigma_1}\right). \tag{9.86}$$

If the receiver is dominated by thermal noise, then Eq. (9.86) reduces to

$$i_{\text{th}} \approx \left(\frac{i_1 + i_0}{2}\right). \tag{9.87}$$

Using Eq. (9.86), we find that

$$Q = \left(\frac{i_1 - i_0}{\sigma_0 + \sigma_1}\right). \tag{9.88}$$

This means that we can rewrite Eq. (9.84) as

$$\text{BER} = \frac{1}{2}\left[\text{erfc}\left(\frac{Q}{\sqrt{2}}\right)\right] \approx \frac{e^{-\frac{Q^2}{2}}}{Q\sqrt{2\pi}}. \tag{9.89}$$

with the last part of this equation valid for $Q > 3$. Squaring Q essentially gives a form of the electrical SNR.

Recall that the receiver sensitivity is defined as the minimum average power needed to keep the BER below a certain value, for example, BER $< 10^{-9}$. Let us now relate our new parameter Q to the incident optical power. Let us assume OOK direct-detection communications using a PIN photodiode. Without loss of generality, we will let $i_0 = 0$, which translates into the zero signal power transmitted for a bit "0". From Eq. (8.100), we know that the electrical SNR for this detection scheme is given by

$$\mathrm{SNR}_{\text{DD-PIN}} \approx \frac{\left(\frac{q\eta}{h\nu}\right)^2 R_L {P_{\text{rec}}}^2}{2qB_e\,\left[\left(\frac{q\eta}{h\nu}\right)P_{\text{rec}} + I_D\right]\,R_L + 4kTF_{\text{rf}}B_e} \approx \frac{R_\lambda^2 {P_{\text{rec}}}^2}{\sigma_1^2}, \tag{9.90}$$

where

$$\sigma_1^2 = \sigma_{\text{sh}}^2 + \sigma_{\text{thermal}}^2, \tag{9.91}$$

$$\sigma_{\text{sh}}^2 = 2q(R_\lambda P_{\text{rec}} + I_D\,)B_e, \tag{9.92}$$

and

$$\sigma_{\text{thermal}}^2 = \frac{4kTF_{\text{rf}}B_e}{R_L}. \tag{9.93}$$

The noise variance for bit "0" is given by

$$\sigma_0^2 = \sigma_{\text{sh0}}^2 + \sigma_{\text{thermal}}^2, \tag{9.94}$$

where

$$\sigma_{\text{sh0}}^2 = 2qI_DB_e. \tag{9.95}$$

Since

$$P_{\text{rec}} = \frac{(P_1 + P_0)}{2} = \frac{P_1}{2}, \tag{9.96}$$

and the thermal noise generally is much larger than the shot noise component, the parameter Q can be written as

$$Q_{\text{DD-PIN}} = \left(\frac{i_1 - i_0}{\sigma_0 + \sigma_1}\right) = \frac{2R_\lambda P_{\text{rec}}}{\sqrt{2q\,(R_\lambda\,P_{\text{rec}} + I_D\,)\,B_e + \sigma_{\text{thermal}}^2} + \sqrt{2qI_DB_e + \sigma_{\text{thermal}}^2}} \tag{9.97}$$

$$\approx \frac{2R_\lambda P_{rec}}{\sqrt{\sigma_{\text{thermal}}^2}\left(1 + \frac{2q\,(2R_\lambda P_{\text{rec}} + I_D\,)B_e}{2\sigma_{\text{thermal}}^2}\right)^{1/2} + \sqrt{\sigma_{\text{thermal}}^2}\left(1 + \frac{2qI_DB_e}{2\sigma_{\text{thermal}}^2}\right)^{1/2}}$$

$$\approx \frac{R_\lambda P_{\text{rec}}\sigma_{\text{thermal}}}{\left(q(R_\lambda P_{\text{rec}} + I_D\,)B_e + \frac{4kTF_{\text{rf}}B_e}{R_L}\right)^{1/2}}.$$

$$\approx \frac{R_\lambda P_{\text{rec}}}{\sqrt{\frac{4kTF_{\text{rf}}B_e}{R_L}}} = \sqrt{\mathrm{SNR}_{\text{Thermal Noise Limited}}}, \tag{9.98}$$

where

$$\mathrm{SNR}_{\text{Thermal Noise Limited}} = \frac{R_\lambda^2 P_{\text{rec}}{}^2}{\frac{4kTF_{\text{rf}}B_e}{R_L}} \tag{9.99}$$

is the thermal-noise-limited electrical SNR.

For an APD photodiode direct-detection system, we have

$$Q_{\text{DD-APD}} \approx \frac{MR_\lambda P_{\text{rec}}}{\sqrt{\frac{4kTF_{\text{rf}}B_e}{R_L}}} = \sqrt{M^2 \mathrm{SNR}_{\text{Thermal Noise Limited}}}. \tag{9.100}$$

The minimum average powers needed to keep the BER below a certain value therefore are

$$P_{\text{rec}} = \begin{cases} \frac{\sigma_{\text{thermal}} Q}{R_\lambda}; & \text{for PIN receivers} \\ \frac{\sigma_{\text{thermal}} Q}{(MR_\lambda)}; & \text{for APD receivers} \end{cases}. \tag{9.101}$$

It is apparent from Eq. (9.101) that $Q_{\text{DD-APD}}$ is a factor of M greater than $Q_{\text{DD-PIN}}$ under thermal noise-limited operations. Here, the parameter M is the average gain factor of M for an APD receiver.

Example 9.9 For $\lambda = 1.55\,\mu\text{m}$, we have $R_\lambda = 1\,\text{A/W}$, $\sigma_{\text{thermal}} = 100\,\text{nA}$, $M = 10$, and $Q = 6$, we have

$$P_{\text{rec}} = \begin{cases} 0.6\,\mu\text{W}(-32.2\,\text{dBmW}); & \text{for the PIN receiver} \\ 60\,\text{nW}(-42.2\,\text{dBmW}); & \text{for the APD receiver} \end{cases}.$$

9.5.2 Receiver Sensitivity for Quantum-Limited Communications

When one is using quantum-limited communications systems, such as the ones using coherent and photon-counting receivers, the signal shot noise is the dominant noise component and the above analysis changes dramatically. Specifically, the electrical SNR now is proportional to the received power and the receiver sensitivity strictly is characterized by the number of received photons. Let us look at this in more detail.

Let us assume an OOK direct-detection receiver. From Eq. (8.101), we know that the electrical SNR for this detection scheme is given by

$$\begin{aligned}\mathrm{SNR}_{\text{DD-PIN}} &\approx \frac{\left(\frac{q\eta}{h\nu}\right)^2 R_L P_{\text{rec}}{}^2}{2qB_e\left[\left(\frac{q\eta}{h\nu}\right)P_{\text{rec}} + I_D\right] R_L + 4kTF_{\text{rf}}B_e} \approx \frac{R_\lambda^2 P_{\text{rec}}{}^2}{2qR_\lambda P_{\text{rec}} B_e} \\ &\approx \frac{R_\lambda P_{\text{rec}}}{2qB_e} = \left(\frac{\eta P_{\text{rec}}}{2h\nu B_e}\right)\end{aligned} \tag{9.102}$$

for signal shot noise-limited communications.

Shannon [20] postulated that the channel capacity for error-free communications is given by

$$C = B_e \log_2(1 + \mathrm{SNR}) = B_e \log_2\left(1 + \frac{P}{N_0 B_e}\right) = B_e \log_2\left(1 + \frac{rE_b}{N_0}\right), \tag{9.103}$$

where

$$r \equiv \text{spectral efficiency (bps/Hz)}$$
$$= \frac{R_b}{B_e} = \log_2(M_{\text{symbol}}). \tag{9.104}$$

In the above equations, C is the capacity (rate in bits per second or bps) over the channel bandwidth B_e (Hz) with average electrical signal power P (watts or W) affected by an AWGN channel with unilateral spectral electrical noise density $N_0/2$ (W/Hz), $E_b = P/R_b$ is the energy per bit (W-s/bit or joules/bit or J/b), R_b is the communications data rate (bps), M_{symbol} is the number of symbols in the system alphabet and the average electrical signal power is $E_b R_b$. The Shannon channel limit is achieved when R_b equals C, the channel capacity. Following community convention, Eq. (9.102) can be rewritten as

$$\text{SNR}_{\text{DD-PIN}} = \left(\frac{r\eta P_{\text{rec}}}{2h\nu R_b}\right) = (\eta r n_0)/2, \tag{9.105}$$

where

$$n_0 = \frac{P_s}{(h\nu R_b)} = \left(\frac{1}{\eta}\right)\frac{E_b}{N_0} \tag{9.106}$$

is the number of transmitted photons/bit (ppb). This implies that the quantum-limited SNR can be written as

$$\text{SNR}_{\text{QL}} = \frac{rE_b}{N_0} = r\eta\left[\frac{P_s}{(h\nu R_b)}\right] \equiv r\eta n_0 \tag{9.107}$$

in most cases.[4] Typically, the average photon count, $\overline{n}_0$, rather than peak photon count, is specified. The average signal photocurrent then is

$$i_1 = 2qr\overline{n}_0 R_b. \tag{9.108}$$

The factor of 2 on the RHS is because the average receiver sensitivity for OOK demodulation is half that of the peak sensitivity since the laser is turned off while transmitting a "0" bit.

$$Q_{\text{DD-PIN}} = \left(\frac{i_1 - i_0}{\sigma_0 + \sigma_1}\right) = \frac{2qr\eta n_0 R_b}{\sqrt{4q^2 r\eta n_0 R_b}} = \sqrt{r\eta\overline{n}_0 R_b} = \sqrt{\overline{m}_0}, \tag{9.109}$$

where

$$\overline{m}_0 = r\eta\overline{n}_0 R_b = r\eta\left[\frac{P_s}{(h\nu R_b)}\right] = \frac{\eta P_s}{(h\nu B_e)}(= \text{SNR}_{\text{QL}}) \tag{9.110}$$

is the average number of received photo-counts (pc) created by the bit "1" signal at the detector.

[4]The one exception is for coherent homodyne receivers, where $\text{SNR}_{QL\text{-}homodyne} = \text{SNR}_{QL}/2$. See Examples 9.10 and 9.11.

The OOK error probability for an ideal signal shot noise-limited, preamplifier optical receiver is approximately given by

$$\mathrm{BER} \approx \frac{1}{2}\left[\mathrm{erfc}\left(\frac{\sqrt{\overline{m}_0}}{\sqrt{2}}\right)\right] \approx \frac{e^{-\frac{\overline{m}_0}{2}}}{\sqrt{2\pi\overline{m}_0}}. \tag{9.111}$$

The 10^{-9}BER corresponds to ~36 pc using either of the two expressions for BER in Eq. (9.111). The receiver sensitivity is ~36 pc/bit. When implemented with NRZ waveforms with equal probability of "1"s and "0"s, the peak power is equal to twice the average power; pulsed RZ waveforms can be used with all the modulations format normally used in optical communications, with the peak-to-average power ratio varying inversely with the duty cycle [13, p. 121].

Caplan noted that the above and the next developments assume that Gaussian noise statistics does not reflect the Poisson statistics of the receiver input signal (see footnote 2 for a similar comment on preamplified noise-limited receivers). However, his numerical comparison between the two PDFs showed that the Rayleigh/Rician PDFs yielded a 10^{-9}BER for $\overline{m}_0 = 38$ pc, while the Gaussian PDF created a 10^{-9}BER for $\overline{m}_0 = 36$ pc [13, pp. 121–122]. From a practical POV, the above is good enough for first-order link budget and receiver sensitivity calculations because of the ease of computing BER using Eq. (9.111).

Example 9.10 Let us look at an OOK (totally coherent) homodyne system. In this case, there is no IF frequency, and laser power presence indicates a "1" and laser power absence denotes a "0." Given we have a LO present at the photodiode in both cases, we have

$$\overline{i_1} \approx i_{\mathrm{LO}} + 2\sqrt{i_{\mathrm{LO}}\, i_s} \tag{9.112}$$

and

$$\overline{i_0} \approx i_{\mathrm{LO}} \tag{9.113}$$

with $\sigma_1^2 = \sigma_0^2 = 2qi_{\mathrm{LO}}B_e$. This means that

$$Q_{\text{OOK-Homodyne}} = \left(\frac{i_1 - i_0}{\sigma_0 + \sigma_1}\right) \approx \frac{i_{\mathrm{LO}} + 2\sqrt{i_{\mathrm{LO}} i_s} - i_{\mathrm{LO}}}{2\sqrt{2qi_{\mathrm{LO}}B_e}} = \sqrt{\frac{i_s}{2qB_e}} = \sqrt{m_0}. \tag{9.114}$$

Therefore, just like in the above direct detection example, a 10^{-9}BER corresponds to ~18 pc. The associated receiver sensitivity is 18 pc/bit. An OOK heterodyne system as a sensitivity of 36 pc/bit, twice the homodyne case [21, p. 118].

Example 9.11 Let us look at a PSK (totally coherent) homodyne system. In this case, there is no IF frequency, and laser power presence indicates a "1" and laser power absence denotes a "0." Given we have a LO present at the photodiode in both cases, we have

$$\overline{i_1} \approx i_{\mathrm{LO}} + 2\sqrt{i_{\mathrm{LO}}\, i_s} \tag{9.115}$$

and

$$\overline{i_0} \approx i_{\mathrm{LO}} - 2\sqrt{i_{\mathrm{LO}}\, i_s} \tag{9.116}$$

with $\sigma_1^2 = \sigma_0^2 = 2qi_{\mathrm{LO}}B_e$. This means that

$$Q_{\text{PSK-Homodyne}} = \left(\frac{i_1 - i_0}{\sigma_0 + \sigma_1}\right) \approx \frac{i_{\mathrm{LO}} + 2\sqrt{i_{\mathrm{LO}} i_s} - i_{\mathrm{LO}} + 2\sqrt{i_{\mathrm{LO}} i_s}}{2\sqrt{2qi_{\mathrm{LO}}B_e}}$$

$$= 2\sqrt{\frac{i_s}{2qB_e}} = 2\sqrt{m_0} \tag{9.117}$$

and a 10^{-9}BER corresponds to 9 pc. The associated receiver sensitivity is 9 pc/bit since each of the two bits must carry an average of 9 pc. A PSK heterodyne system as a sensitivity of 18 pc/bit, twice the homodyne case [21, p. 1118].

Example 9.12 Differentially encoded PSK (DPSK) receivers have gotten considerable attention by the FSOC and TELCO communities because it offers a 3 dB sensitivity improvement over the more commonly used OOK signal modulation formats and a lower peak power to help mitigate nonlinear effects [22, 23]. NRZ-DPSK can be implemented with a constant envelope, which implies the peak power equals the average power. The associated error probability is given by

$$\text{BER} \approx \frac{1}{2}[\text{erfc}(\sqrt{\overline{m}_0})] \approx \frac{e^{-\overline{m}_0}}{2\sqrt{\pi \overline{m}_0}}. \tag{9.118}$$

The 10^{-9}BER corresponds to ~18 pc using either of the two expression for BER in Eq. (9.118). One can attribute the 3 dB receiver sensitivity improvement to the differential encoding, which uses the full energy of each symbol to determine the relative phase for one bit of information [13, p. 122]. Unfortunately, external and internal noise sources can corrupt these two states causing errors to be made.

Example 9.13 As we noted earlier, the above derivation of the BER in the shot-noise limit is not totally valid. Poisson statistics should be used, especially for a small number of photons. Let us look at the few photon limits. For an ideal detector (no thermal noise, no dark current, and $\eta = 1$), Agrawal suggests that bit "0" produces no photo-counts, that is, $\Pr\{1|0\} \equiv 0$, and $\sigma_0 = 0$; errors only occur if bit "1" fails to produce even one photo-count [19]. Recall that the probability of generating m photo-counts is given by

$$\Pr_m = \frac{n_1^m}{m!} e^{-n_1}. \tag{9.119}$$

since $n_1 = m_0(\eta = 1)$ in this case. Since $\Pr\{0|1\} = \Pr_{m=0} = e^{-n_1}$, we have [19]

$$\text{BER} = \frac{1}{2}[\Pr\{1|0\} + \Pr\{0|1\}] = \frac{1}{2}\Pr\{0|1\} = \frac{1}{2}e^{-n_1}. \tag{9.120}$$

If $n_1 = 20$, BER $= 10^{-9}$ using Eq. (9.120), which translates to 10 pc/bit on average.

Example 9.14 Let us look at maximum-likelihood symbol-by-symbol detection of OOK communications in the turbulent channel leveraging Zhu and Kahn [21].

From Section 7.8 in Chapter 7, we find that the covariances of the log-amplitude fluctuation χ for plane and spherical wave are given by Eqs. (7.141) and (7.142), respectively. We know that light traversal through turbules in the channel causes the wave at each interaction, independent, identically distributed phased delays and scatterings. By the Central Limit Theorem, we can write the marginal PDF for log-amplitude fluctuations as

$$f_\chi(\chi) = \frac{1}{\sqrt{2\pi\sigma_\chi^2}} e^{-\frac{(\chi - E[\chi])^2}{2\sigma_\chi^2}}. \tag{9.121}$$

The light intensity is related to the log-amplitude fluctuation χ by the equation

$$I = I_0 e^{-(\chi - E[\chi])}, \tag{9.122}$$

where $E[\chi]$ is the ensemble average of the log-amplitude fluctuation χ. The expectation of the intensity is then

$$E[I] = E\left[I_0 e^{-(\chi - E[\chi])}\right] = I_0\, e^{-2\sigma_\chi^2}. \tag{9.123}$$

This implies that Eq. (9.121) can be rewritten as

$$f_I(I) = \frac{1}{2I\sqrt{2\pi\sigma_\chi^2}} e^{-\frac{[\ln(I) - \ln(I_0)]^2}{8\sigma_\chi^2}}. \tag{9.124}$$

Zhu and Kahn assumed that an intensity modulation/direct detection (IM/DD) FOSC link using OOK coding was dominated by external background shot and thermal noise, both of which were of sufficient magnitude to follow Gaussian statistics. They also assumed a bit interval T_{bit} with a receiver integration time $T_0 \leq T_{\text{bit}}$. The resulting integrated photocurrent is given by

$$i_{\text{pc}} = R_\lambda(I_s + I_{\text{background}}) + i_{\text{thermal}} \approx R_\lambda I_{\text{background}}, \tag{9.125}$$

where I_s and $I_{\text{background}}$ are the received signal and background optical intensities. Both are assumed to be constant during the integration interval. Following the above authors, we express the noise variance to be given by $N_0/2$.

Let us assume that the receiver has knowledge of the marginal distribution of the turbulence-induced fading, but not of the instantaneous channel fading. If we now subtract off the shot noise current out, that is, $i \approx i_{\text{pc}} - R_\lambda I_{\text{background}}$, the resulting signal yields the following conditional probability densities when a transmitted bit is "On" or "Off", respectively:

$$\Pr\{i|\text{Off}\} = \frac{1}{\sqrt{2\pi\left[\frac{N_0}{2}\right]}} e^{-\left(\frac{i^2}{2\left[\frac{N_0}{2}\right]}\right)} = \frac{1}{\sqrt{\pi N_0}} e^{-\left(\frac{i^2}{N_0}\right)} \tag{9.126}$$

$$\Pr\{i|\text{On}\} = \int_{-\infty}^{\infty} \Pr\{i|\text{On}.\ \chi\} f_\chi(\chi)\, d\chi \tag{9.127}$$

$$= \int_{-\infty}^{\infty} \frac{1}{\sqrt{\pi N_0}} f_\chi(\chi) e^{-\left(\frac{(i-R_\lambda I_0\, e^{-(\chi-E[\chi])})^2}{N_0}\right)} d\chi. \tag{9.128}$$

The optimal MAP symbol-by-symbol detector decodes the bit $\hat{s}$ as

$$\hat{s} = \text{argmax}_s \Pr\{i|s\}P(s), \tag{9.129}$$

where $P(s)$ is the probability that a "On" bit or "Off" bit is transmitted [21]. $\Pr\{i|s\}$ is the conditional distribution that if a bit s ("On" or "Off") is transmitted, a photocurrent i will be received. If "On" or "Off" bits are equally probable, or if their *a priori* probabilities are unknown, the symbol-by-symbol maximum likelihood (ML) detector decodes the bits as

$$\hat{s} = \text{argmax}_s \Pr\{i|s\}. \tag{9.130}$$

The likelihood function is

$$\Lambda(i) = \frac{\Pr\{i|\text{On}\}}{\Pr\{i|\text{Off}\}} = \int_{-\infty}^{\infty} f_\chi(\chi) e^{-\left(\frac{(i-R_\lambda I_0 e^{-(\chi-E[\chi])})^2 - i^2}{N_0}\right)} d\chi. \tag{9.131}$$

Now, Zhu and Kahn rewrote the argument of the exponential as

$$-\left(i - R_\lambda I_0 e^{2X-2E[X]}\right)^2 + i^2 = {R_\lambda}^2 E^2[I] \left[2\left(\frac{i}{R_\lambda E[I]}\right)\left(\frac{I_0}{E[I]}\right) e^{2X-2E[X]} - \left(\frac{I_0}{E[I]}\right)^2 e^{4X-4E[X]} \right]$$

$$= {R_\lambda}^2 E^2[I] \left[2\left(\frac{i}{R_\lambda E[I]}\right) e^{-2\sigma_X^2} e^{2X-2E[X]} - e^{-4\sigma_X^2} e^{4X-4E[X]} \right], \tag{9.132}$$

which gives

$$\Lambda(i) = \int_{-\infty}^{\infty} f_\chi(\chi) e^{-\left(\frac{{R_\lambda}^2\, E^2[I]\left[2\left(\frac{i}{R_\lambda\, E[I]}\right) e^{-2\sigma_X^2}\, e^{2X-2E[x]} - e^{-4\sigma_X^2} e^{4X-4E[X]}\right]}{N_0}\right)} d\chi. \tag{9.133}$$

Figure 9.9 displays the ML ratio $\Lambda(i)$ as a function of the normalized photocurrent $\frac{i}{R_\lambda E[I]}$ for various Rytov log-amplitude standard deviations. In this figure, $R_\lambda E[I] = 1$, $E[X] = 0$, and $N_0 = 2 \times 10^{-2}$. This figure shows that the likelihood ratio reduces for increasing photocurrent and increases for increasing Rytov log-amplitude standard deviation.

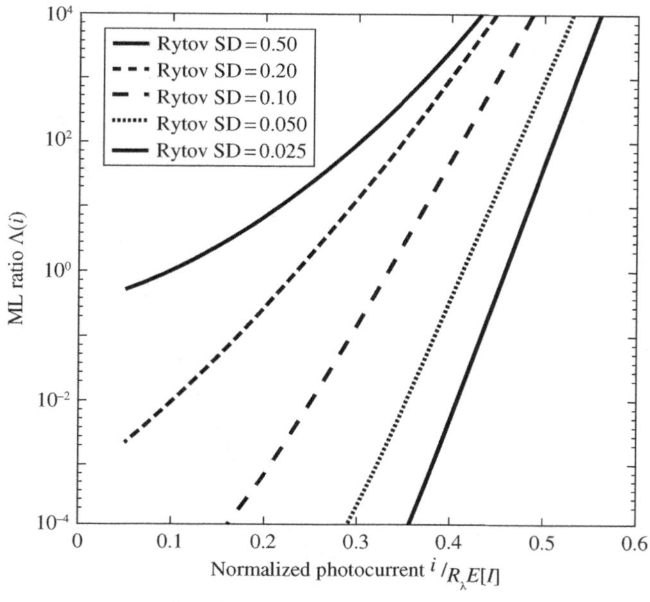

FIGURE 9.9 Maximum likelihood ratio $\Lambda(i)$ as a function of the normalized photocurrent $\frac{i}{R_\lambda E[I]}$ for various Rytov log-amplitude standard deviations.

The BER of OOK can be written as

$$P_e = P(\text{Off})\,P(\text{Bit Error}|\text{Off}) + P(\text{On})\,P(\text{Bit Error}|\text{On}) \tag{9.134a}$$

$$= 0.5\,[P(\text{Bit Error}|\text{Off}) + P(\text{Bit Error}|\text{On})] \tag{9.134b}$$

assuming equal probability of "1"s and "0"s and assuming no intersymbol interference. In this case, we have

$$P(\text{Bit Error}|\text{Off}) = \int_{r_{\text{threshold}}}^{\infty} \Pr(r|\text{off})\,dr \tag{9.135}$$

and

$$P(\text{Bit Error}|\text{On}) = \int_{0}^{r_{\text{threshold}}} \Pr(r|\text{on})\,dr. \tag{9.136}$$

The optimum receiver threshold is selected by finding the threshold that minimizes Eq. (9.134b). Figure 9.10 illustrates an example probability of bit error graph for $\sigma_\chi = 0.15$ and $N_0 = 0.02$. The minimum is clearly seen at a threshold level of 0.31. Figure 9.11 depicts the normalized threshold for ML symbol-by-symbol detection versus the log-amplitude standard deviation of the turbulence for selected values of AWGN. This figure shows that normalized threshold decreases at σ_χ increases for a fixed AWGN level. This is because the turbulence-induced fading increases the

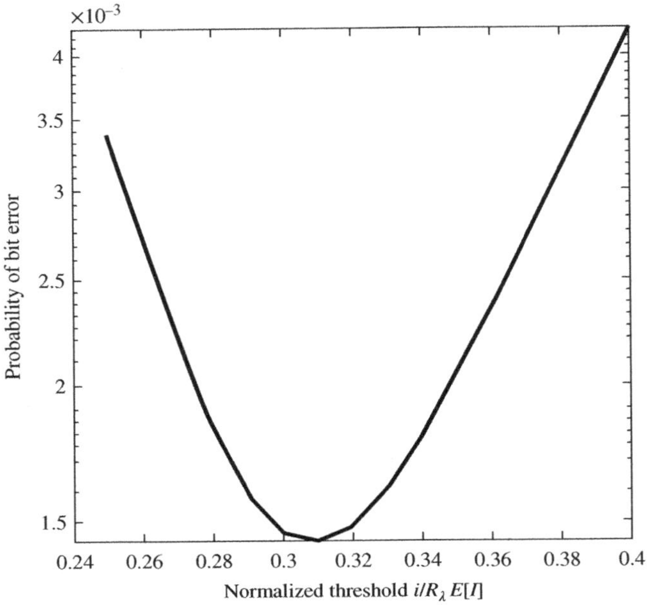

FIGURE 9.10 Example probability of bit error as a function of the normalized threshold for $\sigma_\chi = 0.15$ and $N_0 = 0.02$.

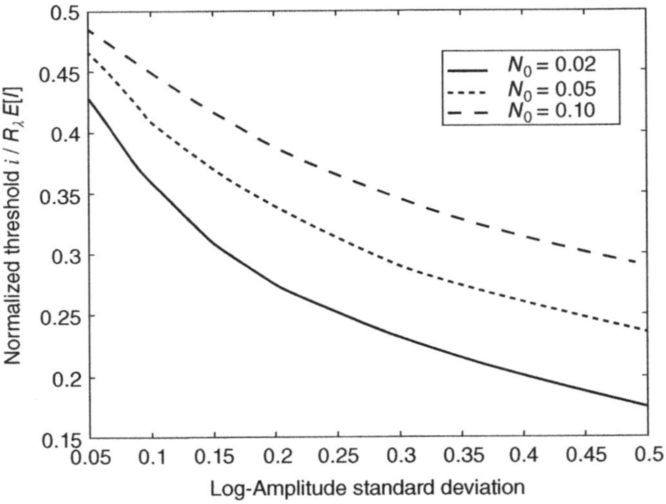

FIGURE 9.11 Normalized threshold for maximum-likelihood symbol-by-symbol detection versus the log-amplitude standard deviation of the turbulence for selected values of AWGN.

fluctuations of "1" signal, while the "0" bit signal level is unchanged [21]. On the other hand, increasing AWGN level increases the threshold level toward 0.5 for all values of σ_{χ}, with the low end moving faster to that limit.

9.6 LASER RADAR (LADAR) AND LIDAR

Coherent visible/infrared imaging is a rapidly emerging product area and there are a multitude of commercial and military applications, with each possessing its own unique digital signal processing approach. For example, in the most direct application of ranging, very high-resolution terrain maps, better known as digital terrain elevation data (DTED) is revolutionizing the map making and area surveys. Figure 9.12 depicts three examples of DTED images at 10, 3, and 1-m ground sample distances. This section deals with the two key system concepts, LAser raDAR (LADAR) and LIght Detection And Ranging (LIDAR).[5] We provide background information on LADAR/LIDAR, then focus on several LADAR concepts, and end by highlighting the governing equation for LIDAR applications.

9.6.1 Background

A LADAR is an electromagnetic system for hard target detection and ranging. It has greater precision over radar because of its wavelength. There are many varieties, with

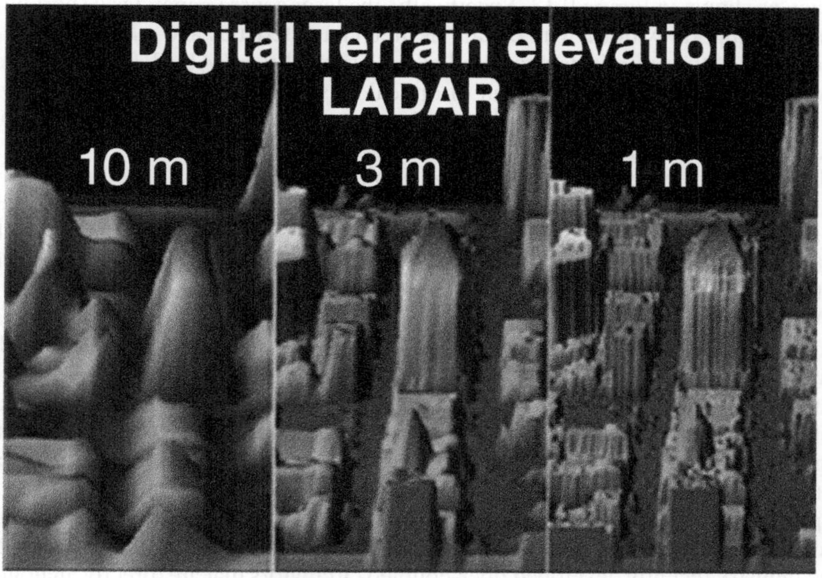

FIGURE 9.12 Examples of DTED.

[5]There are many definitions for acronyms LADAR and LIDAR. For the purpose of this book, these definitions serve well in differentiating between the two capabilities.

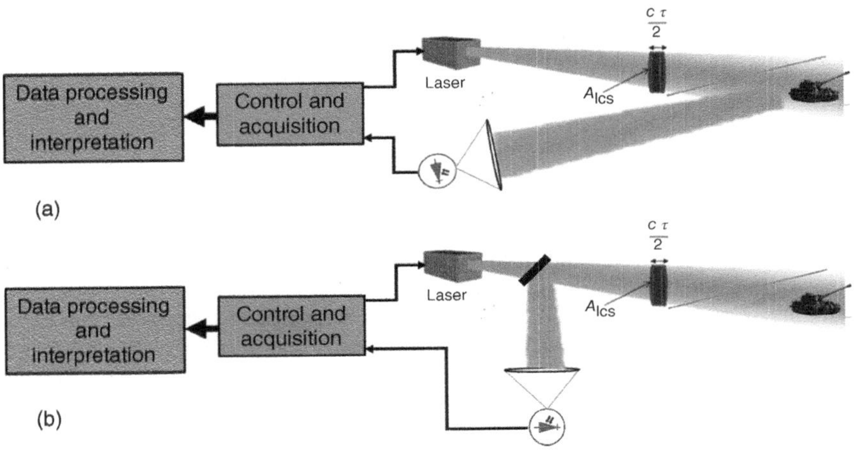

FIGURE 9.13 Block schematics of (a) bistatic and (b) monostatic LADAR/LIDAR systems.

the most popular being coherent, direct detection and photon-counting laser radars. LIDAR systems are remote sensing systems for characterizing environmental elements such as gases, liquids, and terrain [24].

Both of these systems operate by transmitting and receiving electromagnetic energy and then processing to obtain the desired information they were designed to extract, as shown in Figure 9.13. Specifically, the laser source generates a short pulse signal of temporal width τ with certain desired characteristics and uses some sort of optical device to transmit its generated signal. Typically, a LADAR/LIDAR system uses a telescope (or an arrangement of optical lenses) to direct the light beam with minimal spatial and angular spreading to the target, which can be truck or gas cloud, depending on the application.

The simplest system configuration to implement is the bistatic setup shown in Figure 9.13a. Here, separate paths and telescopes are used to transmit the initial signal and receive the return signal created by the target. The main advantage of this configuration is that it minimizes initial signal backscatter of the receiver optics. This configuration is commonly used in LADAR systems, but not always.

An alternate system configuration is to have the same telescope for both the transmitting and receiving light signals, that is, the monostatic configuration illustrated in Figure 9.13b. While using this setup can reduce the size and mechanical complexity of the system, as compared to a bistatic configuration, it does increase the complexity of electronic circuitry and the number of optical subsystems needed. To separate the outgoing and incoming beams, one can use a simple partially reflective-mirror beam splitter, or something more complex. For example, one might use waveplates to rotate the polarization of the outgoing laser beam and a polarization sensitive beam splitter to route the incoming energy into the proper receiver channel. Alternatively, the fiber optic industry recently developed fiber-coupled circulators that are directly analogous to the microwave waveguide and circulator used a transmit/receive switches, and that may the designer's chosen approach [23].

Whether in a mono- or bistatic LADAR/LIDAR, the receiver transforms the incoming return signal energy captured by the telescope into an electrical signal that can be processed to extract the desired information. In particular, a time-gated photodetector is used to convert the incoming return signal photons into photocurrent that goes into the electronic circuitry for processing. Let us look at the type of incoming signal return one might expect.

Since all electromagnetic energy travels at the speed of light c, in free space, the relationship between the range R and the round-trip travel time t is given by

$$t = \frac{2R}{c}. \tag{9.137}$$

Since few terrestrial LADAR applications have round-trip times that even approach seconds, time is usually accounted for in units of nanoseconds. Solving Eq. (9.137) for range, we have

$$R = \frac{ct}{2}. \tag{9.138}$$

At this point, one should note that a LADAR may be viewed as a type optical imaging system and therefore can be characterized in terms of its "resolution." Yes and No. LADAR has a resolution, but its criterion is different from what we discussed earlier for optical imaging systems. The accepted definition for LADAR resolution is the smallest distance separation between two distinct objects illuminated by a LADAR source that can be detected in the signal return. Figure 9.14 illustrates this concept. In this figure, we have defined the target as a step-like structure where the range difference between the two surfaces (A, B) is ΔR. In Figure 9.14a, the laser full beam width $2\theta_{tx}$ is small enough to encompass only one step at a time, which means that a single surface is illuminated by each laser pulse (P_1, P_2). This means that

$$\Delta R = (R_1 - R_2), \tag{9.139}$$

where R_1 is the distance to surface B and R_2 is the distance to surface A.

In this case, the two received signals allow us to write

$$t + \Delta t = \frac{2(R + \Delta R)}{c}, \tag{9.140}$$

where Δt is the time of flight for distance ΔR. If we subtract Eq. (9.137) from the above equation, we obtain

$$\Delta t = \frac{2\Delta R}{c}. \tag{9.141}$$

This type of systems allows us to establish a *range resolution* that can discriminate between different ranged surfaces.

In Figure 9.14b, the laser full beam width $2\theta_{tx}$ is large enough to encompass both surfaces using a single pulse P_1. With a single measured pulse, one cannot determine range resolution between several illuminated objects with a single range gate reading. An example is a laser range finder that makes only a single range measurement of a broad object space per transmitted pulse. However, some systems have processors that can determine multiple ranges from a single return (i.e., "first-pulse, second-pulse" or "first-pulse, last-pulse" logic) that can be useful for applications such as removing foliage from an image to produce "bare-earth" terrain maps.

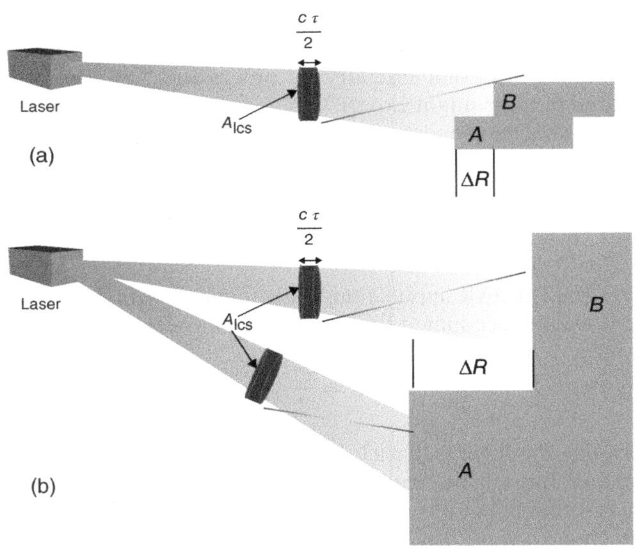

FIGURE 9.14 Range accuracy versus resolution.

LADAR systems often have beam widths smaller than the targets, and the targets can resemble anything from a Lambertian to a specular reflector and often are combinations of both. In general, rough targets scatter the incident light diffusely; smooth targets scatter specularly.

Another target characteristic that LADAR respond to is large area surface roughness (more correctly, the scale of the surface roughness). Measuring this roughness can be of particular interest for applications such as monitoring sea states.

LIDAR system will try to illuminate the target, for example, gas cloud, to get as much backscattered light as possible for the data processing returns into its particle and aerosol constituent parts.

9.6.2 Coherent Laser Radar

Gatt and Henderson have comprehensive article on coherent and direct detection laser radars and their detection statistics [25]. They noted that these two types of laser radar have substantially different detection statistics. Both also are affected by speckle noise and they need to be mitigated. Specifically, speckle noise in the former can be suppressed through some degree of speckle averaging using multiple frequencies or dual polarization. Speckle noise suppression in the latter will occur by temporal averaging of the incoming return signal. Direct detection laser radars also have different modes, for example, continuous wave and photon-counting. This section summarizes their results here, highlighting the three architectures: coherent, continuous direct detection, and photon-counting direct detection. Their performance will be described by their probabilities of detection and false alarm as a function of signal strength, the number of search bins and the level of averaging or accumulation [26]. The reader will see a strong synergism with the previous material.

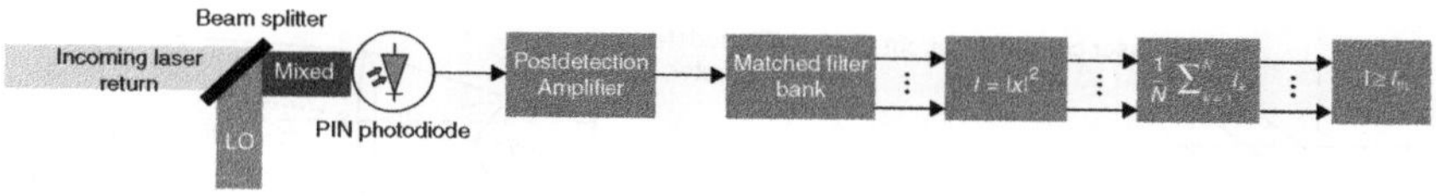

FIGURE 9.15 Coherent detection intensity processor block diagram.

Just like in a communications system, a coherent laser radar beats its weak incoming (return) signal against with a strong LO signal to yield a more robust signal at the detector. Without loss of generality, let us assume a heterodyne laser radar structure. Figure 9.15 shows the envisioned coherent detection intensity processor.

Let $A_s e^{-i2\pi\nu_c t}$ represent the return optical signal field and $A_{\text{LO}} e^{-i2\pi\nu_{\text{LO}} t}$ the optical LO field. Here, ν_c is the return signal's carrier frequency and ν_{LO} the oscillator's carrier frequency. The incoming beam hits beam splitter, reflecting some light out of the beam and letting the rest propagated to photodetector, a PIN photodiode in this case. Simultaneously, a continuous LO beam hits the backside of the same beam splitter allowing some light to be transmitted through and reflecting the remainder to the same photodiode. The optical power collected by the photodetector is the integral of the intensity of the two mixed signals over the detector area, which yields

$$P = |A_s|^2 + |A_{\text{LO}}|^2 + 2\sqrt{\gamma_{\text{he}}}|A_s||A_{\text{LO}}| \cos[2\pi(\nu_c - \nu_{\text{LO}})t + \varphi_c - \varphi_{\text{LO}}] \quad (9.142)$$

$$= P_s + P_{\text{LO}} + 2\sqrt{\gamma_{\text{he}} P_s P_{\text{LO}}} \cos[2\pi\nu_{\text{IF}}\, t + \varphi_c - \varphi_{\text{LO}}] \quad (9.143)$$

with P_s being the return optical signal power, P_{LO} being the LO power, and the frequency $\nu_{\text{IF}} = \nu_c - \nu_{\text{LO}}$ being the IF. If ν_c and ν_{LO} are close in frequency, the IF can be many orders of magnitude smaller than either one.

The parameter γ_{he} is called the *heterodyne mixing efficiency* and is the efficiency with which the LO and signal field interfere on the detector. The two first terms in Eq. (9.143) are the baseband power levels normally used in direct detection. The third term is the IF power level corresponding to the mixed signal created by interference of the signal and LO beams.

The coherent detection photocurrent is obtained by AC coupling the total current. The result is the sum of the IF signal and noise photocurrents being left or

$$i_{\text{IF}} = 2R_\lambda\sqrt{\gamma_{\text{he}} P_s P_{\text{LO}}} \cos[2\pi\nu_{\text{IF}} t + \varphi_c - \varphi_{\text{LO}}] + i_n(t) = i_{\text{IF}}(t) + i_n(t). \quad (9.144)$$

The voltage out of the postdetection amplifier is given by

$$V_{\text{Pre-Amp}} = 2GR_\lambda\sqrt{\gamma_{\text{he}} P_s P_{\text{LO}}} \cos[2\pi\nu_{\text{IF}}\, t + \varphi_c - \varphi_{\text{LO}}]R_L + Gi_n(t)R_L. \quad (9.145)$$

It should be noted that any misalignment between the directions of the two beam washes out the interference term since $\nu_c - \nu_{\text{LO}}$ then will vary sinusoidally with position with the area of the detector.

Example 9.15 For a power detector of dimension d and LO-incoming signal misalignment ψ, Pratt [18, p. 186] showed that the above instantaneous IF signal voltage

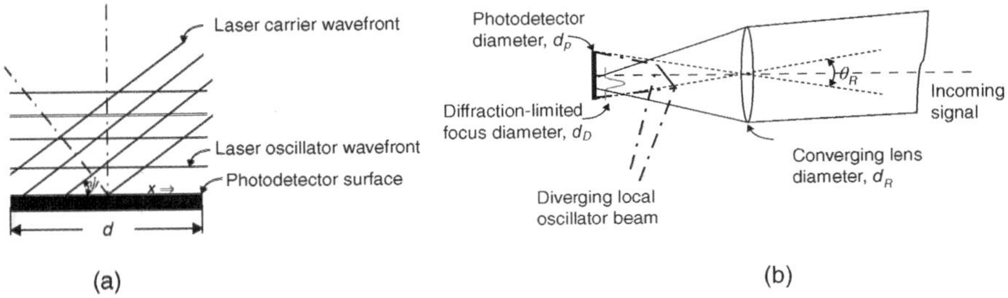

FIGURE 9.16 Spatial orientation of carrier and oscillator signals for (a) a collimated optical receiver and (b) a focusing optical receiver.

can be rewritten as

$$V_{\rm IF} = 2GR_\lambda\sqrt{\gamma_{\rm he}P_sP_{\rm LO}}\cos[2\pi\nu_{\rm IF}t + \varphi_c - \varphi_{\rm LO}]R_L\,\frac{\sin\left(\frac{\omega_c d}{2v_x}\right)}{\left(\frac{\omega_c d}{2v_x}\right)}, \qquad (9.146)$$

where $\omega_c = 2\pi\nu_c$ and

$$v_x = \frac{c}{\sin\psi}. \qquad (9.147)$$

In the above, c is the speed of light. Figure 9.16a depicts the spatial orientation of carrier and LO signals for a collimated optical receiver. Equation (9.146) indicates how the amplitude of the IF voltage is dependent on the misalignment angle ψ.

To keep the interference degradation due to misalignment to 10% or less, the above equation requires that

$$\frac{\sin\left(\frac{\omega_c d}{2v_x}\right)}{\left(\frac{\omega_c d}{2v_x}\right)} \leq 0.1 \text{ or } \omega_c d/2v_x \leq 0.8\,\text{rad}, \qquad (9.148)$$

which means

$$\psi \leq \frac{\lambda}{4d}. \qquad (9.149)$$

Using Eq. (9.149), we find for a 1 μm coherent system with a detector dimension of 1 cm, the angular misalignment must be less than 25 μrad [3, p. 186].

Example 9.16 For an incoming signal return goes through a converging lens telescope as illustrated in Figure 9.16b, Pratt also showed that the mixed signal will create a focused spot on the photodetector surface instead of a collimated beam [3, pp. 186–187]. The result is that the misalignment angle is determined by the receiver field-of-view (FOV) angle. Specifically, he expressed the receiver FOV angle as

$$\theta_R = 2.44\,\lambda\frac{d_p}{(d_d d_R)} \qquad (9.150)$$

with d_p being the photodetector diameter, d_d the diffraction-limited spot diameter, and d_R the receiver aperture diameter (see Figure 9.16b).

Repeating the above example where we have a 10 cm diameter receiver with a 1 cm detector diameter and a 0.01 cm diffraction-limited spot diameter, the angular misalignment must be less than 2.5 mrad [3, p. 186]. The good news is that the misalignment tolerance is relaxed by a factor of 100 compared to the previous example. However, the bad news is that the photomixing only occurs within the diffraction-limited spot. Any light outside this area that also hits the photodetector surface increases the system shot noise, which in turn decreases the SNR. One can minimize this effect by using either an imaging array or image dissection phototube, and only using the pixels that comprise the diffraction-limited spot.

The signal power is related to the time-averaged square of the IF voltage envelope. Mathematically, we have

$$P_{\text{Electrical-IF}} = \frac{V_{\text{IF}}^2(t)}{R_L} = \frac{2\gamma_{\text{he}} G^2 R_\lambda^2 P_s P_{\text{LO}}}{R_L}. \tag{9.151}$$

For a properly designed system, the system noise will be dominated by the LO signal shot noise. We again assume it is a zero-mean Gaussian random process with noise variance

$$\sigma_n^2 = i_n^2(t) = 2G^2 q R_\lambda P_{\text{LO}} B_{\text{IF}} \tag{9.152}$$

with B_{IF} being the IF bandwidth. One of the important metrics of a coherent detection systems is the carrier-to-noise ratio (CNR). Substituting Eqs. (9.141) and (9.142) into Eq. (8.98) gives

$$\text{CNR} = \frac{V_{\text{IF}}^2(t) R_L}{(\sigma_n^2 R_L)} = \frac{(2\gamma_{\text{he}} G^2 P_s P_{\text{LO}})}{(2G^2 q R_\lambda P_{\text{LO}} B_{\text{IF}})}$$

$$= \frac{\left(\gamma_{\text{he}} \left(\frac{q\eta}{h\nu}\right)^2 P_s P_{\text{LO}}\right)}{(q R_\lambda P_{\text{LO}} B_{\text{IF}})} = \frac{\gamma_{\text{he}} \eta P_s}{h\nu B_{\text{IF}}}. \tag{9.153}$$

The above CNR is eight times larger than one would find for subcarrier direct detection system if $\gamma_{\text{he}} \sim 1$ [18, p. 189]. If the system is not LO noise-limited and the thermal, dark current, or external background radiation is appreciable, then the CNR advantage is even greater.

For a matched filter receiver, P_s/B_{IF} is the energy of the return signal, E_s; $E_s/h\nu$ is the number of detector-plane signal photons. Therefore, the matched filter CNR is the number of coherently detected signal photoelectrons, $m_s = \gamma_{\text{he}} \eta E_s/h\nu$. If we define CNR to be the number of signal photoelectrons divided by the effective number of noise photoelectrons, we see that the effective number of noise photoelectrons captured in the integration period $\tau (= 1/B_{\text{IF}})$ is unity [26].

9.6.2.1 Coherent Laser Radar Probability of False Alarm The coherent detection photocurrent is a narrowband random process and can be modeled as

$$i(t) = [i_s(t) + i_n(t)]\, e^{-i\omega_{cc} t}, \tag{9.154}$$

where $i_s(t)$ and $i_n(t)$ now are the complex amplitudes of the signal and noise photocurrents [26]. This means that noise photocurrent can be decomposed into a complex sum of in-phase and quadrature noise components, that is,

$$i_n(t) = i_{n\text{-ip}}(t) - i\, i_{n\text{-q}}(t), \tag{9.155}$$

with each noise component being a zero-mean Gaussian random process with variance σ_n^2. The associated joint PDF for the above two noise components then is a circular complex Gaussian function of the form

$$p_n\{i(t), q\} = \frac{1}{(2\pi\sigma_n^2)} e^{-\frac{(i^2+q^2)}{(2\pi\,\sigma_n^2)}}. \tag{9.156}$$

For a coherent receiver, it is the square magnitude of the photocurrent that is proportional to the optical field intensity. The result is that the intensity noise PDF is an exponential density function

$$p_n\{I\} = \frac{1}{b_n} e^{-\frac{I}{b_n}} \tag{9.157}$$

with $b_n = 2\sigma_n^2$. When averaging is included in the receiver, as illustrated in Figure 9.15, the noised statistics change from an exponential to a gamma (or CHI_{2N}^2) distribution if the averaged signals are statistically independent, which we assume they are. Specifically, we have

$$p_n\{I\} = \Gamma^{-1}(N)\left(\frac{N}{b_n}\right)^N I^{N-1} e^{-\frac{NI}{b_n}}, \tag{9.158}$$

where $\Gamma(N)$ is the gamma function and N is the number of signals averaged. The mean and variance of the above gamma density function are $I = b_n$ and $var[I] = (I - I)^2 = \frac{b_n}{N}$, respectively.

The probability of false alarm then is written as

$$\begin{aligned} Q_{\text{fa}}^N &= \int_{I_{\text{th}}}^{\infty} \Gamma^{-1}(N)\left(\frac{N}{b_n}\right)^N I^{N-1} e^{-\frac{NI}{b_n}}\, dI \\ &= \frac{\Gamma\left(N, \dfrac{NI_{th}}{2\sigma_n^2}\right)}{\Gamma(N)}, \end{aligned} \tag{9.159}$$

where I_{th} is the decision threshold and $\Gamma(\alpha, \beta)$ is the incomplete gamma function defined as

$$\Gamma(\alpha, \beta) = \int_{\beta}^{\infty} t^{\alpha-1} e^{-t}\, dt. \tag{9.160}$$

For a single pulse, that is, no averaging, Eq. (9.159), reduces to

$$Q_{\text{fa}}^1 = e^{-\frac{I_{\text{th}}}{2\sigma_n^2}}. \tag{9.161}$$

9.6.2.2 Coherent Laser Radar Probability of Detection The probability of detection, Q_d, for a coherent laser radar relates to the probability that the detected return signal(s) exceeds a threshold defined by the probability of false alarm developed in the previous section. Gatt and Henderson showed that this parameter was strongly a function of Q_{fa}^N, CNR, N and especially, the target model [25]. The two classic target models used by the community are the (1) diffuse target and (2) perfectly specular model (i.e., smooth or glint) target. Let us discuss the former first.

9.6.2.2.1 Diffuse Target For a diffuse target, the complex signal photocurrent is a circular complex Gaussian random process because of the involved speckle [26]. The joint PDF for the signal is similar to Eq. (9.156), but the difference is that variance of each quadrature component now is σ_s^2 and thus, the total variance is $2\sigma_s^2$. As expected, the signal-plus-noise photocurrent also is a circular complex Gaussian random process. The variance for each component is the sum of the noise and signal variances. Therefore, the intensity PDF for the signal-plus-noise photocurrent is given by

$$p_{sn}\{I\} = \Gamma^{-1}(N)\left(\frac{N}{b_{sn}}\right)^N I^{N-1} e^{-\frac{NI}{b_{sn}}}, \tag{9.162}$$

where $b_{sn} = (2\sigma_s^2 + 2\sigma_n^2) = 2\sigma_n^2\,(1 + 1/\text{CNR})$ since $\text{CNR} = \sigma_s^2/\sigma_n^2$.

For an intensity threshold detector, the probability of detection then is equal to

$$\begin{aligned} Q_d^N &= \int_{I_{\text{th}}}^{\infty} \Gamma^{-1}(N)\left(\frac{N}{b_{sn}}\right)^N I^{N-1} e^{-\frac{NI}{b_{sn}}}\, dI \\ &= \frac{\Gamma\left(N, \frac{NI_{\text{th}}}{[2\sigma_n^2\,(1+1/\text{CNR})]}\right)}{\Gamma(N)}. \end{aligned} \tag{9.163}$$

For a single pulse, Eq. (9.150) reduces to

$$Q_d^1 = e^{-\frac{I_{\text{th}}}{[2\sigma_n^2(1+1/\text{CNR})]}} = (Q_{\text{fa}}^1)^{\frac{1}{(1+1/\text{CNR})}}. \tag{9.164}$$

9.6.2.2.2 Perfectly Specular (Glint) Target For a glint target, the signal photocurrent's complex amplitude is assumed to be a constant A_{glint}, with zero phase for simplicity, and the total photocurrent then is

$$i(t) = A + i_{n\text{-ip}}(t) - i_{n\text{-q}}(t), \tag{9.165}$$

where the quadrature noise components are the same as in the case of the diffuse target. This means that the statistics of the intensity noise and probability of false alarm remain the same. However, the signal-plus-noise statistics are different [26]. In particular, for a single pulse, these statistics follow Rican-square or CHI_2^2 statistics, which means its PDF is given by

$$p_{sn}\{I\} = \frac{1}{(2\pi\sigma_n^2)}\, e^{-\frac{\left(A_{\text{glint}}^2 + I\right)}{(2\pi\sigma_n^2)}}\, I_0\left(\frac{A_{\text{glint}}\sqrt{I}}{(2\pi\sigma_n^2)}\right). \tag{9.166}$$

The density function for the averaged intensity is known to follow the noncentral chi-squared function. Gatt and Henderson used the following definition, which applies to an average rather than accumulation:

$$p_{sn}^N\{I\} = \left[\frac{N}{(2\pi\sigma_n^2)}\right]\left(\frac{I}{A_{\text{glint}}^2}\right)^{\frac{N-1}{2}} e^{-\frac{N\left(A_{\text{glint}}^2+I\right)}{(2\pi\sigma_n^2)}} I_{N-1}\left(\frac{NA_{\text{glint}}\sqrt{I}}{(2\pi\sigma_n^2)}\right). \tag{9.167}$$

Its mean and variance are given by

$$I = A^2 + 2\sigma_n^2 = 2\sigma_n^2\left(1 + \frac{1}{\text{CNR}}\right) \tag{9.168}$$

and

$$\text{var}(I) = \frac{4\sigma_n^2\,(A^2 + \sigma_n^2)}{N}, \tag{9.169}$$

respectively. If we let the probability of detection, then it can be written as

$$Q_d^N = \int_{I_{\text{th}}}^{\infty}\left[\frac{N}{(2\pi\sigma_n^2)}\right]\left(\frac{I}{A_{\text{glint}}^2}\right)^{\frac{N-1}{2}} e^{-\frac{N\left(A_{\text{glint}}^2+I\right)}{(2\pi\sigma_n^2)}} I_{N-1}\left(\frac{NA_{\text{glint}}\sqrt{I}}{(2\pi\sigma_n^2)}\right) dI \tag{9.170}$$

$$= Q_N\left[\sqrt{\frac{NA^2}{\sigma_n^2}}, \sqrt{\frac{NI_{\text{th}}}{\sigma_n^2}}\right] = Q_N\left[\sqrt{2\text{N CNR}}, \sqrt{\frac{NI_{\text{th}}}{\sigma_n^2}}\right], \tag{9.171}$$

where

$$Q_N[\alpha, \beta] = \int_{\beta}^{\infty} z\left(\frac{z}{\alpha}\right)^{N-1} e^{-\frac{(z^2+\alpha^2)}{2}} I_{N-1}(\alpha z)\,dz. \tag{9.172}$$

Equation (9.172) is the definition of the Marcum Q-function [6, p. 113].

9.6.3 Continuous Direct Detection Intensity Statistics

Figure 9.17 shows a block diagram of a continuous direct detection intensity processor. The incoming light is detected, amplified, and then low-pass filtered to remove excess noise. Like in coherent detection, the processed signal then is averaged and compared to a threshold set by preset false alarm probability. The descriptor "continuous" is used here to distinguish it from a "photon-counting" direct detection receiver.

The detector in Figure 9.17 can be a PIN photodiode or a linear-gain detector such as a conventional avalanche photodiode with photoelectron gain G. If P_s is the incoming return signal optical power, then the signal photocurrent is given by

$$i_s(t) = GR_\lambda P_s. \tag{9.173}$$

This photocurrent is a baseband random process, which is directly proportional to the optical field intensity [26]. Similar to what we found for the baseband communications systems, the dominant noise source usually is electrical postdetection amplifier thermal noise, which has the corresponding noise photocurrent $i_n(t)$. The thermal noise again is a zero-mean Gaussian random

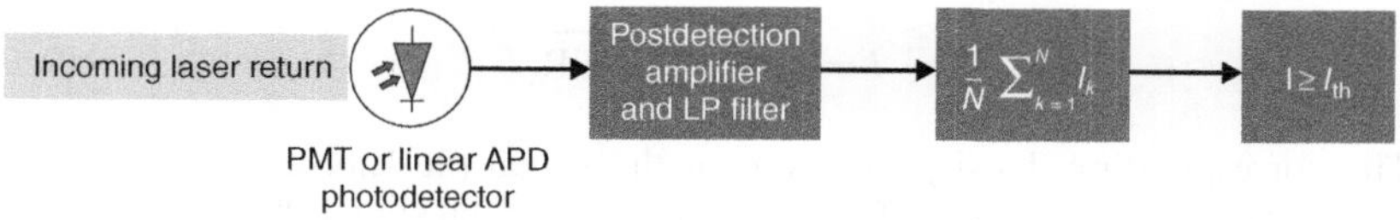

FIGURE 9.17 Continuous direct detection intensity processor block diagram.

process, whose noise variance is given by

$$\sigma_n^2 = N_0 B, \tag{9.174}$$

where N_0 is the unilateral "white noise" spectral density (A^2/Hz).[6] Using Eqs. (9.168) and (9.174), we can write the continuous direct detection receiver SNR as [26]

$$\text{SNR} = \frac{i_s^2(t)}{\sigma_n^2} = \frac{(GR_\lambda P_s)^2}{N_0 B}. \tag{9.175}$$

9.6.3.1 Continuous Direct Detection Probability of False Alarm The PDF of the average noise is Gaussian because the sum of a series of Gaussian random variables remains Gaussian [26]. However, the averaging process reduces the noise power variance by a factor of N, N being the number of the intensities averaged. This means this PDF is given by

$$p_n^N\{I\} = \frac{1}{\sqrt{2\pi\left(\frac{\sigma_n^2}{N}\right)}} e^{-\frac{I^2}{\left(2\pi\left(\frac{\sigma_n^2}{N}\right)\right)}} \tag{9.176a}$$

$$= \sqrt{\frac{N}{2\pi\sigma_n^2}} e^{-\frac{NI^2}{(2\pi\sigma_n^2)}}. \tag{9.176b}$$

Following previous work, it is easy to show that the zero-mean noise process yields a probability of false alarm equal to

$$Q_{fa} = \int_{I_{th}}^{\infty} p_n^N(I)\, dI = 0.5\,\text{erfc}\left(i_{th}\sqrt{\frac{N}{\sigma_n^2}}\right). \tag{9.177}$$

9.6.3.2 Continuous Direct Detection Probability of Detection for a Diffuse Target Gatt and Henderson stated that the density function of the signal intensity follows an exponential density function in the case of single pulse averaging and unit speckle diversity [25]. Previously, we saw the mean of this distribution to be $I = b_n$. Hence, the SNR equals

$$\text{SNR} = \frac{b_n^2}{\sigma_n^2} \tag{9.178}$$

[6] For high levels of signal power (i.e., short range), the signal shot noise can be greater than the thermal noise. Since the photoelectron count level will be very large, the normal Poisson process can be approximated as a Gaussian random process from the Central Limit Theorem.

or

$$I = b_n = \sigma_n \sqrt{\text{SNR}}. \tag{9.179}$$

The corresponding density function for the signal-plus-noise is derived from a convolution of a Gaussian noise PDF with the exponential signal PDF [26]. The result is given by

$$p_n^N\{I\} = \left[2^{-\frac{(N-1)}{2}}\right]\left(\frac{b_n}{N_0}\right)^{-N} \exp\left\{-\frac{NI^2}{(2\sigma_n^2)}\right\}\left(\frac{\sigma_n^2}{N}\right)^{\frac{(N-3)}{2}}$$
$$\times\left[\left(\frac{\sigma_n}{\sqrt{N}}\right)\frac{{}_1F_1\left(\frac{N}{2};\frac{1}{2};\frac{1}{2}\left[\frac{\sqrt{N}\sigma_n}{b_n}-\frac{\sqrt{N}I}{\sigma_n}\right]^2\right)}{\Gamma\left(\frac{N+1}{2}\right)}\right.$$
$$\left.-\left[2^{\frac{(2N-1)}{2}}\right](\sigma_n^2 - b_n I)\Gamma\left(\frac{N+1}{2}\right)\frac{{}_1F_1\left(\frac{N+1}{2};\frac{3}{2};\frac{1}{2}\left[\frac{\sqrt{N}\sigma_n}{b_n}-\frac{\sqrt{N}I}{\sigma_n}\right]^2\right)}{[b_n\sqrt{\pi}\,\Gamma(N)]}\right], \tag{9.180}$$

where ${}_1F_1(a;b;x)$ is the confluent hypergeometric function of the first kind [25, pp. 90–91]. The authors point out that Eq. (9.180) fails to provide adequate numerical accuracy for large values of the noise variance σ_n^2 or small values of the mean signal photocurrent, or SNR < -15 dB [26].

Fortunately, they were able to provide an alternate procedure. That is, they recommended inverse Fast Fourier Transforming the PDF's characteristic function, which is given by

$$Q(-i\omega) = \frac{\exp\left\{-\frac{\omega^2\sigma_n^2}{2N}\right\}}{\left(1 - i\frac{b_n\omega}{N}\right)^N}. \tag{9.181}$$

The probability of detection then is calculated using numerical techniques. The mean and variance of the average signal-plus-noise also can be calculated from the characteristic function, but they indicate that it is more readily derived using moment analysis. In particular, the mean and variance using that latter method are

$$\langle I \rangle = \sigma_n \sqrt{\text{SNR}} \tag{9.182}$$

and

$$\text{var}(I) = \frac{(b_n^2 + \sigma_n^2)}{N} = \frac{\sigma_n^2}{N}(1 + \text{SNR}), \tag{9.183}$$

respectively [26]. The above indicates that signal-plus-noise intensity is equal to the mean signal intensity, which is proportional to $\sqrt{\text{SNR}}$.

9.6.3.3 Continuous Direct Detection Probability of Detection for a Glint Target

For a glint target (or high levels of speckle diversity with high receiver SNR), Gatt and Henderson point out that the signal intensity probability density function is a delta function located at $\sqrt{\text{SNR}}$. The noise PDF is Gaussian with a variance equal to $\frac{\sigma_n^2}{N}$. This implies that the that signal-plus-noise intensity PDF is a Gaussian distribution with a mean of $\sqrt{\text{SNR}}$ and a variance of $\frac{\sigma_n^2}{N}$. For the general case of a nonzero mean Gaussian process, the probability that $x > x_0$ is written as

$$\Pr\{x > x_0\} = \frac{1}{(2\pi\sigma^2)} \int_{x_0}^{\infty} e^{-\frac{(x-u)^2}{(2\sigma^2)}}\, dx = 0.5\,\text{erfc}\left(\frac{(x_0 - u)}{\sqrt{2\sigma^2}}\right), \tag{9.184}$$

which means that the probability of detection of a glint target is given by

$$Q_d = 0.5\,\text{erfc}\left(\frac{(I_{\text{th}} - \sqrt{\text{SNR}})}{\sqrt{2\sigma_n^2}}\right). \tag{9.185}$$

Gatt and Henderson observe that the impact of averaging a glint target return is equivalent to an increase in SNR by a factor of N since the threshold I_{th} scales as $1/N$ [25]. That is, the probability of detection for given Q_{fa}, SNR and N is identical to the single pulse case, except the latter's SNR is multiplied by N.

9.6.4 Photon-Counting Direct Detection Intensity Statistics

Figure 9.18 shows a block diagram of a photon-counting direct detection intensity processor [26]. Key to this receiver is an extremely high-gain detector capable of detecting single photon events. This detector is followed by a gated counter, which reports the number of events (counts) general during a specific counting interval (range gate). Counts from successive laser radar pulse returns are accumulated. The accumulated count then is compared to a threshold value, adjusted to produce a specific probability of false alarm like before.

If E_s is the signal optical energy or integrated signal power, then the number of signal counts is given by

$$K_s = \frac{\eta E_s}{h\nu}. \tag{9.186}$$

The energy E_s is directly proportional to the optical field intensity. Add to the sign are noise counts, K_n. The sources of the noise counts are many. For example, they

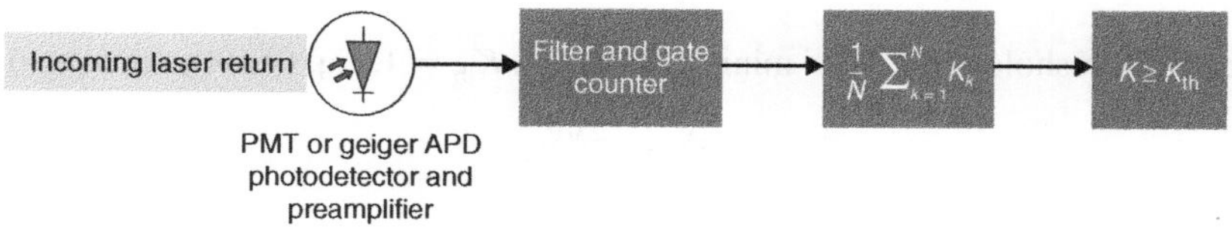

FIGURE 9.18 Photo-counting direct detection intensity processor block diagram.

can come from external background light such as the sun, moon, or stars and dark current as well as the signal's own shot noise. Fortunately, the mean number of noise photons is not important for the detection process because it can be characterized and subtracted from the number of detected photon. However, the fluctuation of the noise and signal counts affects the detectability of the signal photons. We define the SNR for photon-counting direct detection receiver to be

$$\mathrm{SNR} = \frac{K_s^2}{(K_s + K_n)}, \tag{9.187}$$

where K_s and K_n are the mean number of signal and noise photon-counts accumulated in the counting interval.

When $K_n = 0$, the receiver is quantum-noise-limited and like a coherent system, the SNR is proportional to the number of signal photons. Alternately, when $K_n \gg K_s$, the receiver is noise-limited and like the continuous direct detection receiver, the SNR is proportional to the signal-count-squared.

9.6.4.1 Photon-Counting Direct Detection Probability of False Alarm For a photon-counting receiver, the noise power now is a discrete random variable, which follows Poisson statistics as we saw previously. Recall that if we define that k describes the number of photoelectron counts captured in a given time interval, assuming a signal with a stationary mean, then the noise PDF equals

$$p_n\{k\} = \frac{K_n^k e^{-K_n}}{k!}. \tag{9.188}$$

Like the Gaussian process, the Poisson distribution is retained under accumulation [26], that is, the sum of independent Poisson processes results in a random variable that also follows a Poisson distribution. The mean of the accumulated process is the sum of the constituent means; for independent, identically distributed Poisson process, this mean equals NK_n, where N now represents the order of accumulation rather than the number of averages.[7] The accumulated noise PDF for a photon-counting receiver then is

$$p_n^N\{k\} = \frac{(NK_n)^k e^{-NK_n}}{k!}. \tag{9.189}$$

The probability of false alarm is the probability that the Poisson-distributed noise exceeds a threshold, or

$$Q_{\mathrm{fa}}^N = \sum_{k=K_{\mathrm{th}}}^{\infty} p_n^N\{k\} = 1 - \sum_{k=0}^{K_{\mathrm{th}}-1} p_n^N\{k\} = 1 - \frac{\Gamma(K_{\mathrm{th}}, NK_n)}{(K_{\mathrm{th}} - 1)!}. \tag{9.190}$$

When the threshold is set to its minimum, that is, $K_{\mathrm{th}} = 1$, Eq. (9.190) reduces to

$$Q_{\mathrm{fa}}^N = e^{-NK_n}. \tag{9.191}$$

[7] The difference between order of accumulation and average is that an averaged entity is the accumulated value of inputs, divided by the order of accumulation N.

Gatt and Henderson observed that the discrete nature of this counting receiver's threshold creates nonprecise estimate of the probability of false alarm [25].

9.6.4.2 Photon-Counting Direct Detection Probability of Detection-Diffuse Target Gatt and Henderson stated that the probability density function for a discrete random process can be calculated from its continuous counterpart using the Poisson transform

$$p(k) = \int_{-\infty}^{\infty} \left(\frac{x^k e^{-x}}{k!} \right) p(x)\, dx, \tag{9.192}$$

where $p(x)$ is the continuous density function [25]. They point out that the Poisson transform of a continuous distribution is the distribution of a discrete Poisson random variable whose mean, $\bar{x}$, is conditioned on the statistics of the continuous distribution $p(x)$.

For a diffuse target, the averaged signal intensity follows a gamma function. The resulting discrete signal PDF is called a *negative binomial distribution* and is given by

$$p_s^N\{k\} = \binom{N+k+1}{k} \frac{K_s^k}{(1+K_s)^{N+k}}, \tag{9.193}$$

where K_s is the mean number of signal counts in a single accumulation interval. The mean and variance of this distribution equal NK_s and $NK_s(1+K_s)$, respectively. If $N = 1$, Eq. (9.193) reduces to the geometric or Bose–Einstein distribution.

The PDF of the signal-plus-noise case can be evaluated using any one of a number of techniques (e.g., a convolution of the noise PDF with the signal PDF, or by inverse transforming the product of their characteristic functions, or by the Poisson transform integral using the continuous counterpart for $p(x)$). Gatt and Henderson chose to use Goodman's derivation in their paper, which uses the conditional probability integral [27], but with latter's assumed spatial diversity (averaging) parameter M replaced with their order of accumulation N. More specifically they proposed the signal-plus-noise PDF have the form

$$p_{sn}^N\{k\} = \frac{e^{-NK_n}}{(K_s+1)^N (N-1)!} \sum_{j=0}^{k} \frac{(NK_n)^j\, (k+N-j-1)!}{j!(k-j)!} \left[\frac{K_s}{K_s+1} \right]^{k-j}. \tag{9.194}$$

The probability of detection is just the infinite summation of Eq. (9.194), starting at K_{th}, or

$$Q_d^N = \sum_{k=K_{\text{th}}}^{\infty} p_{sn}^N\{k\} = 1 - \sum_{k=0}^{K_{\text{th}}-1} p_{sn}^N\{k\}. \tag{9.195}$$

When the threshold is minimum, that is, $NK_n < Q_{\text{fa}}^N$ such that $K_{\text{th}} = 1$, then the above detection probability becomes

$$Q_d^N = 1 - \frac{e^{-NK_n}}{(1+K_s)^N}. \tag{9.196}$$

In addition, if $NK_n \ll 1$, then Eq. (9.196) reduces to

$$Q_d^N = 1 - (1 + K_s)^{-N}. \tag{9.197}$$

For $N \gg 1$, this expression is equivalent to the threshold crossing probability for a Poisson random process, that is,

$$Q_d^N = 1 - e^{-NK_s}. \tag{9.198}$$

For $N \gg 1$, which implies very little speckle noise, and negligible noise ($NK_n \ll 1$), the number of accumulated photoelectrons required to achieve a given detection probability is given by

$$NK_s = -\ln(1 - Q_d^N). \tag{9.199}$$

9.6.4.3 Photon-Counting Direct Detection Probability of Detection-Glint Target When the target is a glint target, its PDF $p(x)$ is a delta function and the discrete PDF is Poisson distributed with mean NK_s. Therefore, the density function for the signal-plus-noise is also Poisson with mean and variance given by $N(K_s + K_n)$ because the sum of two Poisson processes still is a Poisson process. Therefore, the probability of detection is given by the RHS of Eq. (9.190), but with $N(K_s + K_n)$ for its mean.

9.6.5 LIDAR

Figure 9.19 depicts the concept of operations of LIDAR designed for remote sensing applications. A laser pulse is transmitted into an absorptive, scattering medium and it is situated such that a gated optical receiver can receive scattered light from the volume defined by the expanding laser beam and the receiver field-of-view, that is, receiver FOV overlap function. Inside that volume, the various aerosol and molecule types contribute the backscattered radiation captured by the receiver.

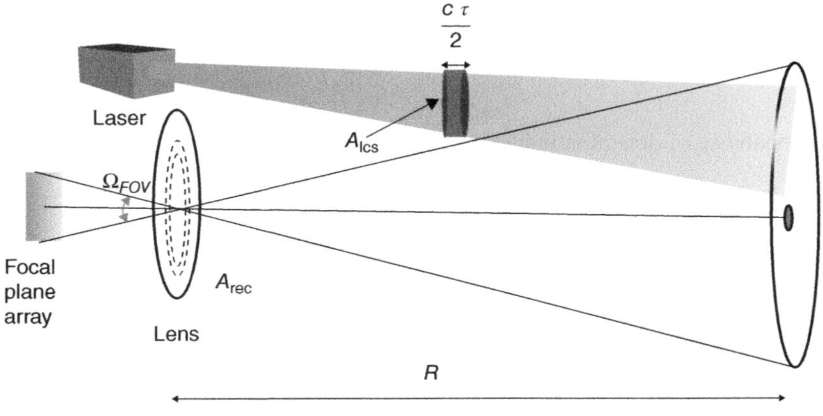

FIGURE 9.19 Concept of operation for LIDAR system.

The basic LIDAR equation for a volume target is given by

$$P_{rec} = \frac{P_t}{A_{lcs}} \sum_i \sigma_i \left(\frac{1}{4\pi R^2}\right) A_{rec}\, \eta\, O(R) L_{atmos}(R), \tag{9.200}$$

where

P_{rec} ≡ Received power
P_t ≡ Transmitted power
A_{lcs} ≡ Laser cross section
A_{rec} ≡ Receiver aperture area
c ≡ Speed of light
τ ≡ Temporal pulse width
R ≡ Range
σ_i ≡ Isotropic scattering cross-section for the ith particle
η ≡ Detector efficiency
$O(R)$ ≡ Receiver FOV Overlap Function

and

$L_{atmos}(R)$ ≡ Atmospheric transmission.

Now, the aerosol and molecular backscatter is the sum of the isotropic scattering cross sections from the various constituents of the sensed atmospheric or ocean column or other scattering channel. Mathematically, this contribution can be written as

$$\sum_i \sigma_i = V \sum_{\substack{\text{unit}\\ \text{volume}}} \sigma_i = A_{lcs}\left(\frac{c\tau}{2}\right) \sum_{\substack{\text{unit}\\ \text{volume}}} \sigma_i, \tag{9.201}$$

where $A_{lcs}(c\tau/2)$ defines the laser beam/receiver spatial-temporal volume. This implies that

$$P_{rec} = P_t\left(\frac{c\tau}{2}\right) \sum_{\substack{\text{unit}\\ \text{volume}}} \sigma_i \left(\frac{1}{4\pi R^2}\right) A_{rec}\eta O(R)\, L_{atmos}(R). \tag{9.202}$$

Now, from Mie scattering theory, we have

$$\sum_{\substack{\text{unit}\\ \text{volume}}} \sigma_i = 4\pi\beta(\pi), \tag{9.203}$$

where $\beta(\pi)$ is the volume scattering function evaluated at $\theta = \pi$ (backscatter term). It is given by

$$\beta(\theta) = \sum_{\substack{\text{unit}\\ \text{volume}}} N_i(R)\, \frac{d\sigma_{i,scat}}{d\Omega}\, (\theta, R)[\text{m}^{-1}\, \text{sr}^{-1}], \tag{9.204}$$

where $N_i(R)$ is the number concentration of the ith particle in the unit volume, and $\left(\frac{d\sigma_{i,scat}}{d\Omega}\right)$ (θ, R) is the differential scattering cross section (m^2/sr) of the ith particle in

Physical process	Backscatter cross section	Mechanism
Mie (aerosol) scattering	10^{-8} to 10^{-10} cm^2/sr	Two photon process; elastic scattering, instantaneous
Resonance fluorescence	10^{-13} cm^2/sr	Two single-photon process (absorption and spontaneous emission); delayed (radiative lifetime)
Molecular absorption	10^{-19} cm^2/sr	Single-photon process
Fluorescence from molecules, liquids, and solids	10^{-19} cm^2/sr	Two single-photon process; inelastic scattering, delayed (lifetime)
Rayleigh scattering	10^{-27} cm^2/sr	Two photon process; elastic scattering, instantaneous
Raman scattering	10^{-30} cm^2/sr	Two photon process; elastic scattering, instantaneous

FIGURE 9.20 Backscatter cross section comparison.

the unit volume at the angle θ and range R [10, 17]. The angle π indicates scattering in backward direction. Figure 9.20 contains a comparison of various backscatter cross sections.

Given this definition, Eq. (9.202) then can be rewritten as

$$P_{\text{rec}} = P_t\left(\frac{c\tau}{2}\right)\left(\frac{\beta}{R^2}\right)A_{\text{rec}}\eta O(R)L_{\text{atmos}}(R) \tag{9.205a}$$

$$= \left[P_t\left(\frac{c\tau}{2}\right)A_{\text{rec}}\,\eta\right]\left(\frac{O(R)}{R^2}\right)[\beta L_{\text{atmos}}(R)]. \tag{9.205b}$$

Equation (9.205b) has divided the equation into three distinct grouping. The first group in the square brackets contains the LIDAR System Constants. The second group in the parenthesis details the range-dependent measurement geometry. The final grouping in the square brackets house the backscatter contribution and atmospheric loss terms.

Not unexpectedly, both the backscatter contribution and atmospheric loss are specifically affected by

(a) Molecular scattering (e.g., gases such as oxygen and nitrogen) → Rayleigh scattering
(b) Particulate scattering (e.g., air pollutants such as dust, sea salt, minerals, pollen, ice, and rain) → Mie scattering (alternately, known as resonant and optical/geometrical optic scattering) [10, 17].

Unfortunately, receiver power measurements alone cannot differentiate between the two contributors. The way out of this predicament is to use the types of scattering effects the aerosols and molecules have on light to determine the types of channel constituents that are present and their concentrations.

In their simplest terms, we have three distinct scattering types:

- *Elastic scattering*: The wavelength (frequency) of the scattered light is the same as the incident light (*Rayleigh and Mie scattering*).
- *Inelastic scattering*: The emitted radiation has a wavelength different from that of the incident radiation (*Raman scattering, fluorescence*).
- *Quasi-elastic scattering*: The wavelength (frequency) of the scattered light shifts (e.g., in moving matter due to Doppler effects).

These scattering effects create the following optical interactions relevant to various LIDAR designs:

- *Rayleigh scattering*: Laser radiation elastically scattered from atoms or molecules with no change of frequency.
- *Mie scattering*: Laser radiation elastically scattered from particulates (aerosols or clouds) of sizes comparable to the wavelengths of radiation with no change of frequency.
- *Raman scattering*: Laser radiation inelastically scattered from molecules with a frequency shift characteristic of the molecule.
- *Resonance scattering*: Laser radiation matched in frequency to that of a specific atomic transition is scattered by a large cross section and observed with no change in frequency.
- *Fluorescence*: Laser radiation matched in frequency to a specific electronic transition of an atom or molecule is absorbed with subsequent emission at the lower frequency.
- *Absorption*: Attenuation of laser radiation when the frequency matched to the absorption band of given molecule.

Using these effects, three basic types of LIDARs have been developed:

- Backscatter LIDARs measure backscattered radiation and polarization.
- Differential absorption LIDAR (DIAL) is used to measure concentrations of chemical species (such as ozone, water vapor, and pollutants) in the atmosphere.
  - Principles: A DIAL LIDAR uses two different laser wavelengths that are selected so that one of the wavelengths is absorbed by the molecule of interest, while the other wavelength is not. The difference in intensity of the two return signals can be used to deduce the concentration of the molecule being investigated
- Doppler LIDAR is used to measure the velocity of a target.
  - Principles: When the light transmitted from the LIDAR hits a target moving toward or away from the LIDAR, the wavelength of the light reflected/scattered off the target will be changed slightly. This is known as a Doppler shift – hence Doppler LIDAR. If the target is moving away from the LIDAR, the return light will have a longer wavelength (sometimes

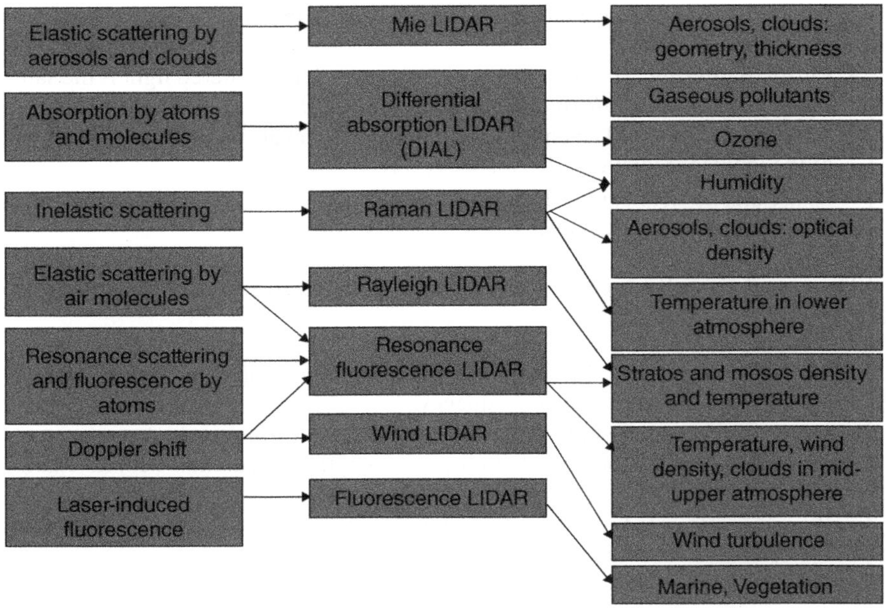

FIGURE 9.21 Examples of LIDAR phenomena, types and application.

referred to as a redshift), and if moving toward the LIDAR, the return light will be at a shorter wavelength (blueshifted). The target can be either a hard target or an atmospheric target – the atmosphere contains many microscopic dust and aerosol particles that are carried by the wind.

Figure 9.21 shows examples of LIDAR phenomena and their associated types of LIDAR(s) and remote sensing applications.

9.7 RESOLVED TARGET DETECTION IN CORRELATED BACKGROUND CLUTTER AND COMMON SYSTEM NOISE

The standard approach for analyzing the optical detection of targets in clutter for the past 50 years was first reported by Helstrom [3] in 1964. It paralleled the signal-plus-additive hypothesis testing developed for RF communications and sensing [6–8]. Received power, and background and system noise standard deviations, were their key performance variables [6–8]. Papers employing these techniques still can be found in the literature today [28–30]. Unfortunately, the signal-plus-noise assumption is inappropriate because a resolved target obscures any background clutter. This can be clearly seen in Figure 9.22, which shows two examples of such a situation [31]. Specifically, we see in these figures that the bar targets, asphalt, airplanes, and/or buildings mask any terrain signature underneath them, and that

FIGURE 9.22 (a) Calibration targets from the photo resolution range, Edwards Air Force Base and (b) Tri-bar array at Eglin Air Force Base, Florida. Imagery from Google Earth [31].

background terrain is found elsewhere. The important distinction here relative to the conventional hypothesis testing is that the noise statistics within the target segment of the image are different than those in the background clutter found elsewhere in the image, although there is a common part, system noise. That is, the signal plus common background and system noise assumption is not valid for the detection of resolved objects in all but a few special cases. This completely changes the two detection hypotheses from the signal-plus-noise and noise-only to signal-only plus common system noise, and clutter-only plus common system noise, respectively. This is the subject of this section.

Let us assume we have an optical image containing a multipixel (resolved) target embedded in a spatially varying background clutter. The target is only subject to system noise and the background is subject to both its spatially varying background clutter noise, and the system noise. The background clutter input vector currents are given by

$$\boldsymbol{\mu}_b = \{\mu_{bn}; n = 1,2, \ldots, N\}. \tag{9.206}$$

The target, pixel input vector currents and other key parameters were defined in a previous section. The two hypotheses in this case are defined as follows:

Hypothesis 0: Background clutter plus system noise

and

Hypothesis 1: Target present plus system noise.

The PDF for Hypothesis H_0 is given by

$$p_0(\boldsymbol{i}) = \left(\frac{1}{2\pi\sigma_T^2}\right)^{\frac{N}{2}} \exp\left[-\sum_{n=1}^{N}\left[\frac{(i_n - \mu_{bn} - \mu_t)^2}{2\sigma_T^2}\right]\right]. \tag{9.207}$$

The PDF for Hypothesis H_1 is given by

$$p_1(\boldsymbol{i}) = \left(\frac{1}{2\pi\sigma_t^2}\right)^{\frac{N}{2}} \exp\left[-\sum_{n=1}^{N}\left[\frac{(i_n - s_n - \mu_t)^2}{2\sigma_t^2}\right]\right]. \tag{9.208}$$

Redefining the current as $m_n = i_n - \mu_{bn} - \mu_t$, the above two PDFs can be rewritten as

$$p_0(\boldsymbol{m}) = \left(\frac{1}{2\pi\sigma_T^2}\right)^{\frac{N}{2}} \exp\left[-\sum_{n=1}^{N}\left(\frac{m_n^2}{2\sigma_T^2}\right)\right] \tag{9.209}$$

and

$$p_1(\boldsymbol{m}) = \left(\frac{1}{2\pi\sigma_t^2}\right)^{\frac{N}{2}} \exp\left[-\sum_{n=1}^{N}\left(\frac{(m_n - \mu_{bn}C_n)^2}{2\sigma_t^2}\right)\right], \tag{9.210}$$

respectively, where σ_b^2 and σ_t^2 are the background and system noise variance, respectively; $\sigma_T^2 = \sigma_b^2 + \sigma_t^2$;

$$m_n \equiv n\text{th pixel value in the image vector } \boldsymbol{m}$$

and

$$C_n \equiv \text{apparent pixel contrast} = \frac{s_n - \mu_{bn}}{\mu_{bn}}. \tag{9.211}$$

Using the above two PDFs, we can write the likelihood ratio as

$$\Lambda(\boldsymbol{m}) = \left(\frac{P_1(\boldsymbol{m})}{P_0(\boldsymbol{m})}\right) = \left(\frac{\sigma_T}{\sigma_t}\right)^N \exp\left[-\sum_{n=1}^{N}\left[\frac{(m_n - \mu_{bn}C_n)^2}{2\sigma_t^2} - \frac{m_n^2}{2\sigma_T^2}\right]\right]. \tag{9.212}$$

Taking the logarithm of Eq. (9.212) gives

$$\ln \Lambda(\boldsymbol{m}) = \ln\left(\frac{P_1(\boldsymbol{m})}{P_0(\boldsymbol{m})}\right) = \frac{N}{2}\ln\left(\frac{\sigma_T^2}{\sigma_t^2}\right) - \sum_{n=1}^{N}\left[\frac{(m_n - \mu_{bn}C_n)^2}{2\sigma_t^2} - \frac{m_n^2}{2\sigma_T^2}\right]$$

$$= \frac{N}{2}\ln\left(\frac{\sigma_T^2}{\sigma_t^2}\right) - \alpha\sum_{n=1}^{N}\left(m_n - \frac{\mu_{bn}C_n}{2\alpha\sigma_t^2}\right)^2 + \sum_{n=1}^{N}\left[\frac{(\mu_{bn}C_n)^2}{4\alpha\sigma_t^4} - \frac{(\mu_{bn}C_n)^2}{2\sigma_t^2}\right] \tag{9.213}$$

with

$$\alpha = \left(\frac{1}{2\,\sigma_t^2} - \frac{1}{2\,\sigma_T^2}\right) = \frac{\sigma_T^2 - \sigma_t^2}{2\sigma_T^2\sigma_t^2} = \frac{\sigma_b^2}{2\sigma_T^2\sigma_t^2} > 0 \tag{9.214}$$

because $\sigma_T^2 > \sigma_b^2 > \sigma_t^2$, in general. Substituting α into Eq. (9.213), we write the equation for the log-likelihood ratio as, after rearranging terms,

$$\ln\Lambda = \frac{N}{2}\ln\left(\frac{\sigma_T^2}{\sigma_t^2}\right) - \alpha\sum_{n=1}^{N}\left(m_n - \frac{\mu_{bn}C_n}{2\alpha\sigma_t^2}\right)^2 + \sum_{n=1}^{N}\frac{(\mu_{bn}C_n)^2\sigma_T^2}{2\sigma_b^2\sigma_t^2} - \sum_{n=1}^{N}\frac{(\mu_{bn}C_n)^2}{2\sigma_t^2}$$

$$= \frac{N}{2}\ln\left(\frac{\sigma_T^2}{\sigma_t^2}\right) - \left(\frac{\sigma_b^2}{2\sigma_T^2\sigma_t^2}\right)\sum_{n=1}^{N}\left(m_n - \frac{\mu_{bn}C_n}{2\alpha\sigma_t^2}\right)^2 + \sum_{n=1}^{N}\frac{(\mu_{bn}C_n)^2}{2\sigma_b^2}$$

or

$$\ln\Lambda - \frac{N}{2}\ln\left(\frac{\sigma_T^2}{\sigma_t^2}\right) - \sum_{n=1}^{N}\frac{(\mu_{bn}C_n)^2}{2\sigma_b^2} = -\left(\frac{\sigma_b^2}{2\sigma_T^2\sigma_t^2}\right)\sum_{n=1}^{N}\left(m_n - \frac{\mu_{bn}C_n}{2\alpha\sigma_t^2}\right)^2. \tag{9.215}$$

Now at this point, most researchers would give up the development because of the negative sign on the RHS of the question. Textbooks such as Helstrom [6, 7] and McDonough and Whalen [8] did not educate the student on what to do in this situation. Stotts and Hoff recognized that this negative test statistic causes one to reverse the inequality signs [32]. Normalizing Eq. (9.213) by $\left(\frac{N\sigma_b^2}{2\sigma_T^2\sigma_t^2}\right)$, we find that the optimum detector is given by

$$q = \frac{1}{N}\sum_{n=1}^{N}\left(m_n - \frac{\mu_{bn}\,C_n\,\sigma_T^2}{\sigma_b^2}\right)^2 \underset{>}{\leq} q_0 \tag{9.216}$$

for deciding between Hypotheses H_1 and H_0. That is, the upper inequality/equal-to decides H_1 and the lower inequality decides H_0. One outcome of this test statistic is that it does not create a peaked ("mountain") matched filter response, but rather an inverted-peak ("trough") matched filter response. This later performance is illustrated in Figure 9.23. The decision level q_0 in Eq. (9.216) is given by

$$q_0 = -\left(\frac{2\sigma_T^2\sigma_t^2}{N\sigma_b^2}\right)\left[\ln\left(\Lambda_0\left(\frac{\sigma_t^2}{\sigma_T^2}\right)^{\frac{N}{2}}\right) - \sum_{n=1}^{N}\frac{(\mu_{bn}C_n)^2}{2\sigma_b^2}\right]. \tag{9.217}$$

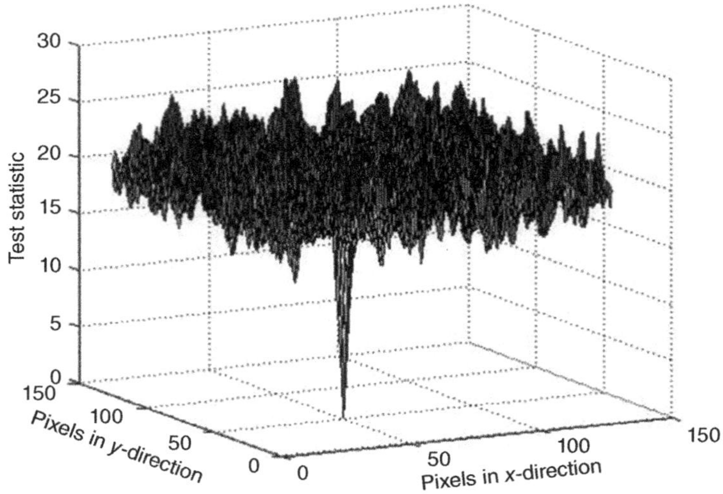

FIGURE 9.23 Matched filter response using the resolved target test statistic.

To implement the Neyman–Pearson form of the optimum detector, we select q_0 to achieve a preassigned false alarm probability rather than using Eq. (9.217).

Let us look for the moment at a test statistics of the form shown in Eq. (9.216), but without the inverse-N term. By letting

$$x_n \equiv \text{zero mean random variable with varance } \sigma^2 \tag{9.218}$$

and

$$A_n \equiv \text{constants,} \tag{9.219}$$

then we define the test statistics to be of the form

$$s = \sum_{n=1}^{N} (x_n + A_n)^2. \tag{9.220}$$

The test statistics in Eq. (9.220) has been analyzed by McDonough and Whalen [8]. By defining the variables $y_n = x_n + A_n$, they showed that the PDF in this case is given by

$$p_s(s)\,ds = \frac{1}{2\sigma^2}\left(\frac{s}{S}\right)^{\frac{(N-2)}{4}} e^{\left[\frac{-(S+s)}{2\sigma^2}\right]} I_{\frac{N}{2}-1}\left(\frac{\sqrt{sS}}{\sigma^2}\right) ds. \tag{9.221}$$

Equation (9.220) is the (unnormalized) *noncentral chi-square density* with N degrees of freedom, where the function $I_\kappa(x)$ is the modified Bessel function of the first kind and of order κ [8, Eq. (4.87), p. 138]. The parameter S is given by the sum of the squares of $\mathcal{E}[y_n] = \mathcal{E}[x_n + A_n]$, or in other words,

$$S = \sum_{n=1}^{N} (\mathcal{E}[x_n + A_n])^2. \tag{9.222}$$

To include the inverse-N term, we define a new random variable $q = \frac{s}{N}$ with probability density function $f_q(q)$ and make the following change of variables,

$$f_q(q) = p_s(Nq)\ |J|, \tag{9.223}$$

where

$$J = \frac{\partial s}{\partial q} = N \tag{9.224}$$

is the Jacobian of the transformation from s to q. This means that Eq. (9.223) equals

$$f_q(s)\,ds = \frac{N}{2\sigma^2}\left(\frac{Nq}{S}\right)^{\frac{(N-2)}{4}} e^{\left[\frac{-(S+Nq)}{2\sigma^2}\right]} I_{\frac{N}{2}-1}\left(\frac{\sqrt{NqS}}{\sigma^2}\right) dq. \tag{9.225}$$

From Eq. (9.222) under Hypothesis H_0, we get

$$S_0 = \left(\frac{\sigma_T^4}{\sigma_b^2}\right)\sum_{n=1}^{N}\frac{\mu_{bn}^2 C_n^2}{\sigma_b^2}. \tag{9.226}$$

However, under Hypothesis H_1, the average of $\mathcal{E}[y_n] = \mathcal{E}[x_n + A_n]$ must include the nonzero average of x_n, which is the term $(s_n - \widehat{\mu}_{bn})$. This implies that the S parameter in this case is given by

$$S_1 = \left(\frac{\sigma_t^4}{\sigma_b^2}\right)\sum_{n=1}^{N}\frac{\mu_{bn}^2 C_n^2}{\sigma_b^2}, \tag{9.227}$$

under Hypothesis H_1.

Given the above, the probability of false alarm for the test statistics in Eq. (9.216) can be written as

$$Q_{\text{fa}} = \int_0^{q_0} f_0(q)\,dq \tag{9.228}$$

$$= \frac{N}{2\sigma_T^2}\int_0^{q_0}\left(\frac{Nq}{S_0}\right)^{\frac{(N-2)}{4}} e^{\left[\frac{-(S_0+Nq)}{2\sigma_T^2}\right]} I_{\frac{N}{2}-1}\left(\frac{\sqrt{S_0 Nq}}{\sigma_T^2}\right) dq. \tag{9.229}$$

If we let $v = \frac{Nq}{\sigma_T^2}$ and define $\lambda_0 = \frac{S_0}{\sigma_T^2}$, we obtain

$$Q_{\text{fa}} = \frac{1}{2}\int_0^{\frac{Nq_0}{\sigma_T^2}}\left(\frac{v}{\lambda_0}\right)^{\frac{(N-2)}{4}} e^{\left[\frac{-(\lambda_0+v)}{2}\right]} I_{\frac{N}{2}-1}(\sqrt{\lambda_0 v})\,dv. \tag{9.230}$$

The integrand of Eq. (9.230) is the standard form of the normalized noncentral chi-squared distribution with N degrees of freedom and noncentrality parameter λ_0

[8, Eq. (4.89), p. 139]. If we define that distribution to be $f_\chi(v; N, \lambda_0)$, then the probability can be written concisely as

$$Q_{\rm fa} = \int_0^{\frac{Nq_0}{\sigma_T^2}} f_\chi(v; N, \lambda_0)\,dv \tag{9.231}$$

with

$$\lambda_0 = \frac{\sigma_T^2}{\sigma_b^2} \sum_{n=1}^{N} \frac{C_n^2}{\left(\frac{\sigma_b^2}{\mu_{bn}{}^2}\right)}. \tag{9.232}$$

Similarly, the probability of detection can be written as

$$Q_d = \int_0^{q_0} f_1(q)\,dq \tag{9.233}$$

$$= \frac{N}{2\sigma_t^2} \int_0^{q_0} \left(\frac{Nq}{S_1}\right)^{\frac{(N-2)}{4}} e^{\left[\frac{-(S_1+Nq)}{2\sigma_t^2}\right]} I_{\frac{N}{2}-1}\left(\frac{\sqrt{S_1 Nq}}{\sigma_t^2}\right) dq \tag{9.234}$$

$$= \int_0^{\frac{Nq_0}{\sigma_t^2}} f_\chi(v; N, \lambda_1)\,dv \tag{9.235}$$

with noncentrality parameter λ_1 given by

$$\lambda_1 = \frac{\sigma_t^2}{\sigma_b^2} \sum_{n=1}^{N} \frac{C_n^2}{\left(\frac{\sigma_b^2}{\mu_{bn}{}^2}\right)}. \tag{9.236}$$

As a final comment, we note that there is an interesting relationship between the above two noncentrality parameters λ_0 and λ_1. Let us define the difference and ratio of these two parameters as the contrast-to-noise ratio (CNR) and the background noise ratio (BNR), respectively, we have

$$\text{CNR} = \lambda_0 - \lambda_1 = \sum_{n=1}^{N} \frac{C_n^2}{\left(\frac{\sigma_b^2}{\mu_{bn}{}^2}\right)} \tag{9.237}$$

and

$$\text{BNR} = \frac{\lambda_0}{\lambda_1} = \frac{\sigma_T^2}{\sigma_t^2} = \frac{\sigma_b^2}{\sigma_t^2} + 1. \tag{9.238}$$

Equation (9.237) is equivalent to the electrical SNR used to characterize detection performance of signal-plus-noise detectors. Its square root is proposed as the correct form of the OSNR as opposed to the version in Reference [32], which cites OSNR as the contrast divided by the standard deviation of the clutter noise σ_b. By replacing

σ_b by σ_b/μ_b, SNR now is dimensionless as it should be, rather than being in inverse photoelectron counts. This equation also shows that any minimum detectable signal using this equation will strongly depend on N, the number of pixels encompassing the target signature. That is, the larger the target, the easier it is to detect. This agrees with the SNR's dependence on target area referenced in the introduction [32].

Rewriting Eq. (9.237) as $\lambda_1 = \lambda_0 - \text{CNR}$ and substituting it into Eq. (9.235), we find that

$$Q_d = \int_0^{\text{BNR}\left(\frac{Nq_0}{\sigma_T^2}\right)} f_\chi(v; N, \lambda_0 - \text{CNR})\,dv. \tag{9.239}$$

The pair of Eqs. (9.231) and (9.239) now defines the false alarm and detection probabilities in terms of CNR and BNR. It is clear from Eq. (9.239) that the higher CNR, the more the PDF under Hypothesis H_1 shifts toward zero and more of the PDF area will be accumulated by the integration process, resulting in a higher detection rate. Alternately, as CNR goes down, the detection probability also goes down, but the exact detection rate will depend upon the BNR.

9.8 ZERO CONTRAST TARGET DETECTION IN BACKGROUND CLUTTER

In the previous development, we assumed a nonzero contrast. One might now ask if the above equations are valid for the situation where $s_n = \mu_{bn}$ for all values of n, the zero-contrast case. Again, the target only will be subject to system noise and the background to both spatially varying background clutter noise, and the same system noise. Letting the pixel contrast equal zero in Eq. (9.216) yields

$$q = \frac{1}{N}\sum_{i=1}^{N} m_i^{2} \underset{>}{\overset{<}{}} \left(\frac{2\sigma_T^2\sigma_t^2}{N\sigma_b^2}\right) \ln\left[\Lambda_0\left(\frac{\sigma_t^2}{\sigma_T^2}\right)^{\frac{N}{2}}\right] = q_0. \tag{9.240}$$

In this situation, the test statistics now is a measure of the data sample variance and is estimating the variability, or "roughness", of the sample set. If the sample is sufficiently smooth, that is, the test is less than the threshold, then the sample is likely a target. However, if the sample is rough, that is, the test is greater than the threshold, the sample is most likely background clutter.

Let us look at the false alarm probability in Eq. (9.230). Again assume that N is even, that is, $N = 2M \neq 0$ where M is an integer. If we further assume the argument of the modified Bessel function is small, then we can write Eq. (9.140) as

$$\begin{aligned} Q_{\text{fa}} &\approx \frac{N}{2\Gamma\left(\frac{N}{2}\right)} \int_0^{q_0} \left(\frac{Nv}{\lambda_0}\right)^{\frac{(N-2)}{4}} e^{\left[\frac{-(\lambda_0+Nv)}{2}\right]} \left(\frac{1}{2}\sqrt{\lambda_0 Nv}\right)^{\frac{N-2}{2}} dv \\ &\to \frac{1}{\Gamma\left(\frac{N}{2}\right)} \int_0^{\frac{Nq_0}{2\sigma_T^2}} w^{\frac{N}{2}-1} e^{-w}\,dw = I\left(\sqrt{\frac{N}{2}}\left(\frac{q_0}{\sigma_T^2}\right), \frac{N}{2}-1\right), \end{aligned} \tag{9.241}$$

as $\lambda_0 \to 0$, assuming the Hypothesis 0 noise variance. Here, $I(u,p)$ is the Pearson's form of the incomplete gamma function given by [9, Eq. (6.5.6), p. 262]

$$I(u,p) = \frac{1}{\Gamma(p+1)} \int_0^{u\sqrt{p+1}} y^p e^{-y}\, dy. \tag{9.242}$$

The integrand of Eq. (9.242) is the well-known chi-squared density with N degrees of freedom [8, Eq. (4.75), p. 135].

Following the same procedure for the probability of detection, we find

$$Q_d \to \frac{1}{\Gamma\left(\frac{N}{2}\right)} \int_0^{\left(\frac{\sigma_T^2}{\sigma_t^2}\right)\left[\frac{Nq_0}{2\sigma_T^2}\right]} w^{\frac{N}{2}-1} e^{-w}\, dw \tag{9.243}$$

$$= \frac{1}{\Gamma\left(\frac{N}{2}\right)} \int_0^{\mathrm{BNR}\left[\frac{Nq_0}{2\sigma_T^2}\right]} w^{\frac{N}{2}-1} e^{-w}\, dw \tag{9.244}$$

$$= I\left(\mathrm{BNR}\left[\sqrt{\frac{N}{2}}\left(\frac{q_0}{\sigma_T^2}\right)\right], \frac{N}{2} - 1\right). \tag{9.245}$$

It should be noted that since the apparent contrast is zero and the target-detection CNR relationship in the previous section does not apply, Eq. (9.245) suggests that BNR takes on the role of the detection SNR, which is also dimensionless. This makes sense because the ratio of the two noise variances will be the deciding factors in the hypothesis test and detection probability. We noted this possible definition in our discussion of various SNR forms.

9.9 MULTISPECTRAL SIGNAL-PLUS-NOISE/NOISE-ONLY TARGET DETECTION IN CLUTTER

Log-likelihood Ratio Test (LRT) algorithms have been applied to multispectral or hyperspectral imagery for almost four decades [9–13, 15, 16, 24, 26]. The reason is that natural clutter from vegetation is characterized by a gray body, and man-made objects, compared with blackbody radiators, emit radiation more strongly at some wavelengths than at others, which makes the optical target detection in clutter a straightforward process. While different log-likelihood ratio approaches have been applied to a variety of surveillance and reconnaissance problems, they appear to have a common framework. One of the most robust examples that still is employed today is the Reed–Xiaoli (RX) algorithm that provided a Generalized Log-likelihood Ratio Test (GLRT) approach to finding targets in clutter [33]. Specifically, it is a generalized hypothesis test formulated by partitioning the received bands into two groups. In one group, targets exhibit substantial coloring in their signatures, but behave either like gray bodies or emit negligible radiant energy in the other group [33–37].

More specifically, the RX algorithm is a GLRT that uses local estimates of the spectral mean and spectral covariance instead of incorporating *a priori* clutter statistics into a detector [34]. In other words, this general framework for ML ratio approach adapt automatically to the local clutter statistics. It satisfies an optimality criterion if locally the spectral data have a multivariate normal probability distribution. The mean and covariance matrix are estimated within a neighborhood of every test pixel. The RX algorithm performs the GLRT based on two windows, a small target window of dimension X_t and a larger background window X_c, as illustrated in Figure 9.5b. There is a region between the windows that is not tested. In operation, the RX algorithm examines each pixel in the multi- or hyperspectral image by performing a GLRT. The output of the RX algorithm is a detection statistics map that can be thresholded. Let us be more specific by focusing on an adaptive RX implementation developed by Yu *et al.* [34].

Their algorithm is a hybrid spectral-spatial matched filter for CFAR detection of small, spectrally selective targets amid a locally Gaussian background spectrum. They assumed that the multispectral sensor creates a set of J-pixel correlated measurement vectors for M band. This allowed them to form a $M \times K$ matrix given by

$$\boldsymbol{J} = [\boldsymbol{j}(1), \boldsymbol{j}(2), \ldots, \boldsymbol{j}(K)], \tag{9.246}$$

which represents K independent pixel observations from M correlated image scenes take from a local window size of $L_x \times L_y = K$ where the clutter is considered stationary. Their measurement vectors of spectral observations in each pixel is given as

$$\boldsymbol{j}(k) = [j_1(k), j_2(k), \ldots, j_M(k)]^T \tag{9.247}$$

for $\{k = 1, 2, \ldots, K\}$. Their spectral covariance $\boldsymbol{R}$ is defined as

$$\boldsymbol{R} = \mathcal{E}\left\{[\boldsymbol{j}(k) - \mathcal{E}\{\boldsymbol{j}(k)\}]\,[\boldsymbol{j}(k) - \mathcal{E}\{\boldsymbol{j}(k)\}]^T\right\}. \tag{9.248}$$

They then postulated a known target signal to be characterized by two vectors:

$$\boldsymbol{b} = [b_1, b_2, \ldots, b_M]^T \tag{9.249}$$

and

$$\boldsymbol{s} = [s(1), s(2), \ldots, s(K)]^T, \tag{9.250}$$

which represents a spectral pattern and spatial pattern, respectively. In addition, they defined

$$\boldsymbol{b}_s = \frac{\boldsymbol{b}}{\|\boldsymbol{b}\|} \tag{9.251}$$

and

$$\boldsymbol{s}^T\boldsymbol{s} = \left\|\boldsymbol{s}^2\right\| = 1. \tag{9.252}$$

The two hypotheses that their adaptive detector will distinguish between are

$$\begin{cases} H_0 : J = \boldsymbol{J}^0 \\ H_1 : J = \boldsymbol{J}^0 + \boldsymbol{b}\boldsymbol{s}^T \end{cases}, \tag{9.253}$$

where $\boldsymbol{J}^0$ is the data samples containing solely residual clutter.

Under these hypotheses, the joint probability density functions of the random variables $\boldsymbol{j}(k)$ for $\{k = 1,2, \ldots, K\}$ are given by

$$p_0(\boldsymbol{X}) = p_0(\boldsymbol{j}(1),\boldsymbol{j}(2), \ldots ,\boldsymbol{j}(K))$$
$$= (2\pi)^{-\frac{KM}{2}}\, |\boldsymbol{R}|^{-K/2} \exp\left\{ -\frac{1}{2}\sum_{k=1}^{K} \boldsymbol{j}^T(k)\boldsymbol{R}^{-1}\boldsymbol{j}(k) \right\} \tag{9.254}$$

and

$$p_1(\boldsymbol{X}) = p_1(\boldsymbol{j}(1),\boldsymbol{j}(2), \ldots ,\boldsymbol{j}(K))$$
$$= (2\pi)^{-\frac{KM}{2}}\, |R|^{-K/2} \exp\left\{ -\frac{1}{2}\sum_{k=1}^{K} (\boldsymbol{j}(k) - \boldsymbol{b}s(n))^T\boldsymbol{R}^{-1}(\boldsymbol{j}(k) - \boldsymbol{b}s(n)) \right\}. \tag{9.255}$$

In the above equations, $|\cdots|$ represents the determinant of the enclosed matrix.

The matched filter weight vector for a target of shape $\boldsymbol{s}$ and spectral distribution $\boldsymbol{b}$ is given by

$$\boldsymbol{w}_{\text{opt}} = \begin{pmatrix} \boldsymbol{R} & \cdots & 0 \\ \vdots & \ddots & \vdots \\ 0 & \cdots & \boldsymbol{R} \end{pmatrix} \begin{pmatrix} \boldsymbol{b}s(1) \\ \vdots \\ \boldsymbol{b}s(K) \end{pmatrix}, \tag{9.256}$$

where the matrix only has nonzero entries down the diagonal and no place else. The output of the multiband matched filter corresponding to the $M \times K$ data sample matrix $\boldsymbol{X}$ has the form

$$\text{MF} = \boldsymbol{b}^T\boldsymbol{R}^{-1}\boldsymbol{X}\boldsymbol{s}. \tag{9.257}$$

The multiband (MB) SNR in the perfectly matched case is

$$(\text{SNR})_{\text{MB}} = \frac{|\mathcal{E}\{\text{MF}|H_1\}|^2}{cov(\text{MF}|H_0)} = \frac{[\boldsymbol{b}^T\boldsymbol{R}^{-1}\boldsymbol{b}\|\boldsymbol{s}\|^2]^2}{\boldsymbol{b}^T\boldsymbol{R}^{-1}\boldsymbol{b}\|\boldsymbol{s}\|^2} = \boldsymbol{b}^T\boldsymbol{R}^{-1}\boldsymbol{b}\|\boldsymbol{s}\|^2. \tag{9.258}$$

Given Eq. (9.258) and a specific false alarm probability PFA, the detection probability for this situation can be written as

$$Q(Q^{-1}(\text{PFA}), \sqrt{(\text{SNR})_{\text{MB}}}), \tag{9.259}$$

where the Q-function is the tail probability of the standard normal distribution. It gives the performance of the matched filter detection statistics in (9.257), which being a linear function of Gaussian observations also is Gaussian. CFAR ROC performance for a given false alarm probability, PFA, is parameterized by the multiband $(\text{SNR})_{\text{MB}}$ parameter given in (9.258).

Equation (9.258) also can be expressed in terms of signal processing gain and single-band (SB) SNR. In particular, we can write

$$(\text{SNR})_{\text{MB}} = G_{\text{MB}}(\text{SNR})_{\text{SB}} \tag{9.260}$$

with G_{MB} is SNR gain obtained through multiband spectral processing and $(\text{SNR})_{\text{SB}}$ is the SNR from single-band processing. For example, if band 1 is selected for single-band processing, then we can write

$$(\text{SNR})_{\text{SB}} = \frac{b_1^2}{\sigma_1^2}\, \boldsymbol{s}^T \boldsymbol{s} \tag{9.261}$$

and

$$G_{\text{MB}} = 1 + \left[\left(\frac{\boldsymbol{b}_2}{b_1}\right)\sigma_1 - \frac{\boldsymbol{V}_{21}}{\sigma_1}\right]^T R_{2|1}^{-1} \left[\left(\frac{\boldsymbol{b}_2}{b_1}\right)\sigma_1 - \frac{\boldsymbol{V}_{21}}{\sigma_1}\right]. \tag{9.262}$$

In the above,

$$\sigma_1^2 = \mathcal{E}\left\{ \left[j_1(k) - \mathcal{E}\{j_1(k)\}\right] \left[j_1(k) - \mathcal{E}\{j_1(k)\}\right]^T \right\} \tag{9.263}$$

and $R_{2|1}^{-1}$ is the conditional covariance matrix of the $(K-1)$-vector

$$\boldsymbol{j}_2(k) \equiv \{j_2(k), j_3(k), \ldots, j_M(k)\}. \tag{9.264}$$

The matrix $\boldsymbol{V}_{21}$ is the cross-covariance matrix of $\boldsymbol{j}_2(k)$ and $\boldsymbol{j}_1(k)$, and is written as

$$\boldsymbol{V}_{21} = \mathcal{E}\left\{ \left[\boldsymbol{j}_2(k) - \mathcal{E}\{\boldsymbol{j}_2(k)\}\right] \left[j_1(k) - \mathcal{E}\{j_1(k)\}\right]^T \right\}. \tag{9.265}$$

Referring to Eq. (9.262), it is clear that $G_{\text{MB}} \geq 1$, which implies that $(\text{SNR})_{\text{MB}} \geq (\text{SNR})_{\text{SB}}$. In other words, multispectral processing is always greater than single-band processing. We will see this effect and how much the SNR improves by adding spectral bands shortly.

As noted earlier, many applications detect targets at a given constant false alarm probability. In general, one may not know the clutter covariance matrix $\boldsymbol{R}$ or spectral signal distribution $\boldsymbol{b}$. This implies that one must estimate these parameters from the observed data. A direct substitution of these estimated parameters into Eq. (9.257) produces a modified matched filter and output. For example, the output of the matched filter using an estimated covariance matrix is given by

$$\widehat{\text{MF}} = \boldsymbol{b}^T \widehat{\boldsymbol{R}}^{-1} \boldsymbol{X}\boldsymbol{s}, \tag{9.266}$$

where $\widehat{\boldsymbol{R}} = (1/K)\boldsymbol{J}^T\boldsymbol{J}$. Equations such as Eq. (9.266), are sometimes called *adaptive matched filters* because they can adapt to an unknown nonstationary clutter background. Unfortunately, there is no straightforward way to make it also have a CFAR property [34]. A GLRT must be employed to remedy this problem.

Let us begin by defining a new scalar amplitude parameter α that creates a modified equation for the scalar amplitude of the target spectral distribution $\boldsymbol{b}$. Specifically, we have

$$\boldsymbol{b} = \alpha \boldsymbol{b}_s = \alpha[b_1, b_2, \ldots, b_M]^T, \tag{9.267}$$

where we have used Eq. (9.251), which is known as the normalized spectral distribution since ${\boldsymbol{b}_s}^T\boldsymbol{b}_s = 1$.

The GLRT for his problem is given by

$$\frac{\max\limits_{\alpha}\{\max\limits_{\boldsymbol{R}} L_1(J;\alpha,\boldsymbol{R})\}}{\max\limits_{\boldsymbol{R}} L_0(J,\boldsymbol{R})} = \begin{cases} \geq l_0, & \text{then } H_1 \\ < l_0, & \text{then } H_0 \end{cases}, \tag{9.268}$$

where $L_1(J;\alpha,\boldsymbol{R})$ and $L_0(J;\boldsymbol{R})$ are the likelihood functions for Hypotheses H_1 and H_0, respectively, and $l_0 \geq 0$ is the threshold of the test.

For the hypothesis test we proposed in Eq. (9.268) with probability density functions given by Eqs. (9.254) and (9.255), we have

$$\max_{\alpha}\{\max_{\boldsymbol{R}} L_1(J;\ \alpha,\boldsymbol{R})\} = \max_{\alpha}\ (2\pi)^{-\frac{KM}{2}}\,|\widehat{\boldsymbol{R}}_{\boldsymbol{\alpha}}|^{-\frac{K}{2}}\exp\left\{-\frac{KM}{2}\right\} \tag{9.269}$$

and

$$\max_{\alpha}\{\max_{\boldsymbol{R}} L_0(J;\ \boldsymbol{R})\} = \max_{\alpha}\ (2\pi)^{-\frac{KM}{2}}\ |\widehat{\boldsymbol{R}}_0|^{-\frac{K}{2}}\ \exp\left\{\frac{-KM}{2}\right\}. \tag{9.270}$$

The well-known maximum likelihood estimates (MLEs) of the unknown covariance matrix $\boldsymbol{R}$ under the respective hypotheses, H_1 and H_0, are

$$\begin{aligned}\widehat{\boldsymbol{R}}_\alpha &= \frac{1}{K}\sum_{k=1}^{K}(\boldsymbol{j}(k) - \alpha\boldsymbol{b}_s s(n))\,(\boldsymbol{j}(k) - \alpha\boldsymbol{b}_s s(n))^T \\ &= (\boldsymbol{J} - \alpha\boldsymbol{b}_s\boldsymbol{s}^T)(\boldsymbol{J} - \alpha\boldsymbol{b}_s\boldsymbol{s}^T)^T\end{aligned} \tag{9.271a}$$

and

$$\widehat{\boldsymbol{R}}_0 = \frac{1}{K}\sum_{k=1}^{K}\boldsymbol{j}(k)\boldsymbol{j}(k)^T = \frac{1}{K}\boldsymbol{J}\boldsymbol{J}^T. \tag{9.271b}$$

Substituting Eqs. (9.271a) and (9.271b) into Eq. (9.268) yields the following GLRT relationship

$$\Lambda(\boldsymbol{J}) = \frac{|\widehat{\boldsymbol{R}}_0|}{\min\limits_{\alpha}|\widehat{\boldsymbol{R}}_{\boldsymbol{\alpha}}|} = \frac{|\boldsymbol{J}\boldsymbol{J}^T|}{\min\limits_{\alpha}|\boldsymbol{F}_{\boldsymbol{\alpha}}|} = \begin{cases} \geq \lambda_0, & \text{then } H_1 \\ < \lambda_0, & \text{then } H_0 \end{cases}, \tag{9.272}$$

where $\lambda_0 = l_0^{\frac{2}{K}}$ is the threshold of the test and

$$F_\alpha = (\boldsymbol{J} - \alpha\boldsymbol{b}_s\boldsymbol{s}^T)(\boldsymbol{J} - \alpha\boldsymbol{b}_s\boldsymbol{s}^T)^T. \tag{9.273}$$

Using the above, Yu *et al.* [35] developed the following decision statistic

$$g(\boldsymbol{J}) = \frac{(\boldsymbol{b}_s^T\boldsymbol{R}^{-1}\boldsymbol{J}\boldsymbol{s})^2}{(\boldsymbol{b}_s^T\boldsymbol{R}^{-1}\boldsymbol{b}_s)\left(1 - \frac{1}{N}\boldsymbol{s}^T\boldsymbol{J}^T\boldsymbol{R}^{-1}\boldsymbol{J}\boldsymbol{s}\right)} \geq g_0 \tag{9.274}$$

with

$$g_0 = K(\lambda_0 - 1). \tag{9.275}$$

The numerator of the test function is the square of the adaptive matched filter output defined in (9.261). The denominator normalizes the adaptive filter output so that its false alarm probability does not depend on the unknown signal amplitude and clutter covariance matrix. The authors further showed that the distribution of $g(\boldsymbol{J})$) under H, depends only on the dimensional parameters J and N. This implies that a CFAR threshold g_0 can be precalculated for any desired false alarm probability. It should be noted that the RX algorithm assumes that the M multispectral image have been preprocessed to create "Gaussian PDFs" [33–37].

The authors developed the following probabilities of detection and false alarm for this problem under CFAR conditions:

$$Q_d = \int_{g_0}^{1} f_{\mathrm{RX}}(q|H_1)\,dr \tag{9.276}$$

and

$$Q_{\mathrm{fa}} = \int_{g_0}^{1} f_{\mathrm{RX}}(r|H_0)\,dr, \tag{9.277}$$

where

$$f_{\mathrm{RX}}(q|H_1) = \int_{g_0}^{1} f(r|q,H_1)f(q|H_1)\,dr, \tag{9.278}$$

$$f(r|q,H_1) = \frac{\Gamma\left(\frac{(K-M+1)}{2}\right)e^{-\left(\frac{aq}{2}\right)}}{\Gamma\left(\frac{K-M}{2}\right)\Gamma\left(\frac{1}{2}\right)}(1-r)^{\frac{(K-M-2)}{2}}r^{-\frac{1}{2}}{}_1F_1\left(\frac{(K-M+1)}{2};\frac{1}{2};\frac{aqr}{2}\right) \tag{9.279}$$

for $0 \le r < 1$,

$$f(q|H_1) = \frac{\Gamma\left(\frac{(K-1)}{2}\right)}{\Gamma\left(\frac{K-M}{2}\right)\Gamma\left(\frac{(M-1)}{2}\right)}(1-q)^{\frac{(M-3)}{2}}q^{\frac{(K-M-2)}{2}} \tag{9.280}$$

for $0 \le q \le 1$, and

$$f_{\mathrm{RX}}(r|H_0) = \frac{\Gamma\left(\frac{(K-M+1)}{2}\right)}{\Gamma\left(\frac{K-M}{2}\right)\Gamma\left(\frac{1}{2}\right)}(1-r)^{\frac{(K-M-2)}{2}}r^{-\frac{1}{2}} \tag{9.281}$$

for $0 \le r < 1$. In Eq. (9.279), ${}_1F_1(a;b;z)$ is confluent hypergeometric function given by

$${}_1F_1(a;b;z) = \frac{\Gamma(b)}{\Gamma(b-a)\Gamma(a)}\int_0^1 e^{zt}t^{a-1}(1-t)^{b-a-1}\,dt, \tag{9.282}$$

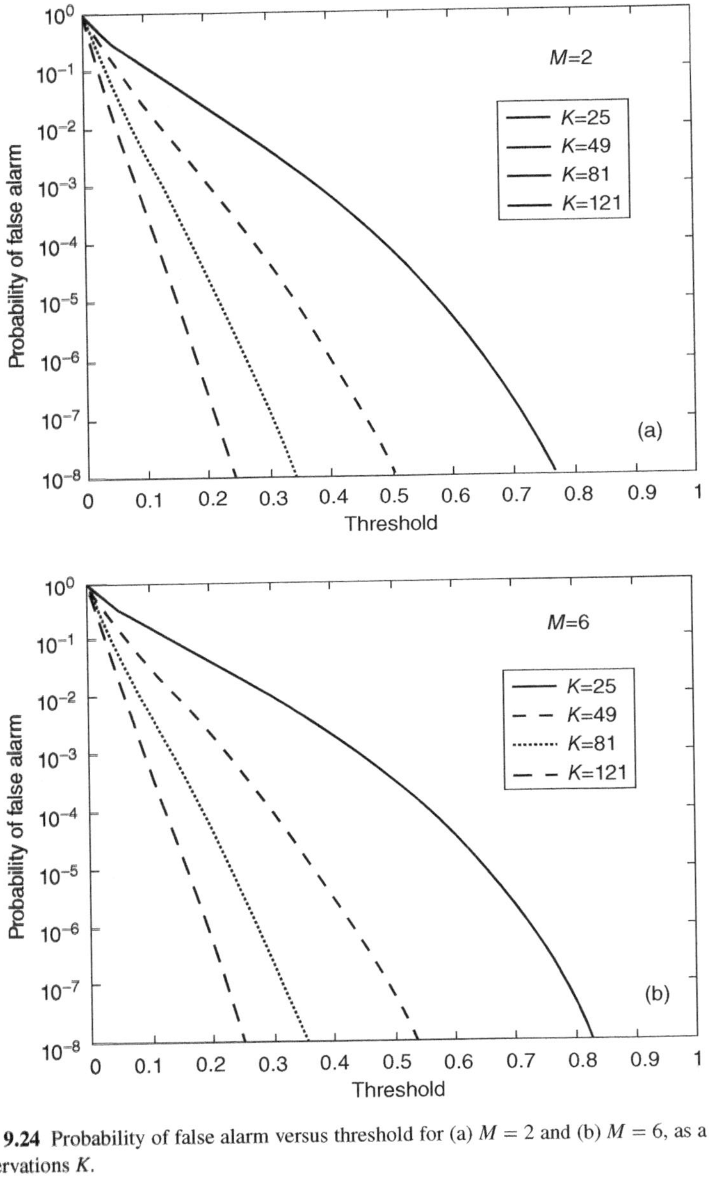

FIGURE 9.24 Probability of false alarm versus threshold for (a) $M = 2$ and (b) $M = 6$, as a function of pixel observations K.

and a is the generalized signal-to-noise ratio (GSNR) defined by

$$a = \text{GSNR} \equiv \boldsymbol{b}_s^T \boldsymbol{R}^{-1} \boldsymbol{b}_s \|\boldsymbol{s}\|^2. \tag{9.283}$$

Figure 9.24a and b plots the probability of false alarm versus threshold g_0 for two spectral band numbers, $M = 2$ and $M = 6$, respectively, as function of pixel observations, $K = 25, 49, 81,$ and 121. Figure 9.25a and b shows the probability of detection versus GSNR for $Q_{\text{fa}} = 10^{-5}$ for $K = 49$ and $K = 81$, respectively, as a function of the number of bands M. Figure 9.26a and b shows the probability of detection versus GSNR for $Q_{\text{fa}} = 10^{-5}$ for $M = 2$ and $M = 6$, respectively, as a function of pixel observations K.

Recall that the probabilities of false alarm and detection for known target in noise using a perfect matched filter in additive white Gaussian noise (AWGN) are equal to

$$Q_{\text{fa}} = 0.5 \,\text{erfc}\left(\frac{r_0}{\sqrt{2}}\right) \tag{9.284}$$

and

$$Q_d = 0.5 \,\text{erfc}\left(\frac{r_0 - \sqrt{\text{GSNR}}}{\sqrt{2}}\right), \tag{9.285}$$

respectively, using this section's notation. The CFAR probability of detection curve for a perfect matched filter and $Q_{\text{fa}} = 10^{-5}$ is included in the last two figures to better illustrate the effects of the numbers of pixel observations and bands, K and M, on Q_d [34]. It is clear in Figures 9.24–9.26 that for a given number of bands M, the probability of false alarm and the CFAR probability of detection both improve as the number of pixel observations increase. The limit is when K goes to infinity, which results in the perfect matched filter curves shown in Figures 9.25 and 9.26 [34]. On the other hand, the CFAR probability of detection for a fixed number of pixel observations decreases if more bands are used to provide the same GSNR. This is because the number of unknown parameters in the covariance matrix $\boldsymbol{R}$ increases as M gets larger [11].

Hoff *et al.* [38, 39] extended to multibands, the two-band weighted difference (additive noise) hypothesis test developed by Stotts [40]. Figures 9.27 and 9.28 show the output SNR of their generalized weighted spectral difference detector using TIMS and SMIFTS image sets, respectively [38]. The first figure indicates that more than 20 dB gain was obtained for detection by processing beyond one spectral band. There was a significant gain of 16 dB in processing just two spectral bands. Since spectral band to band is very highly correlated, two bands processing cancels most of the clutter and additional clutter reference bands provide little help in reducing the clutter variance. (The curves also indicate that the output SNR will gradually level off if more target-reference bands are added.) The second figure confirms these comments using a different data set.

Results such as these suggest that highly correlated, dual-band images provide close to the maximum signal processing gain possible. Adding target and clutter

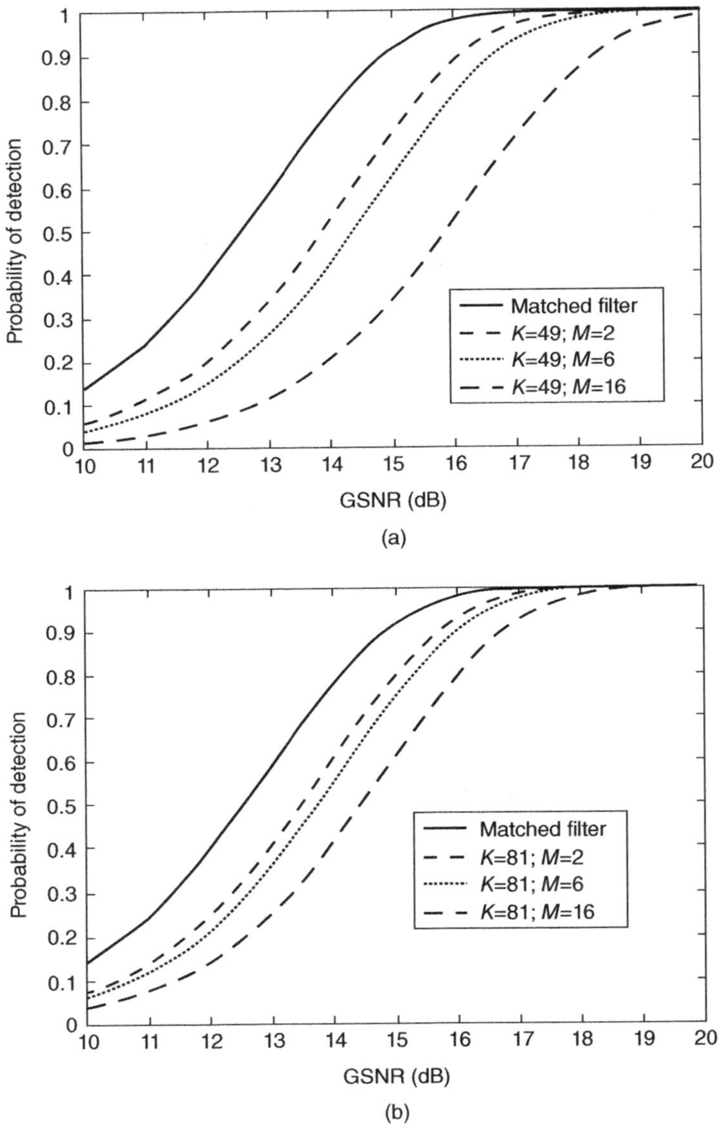

FIGURE 9.25 CFAR probability of detection versus GSNR for (a) $K = 49$ and (b) $K = 81$, as a function of spectral bands M, as compared to that of a perfect matched filter.

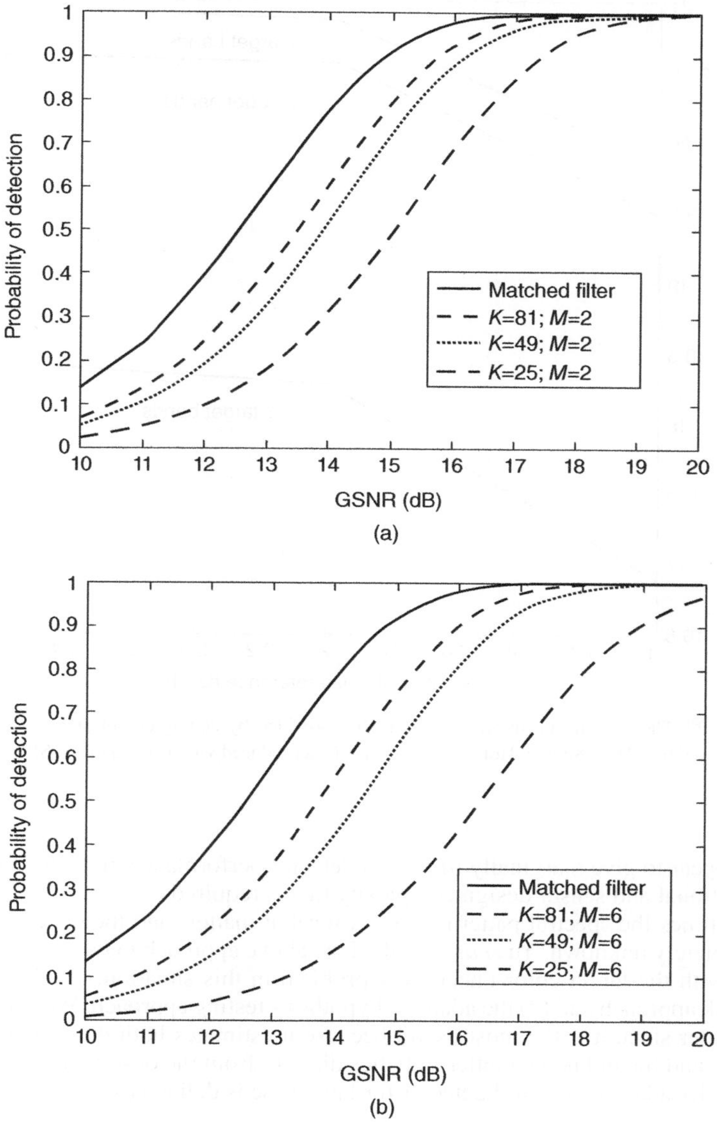

FIGURE 9.26 CFAR probability of detection versus GSNR for (a) $M = 2$ and (b) $M = 6$, as a function of pixel observations K, as compared to that of a perfect matched filter.

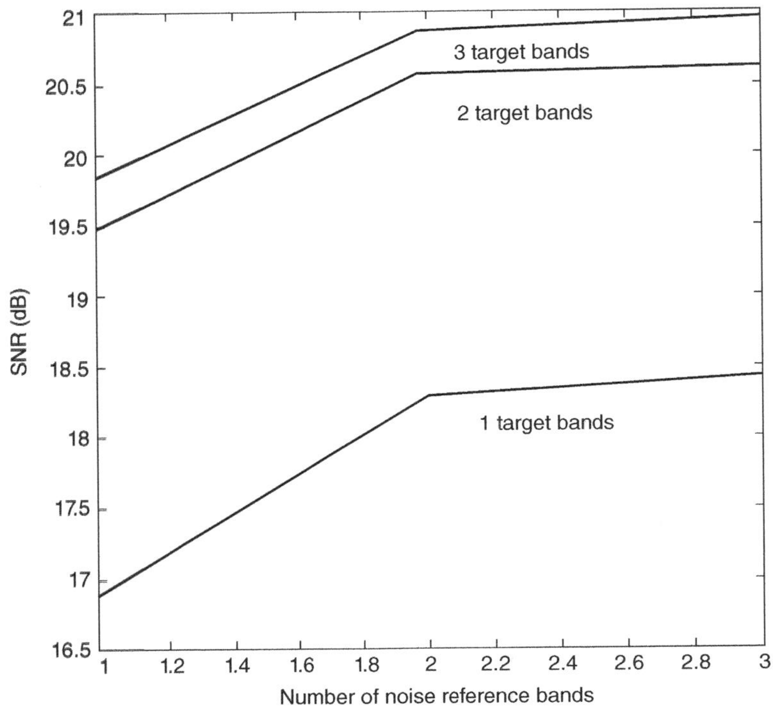

FIGURE 9.27 The incremental improvement in SNR provided by adding clutter and target reference bands collected by TIMS. Source: Hoff *et al.*, 1995 [39]. Reproduced with permission of SPIE.

bands appear to give marginally improved detector performance given the increased computational and sensor design complexity that is required.

Sometimes the spectral pattern $\boldsymbol{b}$ (or $\boldsymbol{b}_s$, for that matter) and the clutter statistics are completely unknown. Yu *et al.* modified the above approach in two specific forms to deal with the target-color uncertainty problem in this situation: (1) the adaptive filter bank approach and (2) the adaptive hypothesis testing approach. We concentrate on the latter since it is the most useful because it estimates both the signal spectral signature and the unknown clutter statistics directly from the observed data [41].

The fully adaptive spectral detector for latter case is defined as

$$r(\boldsymbol{J}) = \frac{(\boldsymbol{Js})^T\,(\boldsymbol{JJ}^T)^{-1}(\boldsymbol{Js})}{\boldsymbol{s}^T\boldsymbol{s}} \geq r_0, \tag{9.286}$$

where r_0 is the detection threshold [41]. This equation only requires knowledge of the target shape vector $\boldsymbol{s}$. Yu *et al.* noted that the MLE of the unknown target spectral vector under Hypothesis H_1 is given by

$$\hat{\boldsymbol{b}} = \frac{(\boldsymbol{Js}\,)}{\boldsymbol{s}^T\boldsymbol{s}}\,. \tag{9.287}$$

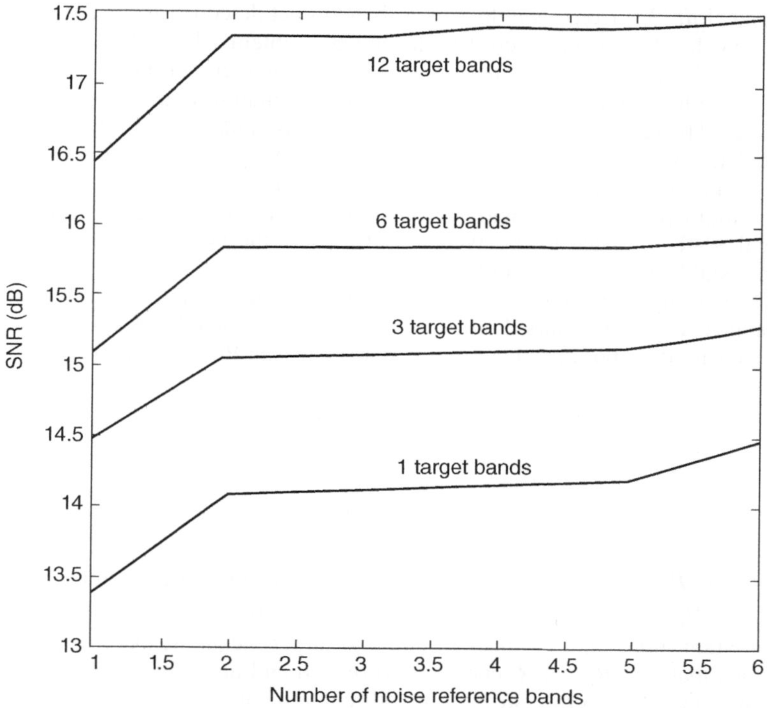

FIGURE 9.28 The incremental improvement in SNR provided by adding clutter and target reference bands collected by SMIFTS. Source: Hoff *et al.*, 1995 [39]. Reproduced with permission of SPIE.

Using $\widehat{\boldsymbol{b}}$ and the estimate of the unknown covariance matrix $\widehat{\boldsymbol{M}} = \frac{(JJ^T)}{N}$, Eq. (9.286) can be rewritten as

$$\widehat{r}(\boldsymbol{J}) = \widehat{\boldsymbol{b}}^T \widehat{\boldsymbol{M}}^{-1} \boldsymbol{J}\boldsymbol{s} \geq \begin{cases} \geq \widehat{r}_0, & \text{then } H_1 \\ < \widehat{r}_0, & \text{then } H_0 \end{cases} \tag{9.288}$$

with $\widehat{r}_0 = N r_0$. Equation (9.288) shows that the fully adaptive spectral detector can be interpreted as a clutter-adaptive spatial-spectral filter that is matched to a ML spectral signature estimate rather than *a priori* specified signature vector $\boldsymbol{b}$ [34]. In other words, this equation says that aside from knowing the approximate spatial extent (constrained by the choice of the dimension for the image prefilter for Gaussianization), no other *a priori* knowledge is required. This fully adaptive approach is widely known as the "RX detector" [41].

9.10 RESOLVED TARGET DETECTION IN CORRELATED DUAL-BAND MULTISPECTRAL IMAGE SETS

The section tackles the problem of resolved target detection in correlated multi-/hyperspectral imagery, beginning with the extension of the work of

Stotts and Hoff [32] from a single- to two-band target detection [42]. This extension focuses on the dual-band spectral replacement problem where the two bands are highly correlated in order to reduce the background clutter noise down to the system noise level, but keep some residual target signature intact, after algorithmic processing. The reason for only looking at two bands is that we found in Section 9.7 that the detector performance is lower as we spread the processing gain across many bands. By getting to system noise-limited operation using just two wavelengths gets the maximum possible gain with minimal operations. It also reconfirms that the electrical SNR is related to the Weber contrast and the normalized noise variance like was established by Stotts and Hoff [32].

In the majority of the optical target detection strategies in multiband imagery, their solution involves the likelihood ratio and the required resolved/unresolved target and noise intensity distributions obey multivariate Gaussian PDFs of the form:

$$H_0 : \boldsymbol{x} \sim N(\bar{\boldsymbol{j}}_n, \boldsymbol{R}_n) \tag{9.289}$$

and

$$H_1 : \boldsymbol{x} \sim N(\bar{\boldsymbol{j}}_t, \boldsymbol{R}_t), \tag{9.290}$$

where $\bar{\boldsymbol{j}}_n$ and $\boldsymbol{R}_n$ and $\bar{\boldsymbol{j}}_t$ and $\boldsymbol{R}_t$ represent the mean vector and covariance matrix for Hypothesis H_k, $k = 0, 1$, respectively. Depending on the problem addressed, $\bar{\boldsymbol{j}}_n$ and $\bar{\boldsymbol{j}}_t$ can contain more than the background and target means, respectively, and their covariance matrices $\boldsymbol{R}_t$ and $\boldsymbol{R}_t$ can have correlated and uncorrelated data. At a minimum, we know that both distributions will contain some sort of system noise that depends on the optical sensor design.

In the general case where $\boldsymbol{R}_n \neq \boldsymbol{R}_t$, the likelihood ratio is given by

$$\Lambda(\boldsymbol{x}) = \frac{|\boldsymbol{R}_n|^{\frac{1}{2}} \exp[-(\boldsymbol{x} - \boldsymbol{j}_t)^T R_t^{-1}(\boldsymbol{x} - \boldsymbol{j}_t)]}{|\boldsymbol{R}_t|^{\frac{1}{2}} \exp[-(\boldsymbol{x} - \boldsymbol{j}_n)^T R_n^{-1}(\boldsymbol{x} - \boldsymbol{j}_n)]} \begin{matrix} > \Lambda_0; \text{to choose } H_1 \\ \leq \Lambda_0; \text{to choose } H_0 \end{matrix}. \tag{9.291}$$

Taking the logarithm and manipulating terms, we obtain the following detection statistics:

$$y = D(\boldsymbol{x}) = (\boldsymbol{x} - \boldsymbol{j}_n)^T \boldsymbol{R}_n^{-1}(\boldsymbol{x} - \boldsymbol{j}_n) - (\boldsymbol{x} - \boldsymbol{j}_t)^T \boldsymbol{R}_t^{-1}(\boldsymbol{x} - \boldsymbol{j}_t) \begin{matrix} > g_0; \text{to choose } H_1 \\ \leq g_0; \text{to choose } H_0 \end{matrix}, \tag{9.292}$$

which compares the Mahalanobis distance of the spectrum under test from the means of the background and target PDFs [43]. In the above, g_0 is the resulting threshold after the terms adjustment. Equation (9.292) is known as the "quadratic detector."

In many of his papers with colleagues, Manolakis noted that the variable $y = D$ is a random variable whose probability density depends on which hypothesis is true. If the two conditional probability densities, $f(q|H_1)$ and $f(q|H_0)$, are known, then the probabilities of detection and false alarm are given by

$$Q_d = \int_{g_0}^{\infty} f(q|H_1) dr \tag{9.293}$$

and

$$Q_{\text{fa}} = \int_{g_0}^{\infty} f_{\text{RX}}(r|H_0)\, dr, \tag{9.294}$$

respectively. They also note that the computation of these two integrals is quite complicated and the ROC curves can only be evaluated not so simply by using Monte Carlo simulation [39] or numerical techniques.

Let us assume Hypothesis H_0 is where we have background clutter plus system noise only in the two images and within N pixel template. The resulting vector is given by

$$\boldsymbol{j}_n = \begin{bmatrix} \boldsymbol{i}_1 - \boldsymbol{b}_1 - \boldsymbol{n}_1 \\ \boldsymbol{i}_2 - \boldsymbol{b}_2 - \boldsymbol{n}_2 \end{bmatrix}, \tag{9.295}$$

where $\boldsymbol{i}_1$ and $\boldsymbol{i}_2$ are the image vectors; $\boldsymbol{b}_1$ and $\boldsymbol{b}_2$ the correlated background clutter vectors (i.e., fixed background structure from trees, grass, roads, etc.) at wavelengths λ_1 and λ_2, respectively, contained in images $\boldsymbol{i}_1$ and $\boldsymbol{i}_2$; and $\boldsymbol{n}_1$ and $\boldsymbol{n}_2$ the mean image vectors for the system noise contained in images $\boldsymbol{i}_1$ and $\boldsymbol{i}_2$, respectively.

The covariance matrix for Hypothesis H_0 is given by

$$\boldsymbol{R}_n = \mathcal{E}\{\boldsymbol{j}_n \boldsymbol{j}_n{}^T\} \tag{9.296}$$

$$= \begin{pmatrix} \sigma_1^2 \boldsymbol{I} & p_{12}\boldsymbol{I} \\ p_{21}\boldsymbol{I} & \sigma_2^2 \boldsymbol{I} \end{pmatrix} \tag{9.297}$$

with

$$\sigma_n^2 \equiv \text{statistical variance of image } \boldsymbol{i}_n \text{ under } H_0,$$

$$= \sigma_{b1}^2 + \sigma_{sn}^2 \approx \sigma_{b1}^2 \tag{9.298}$$

$$\sigma_t^2 \equiv \text{statistical variance of image } i_t \text{ under } H_0,$$

$$= \sigma_{b2}^2 + \sigma_{sn}^2 \approx \sigma_{b2}^2 \tag{9.299}$$

$$\sigma_t^2 \equiv \text{statistical variance of system noise}$$

because one does not see a salt-and-pepper pattern of system noise in good quality imagery. In Eq. (9.297), we have

$$p_{12} = \mathcal{E}\{\boldsymbol{i}_1 \boldsymbol{i}_2{}^T\} = \sigma_{b1}\, \sigma_{b2} \rho = p_{21}. \tag{9.300}$$

In this development, the image background clutter is assumed correlated and the system noise is assumed uncorrelated.

The inverse matrix $\boldsymbol{R}_n^{-1}$ can easily be shown to equal

$$\boldsymbol{R}_n^{-1} = \begin{pmatrix} \sigma_f^2 \boldsymbol{I} & p\boldsymbol{I} \\ p\boldsymbol{I} & \sigma_g^2 \boldsymbol{I} \end{pmatrix} \tag{9.301}$$

with

$$\sigma_f^2 = \frac{1}{\sigma_{b1}^2\,(1-\rho^2)}, \tag{9.302}$$

$$\sigma_g^2 = \frac{1}{\sigma_{b2}^2(1-\rho^2)}, \tag{9.303}$$

$$p = \frac{-\rho}{\sigma_{b1}\sigma_{b2}(1-\rho^2)}. \tag{9.304}$$

and ρ is the correlation coefficient between the two images.

Similarly, we find for Hypothesis H_1 that its resulting vector is given by

$$\boldsymbol{j}_t = \begin{bmatrix} \boldsymbol{i}_1 - \boldsymbol{s}_1 - \boldsymbol{n}_1 \\ \boldsymbol{i}_2 - \boldsymbol{s}_2 - \boldsymbol{n}_2 \end{bmatrix} \tag{9.305}$$

with $\boldsymbol{s}_1$ and $\boldsymbol{s}_{21}$ are the image vectors for our target pixels sampled at wavelengths λ_1 and λ_2, respectively, contained in images $\boldsymbol{i}_1$ and $\boldsymbol{i}_2$,. The covariance matrix for Hypothesis H_1 is equal to

$$\boldsymbol{R}_t = \mathcal{E}\{\boldsymbol{j}_1\boldsymbol{j}_1^T\} \tag{9.306}$$

$$= \sigma_{sn}^2 \begin{bmatrix} \boldsymbol{I} & 0 \\ 0 & \boldsymbol{I} \end{bmatrix}. \tag{9.307}$$

This implies that

$$\boldsymbol{R}_t^{-1} = \sigma_{sn}^{-2} \begin{bmatrix} \boldsymbol{I} & 0 \\ 0 & \boldsymbol{I} \end{bmatrix} \tag{9.308}$$

with $\boldsymbol{I}$ and 0 are the $N \times N$ identity and null matrices, respectively.

Let us remove the signal and system noise-mean image vectors from both images. Then the resulting vector for H_0 becomes

$$\boldsymbol{m}_n = \begin{bmatrix} \boldsymbol{i}_1 - \boldsymbol{b}_1 + \boldsymbol{s}_1 \\ \boldsymbol{i}_2 - \boldsymbol{b}_2 + \boldsymbol{s}_2 \end{bmatrix} \tag{9.309a}$$

$$= \begin{bmatrix} \boldsymbol{i}_1 + \boldsymbol{b}_1\boldsymbol{C}_1 \\ \boldsymbol{i}_2 + \boldsymbol{b}_2\boldsymbol{C}_2 \end{bmatrix}, \tag{9.309b}$$

where

$$\boldsymbol{C}_k = \frac{\boldsymbol{s}_k - \boldsymbol{b}_k}{\boldsymbol{b}_k} \tag{9.310}$$

is the Weber contrast column vector for image $\boldsymbol{m}_0$.

The resulting vector for Hypothesis H_1 is given by

$$\boldsymbol{m}_t = \begin{bmatrix} \boldsymbol{i}_1 \\ \boldsymbol{i}_2 \end{bmatrix} \tag{9.311}$$

and its associated covariance matrix for $\boldsymbol{m}_0$ still is $\boldsymbol{R}_t^{-1}$. Substituting into Eq. (9.291), we have

$$\begin{aligned}
\ln \Lambda &= \ln \left[\frac{p_1(\boldsymbol{i}_2)}{p_0(\boldsymbol{i}_0)}\right] \\
&= \ln \left[\frac{|\boldsymbol{R}_n|^{\frac{1}{2}}}{|\boldsymbol{R}_t|^{\frac{1}{2}}}\right] - \sigma_{sn}^{-2} \left[\boldsymbol{i}_1 \ \boldsymbol{i}_2\right] \begin{bmatrix} \boldsymbol{I} & 0 \\ 0 & \boldsymbol{I} \end{bmatrix} \begin{bmatrix} \boldsymbol{i}_1 \\ \boldsymbol{i}_2 \end{bmatrix} \\
&\quad + \left[\boldsymbol{i}_1 + \boldsymbol{b}_1 \boldsymbol{C}_1 \ \ \boldsymbol{i}_2 + \boldsymbol{b}_2 \boldsymbol{C}_2\right] \begin{pmatrix} \sigma_f^2 \boldsymbol{I} & p\boldsymbol{I} \\ p\boldsymbol{I} & \sigma_g^2 \boldsymbol{I} \end{pmatrix} \begin{bmatrix} \boldsymbol{i}_1 + \boldsymbol{b}_1 \boldsymbol{C}_1 \\ \boldsymbol{i}_2 + \boldsymbol{b}_2 \boldsymbol{C}_2 \end{bmatrix} \\
&= \ln \left[\frac{|\boldsymbol{R}_0|^{\frac{1}{2}}}{|\boldsymbol{R}_1|^{\frac{1}{2}}}\right] - \left(\frac{\sigma_3^2}{\sigma_{sn}^2 \sigma_{b1}^2 (1-\rho^2)}\right) \boldsymbol{i}_1^T \boldsymbol{i}_1 - \left(\frac{\sigma_4^2}{\sigma_{sn}^2 \ \sigma_{b2}^2 \ (1-\rho^2)}\right) \boldsymbol{i}_2^T \boldsymbol{i}_2 \\
&\quad + \frac{(\boldsymbol{i}_1^T \boldsymbol{b}_1 \boldsymbol{C}_1 + \boldsymbol{b}_1^T \boldsymbol{C}_1^T \boldsymbol{i}_1 + \boldsymbol{b}_1^T \boldsymbol{C}_1^T \boldsymbol{b}_1 \boldsymbol{C}_1)}{\sigma_{b1}^2 \ (1-\rho^2)} - \rho \frac{(\boldsymbol{i}_2 + \boldsymbol{b}_2 \boldsymbol{C}_2)^T (\boldsymbol{i}_1 + \boldsymbol{b}_1 \boldsymbol{C}_1)}{\sigma_{b1} \ \sigma_{b2} \ (1-\rho^2)} \\
&\quad - \rho \frac{(\boldsymbol{i}_1 + \boldsymbol{b}_1 \boldsymbol{C}_1)^T (\boldsymbol{i}_2 + \boldsymbol{b}_2 \boldsymbol{C}_2)}{\sigma_{b1} \ \sigma_{b2} \ (1-\rho^2)} \\
&\quad + \frac{(\boldsymbol{i}_2^T \boldsymbol{b}_2 \boldsymbol{C}_2 + \boldsymbol{b}_2^T \boldsymbol{C}_2^T \boldsymbol{i}_2 + \boldsymbol{b}_2^T \boldsymbol{C}_2^T \boldsymbol{b}_2 \boldsymbol{C}_2)}{\sigma_{b2}^2 \ (1-\rho^2)} \begin{matrix} > \ln \Lambda_0; & \text{to choose } H_1 \\ \leq \ln \Lambda_0; & \text{to choose } H_0 \end{matrix},
\end{aligned} \tag{9.312}$$

where

$$\sigma_3^2 = \sigma_{b1}^2 \ (1-\rho^2) - \sigma_{sn}^2 \tag{9.313}$$

and

$$\sigma_4^2 = \sigma_{b2}^2 (1-\rho^2) - \sigma_{sn}^2. \tag{9.314}$$

Rewriting Eq. (9.312), we have

$$\begin{aligned}
g = \Bigg[& i_1^T i_1 + \left(\frac{\sigma_{b1}^2}{\sigma_{b2}^2}\right) i_2^T i_2 + (i_1^T b_1 C_1 + b_1^T C_1^T i_1 + b_1^T C_1^T \ b_1 C_1) \\
& - \rho \left(\frac{\sigma_{b1}}{\sigma_{b2}}\right) (i_2^T \ i_1 + i_2^T b_1 C_1 + b_2^T C_2^T \ i_1 + b_2^T C_2^T b_1 C_1) \\
& - \rho \left(\frac{\sigma_{b1}}{\sigma_{b2}}\right) (i_1^T \ i_2 + i_1^T b_2 C_2 + b_1^T C_1^T \ i_2 + b_1^T C_1^T b_2 C_2) \\
& + \left(\frac{\sigma_{b1}^2}{\sigma_{b2}^2}\right) (i_2^T b_2 C_2 + b_2^T C_2^T i_2 + b_2^T C_2^T \ b_2 C_2) \Bigg] \begin{matrix} > g_0; & \text{to choose } H_1 \\ \leq g_0; & \text{to choose } H_0 \end{matrix},
\end{aligned} \tag{9.315}$$

because we have assumed that $\sigma_{bi}^2(1-\rho^2) \ll \sigma_{sn}^2$ for $i = 1, 2$. Here,

$$g_0 = \sigma_{b1}^2\,(1-\rho^2)\left[\ln\Lambda_0 - \ln\left[\frac{p_1(\boldsymbol{i}_2)}{p_0(\boldsymbol{i}_0)}\right]\right] \tag{9.316}$$

$$\approx \sigma_{b1}^2(1-\rho^2)\ln\left[e^{\frac{A}{(1-\rho^2)}}\right] = \sigma_{b1}^2 A. \tag{9.317}$$

In Eq. (9.317), we adjust $\ln\Lambda_0$ to keep g_0 nonzero for correlation coefficients close to 1. Our assumption $\sigma_{\mathrm{bi}}^2(1-\rho^2) \ll \sigma_{sn}^2$ says that the best weighted difference noise reduction essentially is where we become system noise-limited and the residual clutter is negligible. This condition would be expected for the closely spaced bands [43].

If we further assume that $\rho^2 \cong 1$, with the appropriate adjustments of $\ln\Lambda_0$ to keep g_0 nonzero, then Eq. (9.302) becomes

$$\begin{aligned} g = \Bigg[& \boldsymbol{i}_1^T\boldsymbol{i}_1 + \left(\frac{\sigma_{b1}^2}{\sigma_{b2}^2}\right)\boldsymbol{i}_2^T\boldsymbol{i}_2 + (\boldsymbol{i}_1^T\boldsymbol{b}_1\boldsymbol{C}_1 + \boldsymbol{b}_1^T\boldsymbol{C}_1^T\boldsymbol{i}_1 + \boldsymbol{b}_1^T\boldsymbol{C}_1^T\boldsymbol{b}_1\boldsymbol{C}_1) \\ & - \rho\left(\frac{\sigma_{b1}}{\sigma_{b2}}\right)(\boldsymbol{i}_2^T\boldsymbol{i}_1 + \boldsymbol{i}_2^T\boldsymbol{b}_1\boldsymbol{C}_1 + \boldsymbol{b}_2^T\boldsymbol{C}_2^T\boldsymbol{i}_1 + \boldsymbol{b}_2^T\boldsymbol{C}_2^T\boldsymbol{b}_1\boldsymbol{C}_1) \\ & - \rho\left(\frac{\sigma_{b1}}{\sigma_{b2}}\right)(\boldsymbol{i}_1^T\boldsymbol{i}_2 + \boldsymbol{i}_1^T\boldsymbol{b}_2\boldsymbol{C}_2 + \boldsymbol{b}_1^T\boldsymbol{C}_1^T\boldsymbol{i}_2 + \boldsymbol{b}_1^T\boldsymbol{C}_1^T\boldsymbol{b}_2\boldsymbol{C}_2) \\ & + \left(\frac{\sigma_{b1}^2}{\sigma_{b2}^2}\right)\rho^2(\boldsymbol{i}_2^T\boldsymbol{b}_2\boldsymbol{C}_2 + \boldsymbol{b}_2^T\boldsymbol{C}_2^T\boldsymbol{i}_2 + \boldsymbol{b}_2^T\boldsymbol{C}_2^T\boldsymbol{b}_2\boldsymbol{C}_2)\Bigg] \begin{matrix} > g_0;\text{ to choose } H_1 \\ \leq g_0;\text{ to choose } H_0 \end{matrix} \end{aligned} \tag{9.318}$$

$$= \left|\left(\boldsymbol{i}_1 - \rho\left(\frac{\sigma_{b1}}{\sigma_{b2}}\right)\boldsymbol{i}_2\right) + \left(\boldsymbol{b}_1\boldsymbol{C}_1 - \rho\left(\frac{\sigma_{b1}}{\sigma_{b2}}\right)\boldsymbol{b}_2\boldsymbol{C}_2\right)\right|^2 \begin{matrix} > g_0;\text{ to choose } H_1 \\ \leq g_0;\text{ to choose } H_0 \end{matrix}. \tag{9.319}$$

Normalizing Eq. (9.319), we obtain the following test statistic:

$$\frac{1}{N}\left|\left(\boldsymbol{i}_1 - \rho\left(\frac{\sigma b_1}{\sigma_{b2}}\right)\boldsymbol{i}_2\right) + \left(\boldsymbol{b}_1\boldsymbol{C}_1 - \rho\left(\frac{\sigma_{b1}}{\sigma_{b2}}\right)\boldsymbol{b}_2\boldsymbol{C}_2\right)\right|^2 \begin{matrix} > G_0 \\ \leq G_0 \end{matrix}. \tag{9.320}$$

The $\left(\boldsymbol{i}_1 - \rho\left(\frac{\sigma_{b1}}{\sigma_{b2}}\right)\boldsymbol{i}_2\right)$-term in Eq. (9.320) is analogous to the processing for the side-lobe canceller radar, i.e., weighted-difference equation [44, 45]. It can be shown that the weight $\rho\left(\frac{\sigma_1}{\sigma_2}\right)$ minimizes the difference-image variance in a least-mean-square error sense [44].

Equation (9.320) has the form of the equation

$$\frac{1}{N}\sum_{n=1}^{N}|x_n + A_n|^2, \tag{9.321}$$

which implies from previous work that the probabilities of false alarm and detection for this problem are given by

$$Q_{\text{fa}} = \int_{G_0}^{\infty} f_{\chi}(v; N, \vartheta_0)dq \tag{9.322}$$

and

$$Q_d = \int_{G_0}^{\infty} f_{\chi}(v; N, \vartheta_1)dq, \tag{9.323}$$

respectively, where

$$\vartheta_0 = \frac{\sum_{n=1}^{N} A_n^2(H_0)}{\sigma_T^2} = \frac{\left|b_1 C_1 - \rho\left(\frac{\sigma_1}{\sigma_2}\right) b_2 C_2\right|^2}{\sigma_T^2}, \tag{9.324}$$

$$\vartheta_1 = 4\vartheta_0 \tag{9.325}$$

with

$$\sigma^2 = \sigma_1^2(1 - \rho^2) + \alpha\sigma_{sn}^2. \tag{9.326}$$

(the $\alpha = [1 + \rho(\sigma_1/\sigma_2)]$ multiplier on the system noise variance in Eq. (9.326) comes from increased variance produced by the weighted difference of the two images.)

Following what was done in the last section, we define the CNR to be given by

$$\begin{aligned} \text{CNR} = \vartheta_1 - \vartheta_0 &= \frac{4\left|\left(\boldsymbol{b}_1\boldsymbol{C}_1 - \rho\left(\frac{\sigma_{b1}}{\sigma_{b2}}\right)\boldsymbol{b}_2\boldsymbol{C}_2\right)\right|^2}{(\alpha\, \sigma_{sn}^2)} - \frac{\left|\left(\boldsymbol{b}_1\boldsymbol{C}_1 - \rho\left(\frac{\sigma_{b1}}{\sigma_{b2}}\right)\boldsymbol{b}_2\boldsymbol{C}_2\right)\right|^2}{(\sigma_{b1}^2\, (1-\rho^2) + \alpha\, \sigma_{sn}^2)} \\ &\approx \frac{4\left|\left(\boldsymbol{b}_1\boldsymbol{C}_1 - \rho\left(\frac{\sigma_{b1}}{\sigma_{b2}}\right)\boldsymbol{b}_2\boldsymbol{C}_2\right)\right|^2}{(\sigma_{b1}^2\, (1-\rho^2) + \alpha\sigma_{sn}^2)} - \frac{\left|\left(\boldsymbol{b}_1\boldsymbol{C}_1 - \rho\left(\frac{\sigma_{b1}}{\sigma_{b2}}\right)\boldsymbol{b}_2\boldsymbol{C}_2\right)\right|^2}{(\sigma_{b1}^2\, (1-\rho^2) + \alpha\sigma_{sn}^2)} \end{aligned} \tag{9.327}$$

$$= \frac{3\left|\left(\boldsymbol{b}_1\boldsymbol{C}_1 - \rho\left(\frac{\sigma_{b1}}{\sigma_{b2}}\right)\boldsymbol{b}_2\boldsymbol{C}_2\right)\right|^2}{(\sigma_{b1}^2\, (1-\rho^2) + \alpha\sigma_{sn}^2)}. \tag{9.328}$$

For constant contrast across the target areas, Eq. (9.328) becomes

$$\text{CNR} \approx \left(\frac{NC_1^2}{\left[\frac{(\sigma_{b1}^2\, (1-\rho^2)+\alpha\sigma_t^2)}{b_1^2}\right]}\right)\left[1 - \rho\left(\frac{\sigma_{b1}}{\sigma_{b2}}\right)\left(\frac{b_2 C_2}{b_1 C_1}\right)\right]^2. \tag{9.329}$$

Once again, we see that CNR depends on the Weber contrast-squared divided by the normalized variance, as found in the single channel case [31]. The ratio, $\left(\frac{b_2 C_2}{b_1 C_1}\right)$, is the *color ratio* between the two images. The $[\ldots]^2$ term in Eq. (9.329) always is positive no matter what the value of color ratio is.

9.11 IMAGE WHITENER

As we noted in our discussion on matched filtering, temporal, spatial, or temporal-spatial decorrelation (prewhitening) filtering plays a very useful position in defining an optimum signal detection solution. This section summarizes a first-order image *prewhitening filter* or *whitener* based on the Gram–Schmidt orthogonalization procedure [44, 45]. Other procedures are available [44].

9.11.1 Orthogonal Sets

Assume we have a vector space V that constitutes an inner product space. Further assume that a set of nonzero vectors $\{\boldsymbol{v}_1, \boldsymbol{v}_2, \ldots, \boldsymbol{v}_k\} \in V$ form an orthogonal set if they are orthogonal to each other, that is, $\boldsymbol{v}_i|\boldsymbol{v}_j = 0$ for $i \neq j$, where the symbol $\boldsymbol{x}|\boldsymbol{y}$ represents the *inner product* between vectors $\boldsymbol{x}$ and $\boldsymbol{y}$.

In addition, if all vectors are normalized to one, then we have $\|\boldsymbol{v}_i\| = \langle \boldsymbol{v}_i|\boldsymbol{v}_i\rangle = 1$ for $i = 1,2,\ldots,k$. This implies that $\{\boldsymbol{v}_1, \boldsymbol{v}_2, \ldots, \boldsymbol{v}_k\}$ is an *orthonormal vector set*. It can be shown that any orthogonal set is linearly and V_0 is a finite-dimensional subspace of V [46, 47].

The ramification of the above is that any vector $\boldsymbol{x} \in V$ can be uniquely represented by the vector relation $\boldsymbol{x} = \boldsymbol{w} + \boldsymbol{o}$, where $\boldsymbol{w} \in V_0$ and $\boldsymbol{o} \perp V_0$. The component $\boldsymbol{w}$ is the *orthogonal projection* of the vector $\boldsymbol{x}$ onto the subspace V_0, as illustrated in Figure 9.29. The distance from $\boldsymbol{x}$ to the subspace V_0 is $\|\boldsymbol{o}\|$, where $\boldsymbol{o} = \boldsymbol{x} - \boldsymbol{w}$ is the remainder vector that is orthogonal to $\boldsymbol{w}$.

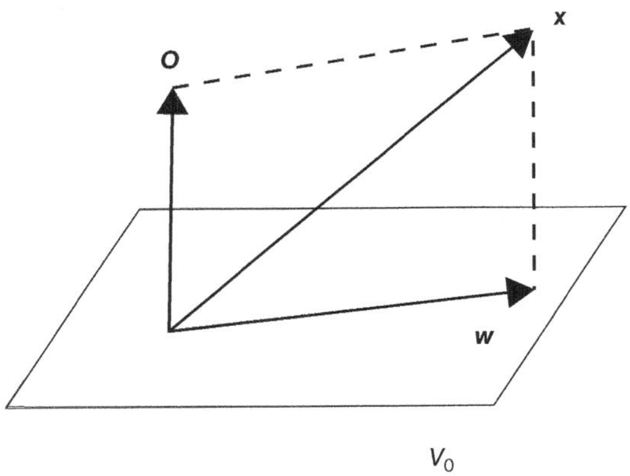

FIGURE 9.29 Vector projections of the components of $\boldsymbol{x}$.

If $\{v_1, v_2, \ldots, v_n\}$ is an orthogonal basis for V_0, then

$$w = \frac{\langle x|v_1\rangle}{\|v_1\|}v_1 + \frac{\langle x|v_2\rangle}{\|v_2\|}v_2 + \cdots + \frac{\langle x|v_n\rangle}{\|v_n\|}v_n \tag{9.330}$$

is the orthogonal projection of the vector x onto the subspace spanned by $\{v_1, v_2, \ldots, v_n\}$ with $\|v_j\| = \sqrt{\langle v_j|v_j\rangle}$. This means that the remainder vector $o = x - w$ is orthogonal to $\{v_1, v_2, \ldots, v_n\}$.

9.11.2 Gram–Schmidt Orthogonalization Theory

Let V be an inner product space. Assume $\{x_1, x_2, \ldots, x_k\}$ is a *basis set* for V. We then can make an orthogonal basis set $\{v_1, v_2, \ldots, v_n\}$ using the following procedure:

$$v_1 = x_1 \tag{9.331}$$

$$v_2 = x_2 - \frac{\langle x_1|v_1\rangle}{\langle v_1|v_1\rangle}v_1 \tag{9.332}$$

$$v_2 = x_3 - \frac{\langle x_3|v_2\rangle}{\langle v_2|v_2\rangle}v_2 - \frac{\langle x_3|v_1\rangle}{\langle v_1|v_1\rangle}v_1 \tag{9.333}$$

$$\vdots$$

$$v_n = x_n - \frac{\langle x_n|v_{n-1}\rangle}{\langle v_{n-1}|v_{n-1}\rangle}v_{n-1} - \cdots - \frac{\langle x_n|v_1\rangle}{\langle v_1|v_1\rangle}v_1. \tag{9.334}$$

Let V be an inner product space. Assume that $\{v_1, v_2, \ldots, v_n\}$ is an *orthogonal basis set* for V. Let

$$w_1 = \frac{v_1}{v_1} \tag{9.335}$$

$$w_2 = \frac{v_2}{v_2} \tag{9.336}$$

$$\vdots$$

$$w_n = \frac{v_n}{v_n}. \tag{9.337}$$

The vector set $\{w_1, w_2, \ldots, w_n\}$ is an *orthonormal basis set* for V. It can be shown that any finite-dimensional vector space with an inner product has an orthonormal basis [45, 46].

In the Gram–Schmidt theory [45], there is a theorem known as the *QR Factorization* that states:

> If A is an $m \times n$ matrix with linearly independent columns, then A can be factored as $A = QR$, where Q is an $m \times n$ matrix whose columns form an orthogonal basis for Col A and R is an $n \times n$ upper triangular invertible matrix with positive entries on its main diagonal.

9.11.3 Prewhitening Filter Using the Gram–Schmidt Process

Recall that we previously define an image vector $\boldsymbol{F}'$ generated by the image matrix $\boldsymbol{F}$ to be

$$\boldsymbol{F}' = \begin{bmatrix} \boldsymbol{F}(1,1) \\ \vdots \\ \boldsymbol{F}(1,N_2) \\ \boldsymbol{F}(2,1) \\ \vdots \\ \boldsymbol{F}(N_1,N_2) \end{bmatrix}. \tag{9.338}$$

Given that, a zero mean vector $\boldsymbol{y}$ of dimension N is assumed to be white if its covariance matrix is the identity matrix $\boldsymbol{I}_N$ is given by

$$\boldsymbol{M}_{yy} = \mathcal{E}(\boldsymbol{y}\boldsymbol{y}^T) = \boldsymbol{I}_N. \tag{9.339}$$

Any set of vectors $\boldsymbol{x}$ can be decorrelated (whitened) by applying some preprocessing stage. The whitening procedure is equivalent to the linear transformation of the vector $\boldsymbol{x}$. Mathematically, we have

$$\boldsymbol{Y} = \boldsymbol{W}\boldsymbol{X}, \tag{9.340}$$

where $\boldsymbol{W}$ is a whitening $N \times N$ matrix.

The above Gram–Schmidt procedure performs whitening by using a simple transformation scheme and as a result decomposes the given matrix $\boldsymbol{x}$ into the product of an orthogonal matrix $\boldsymbol{Q}$ and upper diagonal matrix $\boldsymbol{R}$, that is,

$$\boldsymbol{X} = \boldsymbol{Q}\boldsymbol{R}. \tag{9.341}$$

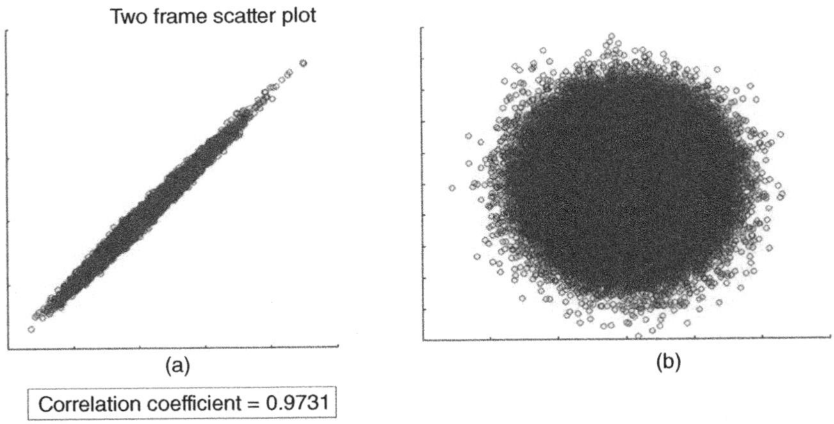

FIGURE 9.30 Scatter plot of two infrared images with system noise taken captured sequentially (a) before and (b) after whitening.

Using this procedure, it can be shown that the whitening filter can be expressed as

$$W = [R^T]^{-1}, \tag{9.342}$$

where $Y^T = W^* X^T$. [44, 45].

Figure 9.30 depicts scatter plots of two sequential infrared images with sensor noise (a) before and (b) after whitening. Figure 9.30a shows a strong correlation between the two images, with their correlation coefficient being 0.9731. One sees some broadening of the 45° "perfect correlation line" by the system noise, for example, signal shot noise and dark current, in the figure. Figure 9.30b shows minimal correlation due to the randomness of the scatter pattern, as represented by its circular shape of the scatter pattern. The reader is recommended to check out reference [44] for other techniques that can be used.

9.12 PROBLEMS

Problem 9.1. Prove the statement on page (5) that the average risk is minimized by choosing the hypothesis for which the conditional risk is the smaller.

Problem 9.2. Derive the conditions that the line of equal risk does not intersect the operating characteristic in the region $0 \ll Q_0 < 1, 0 < Q_d < 1$, and show that they lead to the inequalities

$$C_{11} < C_{01} < C_{00} < C_{10} \text{ (or) } C_{00} < C_{10} < C_{11} < 01.$$

Problem 9.3. Find the Bayes test to choose between the hypotheses, H_0 and H_1, whose prior probabilities are $\frac{5}{8}$ and $\frac{3}{8}$, respectively, when under H_0 the datum x had the PDF

$$p_0(x) = \sqrt{\frac{2}{\pi}}\, e^{\frac{-x^2}{2}}$$

and under H_1, it has PDF

$$p_1(x) = e^{-x},$$

x always being positive. Let the relative costs of the two kinds of errors be equal. Find the minimum average probability of error.

Problem 9.4. The random variables x and y are Gaussian with mean value 0 and variance 1. Their covariance may be 0 or some know positive value $r > 0$. Show that the best choice between these possibilities on the basis of measurement of x and y depends on where the point (x, y) lies with respect to a certain hyperbola in the (x, y)-plane.

Problem 9.5. A sequence of N independent measurements is taken from a Poisson distribution $\{x\}$ whose mean is m_0 under H_0, and m_1 under H_1. On what combination of the measurements should a Bayes test be based, and with what decision level should its outcome be compared, for given prior probabilities $(\xi, 1-\xi)$ and a given cost matrix $\boldsymbol{C}$?

NOTE: The Poisson distribution with mean m assigns a probability $p(x) = \frac{m^x e^{-m}}{x!}$ to the positive integers x and probability 0 to all noninteger values of x.

Problem 9.6. A random variable x is distributed according to the Cauchy distribution,

$$p_1(x) = \frac{m}{\pi(m^2 + x^2)}.$$

The parameter m can take on either of two values, m_0 or m_1, where $m_0 < m_1$. Design a statistical test to decide on the basis of a single measurement of x between the two Hypothesis $H_0(m = m_0)$ and $H_1(m = m_1)$. Use the Neyman–Pearson criterion. For this test, calculated the power $Q_d = 1 - Q_1$ as function of Q_0.

Problem 9.7. Under Hypotheses H_0 and H_1, a random variable has the following probability density functions

$$p_0(x) = 1 - |x|, \;\; |x| < 1, \;\; p_1(x) = \tfrac{(2-|x|)}{4}, \;\; |x| < 2$$
$$= 0, \;\; |x| > 1, \qquad = 0, \;\; |x| > 2.$$

Choosing H_0 when H_1 is true costs twice as much as choosing H_1 when H_0 is true. Correct choices cost nothing. Find the minimax strategy for deciding between the two hypotheses.

Problem 9.8. A choice is made between Hypotheses H_0 and H_1 on the basis of a single measurement x. Under Hypothesis H_0, $x = n$; under Hypothesis H_1, $x = s + n$. Here both s and n are positive random variables with the following PDFs:

$$p_0(x) = be^{-bn} \qquad p_1(x) = ce^{-cn}, \;\; b < c.$$

Calculate the PDFs under Hypotheses H_0 and H_1. Find the decision level on x to yield a given false alarm probability Q_0, and calculate the probability Q_d of correctly choosing Hypothesis H_1.

Problem 9.9. For the logarithm g of the likelihood ratio defined in Eq. (9.36), $e^g = \frac{p_1(x)}{p_0(x)}$. Define the moment generating function of g under Hypotheses H_0 and H_1 by

$$f_j(s) = \mathcal{E}\{e^{sg}|H_j\}, \quad j = 0, 1.$$

Show that $f_1(s) = f_0(s+1)$. Determine $f_0(s)$ for the logarithm of the likelihood ratio in Example (9.3).

Problem 9.10. Derive Eqs. (9.105) and (9.106).

Problem 9.11. Using Eq. (9.108), calculate the value of Q for (a) 10^{-9} and (b) 10^{-12}.

Problem 9.12. Determine the Q expression for preamplifier noise-limited receiver. Hint: Let $i_0 = 0$.

Problem 9.13. Show that if a circular aperture lens of diameter d is used when heterodyning with two match Airy patterns, the equivalent suppression factor for misaligned angles is proportional to

$$\int_{A_d} \left[\frac{J_1(|u|)}{|u|}\right] \left[\frac{J_1(|u-\rho_0|)}{|u-\rho_0|}\right] du,$$

where ρ_0 is the offset distance.

Problem 9.14. Show that

$$Q_{\text{fa}} = \frac{1}{2}\int_0^{\frac{Nq_0}{\sigma_T^2}} \left(\frac{v}{\lambda_0}\right)^{\frac{(N-2)}{4}} e^{\left[\frac{-(\lambda_0+v)}{2}\right]} I_{\frac{N}{2}-1}\left(\sqrt{\lambda_0 v}\right) dv$$

$$\rightarrow \frac{1}{\Gamma\left(\frac{N}{2}\right)}\int_0^{\frac{Nq_0}{2\sigma_T^2}} w^{\frac{N}{2}-1} e^{-w}\, dw = I\left(\sqrt{\frac{N}{2}}\left(\frac{q_0}{\sigma_T^2}\right), \frac{N}{2}-1\right),$$

as $\lambda_0 \to 0$. Here, $I(u,p)$ is the Pearson's form of the incomplete gamma function given by

$$I(u,p) = \frac{1}{\Gamma(p+1)}\int_0^{u\sqrt{p+1}} y^p e^{-y}\, dy.$$

Problem 9.15. Show that

$$\frac{1}{2}\int_0^{q_0} \frac{N}{2}\left(\frac{Nv}{\lambda}\right)^{\frac{(N-2)}{4}} e^{\left[\frac{-(\lambda+Nv)}{2}\right]} I_{\frac{N-2}{2}}\left(\sqrt{\lambda N v}\right) dv$$

$$\rightarrow \frac{N}{\sqrt{8\pi}}\int_0^{q_0} \frac{(Nv)^{\frac{(N-3)}{4}}}{\lambda^{\frac{(N-1)}{4}}} e^{\left[\frac{-(\sqrt{\lambda}-\sqrt{Nv})^2}{2}\right]} dv$$

for large even values of N.

Problem 9.16. Assume we have a 5×5 pixel target with equal signal level at every pixel, and background clutter with equal estimated mean as well. Specifically, we set $s_n = 6$ and $\widehat{\mu}_{bn} = 2$ for all values of n comprising the target. What is the resulting contrast? If we set the total noise variance σ_T^2 equal to 1, with the background noise variance σ_b^2 equal to 0.9 and a system noise variance σ_t^2 equal to 0.1, what is the resulting Contrast? Contrast noise ratio (CNR) in dB? Resulting background noise ratio (BNR) in dB?

Problem 9.17. Let Π be the plane in R^3 spanned by vectors $\boldsymbol{x}_1 = (1,2,2)$ and $\boldsymbol{x}_2 = (-1,0,2)$.

(i) Find an orthonormal basis for Π.

(ii) Extend it to an orthonormal basis for R3.

Hint: Let $\boldsymbol{x}_3 = (0,0,1)$.

Problem 9.18. Find the QR factorization of

$$A = \begin{bmatrix} 1 & 2 \\ 1 & 2 \\ 0 & 3 \end{bmatrix}.$$

REFERENCES

1. Rice, S.O. (1944) The mathematical analysis of random noise. *Bell System Technical Journal*, **23**, 282–332; **24**, 46–256, 1945.
2. Mandel, L. (1959) Fluctuations of photon beams; the distribution of photo-electrons. *Proceedings of the Physical Society*, **74**, 233–243.
3. Helstrom, C.W. (1964) The detection and resolution of optical signals. *IEEE Transactions on Information Theory*, **10** (**4**), 275–287.
4. Lohmann, A.W. (2006) in *Optical Information Processing* (ed. S. Sinzinger), Universitätsverlag Ilmenau, see: http://www.db-thueringen.de/servlets/DerivateServlet/Derivate-13013/Lohmann_Optical.pdf (accessed 14 October 2016)..
5. Papoulis, A. (1968) *Systems and Transforms with Applications in Optics*, McGraw-Hill Book Company, New York.
6. Helstrom, C.W. (1995) *Elements of Signal Detection & Estimation*, Prentice Hall, Englewood Cliffs, New Jersey.
7. Helstrom, C.W. (1968) *Statistical Theory of Signal Detection*, Pergamon Press, New York, Chapter 3..
8. McDonough, R.N. and Whalen, A.D. (1995) *Detection of Signals in Noise*, 2nd edn, Academic Press, Inc., An Imprint of Elsevier, San Diego, CA.
9. Abramowitz, M. and Stegun, I.A. (1970) *Handbook of Mathematical Function with Formulas, Graphs, and Mathematical Tables*, U.S. Government Printing Office, Washington, DC.
10. Karp, S. and Stotts, L.B. (2012) *Fundamentals of Electro-Optic Systems Design: Communications, Lidar, and Imaging*, Cambridge Press, New York.
11. Majumdar, A.K. (2015) in *Advanced Free Space Optics (FSO) A Systems Approach*, Springer Series in Optical Sciences, Volume 186 (ed. W.T. Rhodes), Springer, New York.
12. Alkholidi, A.G. and Altowij, K.S. (2014) in *Contemporary Issues in Wireless Communications*, Chapter 5 (ed. M. Khatib), InTech, Rijeka, Croatia 252 pages. ISBN: ISBN 978-954-51-1732-2
13. Majumdar, A. and Ricklin, J. (2008) *Free Space Laser Communications: Principles and Advances*, Optical and Fiber Communications Series, Springer Science+Business Media, New York.
14. Boroson, D.M. (2005) *Optical Communications, A Compendium of Signal Formats, Receiver Architectures, Analysis Mathematics, and Performance Comparison*, MIT.
15. Zhang, S. (2004) Advanced optical modulation formats in high-speed lightwave systems, Thesis, The University of Kansas.
16. Gagliardi, R.M. and Karp, S. (1995) *Optical Communications*, 2ndWiley Series in Telecommunications and Signal Processing edn, John Wiley and Sons, New York.

17. Karp, S., Gagliardi, R.M., Moran, S.E., and Stotts, L.B. (1988) *OPTICAL CHANNELS: Fiber, Atmosphere, Water and Clouds* Chapter 5, Plenum Publishing Corporation, New York.
18. William, K. (1969) *Pratt, Laser Communications Systems*, John Wiley & Sons, Inc., New York.
19. Agrawal, G.P. (2007) *Optical Communications Systems*, OPT428 Course Note, Source: http://www.optics.rochester.edu/users/gpa/opt428c.pdf (accessed 17 October 2016).
20. Shannon, C.E. (1948) A mathematical theory of communications. *Bell System Technical Journal*, **27**, 379–423 & pp. 623–656.
21. Zhu, X. and Kahn, J.M. (2003) Markov chain model in maximum-likelihood sequence detection for free-space optical communication through atmospheric turbulence channels. *IEEE Transactions on Communications*, **51** (**3**), 509–516. Also, published as a chapter in reference [13].
22. Juarez, J.C., Young, D.W., Sluz, J.E., and Stotts, L.B. (2011) High-sensitivity DPSK receiver for high-bandwidth free-space optical communication inks. *Optics Express*, **19** (**11**), 10789–10796.
23. Bagley, Z.C., Hughes, D.H., Juarez, J.C. *et al.* (2012) Hybrid optical radio frequency airborne communications. *Optical Engineering*, **51**, 055006-1–055006-25. doi: 10.1117/1.OE.51.5
24. Saleh, B.E.A. and Teich, M.C. (2007) *Fundamentals of Photonics*, 2ndWiley Series in Pure and Applied Optics edn, John Wiley & Sons, New York.
25. Gatt, P. and Henderson, S.W., Laser radar detection statistics: A comparison of coherent and direct-detection receivers, *Proc. SPIE 4377*, Laser Radar Technology and Applications VI, pp. 251–162 (2001); DOI: 10.1117/12.440113.
26. Measures, R.M. (1992) *Laser Remote Sensing: Fundamentals and Applications*, Krieger Publishing Company, Malabar, FL 510 pages.
27. Andrews, L.C. (2011) *Field Guide to Special Functions for Engineers*, SPIE Field Guides VOL FG18, John E. Greivenkamp, Series Editor, SPIE Press, Bellingham, WA.
28. Goodman, J.W. (1965) Some effects of target-induced scintillation on optical radar performance. *Proceedings of the IEEE*, **53** (**11**), 1688–1700.
29. Hallenborg, E., Buck, D.L., and Daly, E. (2011) Evaluation of automated algorithms for small target detection and non-natural terrain characterization using remote multi-band imagery. *Proceedings of SPIE*, **8137**, 813711.
30. Kabate, M., Azimi-Sadjadi, M.R., and Tucker, J.D. (2009) *An underwater target detection system for electro-optical imagery data*, in OCEANS 2009, MTS/IEEE Biloxi—Marine Technology for Our Future: Global and Local Challenges, 26–29 October, 2009, pp. 1–8.
31. Fifteen Tri-bar Arrays, Photo Resolution Range, target at Eglin Air Force Base, and at Edwards Air Force Base, CA (from west to east), 2012, 15 images, 8 × 10 inches each, imagery from Google Earth.
32. Stotts, L.B. and Hoff, L.E. (2014) Statistical detection of resolved targets in background clutter using optical/infrared imagery. *Applied Optics*, **53** (**22**), 5042–5052.
33. Smith, S.W. (1997) Special imaging techniquesChapter 25, in *The Scientist and Engineer's Guide to Digital Signal Processing*, California Technical, Carlsbad, CA, pp. 433–434.
34. Reed, I.S. and Yu, X. (1990) Adaptive multiple-band CFAR detection of an optical pattern with unknown spectral distribution. *IEEE Transactions on Acoustics, Speech, and Signal Processing*, **38**, 1760–1770.

35. Yu, X., Reed, I.S., and Stocker, A.D. (1993) Comparative performance analysis of adaptive multispectral detectors. *IEEE Transactions on Signal Processing*, **41**, 2639–2656.
36. Stocker, A.D., Reed, I.S., and Yu, X. (1990) Multidimensional signal processing for electro-optic target detection. *Proceedings of SPIE*, **1305**, 218.
37. Hoff, L.E., Evans, J.R., and Bunney, L.E. (1990) Detection of Targets in Terrain Clutter by Using Multi-Spectral Infrared Image Processing, Naval Ocean Systems Center (NOSC) Report TD 1404.
38. Winter, E. and Schlangen, M.J. (1993) Detection of extended targets using multi-spectral infrared. *Proceedings of SPIE*, **1954**, 173–180.
39. Hoff, L.E., Chen, A.M., Yu, X., and Winter, E.M. (1995) Generalized Weighted Spectral Difference Algorithm for Weak Target Detection in Multiband Imagery. *SPIE Proceedings*, **2561**, 141–152.
40. Hoff, L.E., Chen, A.M., Yu, X., and Winter, E.M. (1995) *Enhanced Classification Performance from Multiband Infrared Imagery*, Proceedings on the Twenty-Ninth Asilomar Conference on Signals, Systems & Computers Proceedings, 30 October-1 November, 1995, Volume 2, pp. 837–841
41. Stotts, L.B. (1988) Moving target indication techniques for optical image sequences, PhD Dissertation, University of California, San Diego.
42. Stotts, L.B. (2015, pp. 103109-1 to 103109-13) Resolved target detection in clutter using correlated, dual-band imagery. *Optical Engineering*, **54** (**10**), 103109. doi: 10.1117/1.OE.54.10.103109
43. Yu, X., Hoff, L.E., Reed, I.S. *et al.* (1997) Automatic target detection and recognition in multi-band imagery: A unified ML detection and estimation approach. *IEEE Transactions on Image Processing*, **6** (**1**), 143–156.
44. Cichocki, A., Osowski, S., and Siwek, K. (2004) *Prewhitening Algorithms of Signals in the Presence of White Noise*, VI International Workshop Computational Problems of Electrical Engineering, Zakopane, pp. 205–208.
45. Leksachenko, V.A. and Shatalov, A.A. (1976) Synthesis of multidimensional 'whitening' filters according to the Gram–Schmidt method. *Radiotekhnika i Elektronika*, **21**, 112–119(Also translated and published in Radio Engineering and Electronic Physics, vol. 21, pp. 92–99, Jan 1976)..
46. Cheney, W. and Kincaid, D. (2009) *Linear Algebra: Theory and Applications*, Jones and Bartlett, Sudbury, Ma, pp. 544–558. ISBN: 978-0-7637-5020-6
47. Harvey Mudd College Math Tutorial on the Gram–Schmidt algorithm, Source: https://www.math.hmc.edu/calculus/tutorials/gramschmidt/gramschmidt.pdf (accessed 17 October 2016).

10

LASER SOURCES

10.1 INTRODUCTION

In Chapter 4, we described blackbody radiation that characterizes light generation from the sun, stars, planets, terrain, reflected solar radiations from objects, and other incoherent optical sources. Because of broadband and generally large angle (without a large lens system) light missions, envisioned optical systems could not compete well with equivalent RF systems in terms of cost, range, system performance, and size, weight, and power (SWaP). An alternative source concept to improve things significantly was proposed in 1917 by Albert Einstein [1], but it took almost 60 years before it became a reality and revolutionize the optics industry. It is the *laser*, which is an acronym for Light Amplification by Stimulated Emission of Radiation. Increased range and system performance by its use emerged immediately to better compete with analogous RF systems. However, it was Fiber Optics Communications (FOC) that created the breakthrough laser technologies, for example, semiconductor lasers, and Erbium-doped fiber lasers and amplifiers, that addressed cost and SWaP, making optics a more competitive and desirable technology option for remote sensing [2–4] and communications [5–7].

As the acronym suggests, the key process for creating this new optical source is *stimulated emission*. Albert Einstein first broached the possibility of stimulated emission in his 1917 paper that investigated the possible interactions between matter and radiation. Specifically, he devised an improved fundamental statistical theory of heat, embracing the quantum of energy. Here is a summary of his theory.

First, he proposed *spontaneous emission*. This is where an electron in an elevated state in isolation can return a lower energy state by emitting a photon wave

Free Space Optical Systems Engineering: Design and Analysis, First Edition. Larry B. Stotts.

Companion website: www.wiley.com\go\stotts\free_space_optical_systems_engineering

packet. Second, the reverse process obviously is possible, which is deemed *photon absorption*, where a photon wave packet is captured by the electron to elevate it to a higher state. To obtain either light emission or absorption, the photon wave packet must have a discrete frequency (or wavelength) that is related to the energy difference between the lower and upper energy states (we will discuss this and its variations in more detail shortly). Third, and the most significant facet, is that if photon wave packet of the same frequency passes in the vicinity of the atom with this elevated state electron, it could stimulate a sympathetic emission of wave packet at the same frequency. Specifically, he postulated that under certain circumstances, a photon wave packet in close proximity to an excited atom could generate a second photon of exactly the same (1) energy (frequency), (2) phase, (3) polarization, and (4) direction of propagation. In other words, a coherent beam resulted. This was hint that something important was possible if things were done right.

Unfortunately, it was not until the 1940s and 1950s that physicists found a use for the concept, even though all that was required to invent a laser was finding the right kind of atom, and adding reflecting mirrors to fortify the stimulated emission process by producing a chain reaction. This did not happen right away. Charles Townes researched radar systems during WWII and then began investigating microwave molecular spectroscopy, a technique that studies the absorption of light by molecules, just after the war. At that time, he noticed that as the wavelength of the microwaves shortened, the more strongly the radiation interacted with the molecules. He then thought it might be possible to develop a device that produced radiation at much shorter wavelengths. He conjectured that the best way to do this was to use molecules to generate the desired frequencies through stimulated emission. After discussing this idea with his colleague, Arthur Schawlow, the latter suggested that a prototype laser could be constructed using a pair of mirrors, one at each end of the lasing cavity, to build up the energy of the confined beam. That is, photon wave packets of specific wavelengths would reflect off the mirrors at each end, thereby traveling back and forth through the lasing medium building beam energy via stimulated emission. Unfortunately, they never built a prototype device, but just published a paper on their concept in the December 1958 issue of the Physical Review. Although they received the patent for their concept 2 years later, it was Theodore Maiman of Hughes Aircraft Company who brought the concept to fruition. Specifically, he demonstrated a 694n ruby laser in 1960, validating their concept. Its first applications were in the areas of Ophthalmology and Dermatology. There are many great books on lasers, and a sampling is provided as Refs [8–18].

In this chapter, we provide an overview of the fundamentals of laser theory, the key source for all communications and remote sensing applications.

10.2 SPONTANEOUS AND STIMULATED EMISSION PROCESSES

This section discusses how laser operations are derived from a collection of atoms experiencing spontaneous and stimulated emissions in the presence of different absorption routes.

10.2.1 The Two-Level System

Let us begin with a two-level atom. As stated above, absorption deals with an electron going from the ground state to an elevated state and emission involving dropping the

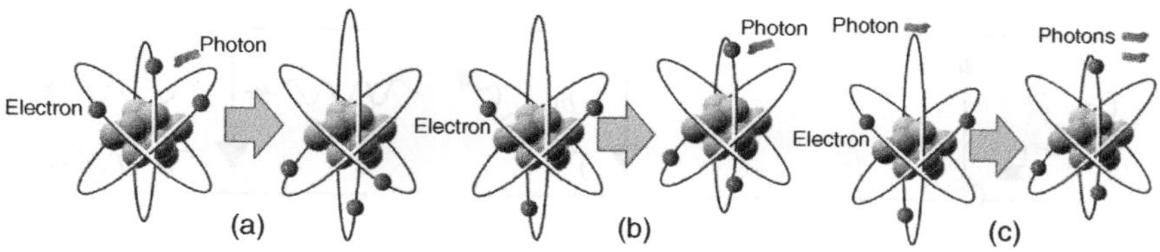

FIGURE 10.1 Bohr model representation of (a) absorption, (b) spontaneous emission, and (c) stimulated emission.

electron from the elevated state to the ground state. The former promotes electron jumping to a higher energy by decreasing the number of photons in the medium. The latter reduces the number of electrons in the elevated state through the emission of photons; this emission requires no other event to create photon. The photon can leave the atom in any direction about the atom. As we found in our discussion of radiation fluctuation of blackbody, or a photoelectron creation from an optical detector, one cannot predict *a priori* when that emission will occur in; it will follow a Poisson distribution in time. If photon of the same frequency passes in the vicinity of the atom with this elevated state electron, we get stimulated emission where the photon will be synchronized with the original photon's trajectory. Figure 10.1 illustrates the Bohr model representation of (a) the absorption process, (b) the spontaneous emission process, and (c) the stimulated emission process. These images show that electrons orbit a nucleus that comprised protons and neutrons. When a photon is absorbed (a), one of the electrons goes from its current orbit (ground state) to a larger diameter orbit (excited state). On spontaneous emission (b), an electron in a higher diameter orbit (excited state) drops down to a lower orbit (ground state) while emitting a photon. Finally, when a photon passes in a close proximity to atom with an electron in the higher orbit state (excited state), the said electron drops down to a lower orbit (ground state) while emitting a photon synchronized to the original close proximity photon.

Today, we use the energy state diagrams based on Quantum Mechanics to describe the above process, which is similar to how we described photoelectron generation in optical detectors. Because of the conservation of energy, this theory states that the energy required to allow electron elevation from the ground states to an elevated state equals

$$E_{\text{elevated}} - E_{\text{ground}} = E_{\text{Photon}} = h\nu = h\frac{c}{\lambda}, \tag{10.1}$$

where ν is the optical frequency of the absorbed or emitted photon wave packet, λ is the associated wavelength to frequency ν, and $c (= \nu\lambda)$ is the speed of light, as before. The above equation relates to the probability that a photon of energy E_{Photon} will excite (or drop) an electron from its initial state to its final state, which will depend on the properties of the two states, for example, spin, parity, and energy difference.[1] Figure 10.2 depicts the same situations in Figure 10.1, but from a quantized energy state POV. If we have many atoms with electrons in an excited state, which is known as a *population inversion* to an elevated state, then spontaneous emission will initially

[1] A detailed discussion of Quantum Mechanics is a book itself and beyond the scope of this introductory text. We therefore will ask the reader to accept the stated premises without the complete details.

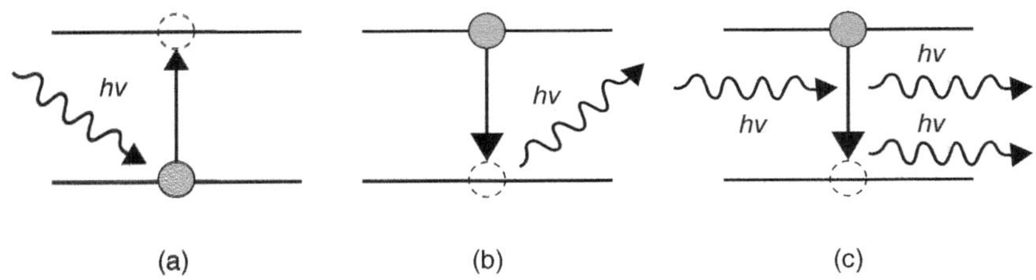

FIGURE 10.2 Quantized energy state representation of (a) absorption, (b) spontaneous emission, and (c) stimulated emission.

occur, but some of the photons will begin creating a number of stimulated photon emissions. If we apply Schawlow's idea of confining atoms between a pair of mirrors, then these synchronized photons grow in a geometric series, that is, 2, 4, 8, 16, 32, and a coherent, high intensity beam forms. The spontaneous emission will continue to occur as random events and will form a glow around the laser medium; its intensity will not increase with time, but just remain in a steady state. Let us now look at the general rate of emission for medium filled with excited atoms.

An excited state can relax to the ground state by (1) spontaneous (radiative) emission at the rate γ_{rad} and (2) by a variety of nonradiative pathway described the rate γ_{nrad}. All of these processes are independent, single-atom processes. The evolution of the excited state population only depends on the number of atoms in the excited state. Mathematically, we can write

$$\frac{dN_e}{dt} = -\gamma_{\text{rad}} N_e - \gamma_{\text{nrad}} N_e = -\gamma_{10} N_e, \tag{10.2}$$

where N_e is the number of electrons in the excited state and γ_{10} represents the total spontaneous relaxation rate from state 2 to state 1.

Let us now look at the transition probabilities of the electrons from an initial state "1" to a final state "2." This can be addressed by perturbation theory. Specifically, if we assume a perturbation $h(t)$ interacting with an atom, a system in energy state "1" makes a transition to energy state "2" with the probability equal to

$$P_{1\to 2} = \frac{1}{h^2} \left| \int_{-\infty}^{t} e^{\frac{iE_{1\to 2} t'}{h}} h(t')\, dt' \right|^2. \tag{10.3}$$

Example 10.1 Assume that we have two-level atom collection whose electrons initially are in energy state "1." If we subject this collection to a harmonic perturbation equal to

$$h(t) = \begin{cases} 0 & \text{for}\, t < 0 \\ 2A_0 \sin 2\pi \nu t & \text{for}\, t > 0 \end{cases}, \tag{10.4}$$

where the frequency of the perturbation ν is close to $\nu_{1\to 2} = \frac{E_{1\to 2}}{h}$. This implies that the transition probability from going to state "1" to state "2" can be written as

$$P_{1\to 2} = \frac{4A_0^2}{h^2} \left| \int_0^t e^{2\pi i \nu_{1\to 2} t'} \left[e^{2\pi i \nu t'} - e^{-2\pi i \nu t'} \right] dt' \right|^2 \tag{10.5}$$

$$\approx \frac{4A_0^2}{(2\pi)^2h^2} \frac{\sin^2(2\pi i(\nu_{1\to2} - \nu)t)}{(\nu_{1\to2} - \nu)^2} \tag{10.6}$$

as the $(\nu_{1\to2} + \nu)$ -term is negligible because of the frequencies involved. Reducing the above for the $P_{2\to1}$ gives the same result. Hence, given the presence of photon can create absorption or stimulated emission with equal likelihood.

Assume now that we have a radiation field and this collection of two-level atoms, which are in thermal equilibrium with each other. In this case, the stimulated emission probability will depend on both the number of electrons in the excited state, which we denote by N_2 and the number of photons incident on the collection of atoms. On the other hand, the spontaneous emission probability also is proportional to the value of the inverted population N_2, but not on the photon density hitting the collection of atoms. This implies that the change per unit time in the population $2 \to 1$ is given by

$$W_{2\to1} = -A_{21}N_2 - B_{21}N_2\,\rho(\nu), \tag{10.7}$$

where A_{21} and B_{21} are the Einstein Coefficients for the transition from $2 \to 1$, and $\rho(\nu)$ denotes the spectral energy density of the isotropic radiation field at the frequency of the transition, that is, Planck's law. The coefficient A_{21} gives the probability per unit time that an electron in state 2 with energy E_2 that will decay spontaneously to state 1 with energy E_1, emitting a photon with the energy $E_2 - E_1 = h\nu$. It is the same as γ_{rad}. The coefficient B_{21} gives the probability per unit time per unit spectral energy density of the radiation field that an electron in state 2 with energy E_2 will decay to state 1 with energy E_1, emitting a photon with an energy $E_2 - E_1 = h\nu$.

The stimulated absorption probability is proportional to the number of the electrons in the ground state N_1 and also the number of photons present. Since there is no such thing as spontaneous absorption, the change per unit time in the population $1 \to 2$ is given by

$$W_{1\to2} = -B_{12}N_1\,\rho(\nu), \tag{10.8}$$

where B_{12} gives the probability per unit time per unit spectral energy density of the radiation field that an electron in state 1 with energy E_1 will absorb a photon with an energy $E_2 - E_1 = h\nu$ and jump to state 2 with energy E_1.

Thus, how do Eqs. (10.7) and (10.8) relate? Given the fact that we have specified that the radiation field and collection of atoms are in thermal equilibrium, these equations for the upward and downward transition rates must balance, or,

$$W_{1\to2} = W_{2\to1}. \tag{10.9}$$

This means that

$$-A_{21}N_2 - B_{21}N_2\,\rho(\nu) = -B_{12}N_1\,\rho(\nu) \tag{10.10}$$

or

$$\frac{N_1}{N_2} = \frac{A_{21} + B_{21}\,\rho(\nu)}{B_{12}\,\rho(\nu)}. \tag{10.11}$$

Planck's law of blackbody radiation at temperature T allows us to write the spectral energy density $\rho(\nu)$ as

$$\rho(\nu) = \frac{8\pi h\nu^3}{c^3}\left[\frac{1}{e^{\frac{(E_2-E_1)}{kT}}-1}\right] = \frac{8\pi h\nu^3}{c^3}\left[\frac{1}{e^{\frac{h\nu}{kT}}-1}\right]. \quad (10.12)$$

Substituting Eq. (10.12) into Eq. (10.11) and rearranging terms, we obtain

$$\frac{N_1}{N_2} = \frac{A_{21}\left(\frac{c^3}{8\pi h\nu^3}\right)\left[e^{\frac{h\nu}{kT}}-1\right]+B_{21}}{B_{12}}. \quad (10.13)$$

The above equation must hold at any temperature, so

$$B_{12} = B_{21} = B, \quad (10.14)$$

and we can write

$$\frac{N_1}{N_2} = \frac{A_{21}\left(\frac{c^3}{8\pi h\nu^3}\right)\left[e^{\frac{h\nu}{kT}}-1\right]+B}{B}. \quad (10.15)$$

Solving for A_{12}, we find that

$$A_{21} = B\left[\frac{N_1}{N_2}-1\right]\frac{\left(\frac{8\pi h\nu^3}{c^3}\right)}{\left[e^{\frac{h\nu}{kT}}-1\right]}. \quad (10.16)$$

Recall that the *Boltzmann's law* describes the frequency distribution of particles in a system over various possible states. Mathematically, the distribution can be expressed as

$$F(E) \propto e^{e^{\frac{h\nu}{kT}}}, \quad (10.17)$$

where E again is the state energy, k Boltzmann's constant, and T the thermodynamic temperature. The *Boltzmann factor* is the ratio of the Boltzmann distribution for two specific states, which results in function dependent only on the energy difference between the two states in thermal equilibrium. Mathematically, it is written as

$$\frac{F(E_1)}{F(E_2)} = e^{e^{\frac{(E_2-e_1)}{kT}}} = e^{\frac{h\nu}{kT}} = \frac{N_1}{N_2}. \quad (10.18)$$

Substituting Eq. (10.18) into Eq. (10.16) yields

$$A_{21} = B\left[\frac{N_1}{N_2}-1\right]\frac{\left(\frac{8\pi h\nu^3}{c^3}\right)}{\left[e^{\frac{h\nu}{kT}}-1\right]} = \left(\frac{8\pi h\nu^3}{c^3}\right)B, \quad (10.19)$$

which is in units of energy density per unit bandwidth. Since $A_{21} = \gamma_{\mathrm{rad}}$, we find that

$$B = \left(\frac{\gamma_{\mathrm{rad}}\, c^3}{8\pi\, h\, \nu^3} \right). \tag{10.20}$$

Given the above, we can write the downward transition rate as

$$\begin{aligned} W_{2\to 1} &= -\left(\frac{8\pi h \nu^3}{c^3}\right) N_2 - \left(\frac{\gamma_{\mathrm{rad}}\, c^3}{8\pi\, h\, \nu^3}\right) N_2 \frac{8\pi h\, \nu^3}{c^3} \left[\frac{1}{e^{\frac{h\nu}{kT}} - 1}\right] \\ &= -\gamma_{\mathrm{rad}} N_2 - N_2 \left[\frac{\gamma_{\mathrm{rad}}}{e^{\frac{h\nu}{kT}} - 1}\right] = -\gamma_{\mathrm{rad}} N_2 \left[1 + \frac{1}{e^{\frac{h\nu}{kT}} - 1}\right] \\ &= -\gamma_{\mathrm{rad}} N_2 \,[1 + \overline{n}_1], \end{aligned} \tag{10.21}$$

where $\overline{n}_1$ is the mean number of photons in a single quantum state, in this case the ground state. In the above,

$$\overline{n}_1 = \frac{1}{e^{\frac{h\nu}{kT}} - 1}$$

is the mean number of photon equation given in Eq. (4.72) that defines the well-known Bose–Einstein distribution. In other words, $W_{2\to 1}$ is proportional to 1 plus the average number of photons incident on the collection of atoms. It is easy to show that

$$W_{2\to 1} = -\gamma_{\mathrm{rad}} N_2 \overline{n}_1. \tag{10.22}$$

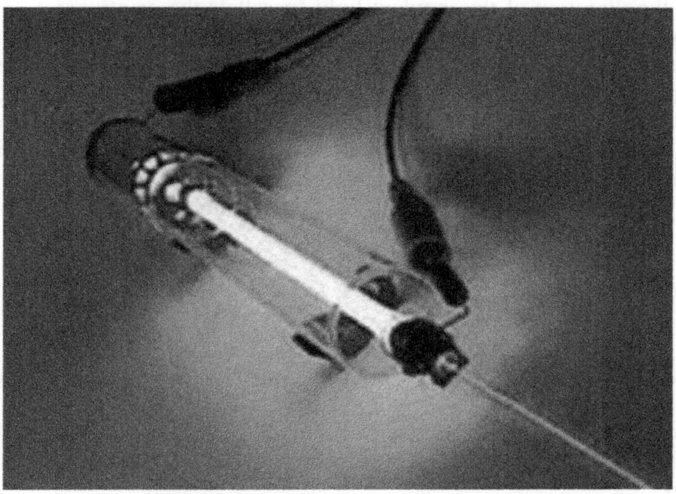

FIGURE 10.3 Insides of a helium–neon gas laser showing the isotropic spontaneous emissions and the directional stimulated emissions within the laser cavity.

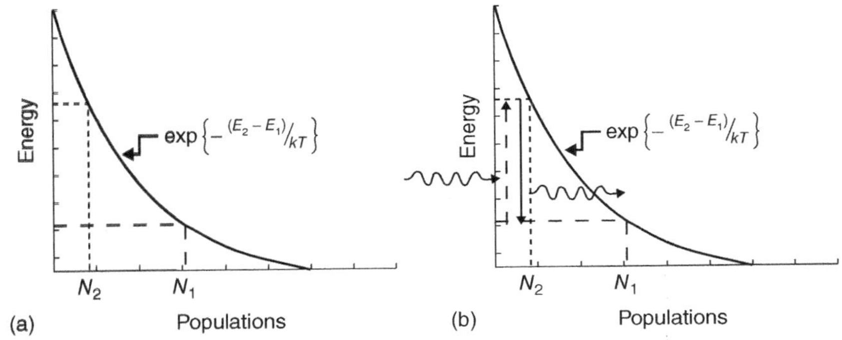

FIGURE 10.4 Graphs of energy versus energy state population following a Boltzmann distribution (a) with and (b) without absorption and stimulated energy diagrams.

Returning to the original rate equation given by Eq. (10.2), we now can write

$$\frac{dN_e}{dt} = -\gamma_{21} N_e - B I_p N_e + B I_p N_g = -\frac{dN_g}{dt} \tag{10.23}$$

$$= -\gamma_{21} N_e - B I_p [N_e - N_g] \tag{10.24}$$

with N_g being the number of electrons in the ground state, n_p is the photon density and where $I_p(= h\nu n_p)$ is the intensity of the photon stream. The first term in Eq. (10.24) represents the emitted photons emitted in all directions, the glow pattern around the laser medium that we cited earlier. The $K_p n_p N_g$ term represents the emitted photons in the direction of the incident light (parallels the cavity axis). Figure 10.3 shows the insides of helium–neon (HeNe) laser that lases at 6328 Å (Red). The light shown on the table holding the laser shows part of the spontaneous emission glow from the laser cavity; the laser beam exiting the laser shown is unidirectional and represents a fraction of the light created by the bidirectional) stimulated emissions (back and forth between the mirrors) in the laser cavity. Therefore, focusing solely on the stimulated emissions activity in the active medium, the number of photons varies according to a reduced version of the above, namely,

$$\frac{dn_p}{dt} = -B I_p [N_g - N_e] = -B I_p \Delta N. \tag{10.25}$$

The solution of Eq. (10.25) shows that the number of photons grows exponentially if $N_e > N_g$.

In thermal equilibrium, the relative populations N_g and N_e for a two-level system are given by the Boltzmann distribution, which is illustrated in Figure 10.4a. This implies that

$$N_e = e^{-\frac{\Delta E}{kT}} N_g < N_g. \tag{10.26}$$

For the visible and infrared spectrum below 1 μm, the Boltzmann factor is $< e^{-45}$, which shows that a population inversion is virtually impossible.

In a steady-state condition, we have

$$\frac{dN_e}{dt} = 0 = -\gamma_{21} N_e - BI_p [N_e - N_g]$$

or

$$N_e = \frac{BI_p}{\gamma_{21} + BI_p} N_g < N_g, \tag{10.27}$$

which is illustrated in Figure 10.4b. That is, you can never get the Boltzmann factor above 50% because the stimulated emission and absorption rates are equal, no net gain. This again shows that a population inversion is impossible.

10.2.2 The Three-Level System

Therefore, how do we get the situation to change to desired population inversion level? In order to achieve $N_e > N_g$, one needs to populate the upper state indirectly, that is, not to use the photon creating the stimulated emissions, and fill it faster than the spontaneous emissions can deplete it. We can illustrate this by taking a high-level look what can happen in a three-level system.

In this case, we couple the ground state ("1") to an intermediate state, say "3," which relaxes to our desire excited electron level ("2"). We now can make the desired $2 \rightarrow 1$ transition making light at the desire frequency (or wavelength). This is depicted in Figure 10.5a. A critical requirement is that the relaxation time for the $2 \rightarrow 1$ transition to be relatively slow (Figure 10.5b), that is, $\tau_{32} \ll \tau_{21}$. Looking at the transition rate equations again, we have

$$\frac{dN_2}{dt} = BI_p N_1 - AN_2 \tag{10.28}$$

and

$$\frac{dN_1}{dt} = -BI_p N_1 + AN_2. \tag{10.29}$$

The first term in each of these equations relates to the absorption process and the second term in each equation relates to spontaneous emission. Let the total number of electrons in all energy states be denoted by N and be given by

$$N \approx N_1 + N_2. \tag{10.30}$$

under the assumption that electrons in energy state "3" decay quickly to energy state "2," but remain there for a relatively long time. If we define

$$\Delta N \approx N_1 - N_2, \tag{10.31}$$

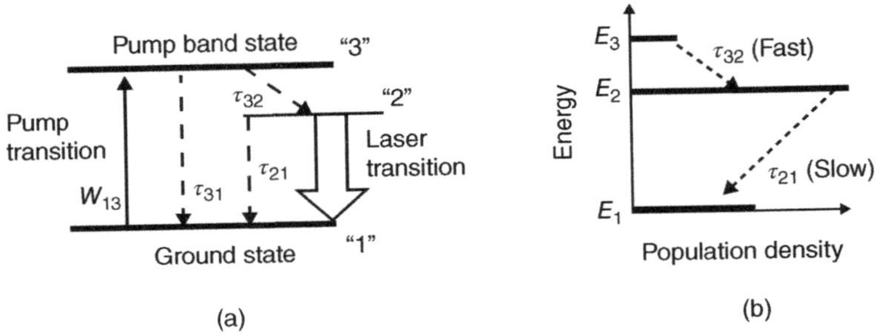

FIGURE 10.5 (a) Energy state diagram depicting pump and laser transition, as well as spontaneous emission between the three-energy states and (b) histogram of energy versus population density highlighting key decay times.

then we write

$$\frac{d\Delta N}{dt} \approx -2BI_pN_1 + 2AN_2 \tag{10.32}$$

$$\frac{d\Delta N}{dt} \approx -BI_pN - BI_p\Delta N + AN - A\Delta N. \tag{10.33}$$

When we now have a steady-state condition, we have

$$\frac{d\Delta N}{dt} = 0 = -BI_pN - BI_p\Delta N + AN - A\Delta N$$

or

$$(A + BI_p)\Delta N = (A - BI_p)N,$$

which implies

$$\Delta N = N\frac{(A - BI_p)}{(A + BI_p)} = N\frac{\left(1 - \left(\frac{B}{A}\right) I_p\right)}{\left(1 + \left(\frac{B}{A}\right) I_p\right)} = N\frac{\left(1 - \left(\frac{I_p}{I_{\text{sat}}}\right)\right)}{\left(1 + \left(\frac{I_p}{I_{\text{sat}}}\right)\right)} \tag{10.34}$$

with

$$\frac{B}{A} = \frac{c^3}{8\pi h \nu^3} \equiv I_{\text{sat}}^{-1}. \tag{10.35}$$

In the above, I_{sat} is called a saturation intensity. This is because if $I_p < I_{\text{sat}}$, then $\Delta N > 0$ and that the ground state has most of the electrons, that is, the ground state is essentially "saturated." However, if $I_p > I_{\text{sat}}$, then $\Delta N < 0$, and population inversion is assured! Hence one can view the gain of the laser medium as proportional to ΔN. However, a significant disadvantage of three-level lasers is the very high pumping levels needed to create the inversion. In particular, for each photon emitted, you lose population in the upper level and gain population in the ground state. This double "hit" does not occur in four-level lasers, which is our next topic.

10.2.3 The Four-Level System

Let us now assume a four-level system and do a more detailed analysis than the above for a three-level system. Figure 10.6 illustrates the various transitions for a four-level system. Let us define R_p as the effective pumping rate for level "2," which involves pumping electrons from the ground state to energy state "3," where they experience a very rapid, nonradiative decay to level "2." Also before, the relaxation time in level "2" is very long compared to that of level "3." Similar to the three-level systems, emission of a photon is derived from the transition from energy state "2" to "1." However, for a four-level system, there now is an additional, quick, nonradiative decay from state "1" to the ground state "0."

The process starts out with state "1" virtually empty and all the electron movement goes to state "3," decaying quickly to create a population inversion in state "2." This essentially can be viewed as a three-level system enhanced by an empty target state. In other words, energy level "1" is not populated for any reasonable amount of time, that is, $N_1 \approx 0$, and essentially we only have to deal with combined populations of the ground and excited laser states, that is, $N \approx N_2 + N_0$.

The transition rate equation for the above situation reduces to just following energy state equation:

$$\frac{dN_2}{dt} \approx BI_pN_0 - AN_2 \tag{10.36}$$

$$\approx BI_p(N - N_2) - AN_2. \tag{10.37}$$

Since $N_1 \approx 0$, we can write Eq. (10.31) as

$$\Delta N \approx N_1 - N_2 \approx -N_2 \tag{10.38}$$

and Eq. (10.37) becomes

$$-\frac{d\Delta N}{dt} \approx BI_p(N + \Delta N) + A\Delta N. \tag{10.39}$$

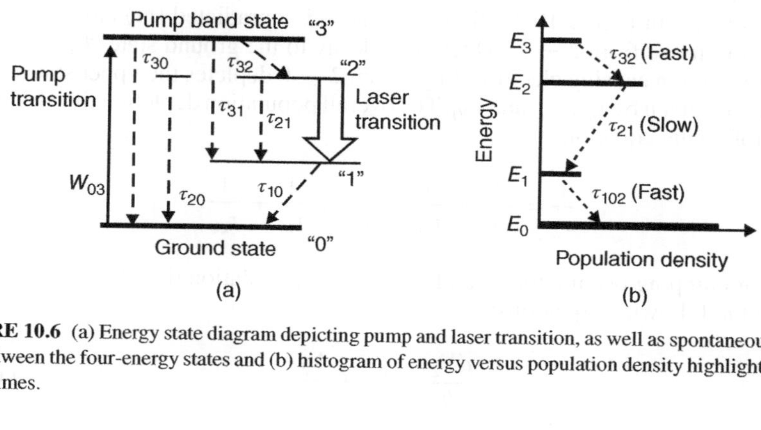

FIGURE 10.6 (a) Energy state diagram depicting pump and laser transition, as well as spontaneous emission between the four-energy states and (b) histogram of energy versus population density highlighting key decay times.

Under steady-state conditions, we have

$$0 \approx BI_p(N + \Delta N) + A\Delta N = (A + BI_p)\Delta N + BI_pN, \tag{10.40}$$

which means that

$$\Delta N \approx \frac{-BI_pN}{(A + BI_p)} = -\left(\frac{B}{A}\right)\left[\frac{I_pN}{\left(1 + I_p\left(\frac{B}{A}\right)\right)}\right] = -N\left[\frac{\left(\frac{I_p}{I_{\text{sat}}}\right)}{\left(1 + \left(\frac{I_p}{I_{\text{sat}}}\right)\right)}\right]. \tag{10.41}$$

Equation (10.41) states that ΔN is negative all the time, which guarantees population inversion is there all the time! This is a great advantage over three-level laser systems. A good example of a four-level laser is a Neodymium-doped Yttrium Aluminum Garnet (Nd : $Y_3Al_{15}O_{12}$ or Nd : YAG) laser, which is the workhorse laser in many remote sensing systems.

10.3 LASER PUMPING

It is apparent from the previous section that we need external pumping to make a laser work. Let us look at pumping in more detail following Piot [19].

10.3.1 Laser Pumping without Amplifier Radiation

Laser pumping moves electrons to the desired excited state "1" either directly or indirectly. This is shown in Figure 10.7a. Our desire is to populate energy level "1." One can populate this state directly from the ground state (left arrow in Figure 10.7a) or indirectly by populating energy state "2" and let spontaneous decays from this state to energy state "1" (see right arrows in Figure 10.7a). We saw in the last section that laser pumping dynamics are described by the rate equations, which quantify the change of population densities N_1 and N_2. We use that approach here as well.

If the medium is unpumped, eventually all atoms we are pushing to the upper states will go to ground state "0," and $N_1 = N_2 = 0$. The processes involved are characterized better in Figure 10.7b. Spontaneous and nonradiated (nr) emissions cause electrons to move from $2 \to 1$, which then decay to the ground state "0," and spontaneous emission causing electrons to go from $2 \to 0$ depletes the upper states. Their lifetimes are given by τ_{sp}, τ_{nr}, and τ_{20}. The overall population depletion time is defined by the following equation:

$$\frac{1}{\tau_2} = \frac{1}{\tau_{21}} + \frac{1}{\tau_{20}} = \frac{1}{\tau_{sp}} + \frac{1}{\tau_{nr}} + \frac{1}{\tau_{20}}. \tag{10.42}$$

When pumping occurs, the rate of increase of population densities is derived by solving the following equations:

$$\frac{dN_2}{dt} = \mathcal{R}_2 + \frac{N_2}{\tau_2} \tag{10.43}$$

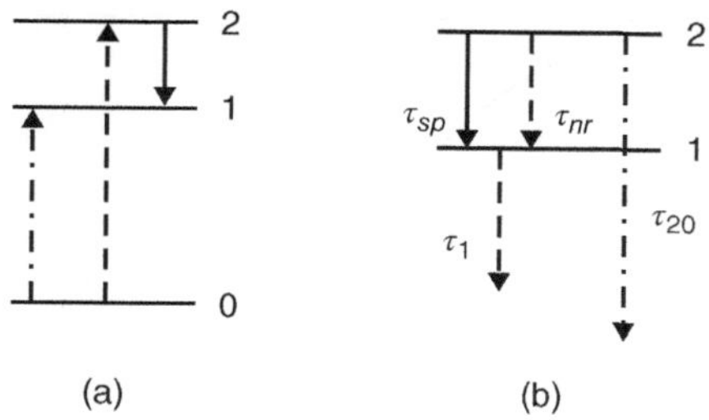

FIGURE 10.7 Illustration of atomic energy states "0," "1," and "2," and decay times.

and

$$\frac{dN_1}{dt} = -\mathcal{R}_1 - \frac{N_1}{\tau_1} + \frac{N_2}{\tau_{21}}. \tag{10.44}$$

In the above equations, the parameters $\mathcal{R}_1$ and $\mathcal{R}_2$ are the rate of pumping atoms out of state 1 and 2, respectively, which are in units of per unit volume per second. The steady-state solution to the above is

$$\Delta N = (N_2 - N_1) = \mathcal{R}_2\tau_2\left(1 - \frac{\tau_1}{\tau_{21}}\right) + \mathcal{R}_1\tau_1 . \tag{10.45}$$

To obtain a large population difference, we need the pump rates to be large, that is, $\mathcal{R}_2, \mathcal{R}_1 \gg 1$, a long τ_2, and a short τ_1 if $\mathcal{R}_1 < \mathcal{R}_2\left(\frac{\tau_2}{\tau_{21}}\right)$. This means that the elevated levels should be pumped strongly and decay slowly, and the ground state should be depleted strongly, which is pointed out in the last section.

If we have $\tau_{21} \ll \tau_{20}$ so that $\tau_2 = \tau_{\text{sp}}$ and $\tau_1 = \tau_{\text{sp}}$, then we have

$$\Delta N = \mathcal{R}_2\tau_{\text{sp}} + \mathcal{R}_1\tau_1 . \tag{10.46}$$

10.3.2 Laser Pumping with Amplifier Radiation

When amplifier radiation and pumping occurs, the rate of increase of population densities are solved from the following equations:

$$\frac{dN_2}{dt} = \mathcal{R}_2 + \frac{N_2}{\tau_2} - (N_2 - N_1)\, W_i \tag{10.47}$$

and

$$\frac{dN_1}{dt} = -\mathcal{R}_1 - \frac{N_1}{\tau_1} + \frac{N_2}{\tau_{21}} + (N_2 - N_1)\, W_i, \tag{10.48}$$

where W_i represents the change per unit time in the population.

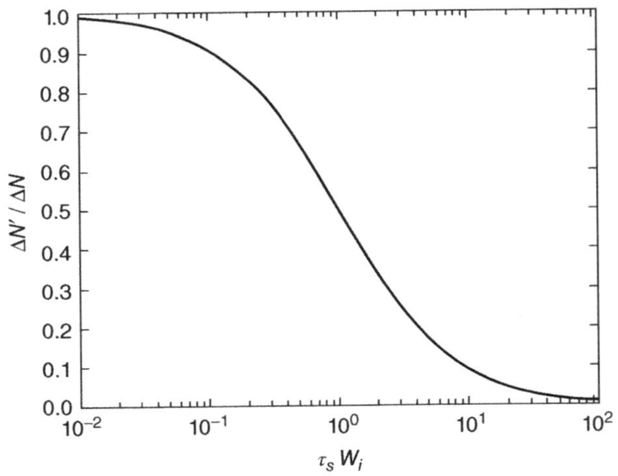

FIGURE 10.8 Plot of $\sum \frac{\Delta N'}{\Delta N}$ as a function of $\tau_s W_i$.

The above equations imply that the population density at the elevated level "2" is increased by absorption from level $1 \rightarrow 2$ and decreased by spontaneous emission from level $2 \rightarrow 1$. The steady-state solution is given by

$$\Delta N' = \frac{\Delta N}{1 + \tau_s W_i} \tag{10.49}$$

with

$$\tau_s = \tau_2 + \tau_1 \left(1 - \frac{\tau_1}{\tau_{21}}\right). \tag{10.50}$$

Figure 10.8 shows plot of $\frac{\Delta N'}{\Delta N}$ as a function of $\tau_s W_i$. From this plot, we see that if $\tau_s W_i \ll 1$ (small signal approximation), then $\Delta N' = \Delta N$. On the other hand, as $\tau_s W_i \rightarrow \infty$, then $\Delta N' \rightarrow 0$.

10.4 LASER GAIN AND PHASE-SHIFT COEFFICIENTS

Let us now use the theory we have developed to derive the laser gain coefficient for the laser medium in our potential laser. This and the following section leverage Refs [19–23].

Referring back to Figure 10.2b, we have an atom initially in "excited" state, E_{elevated}. We saw that for spontaneous emission that the atom decays spontaneously and adds a photon of energy $h\nu$ to an optical stream. This occurs independent of the number of photons already in the said stream. The probability density of this spontaneous emission in units of inverse seconds is given by

$$p_{\text{sp}} = \frac{C}{V}\sigma(\nu), \tag{10.51}$$

where c is the speed of light (m/s), V the volume of the cavity (m^3), and $\sigma(\nu)$ the transition cross section (m^2).

Conversely, Figure 10.2a illustrates the absorption process. Specifically, an atom is initially in the ground state E_{ground} and gets elevated to the "excited" state E_{elevated} by the absorption of a photon of energy $h\nu$. This process is governed by the same law as in spontaneous emission, that is,

$$p_{\text{ab}} = \frac{C}{V}\sigma(\nu). \tag{10.52}$$

However, if there are n photons in the optical stream, the probability is increased by that factor, which implies that the probability of absorption of one photon from a stream with n photons equals

$$p_{\text{ab}}^n = n\frac{C}{V}\sigma(\nu). \tag{10.53}$$

Not surprisingly, the probability density of stimulated emission is the same law that governs spontaneous emission and absorption,

$$p_{\text{st}} = \frac{C}{V}\sigma(\nu), \tag{10.54}$$

which leads to the probability of absorption of one photon from a stream with n photons being given by

$$p_{\text{st}}^n = n\frac{C}{V}\sigma(\nu). \tag{10.55}$$

The total probability density for an atom to emit a photon then is

$$p_{\text{sp}} + p_{\text{st}}^n = c\frac{(n+1)}{V}\sigma(\nu). \tag{10.56}$$

The change per unit time in the population $2 \to 1$ is given by

$$W_{2\to1} = p_{\text{sp}} + p_{\text{st}}^n = c\frac{(n+1)}{V}\sigma(\nu) \tag{10.57}$$

$$\approx n\frac{c}{V}\sigma(\nu) = \Phi\sigma(\nu) \tag{10.58}$$

$$= W_{1\to2} \tag{10.59}$$

for $n \gg 1$ as $p_{\text{ab}}^n = p_{\text{st}}^n$. In Eq. (10.58), Φ is the *photon flux density* (photons/m^2 s).

The transition cross section $\sigma(\nu)$ characterizes the interaction of the atom with the radiation. Integrating $\sigma(\nu)$ over all frequencies defines the *transition (or oscillator) strength S*. Mathematically, we write

$$S = \int_0^\infty \sigma(\nu)d\upsilon. \tag{10.60}$$

Another parameter closely associated with these parameters is the line-shape function $g(\nu)$, which is defined as

$$g(\nu) = \frac{\sigma(\nu)}{S}. \tag{10.61}$$

It is normalized to unity, has units of inverse hertz, and, in general, centered about the resonance frequency ν_0. Its width is approximately the inverse of the resonance bandwidth. This bandwidth comes from collision broadening and Doppler effects. It usually follows a Lorentzian distribution of the form

$$g(\nu) = \frac{\left(\frac{\Delta\nu}{2\pi}\right)}{(\nu - \nu_0)^2 + \left(\frac{\Delta\nu}{2}\right)^2}. \tag{10.62}$$

We will define $\Delta\nu$ shortly.

Given the above statement that the line-shape function is centered around ν_i, the change per unit time in the population $2 \rightarrow 1$ will be centered around that frequency and we rewrite Eqs. (10.57) and (10.59) as

$$W_{2\rightarrow 1} = W_{1\rightarrow 2} = W_i. \tag{10.63}$$

If N_1 and N_2 are the number of atoms in the ground and elevated energy states, respectively, then the average density of absorbed photons (number of photon per unit time per unit volume) is $N_1 W_i$ and the average density of stimulated photons is $N_2 W_i$. The net number of photon gained is

$$NW_i = (N_2 - N_1)W_i. \tag{10.64}$$

The parameter N is the population density difference. If $N > 0$, then this parameter represents a population inversion (more atoms in excited states), which means the medium can act as an amplifier. However, $N < 0$, then this parameter represents an absorber and it cannot do any light amplification, that is, no laser. If $N = 0$, then the medium is transparent. The conclusion is to have $N > 0$, we need an external pump that excites the atoms as discussed in Chapter 9.

Let us assume a cylindrical gain medium of length d. Figure 10.9 depicts photon flux density φ entering and photon flux density $\varphi + d\varphi$ exiting the cylinder of length dz. The incremental photon flux density $d\varphi$ must be the photon flux emitted within the thin slice dz of cylinder. Thus, we have

$$d\varphi = NW_i dz \tag{10.65}$$

or

$$\frac{d\varphi}{dz} = NW_i = N\varphi\sigma(\nu) \tag{10.66}$$

$$= \gamma(\nu)\varphi, \tag{10.67}$$

where $\gamma(n)$ represents the *gain coefficient*, or net gain in the photon flux per unit length, of the medium. From Eq. (10.61), we see that

$$\gamma(\nu) = N\sigma(\nu) = N S g(\nu). \tag{10.68}$$

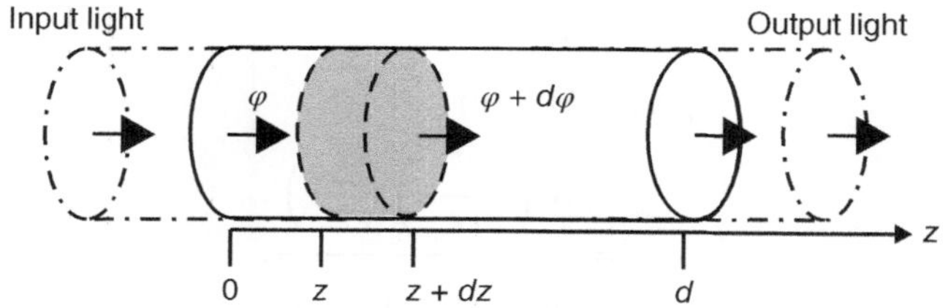

FIGURE 10.9 Illustration of photon flux density entering and exiting a cylinder of length dz.

Recall from Eq. (4.74) that the modal density at frequency ν in a unit volume is given by

$$M(\nu) = \frac{8\pi\nu^2}{c^3}. \tag{10.69}$$

This implies that total probability density of absorption can be written as

$$P_{\text{ab}} = \int_0^\infty \left\{\frac{c}{V}\sigma(\nu)\right\} M(\nu)V\,d\nu = c\int_0^\infty \sigma(\nu)M(\nu)\,d\nu \tag{10.70}$$

$$= c\left(\frac{8\pi\nu^2}{c^3}\right)\int_0^\infty \sigma(\nu)\,d\nu = \left(\frac{8\pi\nu^2}{c^2}\right)S = \left(\frac{8\pi}{\lambda^2}\right)S. \tag{10.71}$$

The lifetime of spontaneous emission is inversely proportional to the above probability density, which implies that

$$\frac{1}{t_{\text{sp}}} = \left(\frac{8\pi}{\lambda^2}\right)S$$

or

$$t_{\text{sp}} = \left(\frac{\lambda^2}{8\pi S}\right). \tag{10.72}$$

The oscillator strength is then

$$S = \left(\frac{\lambda^2}{8\pi t_{\text{sp}}}\right). \tag{10.73}$$

The solution to Eq. (10.67) is

$$\varphi(z) = \varphi(0)e^{\gamma(\nu)z}. \tag{10.74}$$

Then correspondingly the optical intensity is related to the above and can be written as

$$I(z) = h\nu\varphi(z) = h\nu\varphi(0)e^{\gamma(\nu)z} = I(0)e^{\gamma(\nu)z}. \tag{10.75}$$

This last equation implies that $\gamma(\nu)$ also represents the gain in intensity per unit length from the medium. This means that for a medium of length d, the effective gain is given by

$$G(z) = \frac{\varphi(z)}{\varphi(0)} = e^{\gamma(\nu)d}, \tag{10.76}$$

where

$$\gamma(\nu) = NSg(\nu) \tag{10.77}$$

$$= \frac{N\left(\frac{\lambda^2}{8\pi t_{sp}}\right)\left(\frac{\Delta\nu}{2\pi}\right)}{(\nu-\nu_0)^2 + \left(\frac{\Delta\nu}{2}\right)^2}. \tag{10.78}$$

At the center frequency ν_0, we have

$$\gamma(\nu_0) = N\sigma(\nu_0) = \frac{N\left(\frac{\lambda^2}{8\pi t_{sp}}\right)\left(\frac{\Delta\nu}{2\pi}\right)}{\left(\frac{\Delta\nu}{2}\right)^2} = \frac{N\left(\frac{\lambda^2}{4\pi^2 t_{sp}}\right)}{\Delta\nu} \tag{10.79}$$

or

$$\Delta\nu = \frac{\lambda^2}{4\pi^2 t_{sp}\,\sigma(\nu_0)}, \tag{10.80}$$

where $\sigma(\nu_0)$ is peak cross section at the center frequency ν_0. For the special case in which the linewidth $\Delta\nu$ is entirely caused the spontaneous emission lifetime t_{sp} of the excited state, we have

$$\sigma(\nu_0) = \frac{\lambda^2}{4\pi^2}. \tag{10.81}$$

Using Eq. (10.73), we can rewrite Eq. (10.72) as

$$\gamma(\nu) = \frac{\gamma(\nu_0)\left(\frac{\Delta\nu}{2}\right)^2}{(\nu-\nu_0)^2 + \left(\frac{\Delta\nu}{2}\right)^2}, \tag{10.82}$$

which defines the gain coefficient of the medium. Clearly, the gain coefficient is proportional to the Lorentzian distribution like the line-shape function.

The gain of the medium also creates a phase shift in the wave. Consider the electric field associated to the propagating intensity shown in Figure 10.9:

$$E(z) = E(0)e^{\frac{1}{2}\gamma(\nu)z - i\psi(z)}. \tag{10.83}$$

At $z + dz$, the electric field equals

$$E(z+dz) = E(0)e^{\frac{1}{2}\gamma(\nu)(z+dz) - i\psi(z+dz)} \tag{10.84}$$

$$= E(z)e^{\frac{1}{2}\gamma(\nu)dz - i\psi dz}$$

$$= E(z)\left[\frac{1}{2}\gamma(\nu) - i\psi\right]dz. \tag{10.85}$$

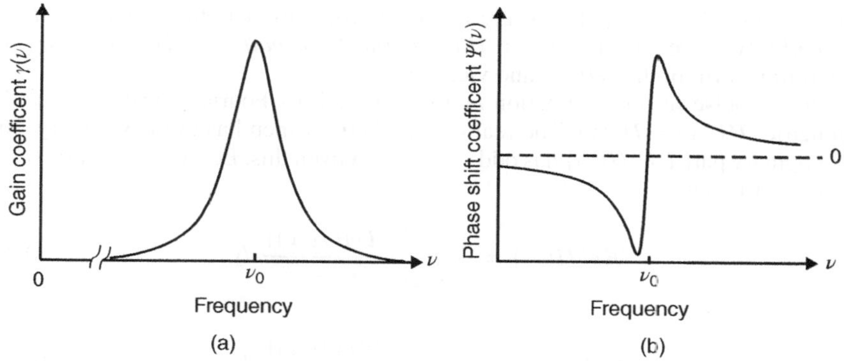

FIGURE 10.10 Spectral plots of (a) gain and (b) phase-shift coefficients for an optical amplifier with a Lorentzian line-shape function.

It can be shown that the real and imaginary parts are related via Hilbert transform and the phase shift is given by

$$\psi(\nu) = \frac{\nu - \nu_0}{\Delta\nu}\,\gamma(\nu). \tag{10.86}$$

The above phase shift is separate from any phase shift created by the laser medium. Figure 10.10 illustrates (a) gain and (b) phase-shift coefficients for an optical amplifier with a Lorentzian line-shape function

Example 10.2 The relationship between real and imaginary parts of Eq. (10.77) can be developed from the transfer function for a forced harmonic oscillator following Saleh and Teich [15, Appendix B].

As we have seen in previous chapters, the impulse response function $h(t)$ of a linear shift-invariant causal system must vanish for $t < 0$, since the system's response cannot begin before the application of the input. The function $h(t)$ is therefore not symmetric and its Fourier transform, the transfer function $\mathcal{H}(\nu)$, must be complex. It can be shown that if $h(t) = 0$ for $t < 0$, then the real and imaginary parts of $\mathcal{H}(\nu)$, denoted $\Re e\{\mathcal{H}(\nu)\}$ and $\mathcal{I}m\{\mathcal{H}(\nu)\}$, respectively, are related by

$$\Re e\{\mathcal{H}(\nu)\} = \frac{1}{\pi}\int_{-\infty}^{\infty} \frac{\mathcal{I}m\{\mathcal{H}(\nu)\}}{s - \nu}\,ds \tag{10.87}$$

and

$$\mathcal{I}m\{\mathcal{H}(\nu)\} = \frac{1}{\pi}\int_{-\infty}^{\infty} \frac{\Re e\{\mathcal{H}(\nu)\}}{s - \nu}\,ds, \tag{10.88}$$

where the Cauchy principal values of the integrals are to be evaluated, that is,

$$\int_{-\infty}^{\infty} = \int_{-\infty}^{\nu-\Delta} + \int_{\nu+\Delta}^{\infty}; \quad \Delta > 0. \tag{10.89}$$

Equations (10.87) and (10.88) are *Hilbert transforms*. The functions $\Re e\{\mathcal{H}(\nu)\}$ and $\mathcal{I}m\{\mathcal{H}(\nu)\}$ are said to form a *Hilbert transform pair*, $\mathcal{I}m\{\mathcal{H}(\nu)\}$) being the Hilbert transform of $\Re e\{\mathcal{H}(\nu)\}$ and vice versa.

If the impulse response function $h(t)$ is also real, its Fourier transform must be symmetric, $\mathcal{H}(-\nu) = \mathcal{H}^*(\nu)$. The real part $\Re e\{\mathcal{H}(\nu)\}$ then has even symmetry, and the imaginary part $\mathcal{I}m\{\mathcal{H}(\nu)\}$ has odd symmetry. Given this, Eqs. (10.78) and (10.79) can be rewritten as

$$\Re e\{\mathcal{H}(\nu)\} = \frac{2}{\pi}\int_{-\infty}^{\infty} s\,\frac{\mathcal{I}m\{\mathcal{H}(\nu)\}}{s^2-\nu^2}\,ds \tag{10.90}$$

and

$$\mathcal{I}m\{\mathcal{H}(\nu)\} = \frac{1}{\pi}\int_{-\infty}^{\infty} \nu\,\frac{\Re e\{\mathcal{H}(\nu)\}}{\nu^2-s^2}\,ds. \tag{10.91}$$

Equations (10.90) and (10.91) are known as the *Kramers–Kronig relations*.

Consider a harmonic oscillator described by the equation:

$$\left[\frac{d^2}{dt^2} + \sigma\frac{d}{dt} + \omega_0^2\right] f_1(t) = f_2(t). \tag{10.92}$$

Equation (10.92) describes a harmonic oscillator with displacement $f_1(t)$ under an applied force $f_2(t)$, where ω is the resonance angular frequency and σ is a coefficient representing the damping effects. Let us assume and external harmonic force and displacement of the forms

$$f_2(t) = e^{-2\pi i\nu t} \tag{10.93}$$

and

$$f_1(t) = \mathcal{H}(\nu)\,e^{-2\pi i\nu t}. \tag{10.94}$$

The transfer function $\mathcal{H}(\nu)$ of this system may be obtained by Eqs. (10.93) and (10.94) into Eq. (10.92), which yields

$$\mathcal{H}(\nu) = \left(\frac{1}{2\pi}\right)^2 \frac{1}{\nu_0^2 - \nu^2 - i\nu\Delta\nu} \tag{10.95}$$

$$= \left(\frac{1}{2\pi}\right)^2 \frac{1}{(\nu_0^2-\nu^2)^2 + (\nu\Delta\nu)^2}[(\nu_0^2-\nu^2) + i\nu\Delta\nu], \tag{10.96}$$

where $\nu_0 = \frac{\omega_0}{2\pi}$ is the resonance frequency, and $\Delta\nu = \frac{\sigma}{2\pi}$. This implies that

$$\Re e\{\mathcal{H}(\nu)\} = \left(\frac{1}{2\pi}\right)^2 \frac{(\nu_0^2-\nu^2)}{(\nu_0^2-\nu^2)^2 + (\nu\Delta\nu)^2} \tag{10.97}$$

and

$$\mathcal{I}m\{\mathcal{H}(\nu)\} = \left(\frac{1}{2\pi}\right)^2 \frac{\nu\Delta\nu}{(\nu_0^2-\nu^2)^2 + (\nu\Delta\nu)^2}. \tag{10.98}$$

Since the system is causal, $\Re e\{\mathcal{H}(\nu)\}$ and $\mathcal{I}m\{\mathcal{H}(\nu)\}$ satisfy the Kramers–Kronig relations. When $\nu_0 \gg \Delta\nu$, $\Re e\{\mathcal{H}(\nu)\}$, and $\mathcal{I}m\{\mathcal{H}(\nu)\}$ are narrow functions centered about ν_0. For $\nu = \nu_0$, we have $\nu_0^2 - \nu^2 \approx 2\nu_0(\nu_0 - \nu)$, which allows us to write

$$\mathcal{I}m\{\mathcal{H}(\nu)\} = \left(\frac{1}{2\pi}\right)^2 \frac{\nu\Delta\nu}{(\nu_0^2 - \nu^2)^2 + (\nu\Delta\nu)^2} \approx \left(\frac{1}{2\pi}\right)^2 \frac{\nu_0\Delta\nu}{4\nu_0^2(\nu_0 - \nu)^2 + \nu_0^2(\Delta\nu)^2}$$
$$\approx \left(\frac{1}{2\pi}\right)^2 \frac{\left(\frac{\Delta\nu}{4\nu_0}\right)}{(\nu_0 - \nu)^2 + \left(\frac{\Delta\nu}{2}\right)^2}. \tag{10.99}$$

We also see from Eq. (10.99) that

$$\Re e\{\mathcal{H}(\nu)\} = \left(\frac{1}{2\pi}\right)^2 \frac{(\nu_0^2 - \nu^2)}{(\nu_0^2 - \nu^2)^2 + (\nu\Delta\nu)^2}$$
$$\approx \left(\frac{1}{2\pi}\right)^2 \left[\frac{2\nu_0(\nu_0 - \nu)}{\nu_0\Delta\nu}\right] \frac{\nu\Delta\nu}{(\nu_0^2 - \nu^2)^2 + (\nu\Delta\nu)^2}$$
$$\approx \left[\frac{2(\nu_0 - \nu)}{\Delta\nu}\right] \mathcal{I}m\{\mathcal{H}(\nu)\}. \tag{10.100}$$

The real part of $\mathcal{H}(\nu)$ contributes to the index of refraction, which in turns creates a phase shift of the propagating wave. On the other hand, the imaginary part $\mathcal{H}(\nu)$ is the gain or absorption loss. Thus, like a forced oscillator, a laser medium creates a net change in the amplitude of the field as well as a phase change.

10.5 LASER CAVITY GAINS AND LOSSES

Let us consider a three-level medium. As we saw in Section 10.2, when one has a large number of atoms in the ground state of a medium irradiated with light of the appropriate energy, the light will be absorbed following a Beer's-type law, based on the form of the equations cited. Namely, the absorbed light through the medium will be given by

$$I = I_0 e^{-\sigma_{21} N_1 d} \tag{10.101}$$

with I_0 being the intensity of the beam before entering the medium, σ_{21} being the cross section for absorption or emission between ground and elevated states, and N_1 being the population density of atoms residing in the ground state (number of atoms per unit volume). Equation (10.101) follows the same type of exponential decay shown by the solid curve in Figure 10.4. Obviously, the product $\sigma_{21}N_1$ represents an "absorption coefficient."

From the same discussion, it is a straightforward conclusion that stimulated emission would follow the same law but with the sign of the exponent in Eq. (10.101) being positive. Thus, the equation for stimulation emission light production is given by

$$I = I_0 e^{\sigma_{21} N_2 d}, \tag{10.102}$$

where we now have the population density N_2 in the expression along with the appropriate cross section σ_{21}. The resulting light intensity through the medium then will be given by

$$I = I_0 e^{\sigma_{21}(N_2 - N_1)d} \tag{10.103}$$

and photon production (lasing) will occur only if $N_2 > N_1$, that is, population inversion. Typical gains from a single pass range from 2% to 1000%, depending on the laser type [24]. For many applications, this growth is insufficient and multiple path resonators are used to increase the laser power to the desired level. The flat mirror cavity is the simplest structure for this purpose. However, there comes a limit. Somewhere between 2 passes (dye lasers) and 500 passes (HeNe lasers), the propagating beam no longer has sufficient atoms in the elevated state to continue the photon production indefinitely [24]. This is the condition known as *beam saturation*, with the resulting intensity called the *saturation intensity*. When it reaches this intensity, the laser will stabilize to a specific value as long as the laser pumping from 2 → 1 continues. In other words, the conversion from pump power to laser photons reaches an equilibrium balance [15, 24]. The condition of saturation occurs when the exponent of Eq. (10.103) reaches a value of approximately 12 [24]. The resulting growth factor then is

$$e^{\sigma_{21}(N_2 - N_1)d} = e^{g_{21} d} = e^{12} = 1.63 \times 10^5, \tag{10.104}$$

where

$$g_{21} = \sigma_{21}(N_2 - N_1) \tag{10.105}$$

is the *small-signal-gain coefficient* [15, 24].

Another limiting factor is losses within the cavity like from imperfect reflectivity of the end mirrors or from losses derived from any other optical elements housed in the cavity. There are two expressions for the threshold condition necessary for a laser beam to occur, based on the laser cavity characteristics. For our simple laser structure, he stated that the threshold condition for the gain coefficient g is given by

$$g = \frac{1}{2d} \ln\left(\frac{1}{R^2}\right), \tag{10.106}$$

where d again is the cavity length and R the identical reflectivity of the two mirrors, assuming no other losses in the cavity [15, 24]. If $g_{21} > g$, then a laser beam will be produced.

For a more complex laser cavity in which the mirrors have different reflectivities R_1 and R_2, and L_1 and L_2 represent other losses within the cavity, the expression for the threshold gain g is given as

$$g = \frac{1}{2d} \ln\left(\frac{1}{R_1 R_2 (1 - L_1)(1 - L_2)}\right) + \alpha, \tag{10.107}$$

where α accounts for the absorption loss within the laser medium. Similarly, if $g_{21} > g$, then a laser beam will be produced.

Example 10.3 Assume a helium neon (HeNe) laser with a cavity length of 20 cm, mirror reflectivities of $R_1 = 0.999$ and $R_2 = 0.99$; cavity losses of $L_1 = L_2 = 0.002$; and $\alpha = 0.0$. The threshold gain g then equals

$$\begin{aligned} g &= \frac{1}{2d} \ln\left(\frac{1}{R_1 R_2 (1 - L_1)(1 - L_2)}\right) + \alpha \\ &= \frac{1}{2 \times 20\text{ cm}} \ln\left(\frac{1}{(0.999)(0.99)(1 - 0.002)(1 - 0.002)}\right) \\ &= 3.76 \times 10^{-4}\,\text{cm}^{-1}. \end{aligned}$$

The product $gd = 0.00753$ and one gets 0.7% more light per pass.

The useful power from the laser is obtained making one of the end flat mirrors partially transmissive so a fraction of the beam "leaks out." The initial gain in the amplifier must be greater than the loss of the transmitting mirror (plus other mirror and cavity losses) or the beam will not develop. A simple expression for the optimum mirror transmission $T_{2\text{-opt}}$ is possible for the second mirror in terms of the small-signal-gain coefficient g, the actual amplifier length d, and the absorption loss $L_{\text{single pass}}$ (averaged over a single pass from one mirror to the other). In particular, we have, namely [15, 24].

$$T_{2\text{-opt}} = \sqrt{g d L_{\text{single pass}}} - L_{\text{single pass}}. \tag{10.108}$$

Example 10.4 Assume the helium neon (HeNe) laser characteristics in Example 10.3 and that the gain is 10 times the threshold value or

$$g = 10 \times (3.76 \times 10^{-4}\,\text{cm}^{-1}) = 3.76 \times 10^{-3}\,\text{cm}^{-1}.$$

The optimum mirror transmittance then equals

$$\begin{aligned} T_{2\text{-opt}} &= \sqrt{(3.76 \times 10^{-3}\,\text{cm}^{-1})(20\text{ cm})(0.002)} - (0.002) \\ &= 1.03 \times 10^{-2} \approx 1\%, \end{aligned}$$

assuming $L = 0.5\,(L_1 + L_2)$.This implies that the reflectivity of the second mirror equals 98.96% or approximately 99%,which essentially is the second mirror reflectivity we assumed in Example 10.3.

The maximum laser beam output intensity I_{max} emitted from the output mirror also can be estimated in terms of the saturation intensity I_{sat}, $T_{2\text{-opt}}$, and the average absorption loss per pass L. Specifically, we have [24]

$$I_{\text{max}} = \left(\frac{T_{2\text{-opt}}^2}{2L}\right) I_{\text{sat}}. \tag{10.109}$$

Example 10.5 Using the helium neon (HeNe) laser characteristics in Example 10.4, we have

$$I_{max} = \left(\frac{(0.0103)^2}{[0.004]} \right)^2 (6.2\,\mathrm{W/cm^2}) = 164\,\mathrm{mW/cm^2},$$

assuming $I_{sat} = 6.2\ \mathrm{W/cm^2}$. This is on the high end of the power spectrum for a HeNe laser [24].

10.6 OPTICAL RESONATORS

Up to this point, we have assumed that a laser has a layout as depicted in Figure 10.11. In this figure, we see that we have sort of laser medium, for example, gas, crystal dye, centered axially between two flat mirrors; one is totally reflective and the other is partially reflected, say 95%. The laser medium is pumped by some external sources such as laser diode bars and flash lamps to provide the population inversion necessary for laser beam creation as described in the last section. The exact design depends on wavelength, external CW or pulse light pumping, desired power level, and so on. This configuration implies a light beam multimodal both in time and space. In most applications, one would like to configure the laser to produce light at the greatest joules or watts per meter-squared (fluence) level the design dictates, that is, single-mode operation. This means that the laser medium and pump source are placed in *optical resonator or cavity* to facilitate laser mode selection.

An optical resonator is analogous to the electronic resonant circuit. More specifically, it is an optical transmission system with feedback, or *optical circulator*. Figure 10.7 represents a *Fabry–Perot resonator*, which comprised two parallel mirrors, which is the simplest configuration one can use. However, many alternate resonator configurations have found great utility. These include spherical mirror, ring, and two- and three-dimensional rectangular cavity as optical resonant structures.

The following sections discuss some of the approaches to optical cavity design and their analysis.

10.6.1 Planar Mirror Resonators – Longitudinal Modes

Let us again assume a monochromatic wave of frequency ν of the form

$$u(\boldsymbol{r}, t) = \mathrm{Re}\{A(\boldsymbol{r})\, e^{-i2\pi\nu t}\}, \tag{10.110}$$

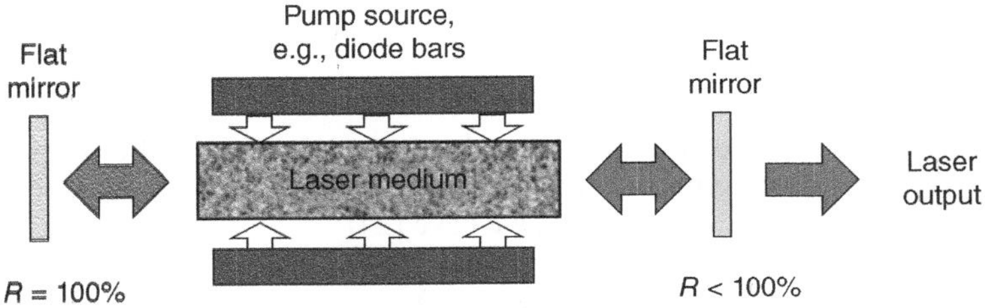

FIGURE 10.11 Basic laser layout.

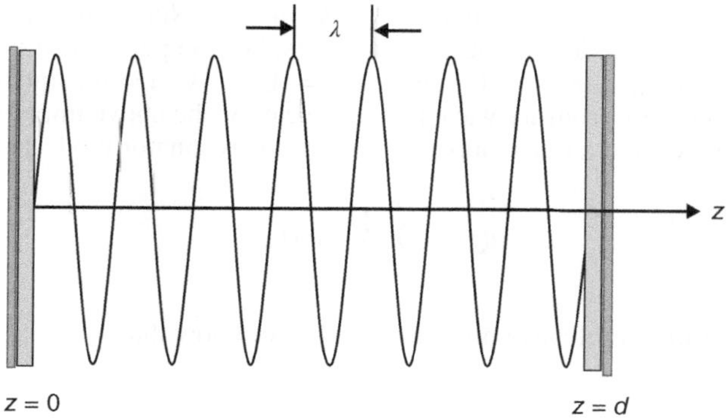

FIGURE 10.12 Depiction of the complex amplitude signal within a flat mirror resonator.

which represents the transverse component of the electric field. The complex amplitude satisfies the stationary Helmholtz equation given in Eq. (2.21),

$$\nabla^2 u(\boldsymbol{r}) - k^2 u(\boldsymbol{r}) = 0$$

with $k = \frac{2\pi v}{c}$ being the wave number and c being the speed of light given by

$$c = \frac{1}{\sqrt{\mu_0 \varepsilon_0}} = 2.99792458 \times 10^8 \text{ m/s}.$$

In materials with permittivity, ε_0, and relative permeability, μ, the phase velocity of light was given in Eq. (2.10) as

$$v_p = \frac{c}{n} = \frac{1}{\sqrt{\mu \varepsilon_0}}.$$

Here, n is the medium's refractive index. The plane wave resonator modes are the solutions to the stationary Helmholtz equation under the appropriate boundary conditions. For a lossless flat mirror resonator of length d, the transverse component of the electric field vanishes at the mirror surfaces, which means that $A(\boldsymbol{r})$ equals zero at in z-plane locations $z = 0$ and $z = d$, as depicted in Figure 10.12.

The standing wave shown in Figure 10.12 satisfies the stationary Helmholtz equation and is written as

$$u(z) = A \sin(kz), \tag{10.111}$$

where is a constant and vanishes at $z = 0$ and $z = d$ if $k = \frac{q\pi}{d} = k_p$, with q being an integer. This means that Eq. (10.111) needs to be rewritten as

$$u_q(z) = A_q \sin(k_q z), \tag{10.112}$$

where $q = 1, 2, 3, \ldots$. This equation says that the flat mirror radiator only allows certain set of standing waves, or *modes*, to exist in the cavity defined by the two mirrors. The parameter q is known as the *mode number*. Note that negative values of q do not reflect independent modes, just a reversal of the peaks, that is, $\sin(k_{-q}z) = \sin(-k_q z) = -\sin(k_q z)$, and $q = 0$ yields $u(z) = A_0 \sin(k_0 z) = A_0 \sin(0) = 0$, a zero energy signal. For any arbitrary wave inside the cavity, the above implies that we can compose this wave into a linear superposition of the resonator modes, or

$$u(z) = \sum_{q=1}^{\infty} A_q \sin(k_q z). \tag{10.113}$$

From our definition of wave number $k = \frac{2\pi\nu}{c}$, we know that

$$k_q = 2\pi \frac{\nu_p}{c} = q\frac{\pi}{d}$$

or

$$\nu_q = q\frac{c}{2d}, \tag{10.114}$$

for $q = 1, 2, 3, \ldots$. Equation (10.114) represents the *resonant frequencies* of the cavity. The frequency spectrum of the cavity is shown in Figure 10.13. It is clear that resonant frequencies are separated by a constant frequency difference, $\nu_F = \frac{c}{2d}$, known as the *free spectral range*. For a cavity of length $d = 50$ cm, the free spectral ranges equal 300 MHz.

The wavelengths associated with the frequency definition in Eq. (10.46) are given by

$$\lambda_q = \frac{c}{\nu_q} = \frac{2d}{q}. \tag{10.115}$$

The round-trip distance in the resonator must be precisely equal an integer number of wavelengths as suggested by Figure 10.13, that is, $2d = q\nu_q$ for $q = 1, 2, 3, \ldots$. Since a number of laser mediums may not have unity refractive index, the speed of light in the medium will be the phase velocity $v_p = \frac{c}{n}$ and λ_q will be the wavelength in that medium, or

$$\lambda_q = \frac{\nu_p}{\nu_q} = \frac{2d}{nq}. \tag{10.116}$$

The number of modes per unit frequency is the inverse of the spectral free range, or

$$\text{Number of modes} = \frac{1}{\nu_F} = \frac{2d}{c} \tag{10.117}$$

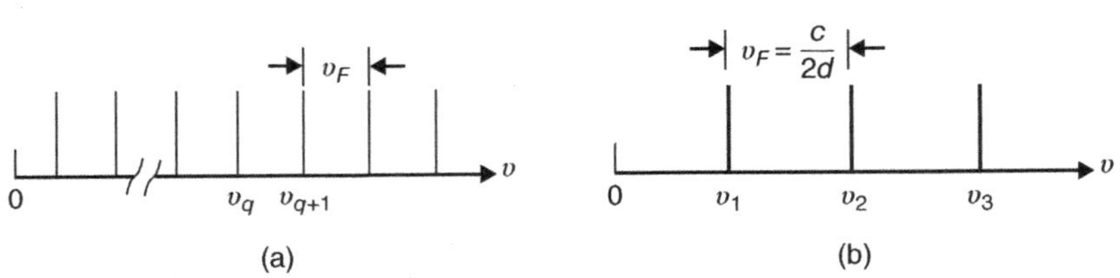

FIGURE 10.13 (a) Example resonator frequency spectrum and (b) an expanded view of its first three frequencies.

in each of the polarization allowed by the resonator. The density of modes $M(\nu)$ is the number of modes per unit frequency per unit length, accounting for both polarizations, therefore is

$$M(\nu) = \frac{4}{c}. \tag{10.118}$$

The number of modes in a resonator of length d in the frequency interval $\Delta\nu$ is thus

$$d\,\Delta\nu\,M(\nu) = \left(\frac{4d}{c}\right)\Delta\nu. \tag{10.119}$$

This number represents the number of degrees of freedom for the optical waves exiting in the cavity.

Let us assume the complex round-trip amplitude attenuation factor is given by

$$L_{\text{cavity}} = |r|e^{i\varphi}. \tag{10.120}$$

The intensity of the light in the cavity is given by[2]

$$I = |u|^2 = \frac{|u_0|^2}{|1 - |r|e^{i\varphi}|^2} = \frac{I_0}{1 + |r|^2 - 2\,|r|\,\cos\varphi}, \tag{10.121}$$

which can be rewritten as

$$\begin{aligned} I &= \frac{I_0}{1 + |r|^2 - 2|r| + 2|r| - 2\,|r|\,\cos\varphi} = \frac{I_0}{(1 - |r|)^2 + 2|r|(1 - \cos\varphi)} \\ &= \frac{I_0}{(1 - |r|)^2}\left[\frac{1}{1 + \left(\frac{2|r|}{(1 - |r|)^2}\right)(1 - \cos\varphi)}\right] = \frac{I_{\max}}{1 + \left(\frac{2\mathcal{F}}{\pi}\right)^2 \sin^2\left(\frac{\varphi}{2}\right)}, \end{aligned} \tag{10.122}$$

where

$$I_{\max} = \frac{I_0}{(1 - |r|)^2} \tag{10.123}$$

and

$$\mathcal{F} = \frac{\pi\sqrt{|r|}}{(1 - |r|)}. \tag{10.124}$$

[2] In general, the finesse is given by

$$\mathcal{F} = \frac{\pi\,(R_1R_2)^{1/4}}{\left(1 - (R_1R_2)^{1/4}\right)},$$

where R_1 and R_2 are the mirror reflectivity of mirrors 1 and 2, respectively.

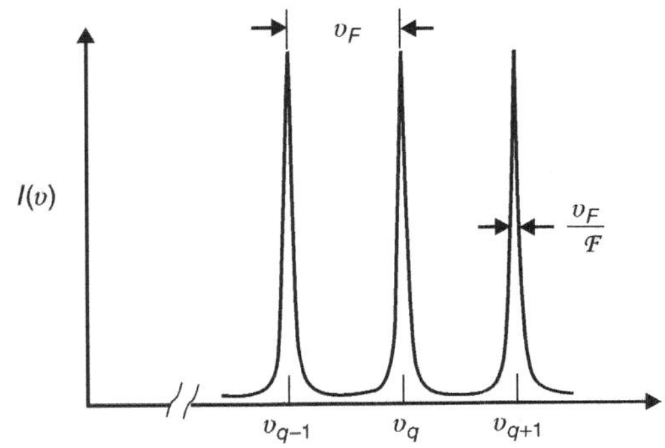

FIGURE 10.14 Mode spectrum for a lossy resonator.

In the above, $I_0 = |u_0|^2$ is the intensity of the initial wave and $\mathcal{F}$ is the *finesse* of the resonator. The phase φ can be shown to be equal to [15, p. 372]

$$\varphi = \frac{2\pi \nu d}{c}, \tag{10.125}$$

which implies that the intensity becomes

$$I = \frac{I_{max}}{1 + \left(\frac{2\mathcal{F}}{\pi}\right)^2 \sin^2\left(\frac{\pi\nu}{\nu_F}\right)}. \tag{10.126}$$

Figure 10.14 illustrates a sample mode spectrum for lossy resonator, that is, $|r| \neq 0$. The minimum intensity at the midpoints between the intensity peaks is equal to

$$I_{min} = \frac{I_{max}}{1 + \left(\frac{2\mathcal{F}}{\pi}\right)^2}. \tag{10.127}$$

In general, when the system is lossy, but the finesse is large ($\mathcal{F} \gg 1$), the spectral response of the resonator is sharply peaked about the resonance frequencies and $\frac{I_{min}}{I_{max}}$ is small, as illustrated in Figure 10.13. [Figure 10.12 illustrates the spectral response when the resonator is lossless ($\mathcal{F} = \infty$).] The FWHM of the resonance peak is equal to

$$\delta\nu \approx \frac{\nu_F}{\mathcal{F}}. \tag{10.128}$$

In summary, the spectral response of a Fabry–Perot optical resonator has intensity peaks spaced at separations given by the free spectral range and with spectral widths described by Eq. (10.128).

Generally, a laser does not have an infinite number of laser lines to choose from. The gain coefficient specifically limits the number of lines the laser will be support. Figure 10.15a depicts the effect on the available longitudinal modes and their

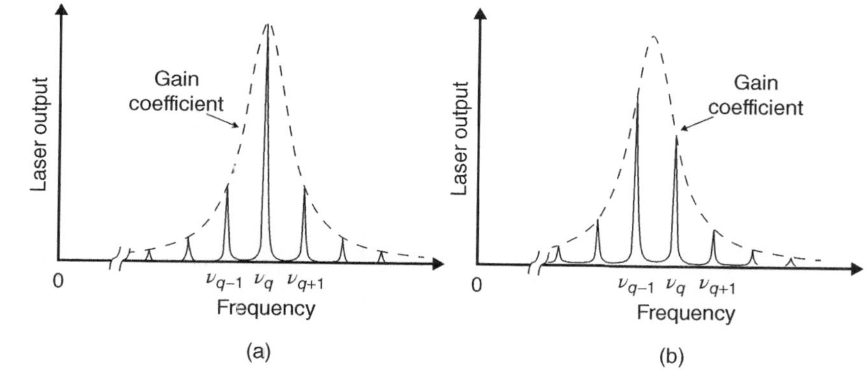

FIGURE 10.15 Effect on several longitudinal mode laser output by gain coefficient of the laser medium.

potential output by the laser medium's gain coefficient when a mode and the peak gain coincide. In this example, the center frequency has maximum gain and the adjacent lines get decreasing gain as the separation from the center increases. Sometimes, the modes and the center frequency are not collocated. Figure 10.15b shows an example where the available longitudinal modes and the peak gain coefficient do not exactly coincide. When multiple longitudinal modes are present in a laser cavity, the output of the laser exhibits mode beating. This is due to the interaction of the modes in time and results in spikes in the temporal profile of the laser output.

As a final comment, another quantity used to describe cavities is the center frequency divided by the full width at half maximum of the transmission peak. It is called the Quality Factor Q, which is given by

$$Q = \frac{\nu_0}{\Delta\nu_{\text{FWHM}}} = \frac{2\pi nd}{\lambda_0}\frac{\pi\left(R_1R_2\right)^{1/4}}{\left(1-\left(R_1R_2\right)^{1/4}\right)}. \tag{10.129}$$

10.6.2 Planar Mirror Resonators – Transverse Modes

A planar mirror resonator exhibits unique transverse modes as well as the specific longitudinal modes described above. The round-trip phase change in the resonator generally is a function of those transverse field distributions. Specifically, the resonator modes must reproduce shape at each point on each pass, with the round-trip phase shift being a multiple of 2π. Fortunately, the solution to the wave equation for this resonator can be expanded into Hermite–Gaussian modes, which are given by

$$E(x,y,z) = E_0\left(\frac{w_0^2}{w^2(z)}\right)H_n\left(\frac{\sqrt{2}x}{w(z)}\right)H_m\left(\frac{\sqrt{2}y}{w(z)}\right)e^{-\left[\frac{(x^2+y^2)}{w^2(z)}\right]}$$
$$\times e^{-i\left[k\left(\frac{(x^2+y^2)}{2R(z)}\right)-(1+n+m)\eta(z)\right]}, \tag{10.130}$$

where

$$w(z) = w_0 \sqrt{1 + \frac{z^2}{z_0^2}}, \tag{10.131}$$

$$z_0 = \frac{\pi w_0^2}{\lambda}, \tag{10.132}$$

$$R(z) = z \left[1 + \frac{z_0^2}{z^2} \right], \tag{10.133}$$

and

$$\eta(z) = \tan^{-1} \left(\frac{z}{z_0} \right). \tag{10.134}$$

In Eq. (10.130), the functions $H_k(x)$ are the Hermite Polynomial defined by the recurrence relation

$$H_{l+1}(x) = 2x H_l(x) - 2l H_{l-1}(x), \tag{10.135}$$

with $H_0(x) = 1$. The transverse modes described by Eq. (10.130) are called transverse electromagnetic wave (TEM_{mnq}) modes. The first two subscripts, m and n, describe the amplitude distribution transverse to the optic axis, while the third subscript q is the axial mode number having the same definition as in Eq. (10.114). Figure 10.16 shows the irradiance patterns for the four lowest transverse modes TEM_{mn} of the optical resonator, assuming the same value of q. It is clear from this figure that the higher-order

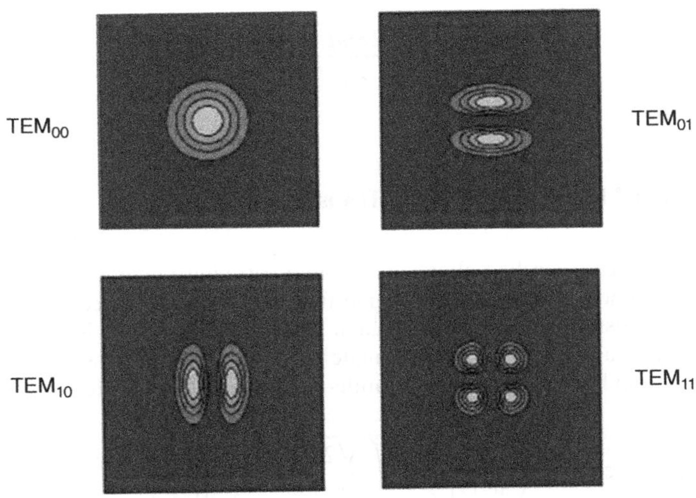

FIGURE 10.16 The irradiance distribution for the four lowest transverse modes TEM_{mnq} (same q) of the optical resonator. The subscripts m and n are identified as the number of nodes along two orthogonal axes perpendicular to the axis of the optical cavity.

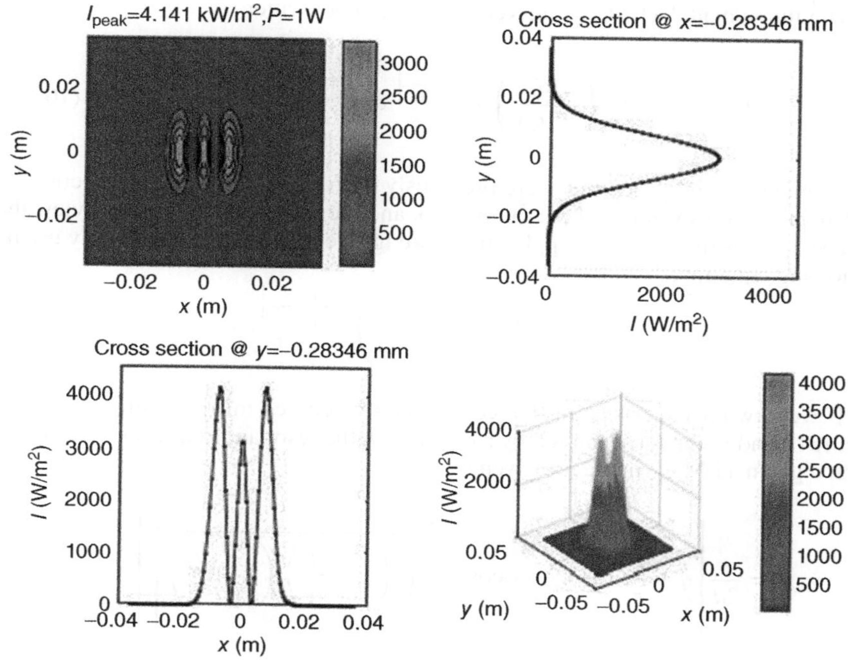

FIGURE 10.17 Example irradiance distribution for the TEM_{20} transverse mode (same q) of the optical resonator.

transverse modes exhibit irradiance distribution with one or more nulls. This means that the subscripts m and n really indicate the number of nulls along two orthogonal axes perpendicular to the resonator axis. This implies that if a small optical receiver is centered on the propagation axis, many of these irradiance distributions, for example, TEM_{10}, TEM_{01}, and TEM_{11}, have holes in the center of the pattern and negligible light will be received. This is not a desirable situation for any system design. For the few that do have energy patterns centered on the said axis, such as the TEM_{20} irradiance pattern shown in Figure 10.17, a lot of the total laser energy is outside of the center pattern. This also is not a desirable situation for some laser applications. Thus, the fundamental mode TEM_{00q} has a very desirable irradiance distribution that would be useful in any laser application since it has with no nulls and has all of its power centered around the propagation axis. Not unexpectedly, this distribution will experience the least diffraction loss and angular divergence, which allow the highest receiver power and smallest possible focused spot, respectively. There are practical examples of laser running in several transverse modes such as laser designators where high energy output is traded for less perfect beam quality. In these lasers, many transverse modes are allowed to fill up the mode volume of the lasing medium, allowing extraction of the pump light more efficiently and lowering the peak power at the output surface. Unstable cavities also are used to increase the mode size in this type of laser.

The fundamental TEM_{00q} transverse mode can be written as

$$E(x,y,z) = E_0\left(\frac{w_0^2}{w^2(z)}\right) e^{-\left[\frac{r^2}{w^2(z)}\right]} e^{-i\left[k\left(\frac{r^2}{2R(z)}\right)-\eta(z)\right]}. \tag{10.136}$$

Its intensity and properties were previously derived and discussed in Section 7.7 of Chapter 7, for example, see Example 7.6, and the reader is directed to review that material at this time. In most applications, we ignore the phase term and only use the following equation:

$$E(x,y,z) = E_0\left(\frac{w_0^2}{w^2(z)}\right) e^{-\left[\frac{r^2}{w^2(z)}\right]}. \tag{10.137}$$

If we now have a resonator that comprised two curved mirrors with radii of curvature c_1 and c_2, again separated by a distance d, the resonant frequencies associated with a given TEM_{mnq} mode are given by

$$\nu_{mnq} = \frac{c}{2d}\left[q + \frac{1}{\pi}(1+n+m)\cos^{-1}\left\{\sqrt{\left(1-\frac{d}{c_1}\right)\left(1-\frac{d}{c_2}\right)}\right\}\right]. \tag{10.138}$$

10.7 THE ABCD MATRIX AND RESONATOR STABILITY

A stable optical cavity consists of two or more optical elements (usually mirrors) that causes the rays to eventually replicate themselves. An example is shown in Figure 10.18a. The initial ray is reflected by the curved mirrors and circulates back on itself. They have low diffraction loss a smaller mode volume. This configuration is useful for low-gain, low-volume lasers.

An unstable optical cavity consists of two or more optical elements that cause the rays to eventually not replicate themselves. The initial ray is reflected by the non-symmetric curved mirrors and each mirror reflection takes the ray further from the optic axis, resulting in high diffraction losses. An example is shown in Figure 10.18b. This configuration is most useful in high-gain, high-volume lasers. Unstable resonator modes are not described by Hermite–Gaussian functions and their intensity profile is often generated numerically. Special techniques such as using radially graded reflectivity mirrors in the resonator are sometimes used to optimize unstable resonators.

The choice of cavity geometry affects the mode size and as we saw in the transverse mode section, power will not be extracted from areas outside the mode [25]. This implies that laser power may go to waste with the wrong design. In addition, undesired parasitic modes may occur. The laser designer needs to design the cavity so that the mode overlaps the pumped region of the gain medium, as illustrated in Figure 10.19 [25].

Reference [25] discussed the ABCD matrix approach to analyzing optical systems, which has been summarized in Appendix D. Following Galvin and Eden [22], we show how this technique can be applied to the resonator stability.

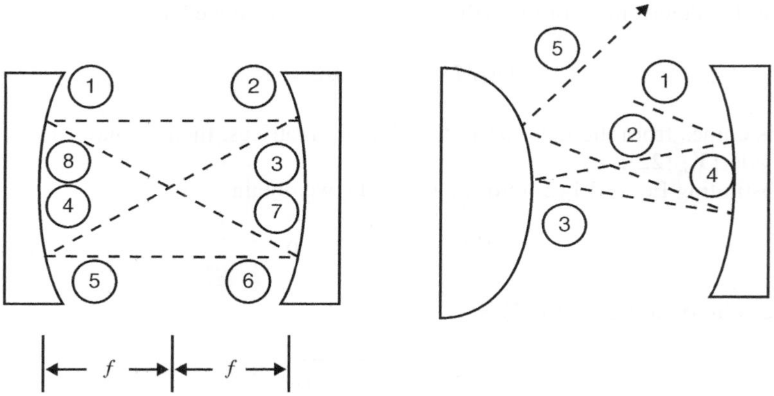

(a) (b)

FIGURE 10.18 Example configurations of (a) stable and (b) unstable resonators.

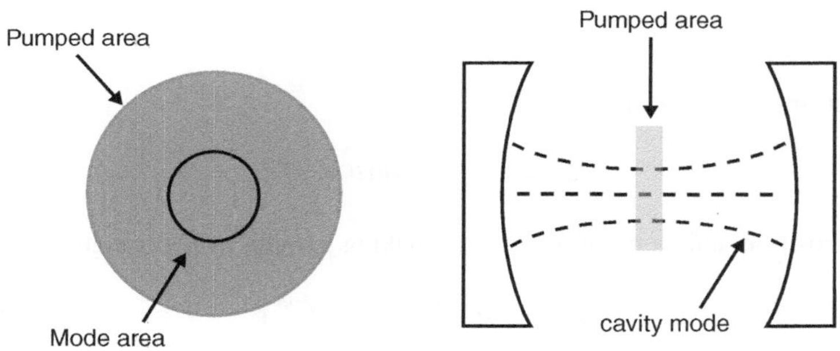

(a) (b)

FIGURE 10.19 Relationship between laser pump region and cavity mode structure.

They first cast the stability of a cavity as an eigenvalue problem using the ABCD matrix formulism. Specifically, they wrote

$$\begin{bmatrix} A & B \\ C & D \end{bmatrix} \begin{bmatrix} \rho \\ \rho' \end{bmatrix} = \zeta \begin{bmatrix} \rho \\ \rho' \end{bmatrix}, \tag{10.139}$$

where ζ represents the eigenvalues of the ABCD matrix. Multiplying out Eq. (10.138), we have

$$(A - \zeta)(D - \zeta) - BC = 0 \tag{10.140}$$

or

$$\zeta^2 (A + D)\zeta + AD - BC = 0. \tag{10.141}$$

Now, the determinant of the ABCD matrix is unity, so we have

$$\det\{\boldsymbol{M}\} = AD - BC = 1. \tag{10.142}$$

This comes from the fact that if $\boldsymbol{X}$ and $\boldsymbol{Y}$ are matrices, then we have $\det\{\boldsymbol{XY}\} = \det\{\boldsymbol{X}\}\det\{\boldsymbol{Y}\}$ [22].[3]

Substituting Eq. (10.142) into Eq. (10.141), we obtain

$$\zeta^2 (A + D)\zeta + 1 = 0. \tag{10.143}$$

The solution to Eq. (10.143) is

$$\zeta = \xi \pm \sqrt{\xi^2 - 1}, \tag{10.144}$$

where

$$\xi = \frac{(A + D)}{2}. \tag{10.145}$$

Assume that $|\xi| \leq 1$ for the moment. Then we can write

$$\xi = \cos\theta, \tag{10.146}$$

which means we can write

$$\zeta_\pm = \cos\theta \pm i\sin\theta = e^{i\theta}. \tag{10.147}$$

The position and slope of a ray after making j round trips through the cavity is then

$$\boldsymbol{\rho}_j = c_+\boldsymbol{\rho}_+ e^{ij\theta} + c_-\boldsymbol{\rho}_- e^{-ij\theta}, \tag{10.148}$$

where $\boldsymbol{\rho}_+$ and $\boldsymbol{\rho}_-$ are the eigenvectors of the ABCD matrix.

Alternately, Eq. (10.148) can be rewritten as

$$\boldsymbol{\rho}_j = \boldsymbol{\rho}_{\cos}\cos(j\theta) + \boldsymbol{\rho}_{\sin}\sin(j\theta). \tag{10.149}$$

This equation implies that the maximum displacement is bounded by $\boldsymbol{\rho}_{\cos} + \boldsymbol{\rho}_{\sin}$. In addition, because it is the sum of sines and cosines, the position will periodically repeat itself.

Thus, our assumption that our assumption of $|\xi| \leq 1$ indicates that the cavity is stable.

This inequality also can be written as

$$-1 \leq \frac{(A + D)}{2} \leq 1 \tag{10.150}$$

[3]This assumes that matrix $\boldsymbol{M}$ represents an optical system surrounded by the same medium. If the media on the two sides have refractive indices n_{in} and n_{out}, then $\det\{\boldsymbol{M}\} = n_{\text{out}}/n_{\text{in}}$.

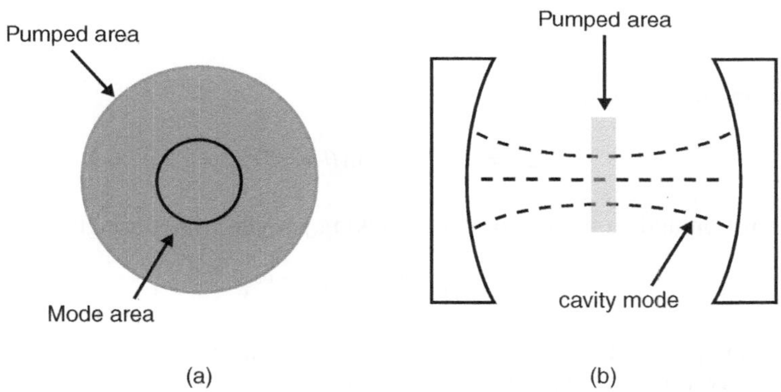

or

$$0 \leq \frac{(A+D+2)}{4} \leq 1. \tag{10.151}$$

Given the above, it will not be too surprising that $|\xi| > 1$ indicates an unstable resonator. The analysis is as follows: If $|\xi| > 1$, then $\zeta = \xi \pm \sqrt{\xi^2 - 1}$ is both real and positive. This means that after j passes through the cavity, the solution vector is given by

$$\rho_j = c_+ \rho_+ e^{j\zeta_+} + c_- \rho_- e^{j\zeta_-}, \tag{10.152}$$

which shows a solution that grows without bounds and never finds an equilibrium.

10.8 STABILITY OF A TWO-MIRROR RESONATOR

Let us now analyze a laser resonator that comprised two spherical mirrors with radii R_1 and R_2 separated by a distance d, as shown in Figure 10.20a. Assume that the input and output environments are the same. Using the results from Appendix D, we can write the single round-trip ABCD matrix for this resonator as

$$\begin{bmatrix} A & B \\ C & D \end{bmatrix} = \begin{bmatrix} 1 & 0 \\ -\frac{2}{R_1} & 1 \end{bmatrix} \begin{bmatrix} 1 & d \\ 0 & 1 \end{bmatrix} \begin{bmatrix} 1 & 0 \\ -\frac{2}{R_1} & 1 \end{bmatrix} \begin{bmatrix} 1 & d \\ 0 & 1 \end{bmatrix} \tag{10.153}$$

$$= \begin{bmatrix} 1 - \frac{2d}{R_1} & 2d\left(1 - \frac{d}{R_2}\right) \\ -\frac{2}{R_1} - \frac{2}{R_2} + \frac{4}{(R_1 R_2)} & 1 - \frac{4d}{R_1} - \frac{2d}{R_2} + \frac{4d^2}{(R_1 R_2)} \end{bmatrix} \tag{10.154}$$

$$= \begin{bmatrix} 2g_2 - 1 & 2dg_2 \\ \frac{2}{d}[2g_1g_2 - g_1 - g_2] & 4g_1g_2 - 2g_2 - 1 \end{bmatrix}, \tag{10.155}$$

where

$$g_1 = 1 - \frac{d}{R_1} \tag{10.156}$$

and

$$g_2 = 1 - \frac{d}{R_2}. \tag{10.157}$$

If we recall the ABCD matrix for a thin lens in Appendix D, that is, Eq. (D.33), it is apparent that the first and third matrix in Eq. (10.153) are the same if the focus length of the lens is one half the curvature of the spherical mirror, that is, $f = \frac{R}{2}$. This implies that our cavity is equivalent to the lens system in Figure 10.20b under that condition.

After the kth roundtrip within the optical resonator, the ray is given by

$$\hat{\rho}_k = \mathbf{M}^1 \cdot \hat{\rho}_{k-1} = \mathbf{M}^2 \cdot \hat{\rho}_{k-2} = \mathbf{M}^k \cdot \hat{\rho}_0. \tag{10.158}$$

This equation also suggests that the cavity under k roundtrips is equivalent to a periodic lens system shown in Figure 10.20b repeated k times.

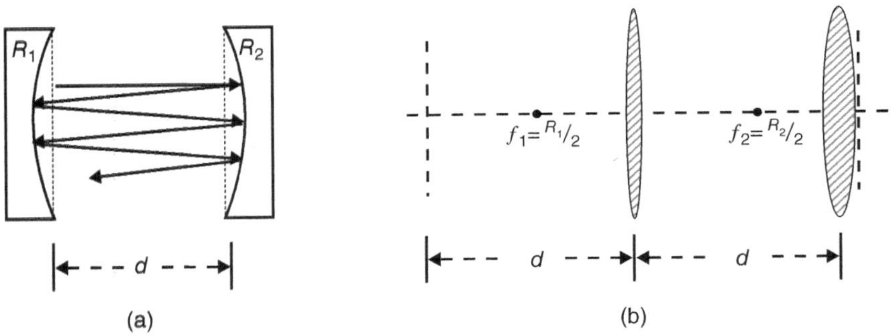

FIGURE 10.20 (a) Optical cavity formed by two spherical mirrors and (b) two lens system.

Now for a two-mirror resonator, the parameter ξ derived in the last section under a single pass condition equals

$$\xi = \frac{(A+D)}{2} = \frac{1}{2}[2g_2 - 1 + 4g_1g_2 - 2g_2 - 1] \tag{10.159}$$

$$= \frac{1}{2}[4g_1g_2 - 2] = 2g_1g_2 - 1. \tag{10.160}$$

For the cavity to be stable, we require

$$0 \leq \frac{(A+D+2)}{4} = g_1g_2 \leq 1. \tag{10.161}$$

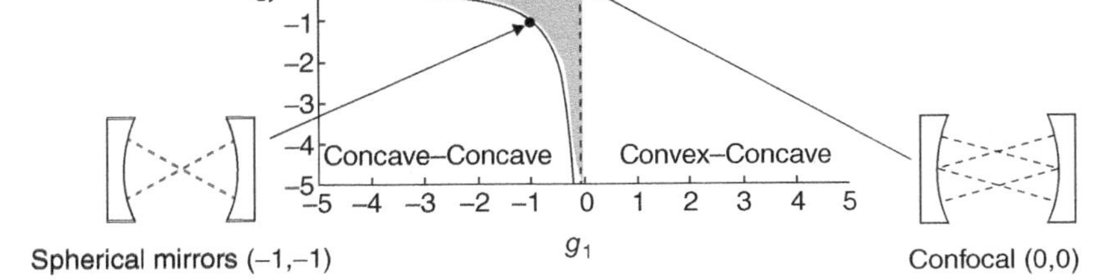

FIGURE 10.21 Plot of g_1 versus g_2 showing regions of stability and instability with example resonator structures.

Figure 10.21 plots g_1 versus g_2 showing the regions of stability and instability. In addition, with example, resonator design points are illustrated.

10.9 PROBLEMS

Problem 10.1. The ratio of the spontaneous to stimulated emission rates is given by

$$\frac{R_{\text{Spon}}}{R_{\text{Stim}}} = N_1 e^{\frac{h\nu}{kT}} - 1.$$

What is the ratio at $\lambda = 600$ nm for a tungsten lamp operating at 2000 K?

Problem 10.2. Assume that we have a lamp that has a radiance of 95 W/(cm^2 sr) at $\lambda = 546$ nm. What is the radiance of a 1 W Argon laser at $\lambda = 546$ nm, assuming a diffraction-limited beam? Which is larger and by now much?

Problem 10.3. Consider a lower energy level situated 200 cm^{-1} from the ground state. There are no other energy levels nearby. Determine the fraction of the population found in this level compared to the ground state population at a temperature of 300 K.

HINT: The conversion from cm^{-1} to joules is given by

$$E(\text{joules}) = 100\, hc\, E(\text{cm}^{-1}).$$

Problem 10.4. Find the FSR, Q, and F of the cavity shown as follows.

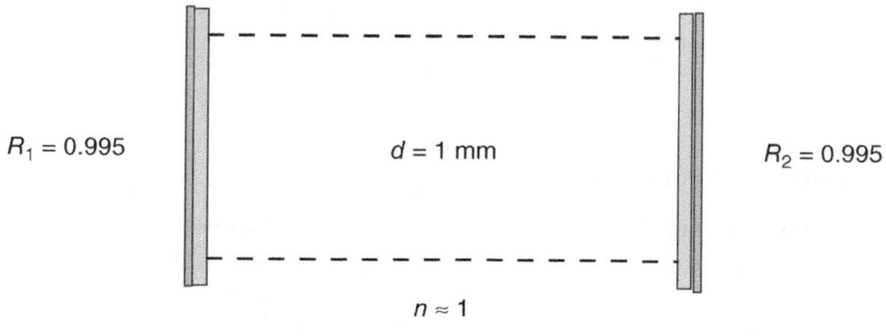

Assume that the wavelength of light equals 1 μm.

Problem 10.5. Find the FSR, Q, and F of the cavity shown as follows.

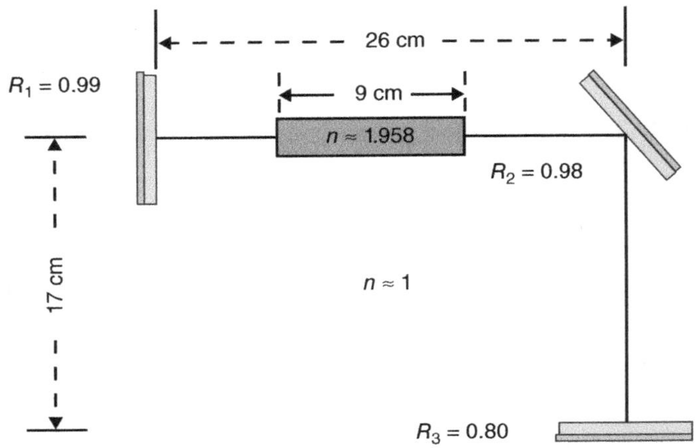

Assume that the wavelength of light equals 1 μm and $\alpha = 0.001\ \text{cm}^{-1}$.

HINT: If we have a cavity with a nonuniform index of refraction, then the resonance frequency of the cavity then is given by

$$\nu_q = q\frac{c}{\int_{\text{Round Trip}} n(z)dz}$$

and a photon surviving a round trip in the cavity is given by

$$S = \left(R_1 e^{-\int_0^{L_0} \alpha(z)dz}\right)\left(R_2 e^{-\int_{L_0}^{0} \alpha(z)dz}\right),$$

where L_0 is the length of the absorptive medium.

Problem 10.6. The amplifying medium of a helium–neon laser has an amplification spectral band equal to $\Delta\nu = 1$ GHz at $\lambda = 632.8$ nm. For simplicity, the spectral profile is assumed to be rectangular. The linear cavity is 30 cm long. Calculate the number of longitudinal modes that can oscillate in this cavity.

Problem 10.7. Assume that we have a CO_2 laser that has a bandwidth of $\Delta\nu = 1$ GHz at = 10.6 μm . For simplicity, the spectral profile is assumed to be rectangular. The length of the cavity is equal to 1 m. Calculate the number of longitudinal modes that can oscillate in this cavity. At what distance must the cavity mirror be placed so that at least one of the modes falls in the amplification band?

Problem 10.8. Assume an InGaAsP–InP laser diode that has a resonator cavity equal to 250 μm. The peak radiation is at $\lambda = 1.55$ μm. The refractive index of InGaAsP is 4. The optical gain bandwidth (as measured between the 50% intensity points) is assumed for this problem to be 2 nm. What is the mode integer of the radiation peak? What is the separation $\Delta\lambda$ between the modes of the cavity? How many of the modes are within the gain band of the laser? What is the reflection coefficient and reflectance at the ends of the resonator cavity, which we will assume at the ends of the InGaAsP crystal? The beam divergence full angle is 5° in the x-direction and 20° in the y-direction. Estimate the x and y dimensions of the cavity?

HINT: Assume a Gaussian beam with the waist located at the output of the cavity where we assume that its waist size approximates the x and y dimensions of the cavity.

Problem 10.9. Verify that for $m = n = 0$, with plane-parallel mirrors ($c_1 = c_2 = \infty$) in a cavity of length d, the resonant longitudinal modes are given by Equation (10.114).

Problem 10.10. Calculate the gap in frequency between two longitudinal modes in a linear cavity with a length of 300 mm.

Problem 10.11. A helium–neon laser emitting at 632.8 nm light makes a spot with a radius equal to 100 mm at e^{-2} at a distance of 500 m from the laser. What is the radius of the beam at the waist (considering the waist and the laser are in the same plane)?

HINT: Assume a Gaussian laser beam.

Problem 10.12. Find the output ray of the system shown as follows when the input ray is characterized by $\rho = 0.1$ cm and $\rho' = -0.1$ cm^{-2}.

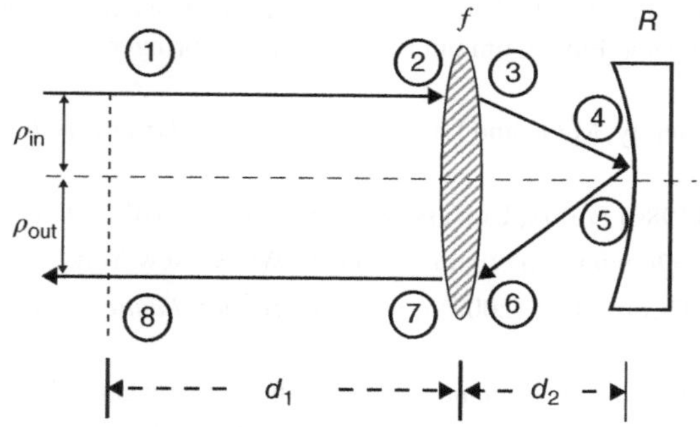

assuming $d_1 = 7$ cm, $d_2 = 5$ cm, $f = 200$ cm, and $R = 400$ cm.

Problem 10.13. Determine the minimum radius of curvature of the two mirrors to ensure the following cavity is stable:

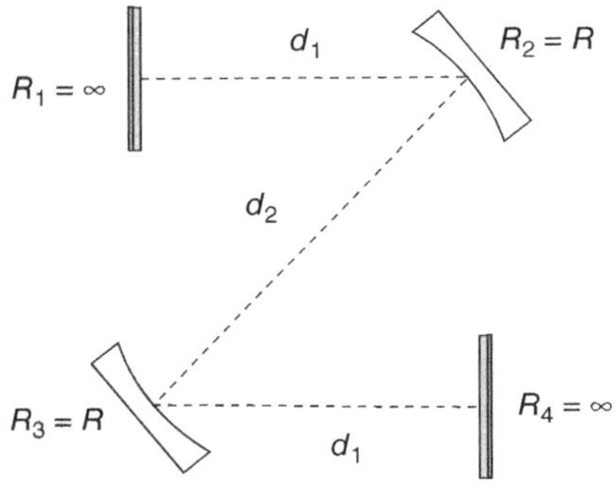

REFERENCES

1. Einstein, A. (1917) Zur Quantentheorie der Strahlung (On the quantum theory of radiation). *Physika Zeitschrift*, **18**, 121–128.
2. Chen, J. and Toth, C.K. (2009) *Topographic Laser Ranging and Scanning: Principles and Processing*, CRC Press, Boca Raton, FL, 616 pages.
3. Weitkamp, C. (ed.) (2005) *Lidar: Range-Resolved Optical Remote Sensing of the Atmosphere*, Springer Series in Optical Sciences, 456 pages, Springer, New York.
4. Measures, R.M. (1992) *Laser Remote Sensing: Fundamentals and Applications*, Krieger Publishing Company, Malabar, FL, 510 pages.
5. Majumdar, A.K. (ed.) (2014) *Advanced Free Space Optics (FSO): A Systems Approach*, Springer, New York.
6. Karp, S. and Stotts, L.B. (2012) *Fundamentals of Electro-Optic Systems Design: Communications, Lidar, and Imaging*, Cambridge Press, New York.
7. Majumdar, A. and Ricklin, J. (2008) *Free Space Laser Communications: Principles and Advances*, Optical and Fiber Communications Series, Springer Science+Business Media, New York.
8. Sargent, M. III, Scully, M.O., and Lamb, W.E. Jr. (1974) *Laser Physics*, Addison-Wesley, Reading, MA.
9. Siegman, A.E. (1986) *Lasers*, University Science Books, Mill Valley, CA.
10. Yariv, A. (1989) *Quantum Electronics*, 3rd edn, Wiley, New York.
11. Cagnac, B. and Faroux, J.-P. (2002) *Lasers. Interaction Lumière-Atomes*, EDP Sciences, Les Ulis.
12. Silvfast, W.T. (2004) *Laser Fundamentals*, Cambridge University Press, Cambridge.
13. Dangoisse, D., Hennequin, D., and Zehnlé, V. (2004) *Les Lasers*, 2nd edn, Dunod, Paris.
14. Koechner, W. (2006) *Solid-State Laser Engineering*, Springer, New York.
15. Saleh, B.E.A. and Teich, M.C. (2007) *Fundamentals of Photonics*, 2nd, Wiley Series in Pure and Applied Optics edn, John Wiley & Sons, New York.
16. Träger, F. (ed.) (2007) *Handbook of Lasers and Optics*, Springer, New York.
17. Delsart, C. (2008) *Lasers et Optique Non-linéaire*, Ellipses, Paris.

18. Grynberg, G., Aspect, A., and Fabre, C. (2010) *Introduction to Quantum Optics: From the Semi-classical Approach to Quantized Light*, Cambridge University Press, Cambridge.
19. Piot, P. (2008) *Physics 630 Lecture Notes*, Lecture 17, nicadd.niu.edu/~piot/phys_630/Lesson17.pdf (accessed 26 October 2016).
20. Sheehy, B. (2013) *Laser Basics*, Session 1, US Particle Accelerator School, http://uspas.fnal.gov/materials/13Duke/Session1_LaserBasics.pdf (accessed 26 October 2016).
21. Li, Y. (2008) *Laser Basics*, Session 1, US Particle Accelerator School, http://uspas.fnal.gov/materials/08UMD/Lecture1.pdf (accessed 26 October 2016).
22. Galvin, T. and Eden, G. *Optical Resonator Modes*, ECE 455 Optical Electronics Lecture Notes, https://courses.engr.illinois.edu/ece455/Files/Galvinlectures/02_CavityModes.pdf (accessed 26 October 2016).
23. Franks, L.E. (1969) *Signal Theory*, Prentice-Hall, Englewood Cliffs, NJ.
24. Silfvast, W.T. (2003) Lasers, in *Fundamentals of Photonics* (eds A. Guenther, L.S. Pedrotti, and C. Roychoudhuri), Lecture Notes, Module 5, Materials developed under Project STEP (Scientific and Technological Education in Photonics) by The University of Connecticut and CORD funded by National Science Foundation.
25. *Laser Light: Physics 5734/4734*, Lecture Notes, Physics Department, University of Arkansas, http://physics.uark.edu/lasers/LPCH01.pdf (accessed 26 October 2016).

APPENDIX A

STATIONARY PHASE AND SADDLE POINT METHODS

A.1 INTRODUCTION

The intent of this appendix is to provide a simple approximate solution for the integral

$$\int_{A}^{B} g(x) e^{ikf(x)}\, dx. \tag{A.1}$$

A.2 THE METHOD OF STATIONARY PHASE

In one dimension, the solution can be found by reducing the Eq. (A.1) to the Fresnel Integral

$$F = \int_{-\infty}^{\infty} e^{iax^2} dx = \sqrt{\frac{\pi}{2a}}(1+i). \tag{A.2}$$

To understand the solution to come, let us look at the real and imaginary parts of the integrand of

$$F = \sqrt{\frac{\pi}{2a}}(1+i).$$

The main contribution to the real part of F

$$F = \int_{-\infty}^{\infty} \cos(ax^2) dx = \sqrt{\frac{\pi}{2a}} \tag{A.3}$$

comes from the interval, $-\sqrt{\frac{\pi}{2a}} < x < \sqrt{\frac{\pi}{2a}}$, and the rest cancels out because of the oscillations of the cosine function (see Figure A.1a).

Free Space Optical Systems Engineering: Design and Analysis, First Edition. Larry B. Stotts.

Companion website: www.wiley.com\go\stotts\free_space_optical_systems_engineering

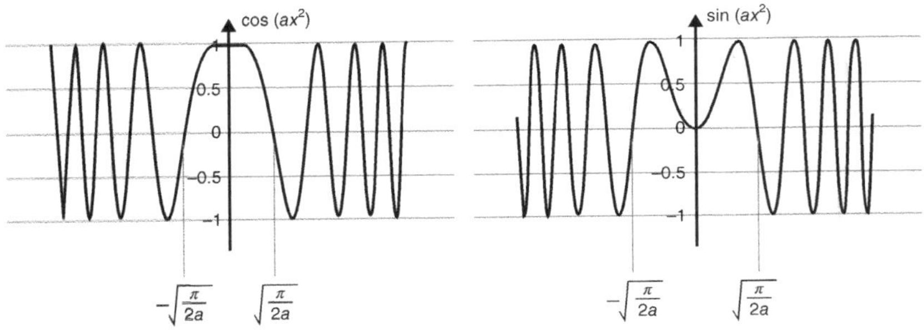

FIGURE A.1 Oscillatory nature of the (a) Cosine function and (b) Sine function.

Referring to Figure A.1b, it is plausible that the imaginary part of F is given by

$$F = \int_{-\infty}^{\infty} \sin(ax^2)dx \approx \sqrt{\frac{\pi}{2a}} \tag{A.4}$$

over the same interval $-\sqrt{\frac{\pi}{2a}} < x < \sqrt{\frac{\pi}{2a}}$, with the rest canceling out because of the oscillations of the sine function.

Looking again at Eq. (A.1), let us see how it will assume the form of a Fresnel Integral. The exponent $kf(x)$ might vary rapidly over most of the x-regime, but let us assume that it is "stationary" around $x = x_0$. This means that the first derivate $\frac{\partial f}{\partial x}$ equals 0 at $x = x_0$ and that we can approximate function $f(x)$ by the equation

$$f(x) \approx f(x_0) + \frac{1}{2!}\frac{\partial^2 f(x_0)}{\partial x^2}. \tag{A.5}$$

Given the above, the result of the integration will depend on

(a) $g(x)$
(b) $\cos[kf(x)]$
(c) the width of the unusually wide maximum of $\cos[kf(x)]$ at $x = x_0$.

Item (c) in the above list will be narrow if the bend of $f(x)$ at $x = x_0$ is sharp, which depends on $\frac{\partial^2 f(x_0)}{\partial x^2}$. The function $g(x)$ should change only a little during one oscillation, which is achieved by a large k (see Figure A.2). If $f(x)$ is stationary only once within an interval $A < x < B$, then we will have a contribution only from there. If so, it does not make a difference to the answer if we extend the integration limits from (A, B) to $(-\infty, \infty)$ as long as $kf(x)$ does not have another stationary point outside the interval (A, B) .

Given the above is true, we have

$$\int_{A}^{B} g(x)\, e^{ikf(x)}dx \approx \int_{-\infty}^{\infty} g(x)\, e^{ikf(x)}dx \approx \int_{-\infty}^{\infty} g(x) e^{ik\left[f(x_0)+\frac{1}{2!}\frac{\partial^2 f(x_0)}{\partial x^2}\right]}dx. \tag{A.6}$$

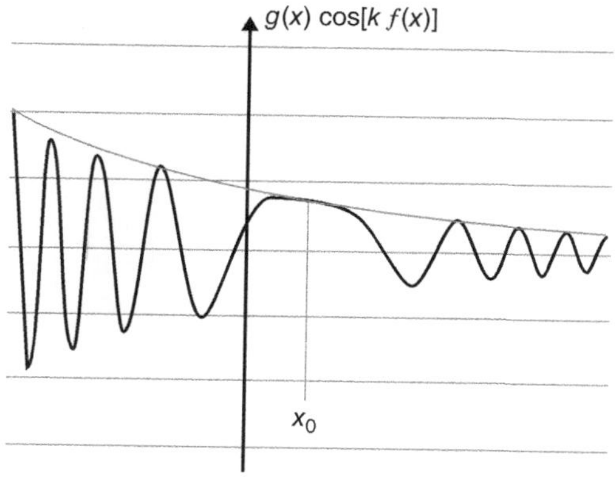

FIGURE A.2 Notional plot of $g(x)\cos[kf(x)]$.

Expanding $g(x)$ into a Taylor Series and integrating, we have

$$\int_A^B g(x)e^{ikf(x)}dx \approx \sqrt{\frac{2\pi}{\left(\frac{\partial^2 f(x_0)}{\partial x^2}\right)}}g(x_0)e^{ikf(x_0)+\frac{\pi}{4}}. \tag{A.7}$$

If $\frac{\partial f}{\partial x}$ has more than one zero, then

$$\int_A^B g(x)e^{ikf(x)}dx \approx \sum_n \sqrt{\frac{2\pi}{\left(\frac{\partial^2 f(x_n)}{\partial x^2}\right)}}g(x_n)e^{ikf(x_n)+\frac{\pi}{4}}. \tag{A.8}$$

A.3 SADDLE POINT METHOD

This method is essentially the same as the method of stationary phase, except it applies to the two-dimensional version of the previous integral. That is, we are interested in the approximate solution to the integral

$$\int_{A_x}^{B_x}\int_{A_y}^{B_y} g(x,y)\,e^{ikf(x,y)}dx\,dy. \tag{A.9}$$

Following in essence the same development as above, we can show that

$$\int_{A_x}^{B_x}\int_{A_y}^{B_y} g(x,y)\,e^{ikf(x,y)}\,dx\,dy \approx \frac{2\pi g(x_0,y_0)\,e^{ikf(x_0,y_0)+\frac{i\pi}{2}}}{k\sqrt{\left(\frac{\partial^2 f(x_0,y_0)}{\partial x^2}\right)\left(\frac{\partial^2 f(x_0,y_0)}{\partial y^2}\right)-\left(\frac{\partial^2 f(x_0,y_0)}{\partial x\partial y}\right)}}, \tag{A.10}$$

where x_0 and y_0 are solutions to the equations $\frac{\partial f}{\partial x}=0$ and $\frac{\partial f}{\partial y}=0$, respectively.

APPENDIX B

EYE DIAGRAM AND ITS INTERPRETATION

B.1 INTRODUCTION

In communications, an *eye diagram* is used to visually assess the performance of a system in operation.

B.2 EYE DIAGRAM OVERVIEW

It is called an eye diagram, or eye pattern, because the pattern looks like a series of eyes between a pair of rails for several types of coding schemes. It is created by taking the time domain signal and overlapping the traces for a certain number of symbols. If we are sampling a signal at a rate of 10 samples per second and we want to look at two symbols, then we would cut the signal every 20 samples and overlap them.

In practice, it is created on an oscilloscope display by repetitively sampling the digital signal from a receiver and placing it into the vertical input of the scope while the data rate triggers the horizontal sweep.

The overlapped signals show us a lot of useful information, as is depicted in Figure B.1:

- Eye opening quality (height, peak-to-peak) ↔ Additive noise/intersymbol interference
- Eye overshoot/undershoot ↔ Peak distortion due to interruption in the signal path
- Eye width ↔ Timing synchronization and jitter effects.

Figure B.2 shows some eye diagrams for single and multiple trigger sweeps.

Free Space Optical Systems Engineering: Design and Analysis, First Edition. Larry B. Stotts.

Companion website: www.wiley.com\go\stotts\free_space_optical_systems_engineering

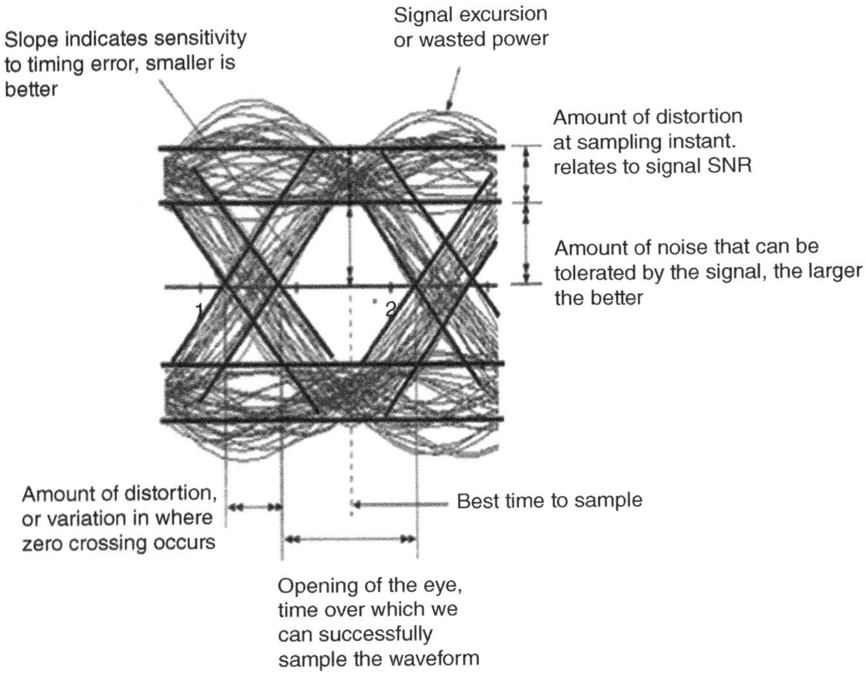

FIGURE B.1 Illustration of an eye diagram and its interpretation. Source: Reproduced by Permission by Originator, Charan Langton.

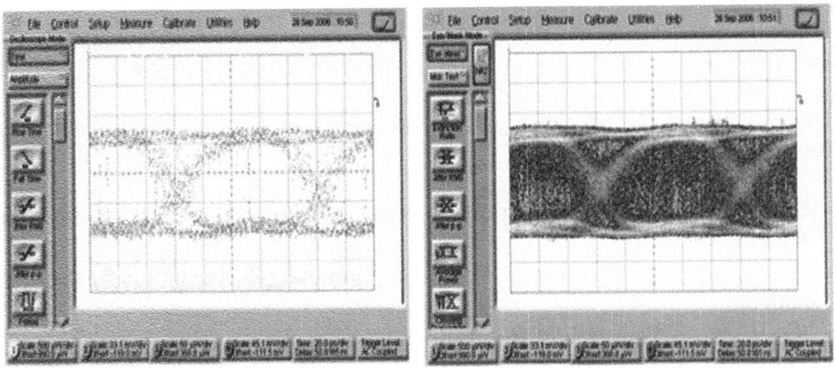

(a) (b)

FIGURE B.2 Example of an eye diagram from (a) single trigger of recovered 10 Gbps eye, 600 ms persistence and (b) long persistence of recovered 10 Gbps eye.

APPENDIX C

VECTOR-SPACE IMAGE REPRESENTATION

C.1 INTRODUCTION

This appendix develops a vector-space algebra formalism for representing discrete image fields as vectors, both easing mathematical analysis and allowing leveraging RF radar and communication algorithms [1].

C.2 BASIC FORMALISM

Figure C.1 illustrates a basic optical surveillance and/or reconnaissance sensing scenario. A continuous radiance field $F(x, y; t)$, containing both a target (here, a truck) and clutter (terrain) radiative sources, is imaged by a lens system onto a two-dimensional optical focal plane array. The detector array creates a sampled output image $\{F(m, l); 1 \leq m \leq N_1; 1 \leq l \leq N_2\}$. Within each sample *picture element* (*pixel*), one has the number of photoelectrons (or counts) derived from the radiance field $F(x, y; t)$ with the equivalent ground area that is captured during the integration period T_1. It is subject to various internal optical receiver noise sources that we reviewed in Chapter 8. This output image is fed either into system hardware for immediate processing or into a storage medium that will be hooked into processing hardware at a later time. For purposes of statistical derivations and analysis, it is often convenient to convert the image matrix to vector form by column (or row) scanning F, and then stringing the elements together in a long vector, that is, we define the image vector $\boldsymbol{F}'$ generated by the image matrix $\boldsymbol{F}$ as

$$\boldsymbol{F}' = \begin{bmatrix} F(1, 1) \\ \vdots \\ F(1, N_2) \\ F(2, 1) \\ \vdots \\ F(N_1, N_2) \end{bmatrix}. \tag{C.1}$$

Free Space Optical Systems Engineering: Design and Analysis, First Edition. Larry B. Stotts.

Companion website: www.wiley.com\go\stotts\free_space_optical_systems_engineering

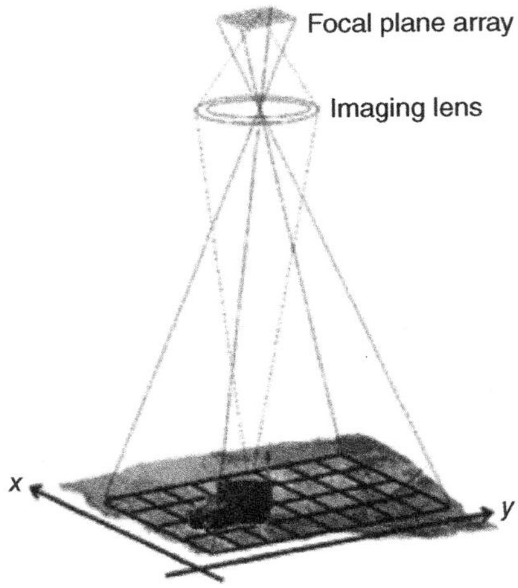

FIGURE C.1 Illustrative optical surveillance and reconnaissance geometry.

Pratt proposed that an equivalent scanning operation can be expressed in quantitative form by the use of an $N_2 \times 1$ operational vector $\boldsymbol{v}_n$ and an $N_1 * N_2 \times N_2$ matrix $\boldsymbol{N}_n$ defined as

$$\boldsymbol{v}_n = \begin{bmatrix} 0 \\ \vdots \\ 0 \\ 1 \\ 0 \\ \vdots \\ 0 \end{bmatrix} \begin{matrix} 1 \\ \vdots \\ n-1 \\ n \\ n+1 \\ \vdots \\ N_2 \end{matrix} \tag{C.2}$$

and

$$\boldsymbol{N}_n = \begin{bmatrix} \mathbf{0} \\ \vdots \\ \mathbf{0} \\ \mathbf{1} \\ \mathbf{0} \\ \vdots \\ \mathbf{0} \end{bmatrix} \begin{matrix} 1 \\ \vdots \\ n-1 \\ n \\ n+1 \\ \vdots \\ N_2 \end{matrix} \tag{C.3}$$

[1].

Then, the vector representation of the image matrix $\boldsymbol{F}$ is given by the stacking operation

$$\boldsymbol{F}' = \sum_{n=1}^{N_2} \boldsymbol{N}_n \boldsymbol{F} \boldsymbol{v}_{n.} \tag{C.4}$$

The vector in Eq. (C.4) extracts the nth column from $\boldsymbol{F}$ and the matrix places this column into the nth segment of the vector $\boldsymbol{F}'$. Thus, $\boldsymbol{F}'$ contains the column-scanned elements of $\boldsymbol{F}$. The inverse operation is given by

$$\boldsymbol{F} = \sum_{n=1}^{N_2} \boldsymbol{N}_n^T \boldsymbol{F}' \boldsymbol{v}_n^T. \tag{C.5}$$

In summary, we can easily convert images into a two-dimensional array and vice versa using the matrix-to-vector operator of Eq. (C.4) and the vector-to-matrix operator of Eq. (C.5), respectively.

REFERENCE

1. Pratt, W.K. (2007) *Digital Image Processing*, 4th edn, John Wiley and Sons, pp. 121–123.

APPENDIX D

PARAXIAL RAY TRACING – ABCD MATRIX

D.1 INTRODUCTION

As we found in many of the chapters, the first-order analysis of light propagation through any medium or system involves light rays that stay close to the optical axis, usually taken to be the z-axis, of the system. Such rays are called *paraxial rays*. These rays obey the following trigonometric approximations:

$$\tan\theta \approx \theta \approx \sin\theta. \tag{D.1}$$

This approximation is accurate to within 2% for angles less than 25°.

Using this approximation, one can completely characterize an optical system or medium using a 2×2 ray-transfer matrix relating to the position and inclination of the transmitted ray after modification of the original ray by the elements of the system/medium [1–3]. This appendix provides the background theory of this approach following references [1].

D.2 BASIC FORMALISM

Figure D.1 shows a paraxial ray making a small angle with the optical axis. This ray has an arc length of ds that is approximating equal to dz, which results in the following propagation equation:

$$\nabla n = \frac{d}{dz}\, n\, \frac{d\boldsymbol{r}}{dz}. \tag{D.2}$$

In Eq. (D.2), $n = n(\boldsymbol{r})$ is the refractive index of the element/medium, $\boldsymbol{r} = \rho\, e_\perp + z\, e_\parallel$ and ρ is the lateral displacement of the ray from the z-axis. Since most optical systems

Free Space Optical Systems Engineering: Design and Analysis, First Edition. Larry B. Stotts.

Companion website: www.wiley.com\go\stotts\free_space_optical_systems_engineering

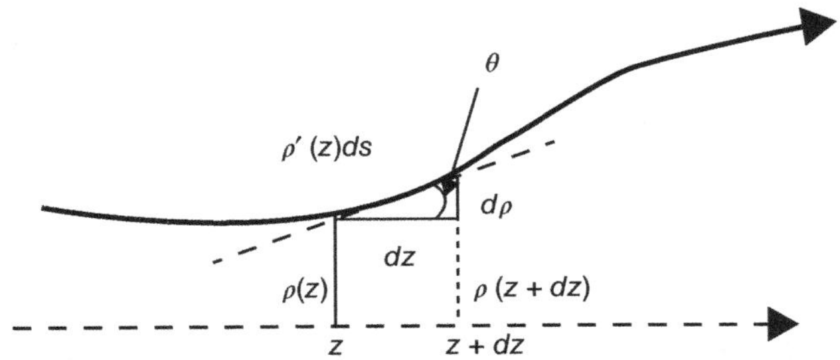

FIGURE D.1 Example of a paraxial ray making a small angle with the optical axis.

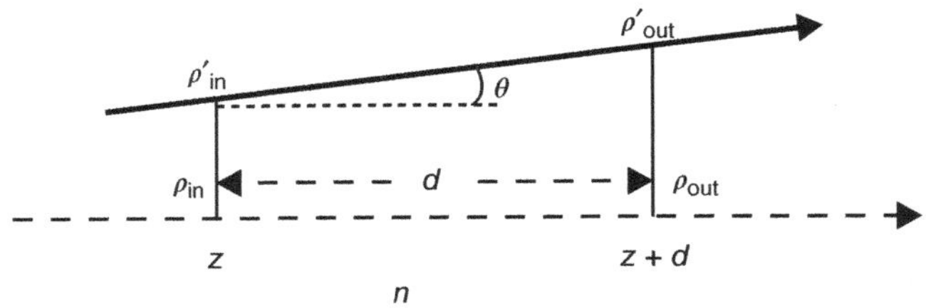

FIGURE D.2 Propagation in a homogeneous medium of length *d*.

and mediums have cylindrical symmetry, we have

$$e_{\perp} \cdot \nabla n = \frac{d}{dz} n \frac{d\rho}{dz} \tag{D.3}$$

This equation determines a propagating ray, given the lateral displacement $\rho_0 = \rho(z_0)$ and a slope defined as

$$\rho_0' \equiv \left. \frac{d\rho}{dz} \right|_{z=z_0} . \tag{D.4}$$

This situation and definitions for a homogeneous medium of length d are illustrated in Figure D.2.

Example D.1 In a homogeneous medium where $n \equiv$ constant, Eq. (D.3) reduces to

$$\frac{d}{dz} n \frac{d\rho}{dz} = 0 \tag{D.5}$$

Integrating this equation from z_0 to z, we find that the ray is a straight line characterized by the following slope and position:

$$\rho'(z) = \rho_0' \tag{D.6}$$

and

$$\rho(z) = \rho_0 + \rho_0' \, (z - z_0), \tag{D.7}$$

respectively.

If the ray goes from one medium to another, then we have

$$\rho_2'(z) = \left(\frac{n_1}{n_2}\right)\rho_1' \tag{D.8}$$

and

$$\rho_2(z) = \rho_1 + \left(\frac{n_1}{n_2}\right)\rho_1' z, \tag{D.9}$$

assuming that the interface between the two mediums is at $z_0 = 0$. Equation (D.8) is just Snell's law in the paraxial approximation that we were introduced to in Chapter 3. The second equation gives the ray displacement in the second medium in terms of the ray displacement ρ_1 and slope $\rho_2'(z) = \left(\frac{n_1}{n_2}\right)\rho_1'$ at the boundary just inside the second medium.

Given the above, it appears that a ray is completely specified by its lateral displacement ρ from the optical axis and its slope $\rho' = \frac{d\rho}{dz}$ under paraxial conditions. If we introduce a column matrix with ρ and ρ' its elements, which is written as

$$\hat{\rho} \equiv \begin{bmatrix} \rho \\ \rho' \end{bmatrix}, \tag{D.10}$$

we can introduce a 2×2 ray transfer function matrix of the form

$$\boldsymbol{M} \equiv \begin{bmatrix} A & B \\ C & D \end{bmatrix}. \tag{D.11}$$

To find the position and slope of a ray after propagating through an optical system/medium, we solve the following equation[1]:

$$\hat{\rho}_2 = \begin{bmatrix} \rho_2 \\ \rho_2' \end{bmatrix} = \boldsymbol{M} \cdot \hat{\rho}_1 \equiv \begin{bmatrix} A & B \\ C & D \end{bmatrix} \begin{bmatrix} \rho_1 \\ \rho_1' \end{bmatrix}. \tag{D.12}$$

For most applications, we just need to know the transformation properties of three basic elements: (1) free propagation in a homogeneous medium of length d and refractive index n, (2) reflection from a curved surface of radius of curvature R, and (3) refraction at a curved interface (radius of curvature R) between two media with refractive indices n_1 and n_2 when a ray is incident from medium n_1. Matrices for most other elements or complex systems can be derived from these equations.

D.2.1 Propagation in a Homogeneous Medium

Consider again a ray propagating from plane z to plane $z + d$ in a homogeneous medium of refractive index n, as defined in Example D.1. The solution we found in

[1]Different books, for example, Siegman [4], use different conventions for ABCD matrices than defined here. Some switch the position of r and r_0. Another possible convention is to multiply the slope by the refractive index. The reader is advised to pay attention to the form that is utilized.

the example suggests that we can write the solution in terms of the ABCD matrix as

$$\begin{bmatrix} \rho_2 \\ \rho_2' \end{bmatrix} = \begin{bmatrix} A & B \\ C & D \end{bmatrix} \begin{bmatrix} \rho_1 \\ \rho_1' \end{bmatrix} = \begin{bmatrix} 1 & d \\ 0 & 1 \end{bmatrix} \begin{bmatrix} \rho_1 \\ \rho_1' \end{bmatrix}. \tag{D.13}$$

D.2.2 Propagation Against a Curved Interface

Let us now turn to a ray propagating into a reflective curved interface. Figure D.3 illustrates this situation. Let R represent the curvature of this mirror surface, which is assumed to be concave mirror (reflecting surface). According to the laws of reflection, the angles of incidence and reflection of the incident ray are equal, that is, $\theta_1 = \theta_2$, which leads to

$$\rho_{\text{out}} = \rho_{\text{in}}, \tag{D.14}$$

and

$$\alpha_{\text{out}} - \varphi = \varphi - \alpha_{\text{in}} \tag{D.15}$$

or

$$\alpha_{\text{out}} = 2\varphi - \alpha_{\text{in}}, \tag{D.16}$$

referring to Figure D.3. Using these facts, the input and output ray slopes are given by

$$\rho'_{\text{in}} = \alpha_{\text{in}} \tag{D.17}$$

and

$$\rho'_{\text{out}} = -\alpha_{\text{out}}, \tag{D.18}$$

respectively. From this figure, we also see

$$\varphi = \frac{r_{\text{in}}}{R}. \tag{D.19}$$

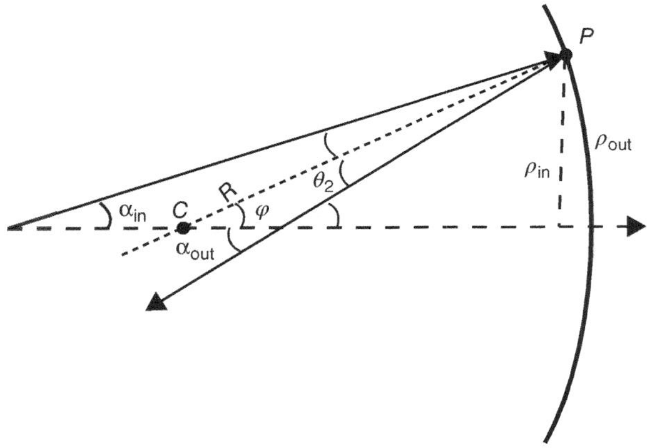

FIGURE D.3 Propagation into a curved mirror.

The above allows us to write that the exit ray slope as

$$\rho'_{out} = -2\left(\frac{r_{in}}{R}\right) + \alpha_{in}, \tag{D.20}$$

The transfer function matrix for a reflecting surface then is

$$\begin{bmatrix} A & B \\ C & D \end{bmatrix} = \begin{bmatrix} 1 & 0 \\ -\dfrac{2}{R} & 1 \end{bmatrix}. \tag{D.21}$$

D.2.3 Propagation into a Refractive Index Interface

This section addresses a ray propagating into a curved interface between two mediums of refractive index n_1 and n_2. Figure D.4 illustrates this situation. Let R represent the curvature of this curved interface between the two mediums, which is assumed to be convex from the perspective of the first medium.

From Figure D.4, we see that the input and output displacements are the same, that is,

$$\rho_{out} = \rho_{in}, \tag{D.22}$$

From Snell's law, we have

$$n_1\theta_1 \approx n_2\theta_2, \tag{D.23}$$

or

$$n_1(\alpha_{in} + \varphi) = n_2(\varphi + \alpha_{out}) \tag{D.24}$$

or

$$\alpha_{out} = -\left[\frac{n_2 - n_1}{n_2}\right]\varphi + \frac{n_1}{n_2}\alpha_{in}. \tag{D.25}$$

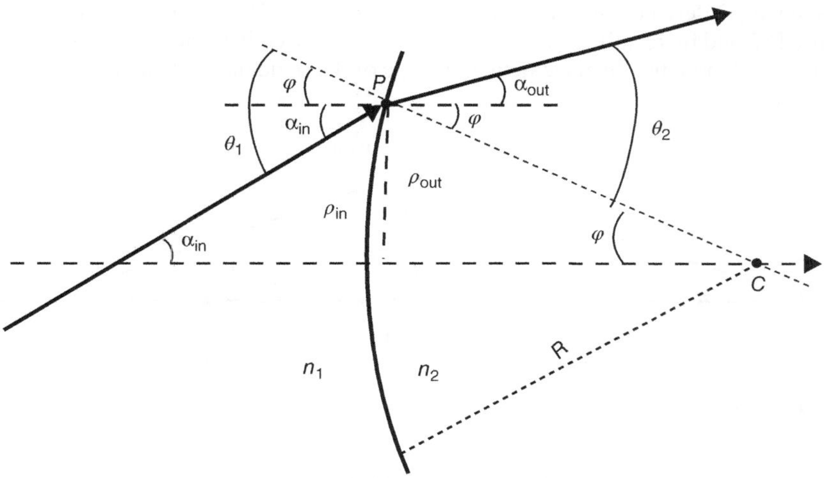

FIGURE D.4 Propagation into a refractive index interface.

Referring to Figure D.4 again, this means that the input and output ray slopes are given by

$$\rho_2' = -\left[\frac{n_2 - n_1}{n_2}\right]\frac{\rho_{\text{in}}}{R} + \frac{n_1}{n_2}\rho_1'. \tag{D.26}$$

The transfer function matrix for a refractive index interface then is

$$\begin{bmatrix} A & B \\ C & D \end{bmatrix} = \begin{bmatrix} 1 & 0 \\ -\dfrac{1}{R}\left[\dfrac{n_2 - n_1}{n_2}\right] & \dfrac{n_1}{n_2} \end{bmatrix}. \tag{D.27}$$

Example D.2 Figure D.5 shows a simple lens with focal length f. For a thin lens, we can derive the resulting transfer function matrix by multiplying the ABCD matrix for each of the two surfaces, which is given by

$$\begin{bmatrix} A & B \\ C & D \end{bmatrix} = \begin{bmatrix} 1 & 0 \\ -\dfrac{1}{R_2}\left[\dfrac{n_1 - n_2}{n_1}\right] & \dfrac{n_2}{n_1} \end{bmatrix}\begin{bmatrix} 1 & 0 \\ -\dfrac{1}{R_1}\left[\dfrac{n_2 - n_1}{n_2}\right] & \dfrac{n_1}{n_2} \end{bmatrix}. \tag{D.1}$$

$$= \begin{bmatrix} 1 & 0 \\ -\left[\dfrac{n_2 - n_1}{n_1}\right]\left(\dfrac{1}{R_1} - \dfrac{1}{R_2}\right) & 1 \end{bmatrix}. \tag{D.2}$$

$$= \begin{bmatrix} 1 & 0 \\ -\dfrac{1}{f} & 1 \end{bmatrix}. \tag{D.3}$$

Reference [1] derived the ABCD matrix of a thin lens of focal length f directly with Figure D.5 and by recalling the thin lens formula we used in Chapter 3. As before, let ρ_{in} and ρ'_{in} denote the displacement and slope of the incident ray just before the lens,

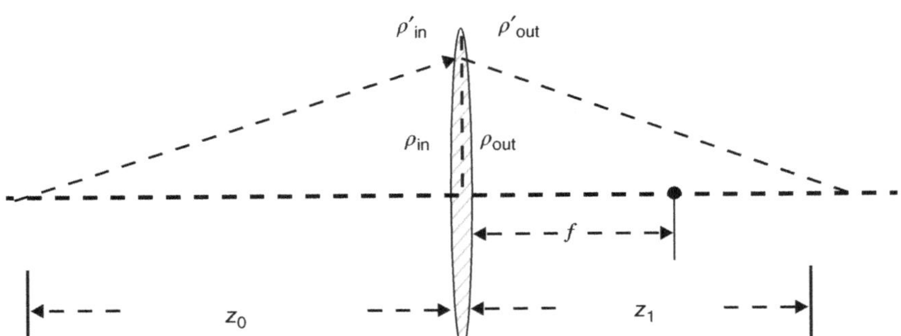

FIGURE D.5 Simple lens system.

and ρ_{out} and ρ'_{out} their values just after the lens. Because the lens is thin, the distance from the axis has no chance to change while light propagates through the lens and we have

$$\rho_{out} = \rho_{in}. \tag{D.31}$$

To determine the slope after the lens, we assume that the incident ray is coming from the axial point z_0 and the emergent ray travels towards the point z_1. Recall that these distances are related by the thin lens formula given in Eq. (3.24), that is, we have

$$\frac{1}{z_0} + \frac{1}{z_1} = \frac{1}{f}.$$

Multiplying both sides by ρ_{in} and noting that $\frac{\rho_{in}}{z_0} = \rho'_1$ and $\frac{\rho_{in}}{z_1} = -\rho'_1$ (the emergent ray has negative slope) and rearranging the terms we find the slope of the emergent ray

$$\rho'_{out} = \rho'_{in} - \frac{\rho_{in}}{f}. \tag{D.32}$$

Using Eqs. (D.31) and (D.32), we find that

$$\begin{bmatrix} \rho_2 \\ \rho'_2 \end{bmatrix} = \begin{bmatrix} A & B \\ C & D \end{bmatrix} \begin{bmatrix} \rho_1 \\ \rho'_1 \end{bmatrix} = \begin{bmatrix} 1 & 0 \\ -\frac{1}{f} & 1 \end{bmatrix} \cdot \begin{bmatrix} \rho_1 \\ \rho'_1 \end{bmatrix}, \tag{D.33}$$

which is the same ABCD matrix found in Eq. (10.31).

Example D.3 The ABCD matrix for propagation through a plate of a homogeneous medium with refractive index n and length d is

$$\begin{bmatrix} A & B \\ C & D \end{bmatrix} = \begin{bmatrix} 1 & \frac{d}{n} \\ 0 & 1 \end{bmatrix}. \tag{D.34}$$

Example D.4 Figure D.6 depicts a multielement optical system. As noted earlier, once these basic matrices are known for the various elements of an optical system, we can calculate the overall ABCD matrix for any system of optical element through the multiplication of their individual ABCD matrices [1]. For example, the ABCD matrix of the systems shown in Figure D.6 (i.e., a thin lens, followed by free space and a dielectric slab of refractive index n, then free space and finally, a thick lens). We have labeled the various optical elements as $1, 2, 3, \ldots, 6$ in the order in which they are encountered. This means that the overall transfer matrix can be written as

$$\boldsymbol{M}_{overall} = \boldsymbol{M}_6 \cdot \boldsymbol{M}_5 \cdot \boldsymbol{M}_4 \cdot \boldsymbol{M}_3 \cdot \boldsymbol{M}_2 \cdot \boldsymbol{M}_1, \tag{D.35}$$

where

$$\boldsymbol{M}_k \equiv \begin{bmatrix} A & B \\ C & D \end{bmatrix}_k \tag{D.36}$$

is the ABCD matrix for kth element in the optical system train. Note that the matrices are written from right to left in the order in which they are encountered by the ray.

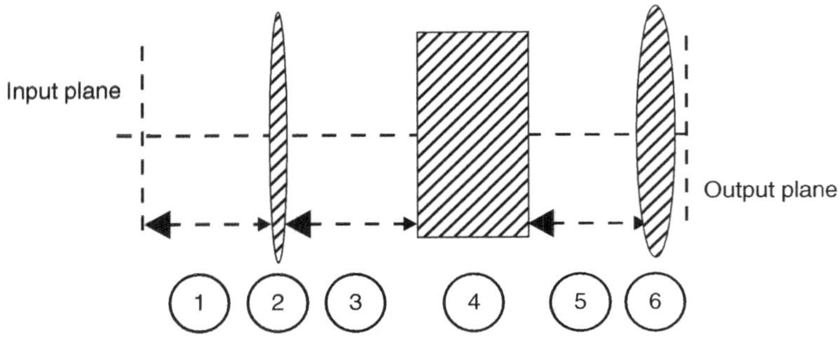

FIGURE D.6 Multielement optical system.

REFERENCES

1. *Laser Light: Physics 5734/4734*, Lecture Notes, Physics Department, University of Arkansas, http://physics.uark.edu/lasers/LPCH01.pdf (accessed 28 October 2016).
2. Galvin, T. and Eden, G., *Optical Resonator Modes*, ECE 455 Optical Electronics Lecture notes; https://courses.engr.illinois.edu/ece455/Files/Galvinlectures/02_CavityModes.pdf (accessed 28 October 2016).
3. Saleh, B.E.A. and Teich, M.C. (2007) *Fundamentals of Photonics*, 2nd edn, Wiley Series in Pure and Applied Optics, John Wiley & Sons, New York.
4. Siegman, A.E. (1986) *Lasers*, University Science Books, Mill Valley, CA.

INDEX

Free Space Optical Systems Engineering: Design and Analysis, First Edition. Larry B. Stotts.

Companion website: www.wiley.com\go\stotts\free_space_optical_systems_engineering

Fundamentals of Infrared and Visible Detector Operation and Testing, 2nd Edition
by John David Vincent, Steve Hodges, John Vampola, Mark Stegall, Greg Pierce

Statistical Optics, 2nd Edition
by Joseph W. Goodman

Optomechanical Systems Engineering
by Keith J. Kasunic

Wavelength Division Multiplexing: A Practical Engineering Guide
by Klaus Grobe, Michael Eiselt

Nematicons: Spatial Optical Solitons in Nematic Liquid Crystals
by Gaetano Assanto

Nonlinear Optics: Phenomena, Materials and Devices
by George I. Stegeman, Robert A. Stegeman

Introduction to Adaptive Lenses
by Hongwen Ren, Shin-Tson Wu

Computational Lithography
by Xu Ma, Gonzalo R. Arce

Optics of Liquid Crystal Displays, 2nd Edition
by Pochi Yeh, Claire Gu

Building Electro-Optical Systems: Making It all Work, 2nd Edition
by Philip C. D. Hobbs

Ultrafast Optics
by Andrew Weiner

Photonic Crystals, Theory, Applications and Fabrication
by Dennis W Prather, Ahmed Sharkawy, Shouyuan Shi, Janusz Murakowski, Garrett Schneider

Physics of Photonic Devices, 2nd Edition
by Shun Lien Chuang

Optical Shop Testing, 3rd Edition
by Daniel Malacara (Editor)

Fundamentals of Photonics, 2nd Edition
by Bahaa E. A. Saleh, Malvin Carl Teich

Liquid Crystals: Physical Properties and Nonlinear Optical Phenomena, 2nd Edition
by Iam-Choon Khoo, Bahaa E. A. Saleh (Series Editor)

Diffraction, Fourier Optics and Imaging
by Okan K. Ersoy

Infrared System Engineering
by Richard D. Hudson, Jr.

Optical Signal Processing
by Anthony VanderLugt

Optical Waves in Layered Media
by Pochi Yeh

Fabrication Methods for Precision Optics
by Hank H. Karow

Fundamentals of Optical Fibers, 2nd Edition
by John A. Buck

Foundations of Image Science
by Harrison H. Barrett, Kyle J. Myers

Optical Waves in Crystals: Propagation and Control of Laser Radiation
by Amnon Yariv, Pochi Yeh

Optical Detection Theory for Laser Applications
by Gregory R. Osche

Elements of Photonics, Volume I, In Free Space and Special Media
by Keigo Iizuka

Elements of Photonics, Volume II, For Fiber and Integrated Optics
by Keigo Iizuka

Elements of Photonics, 2 Volume Set
by Keigo Iizuka

The Physics and Chemistry of Color: The Fifteen Causes of Color, 2nd Edition
by Kurt Nassau

The Fractional Fourier Transform: with Applications in Optics and Signal Processing
by Haldun M. Ozaktas, Zeev Zalevsky, M. Alper Kutay

Statistical Optics
by Joseph W. Goodman

Color Science: Concepts and Methods, Quantitative Data and Formulae, 2nd Edition
by Günther Wyszecki, W. S. Stiles

Infrared Detectors and Systems
by E. L. Dereniak, G. D. Boreman

Fiber Optic Smart Structures
by Eric Udd (Editor)

Introduction to Photorefractive Nonlinear Optics
by Pochi Yeh

Fundamentals of Infrared Detector Operation and Testing
by John David Vincent

Linear Systems, Fourier Transforms, and Optics
by Jack D. Gaskill

Printed and bound by CPI Group (UK) Ltd, Croydon, CR0 4YY
17/09/2026
14964322-0002